8th International Symposium on the Science and Technology of Light Sources

Greifswald, Germany
30 August – 3 September 1998

ISBN: 978-1-5108-4112-3

Printed from e-media with permission by:

Curran Associates, Inc.
57 Morehouse Lane
Red Hook, NY 12571

Some format issues inherent in the e-media version may also appear in this print version.

Printed by Curran Associates, Inc. (2017)

For permission requests, please contact Foundation for the Advancement of the Science & Technology of Light Sources (FAST-LS) at the address below.

Foundation for the Advancement of the Science & Technology of Light Sources
FAST-LS
Belmayne House
99 Clarkehouse Road
Sheffield, United Kingdom
S10 2LN

www.fast-ls.org

Additional copies of this publication are available from:

Curran Associates, Inc.
57 Morehouse Lane
Red Hook, NY 12571 USA
Phone: 845-758-0400
Fax: 845-758-2633
Email: curran@proceedings.com
Web: www.proceedings.com

Symposium Proceedings

The 8th International Symposium
on the Science & Technology of

LIGHT SOURCES

LS-8

TOME 1

Greifswald - Germany
30th August - 3rd September 1998

Symposium Proceedings

The 8th International Symposium
on the Science & Technology of

LIGHT SOURCES

LS-8

TOME 2

Greifswald - Germany
30th August - 3rd September 1998

Proceedings of

The 8th International Symposium on the Science & Technology of Light Sources (LS-8)

Greifswald - Germany

30.8. - 3.9.1998

Invited Papers

and Poster Papers

Editor: Gerhard Babucke

Organised by:

Institute for Low-Temperature Plasma Physics Greifswald (INP)
Robert-Blum-Str. 8-10
17489 Greifswald
Germany

and

Department of Physics
University of Greifswald
Domstr. 10a
17489 Greifswald
Germany

Printed in Germany by:	KIEBU-DRUCK GmbH Ziegelhof 27 17489 Greifswald
Distributed by:	LS-8 Organising Committee c/o Dr. G. Babucke Institute for Low-Temperature Plasma Physics Robert-Blum-Str. 8-10 17489 Greifswald

PREFACE

This volume contains both the full texts of the invited papers and the abstracts of the accepted contributed posters that are to be presented on the 8th International Symposium on the Science and Technology of Light Sources (LS-8) being held in Greifswald, Germany, from 31st August to 3rd September 1998.

The LS conferences are the only international meetings which are focussed on the science and technology of light sources. They bring together scientists and engineers from industry, universities and research establishments involved in their development and production. Previous symposia were held in Loughborough (1975; 90 delegates), Enschede (1979; 126), Toulouse (1983; 150), Karlsruhe (1986; 198), York (1989; 250), Budapest (1992; 150) and most recently in Kyoto (1995; 278). Each of these conferences had a significant impact on the development of light sources. The Greifswald· symposium is the first in the series of conferences to be held in the reunited Germany. There, 128 contributed papers will be presented, and the number of participants is expected to be over 230 from 27 countries of three continents.

The format of the Greifswald symposium has followed the tradition of the series of LS meetings. Six main sessions contained in the scientific programme of LS-8 comprise a long poster session bracketed by two Invited lectures and two Landmark Papers, only the session following the Opening ceremony consists of two Invited lectures and four Landmark papers. Sunday night two lectures of more general interest will be presented by two distinguished scientists (‚Exotic gas discharges‘ - experimental lecture by A. Rutscher, not printed in the proceedings, and ‚Existing state and future perspectives of education in the field of light sources and its related technologies in university‘, presented by T. Yamaya).

The Landmark papers included in the programme were selected from the Contributed Papers by the Organising Committee. Selection has been made with regard to novelty, timeliness, quality and significance to the general programme of the Symposium. In addition, special consideration was given to the papers of younger scientists.

Each paper in the volume has been assigned a letter designation according to the list of topics in addition to the category of Invited and Landmark Papers and a further number of the position in the running order as planned in the programme.

Johannes Conrads

Chairman LS-8

June 1998

COMMITTEES

International Organising Committee

Dr. R. Bergman	Lighting Technology Division, GEL, USA
Prof. J. Conrads	Institute for Low-Temperature Plasma Physics Greifswald, Germany
Dr. R. Devonshire	University of Sheffield, UK
Prof. K. Guenther	OSRAM GmbH, Germany
Dr. W. van den Hoek	Philips Lighting B.V., The Netherlands
Dr. A.G. Jack †	Philips Lighting B.V., The Netherlands
Dr. S. Kamiya	Consultant, Japan
Prof. M. Neiger	Lighting Technology Institute, University of Karlsruhe, Germany
Dr. E. Smolka	Heraeus Noblelight GmbH, Germany
Dr. D.O. Wharmby	SLI Lighting Ltd., UK
Prof. G. Zissis	Centre de Physique des Plasmas et des Applications, France

Corresponding Committee

Dr. B.M. Ditchek	Lamp & Systems Development, OSRAM, USA
Prof. D. Karabourniotis	University of Crete, Greece
Prof. J.E. Lawler	University of Wisconsin, USA
Prof. P. Ramakrishnan	Indian Institute of Technology, India
Mr. Y. Yuge	Toshiba Lighting and Technology Corporation, Japan
Prof. S. Zhu	Fudan University, PR China

Advisory Committee

Prof. L. Bartha	Research Institute for Technical Physics, Hungary
Prof. Z. Cai	Fudan University, PR China
Prof. J.-J. Damelincourt	Paul Sabatier University, France
Prof. R. Itatani	Niihama National College of Technology, Japan
Prof. H.-P. Popp	Lighting Technology Institute, University of Karlsruhe, Germany
Dr. J.F. Waymouth	Consultant, USA

Local Organising Committee, Greifswald, Germany

Prof. J. Conrads	Chairman	Institute for Low-Temperature Plasma Physics
Dr. G. Babucke	Secretary	Institute for Low-Temperature Plasma Physics
Prof. H. Deutsch	Scientific Advisor	Dept. of Physics, Ernst-Moritz-Arndt University
Ms. I. Giannikos	Assistant	Institute for Low-Temperature Plasma Physics
Dr. H. Hess	Scientific Advisor	Institute for Low-Temperature Plasma Physics
Prof. R. Hippler	Scientific Advisor	Dept. of Physics, Ernst-Moritz-Arndt University
Dr. H. Kersten	Scientific Advisor	Dept. of Physics, Ernst-Moritz-Arndt University
Dr. E. Kindel	Scientific Advisor	Institute for Low-Temperature Plasma Physics
Ms. C. Krcka	Assistant	Institute for Low-Temperature Plasma Physics
Ms. B. Lindemann	Assistant	Institute for Low-Temperature Plasma Physics
Mr. H. Schoepp	Treasurer	Institute for Low-Temperature Plasma Physics

Alexander G. Jack (1944-1997)

Dr. Alexander G. (Sandy) Jack died suddenly on November 12th 1997. He had been struggling to recover from a stroke which he suffered six weeks before. Since the eighties he was involved in the LS Organizing Committees and made a substantial contribution to the continuing success of the LS series of conferences.

A native of Scotland and educated in electrical engineering at the Universities of Edinburgh and Cambridge, Sandy Jack spent his whole professional career within the Research and Development of Philips Lighting in the Netherlands. Starting in 1969, he was first involved in the scientific aspects of lamp development. A number of papers give evidence of his excellent work on high-pressure as well as low-pressure gas discharge lamps. His paper, written together with M. Koedam, on the Energy Balance of High-Pressure Discharge Lamps (J.IES 1974), is always referred to and can be considered as a 'classical' contribution in the field of gas discharge lamps.

The qualities of Sandy Jack, however, covered a wider scope. Various management functions were entrusted to him. In recent years he served as the R&D Strategy and Planning Manager of the Central Development Laboratories.

In Sandy Jack the lighting community has lost a good friend and colleague. As a scientist he was very sound ; as a manager he was open-minded and always trying to achieve harmony. It was a privilege to work with him.

W. J. van den Hoek

Sessions and Timetable

IL: Invited Lecture (No., place) LL: Landmark Lecture (No., place)

CC: Conference Centre A: Aula Postersessions: Mensa

Time	Su 30.08.1998	Mo 31.08.1998	Tu 01.09.1998	We 02.09.1998	Th 3.9.98
09:00		Opening ceremony A	IL 06 CC	IL 10 CC	IL 14 CC
09:30		09:40 IL 02 A	LL 08 CC	LL 12 CC	LL 14 CC
09:50		10:10 LL 02 A	Postersession 2 + Coffee	Postersession 4 + Coffee	Postersession 6 + Coffee
		10:30 Break Audi Max Press conference Physics lecture hall			
11:15		LL 03 A			
11:35		LL 04 A			
11:55		LL 05 A	LL 09 CC	LL 13 CC	LL 15 CC
12:15		IL 03 A	IL 07 CC	IL 11 CC	IL 15 CC
12:45		Lunch	Lunch	Lunch till 14:00 !	Closing Conference
14:15	Registration Physics lecture hall University	IL 04 CC	IL 08 CC	14:00 IL 12 CC	
14:45		LL 06 CC	LL 10 CC	14:30 Postersession 5 + Coffee	
15:05		Postersession 1 + Coffee	Postersession 3 + Coffee	16:00 IL 13 CC	
				16:30 Social Programme 'Motor Vessel Tour'	
17:10		LL 07 CC	LL 11 CC		
17:35		IL 05 CC	IL 09 CC		
18:00	IL 01 Physics lecture hall	'COST'-Meeting	Workshop 'Thermionic Cathodes' CC		
19:00	LL 01 Physics lecture hall				
20:00	Get together 'Alter Fritz'	Mayor's Reception 'Town hall'	Organising Committee meeting: Hotel	End < 24:00	

POSTER SESSION

Date / Topic	Session 1	Session 2	Session 3	Session 4	Session 5	Session 6
	Mo 15:05-17:10	Tu 09:50-11:55	Tu 15:05-17:10	We 09:50-11:55	We 14:30-16:00	Th 09:50-11:55
A	1-4	5-7	8-10	11-14	15-17	18-20
B	1, 2	3-5	6-8			
C	1-4	5-8	9-11	12-14	15-17	18-21
D	1, 2	3	4	5		
E	1	2	3			
F	1, 2	3, 4	5			
G	1, 2	3	4			
H				1, 2	3-5	6, 7
I	1	2, 3	4, 5	6		7, 8
K	1, 2	3, 4	5-7	8-10	11, 12	13-16
L	1, 2	3-5	6-8	9-11	12-14	15-17
M	1, 2	3, 4	5-7	8, 9	10-12	13, 14

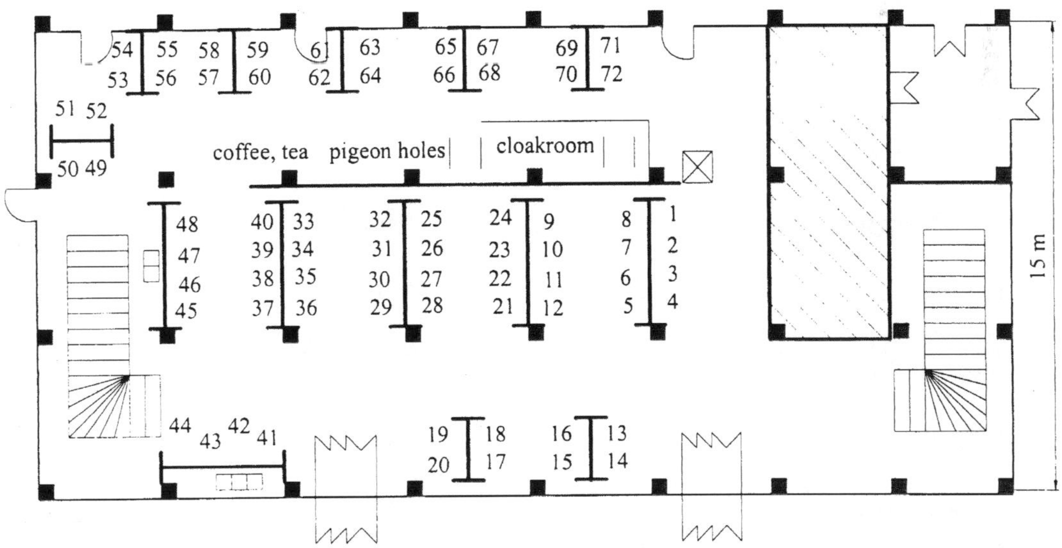

Scientific Programme

Texts of Invited and Poster Papers

Topics of LS-8

INVITED LECTURES (IL)

LANDMARK LECTURES (LL)

Contributed Posters (according to the list of topics)

Invited Lectures

IL 02

ENDURA: A NEW HIGH OUTPUT ELECTRODELESS FLUORESCENT LIGHT SOURCE

Valery Godyak and Jack Shaffer

OSRAM SYLVANIA Development Inc.
71 Cherry Hill Drive, Beverly, MA 01915, USA

INTRODUCTION

In this lecture we describe a new type of high output, high efficiency electrodeless fluorescent lamp ENDURA™ (ICETRON™ in USA) developed at OSRAM SYLVANIA. The lamp operates at 250 kHz and produces 12,000 lm at a system power of 150 W. Utilization of low frequency results in significantly reduced EMI and ballast complexity. Today ENDURA is the most powerful and efficient fluorescent RF lamp on the market.

Short history

The idea of RF lighting (Tesla, 1891) as well as the first patent on an RF lamp driven with a mechanical RF generator (Hewitt, 1907) appeared long before the first fluorescent and HID lamps became a reality. It took a century for the first commercial RF lamps of Matsushita, 1991 and Philips, 1991 to come to the market and an RF lamp to be viewed as the light source of the future.
There are a few reviews dealing with the history, basic principles, and development of commercial RF lighting (Netten and Verheij, 1991; Shinomiya et al, 1991; Warmby, 1993, 1997; Godyak, 1993; Shaffer and Godyak, 1998).

Advantages of electrodeless lamps

The most attractive feature of RF lighting (which has long been recognized) is the absence of electrodes that are the main factor in limiting lamp life, shape, efficacy and wattage. Suffice to say that electrodes of fluorescent lamps limit their life to 5,000 – 20,000 hours, cannot withstand high discharge current (usually less than 1.5 A), and require (for reasonable long life) a gas pressure which is an order of magnitude higher than that corresponding to maximal lamp efficacy. Inherent to all electroded discharges are the cathode and anode voltage drops with extremely inefficient electric power conversion to radiation. Thus, the lighting discharges should be long and slim so that the total discharge voltage does not considerably exceed the voltage drop across the radiating positive column. In RF lamps the discharge current can flow without electrodes, along a closed path in a plasma. This opens practically unlimited opportunities for lamp shapes, sizes, gas fills, and wattage. But the principal factor that delayed the commercialization of electrodeless RF lamps was that necessary electronics for reliable, low cost generation of RF power were not available earlier. Only in the last decade has progress in semiconductor electronics components and in power switching technology made feasible the commercial realization of RF lamps with compact, highly efficient and reasonable cost RF power supplies (ballasts).

RF discharge types

There are three basic ways to energize RF light sources (here we will consider only RF fluorescent lamps) corresponding to different type of interaction of an electromagnetic field with a bounded plasma and corresponding to different types of RF discharges (Lieberman and Lichtenberg, 1994). They are:

- Capacitive RF discharges where the plasma RF current path is closed by displacement currents in the RF electrode sheaths (whether the RF electrodes are inside or outside the discharge vessel). Capacitive RF discharges are excited by an electrostatic RF E-fields which have wave lengths λ that are much larger than the discharge size d, ($\lambda >>$ d).
- Inductive RF discharges where the plasma RF current is induced by an RF magnetic H-field and an RF plasma current forms a closed discharge path within the plasma. The frequency of excitation of inductive discharges satisfies the condition ($\lambda >>$ d).
- RF wave discharges are maintained by an electromagnetic wave that is incident on the plasma surface or propagates along it and whose length is comparable to the plasma size ($\lambda \sim$ d). Wave discharges are usually maintained by microwave power sources at frequencies close to or higher than 1000 MHz. In a surface wave discharge the plasma column works as a slow wave structure making the length of the propagating wave much shorter than that in a vacuum, and the frequency can be much smaller (10–100 MHz).

Choice of frequency and discharge type

Generally, it is easier to maintain an RF plasma at higher frequency, but the higher the frequency, the more costly and less efficient is RF power production. Also there are more problems with electromagnetic interference (EMI) for discharges with open surfaces such as light sources. One of the first RF fluorescent lamps based on a capacitive discharge energized with 915 MHz (Proud and Smith, 1980) has been continually operating in our Research Center since that date. However, the low efficiency and complicated microwave power supply does not give any chance for commercialization of such an RF lamp.

Attempts to drive compact capacitive RF fluorescent lamps at a frequency of 13.56 MHz, allocated for industrial application, have shown that an essential portion of RF power is spent for ion acceleration in the near electrode RF sheath (Godyak et al, 1991). This makes such lamps inefficient at an RF power exceeding a few watts. At 2.65 MHz, specifically allocated for lighting application, the portion of RF power spent for ion acceleration in a capacitive RF discharge is even greater.

With a limited choice of available frequencies for industrial and residential applications, the inductively coupled plasma source seems to be the most practical construction for commercially viable lamps (ones with light output over 1000 lumen). To our knowledge, all present efforts to make RF fluorescent lamps (including those on the market) are based on application of inductively coupled plasmas. All (except Everlight of Matsushita) follow the concept, patented more than 60 years ago (Bethenod et al, 1936) of a re-entrant cavity housing an inductor coil with a ferrite or air core. Such an RF lamp topology hides the RF coil and allows for a compact, bulb shape lamp with free external surface that eventually makes this lamp an ideal substitution for spherical-like incandescent lamps.

The utilization of relatively high RF frequencies in existing inductive lamps (2.65 and 13.56 MHz) requires significant efforts to suppress EMI. EMI is created by the RF lamp itself, as well as that originated by RF power electronics working in the switching mode and accompanied with a large spectrum of RF harmonics (Wharmby, 1997). Another problem is the localization of the RF power deposition around the re-entrant cavity near the lamp mid-plane, resulting in local wall overheating. The heat pipe (Netten and Verheij, 1991) can reduce the problem, but the heat localization inherent to the re-entrant lamp topology puts a limitation on the lamp durability, power, and maximum lumen output.

LOW FREQUENCY INDUCTION LAMP

New is old (long forgotten)

The first successful attempt to energize an electrodeless fluorescent lamp with low (100–500 kHz) frequency was made by Anderson, 1970. He applied the idea of the ring inductive discharge (widely used to maintain plasma in toroidal devices for plasma fusion) to a standard T12 fluorescent lamp with no electrodes and closed upon itself. The tube ring was penetrated with a ferrite core having a primary winding connected to an RF power source, thus, forming a transformer with the discharge being a one turn secondary winding. Operation of this lamp required a huge amount of ferrite material, yet an essential portion of RF power was dissipated in the ferrite core making this lamp impractical. Up until now, Anderson's low frequency approach has been considered among lighting specialists as a "dead end" and unworthy of serious attention.

Reinventing Anderson's lamp

Significant progress in the understanding of RF discharges gained in the last decade, driven mainly by the needs of plasma processing technology for manufacturing of semiconductor chips, allowed reconsideration of Anderson's approach and finding the way to it successful realization.

The key point in creating an efficient light source based on an RF discharge is to minimize the RF power loss in the RF coupler (inductor) which for Anderson's type of RF lamp is due to losses in the ferrite core. This loss depends on the particular ferrite material, its size, geometry, working frequency, and different characteristics of the driven discharge. The analysis of power transfer efficiency in this kind of inductive lamp shows that the correct choice of discharge current has a crucial effect on the ferrite core loss.

This can best be understood from the following considerations. Low pressure gas discharges have a negative voltage/current characteristic, so that the discharge voltage V_d is related to the discharge current I_d such that: $V_d \propto I_d^{-k}$ where $k < 1$. Ferrite core loss P_c is a steep function of the core magnetic induction B, $P_c \propto B^n$ where $n > 2$. The induction B, in turn, is proportional to the discharge voltage V_d, thus: $P_c \propto V_d^n \propto I_d^{-kn}$. Accounting for the discharge power $P_d = I_d V_d \propto I_d^{1-k}$, one can obtain the relative core loss as following: $P_c/P_d \propto I_d^{-[k(n-1)+1]}$. For discharges typical of fluorescent lamps, k=0.3 and for low frequency ferrite materials n=2.8, so one obtains $P_c/P_d \propto I_d^{-1.5}$. Thus, for a given ferrite core (with a fixed temperature) driving a discharge with given geometry and gas pressure, increasing the discharge current from 0.5 A (as per Anderson) to 7.1 A (in ENDURA lamp) provides a reduction of core loss by $(7.1/0.5)^{1.5} = 54$ times! This is qualitatively shown in Fig. 1 where the discharge and core loss powers are given as functions of the discharge current. Amazingly, the more power the ferrite core delivers to the discharge, the less power loss occurs in the core itself. This unusual feature is a consequence of the negative voltage/current characteristic of the discharge and its strong coupling with the primary winding due to a closed magnetic core of high permeability. The heating of the core and/or neutral gas in the discharge volume results in increasing core loss at very high discharge intensity.

Increased power transfer efficiency, however, is not a sufficient condition to yield an efficacious lamp. To make an efficacious lamp it is also important to optimize the discharge itself to obtain maximum in UV conversion at high lamp loading. That is achieved with reduction of buffer gas pressure to a few hundred of mTorr. At standard fluorescent lamp gas pressure of about 2.5 Torr the UV conversion at high discharge current becomes very poor.

Lamp construction

The principal of operation and external appearance of this new lamp are similar to the early work done by Anderson, 1970.

The lamp is of a closed loop tubular configuration and is driven through externally mounted two toroidal ferrite cores as it shown in Fig. 2. The lamp is fabricated from 54 mm diameter glass tubing coated inside with tri-phosphor mixture and filled with krypton gas at a pressure of 250 mTorr. The length of the lamp body is 350 mm and the length of the discharge path is 720 mm. A quantity of a bismuth-indium-mercury amalgam is located within a glass appendage to control the mercury vapor pressure during lamp operation. Also, supported from a wire extending inward from the appendage, is a thin metal foil coated with indium metal to assist lumen run-up at the time of turn-on. Two toroidal ferrite cores, made of Siemens N87 material, have an inside diameter of 42 mm, an outside diameter of 65 mm and a length of 18 mm. The ferrite cores are provided with a white coating to reduce light absorption. The cores are cut into two halves to permit assembly around the lamp tubing. Each ferrite core is provided with an 18 turn winding of wire with white colored insulation. The core halves are held together in the aluminum mounting brackets by flat stainless steel springs.

IL 02

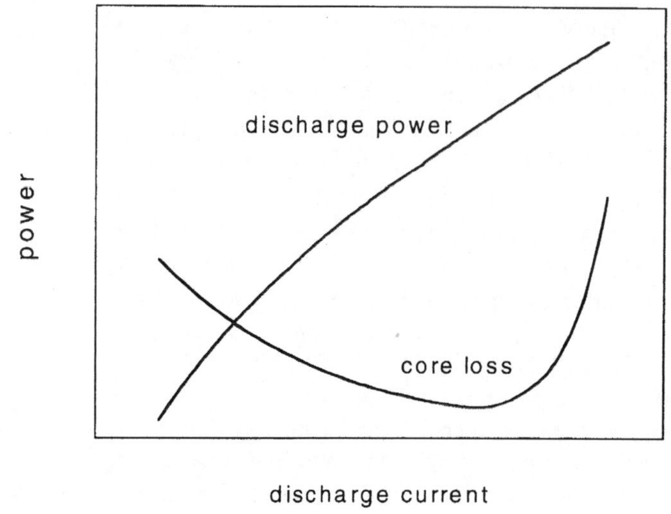

Fig.1. A qualitative diagram of discharge power and core loss versus discharge current. Raise of the core loss at high discharge current is due to core heating at high discharge intensity

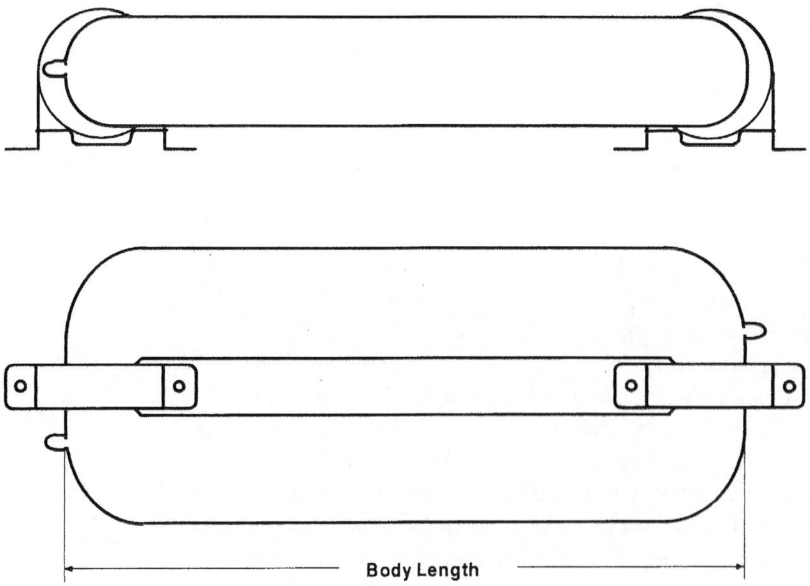

Fig. 2. Schematic representation of the lamp construction

Lamp characteristics

There are three commercial versions of ENDURA lamp today:

- 150 W system power, 350 mm body length, 12,000 lm output
- 100 W system power, 250 mm body length, 8,000 lm output
- 150 W system power, 250 mm body length, 11,000 lm output

All of them have the same tube diameter, similar color temperature and CRI. In Table 1 the characteristics of 350mm lamp at different discharge current are given together with characteristics of the Anderson's lamp. A dramatic decrease in the relative core loss is seen when operating at large current.

Lamp Parameters:	ENDURA at normal regime	ENDURA at low current	Anderson's Lamp
Vessel diameter	54 mm	54 mm	38 mm
Discharge length	72 cm	72 cm	89 cm
Discharge current	7.1 A	0.50 A	0.5 A
Buffer gas	krypton	krypton	argon
Buffer gas pressure	0.25 Torr	0.25 Torr	2.5 Torr
Discharge power, P_d	138 W	18.6 W	32 W
Core loss, P_c	2.8 W	11.0 W	10 W
Rel. Core loss, P_c/P_d	2 %	59 %	31 %
Core mass	320 g	320 g	1,240 g
Lumen output	12,280 lm	1,860 lm	unknown

Table 1. A comparison of electrodeless fluorescent lamp performance when operated at different loading.

Detailed technical characteristics of different versions of ENDURA lamp are given elsewhere (Shaffer and Godyak, 1998). Here we present some light characteristics of the ENDURA system.

Fig. 3 shows a typical run-up curve for the lamp after 16 hours of off time at 25°C ambient temperature. Under these conditions, lumen output reaches more than 90% of peak after 10 seconds. Subsequent lumen variations are measurable but not noticeable to the eye. For very short off periods, the light output returns almost immediately to its equilibrium value.

IL 02

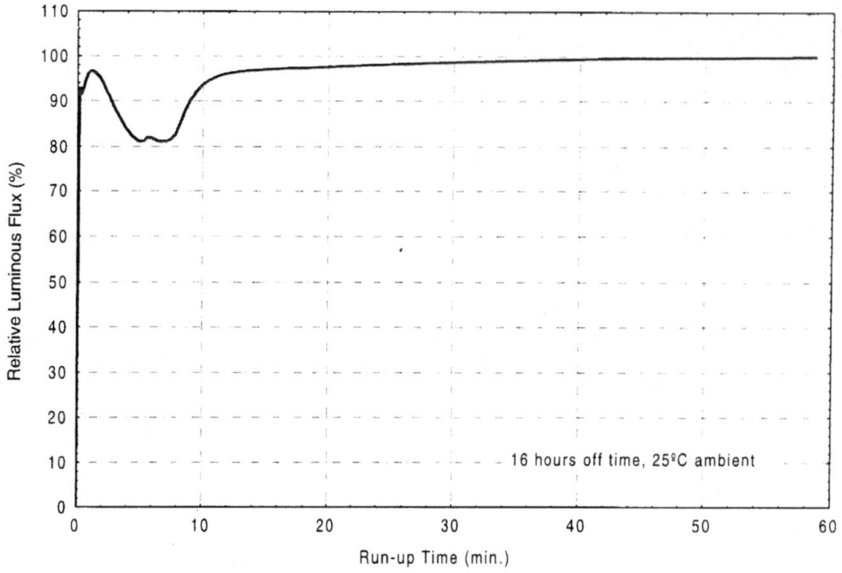

Fig.3. Lumen run-up characteristic.

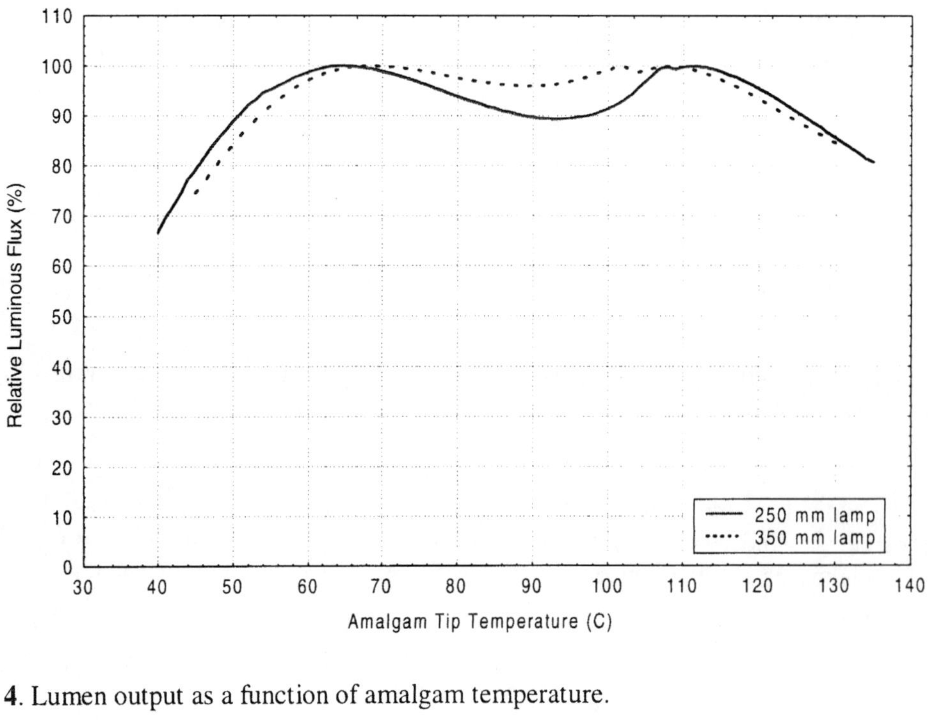

Fig. 4. Lumen output as a function of amalgam temperature.

The main amalgam stabilizes the mercury pressure over a wide range of amalgam temperature as shown in Fig. 3. Lumen output is at or above 90% of peak in the temperature interval between about 52 and 125°C. The corresponding range of ambient temperature over which lumen output exceeds 90% depends on fixture design.

The luminous intensity distribution is generally what would be expected from the lamp configuration and is shown in **Fig. 5** is for the 250mm lamp, the distribution pattern is similar for both lamp lengths.

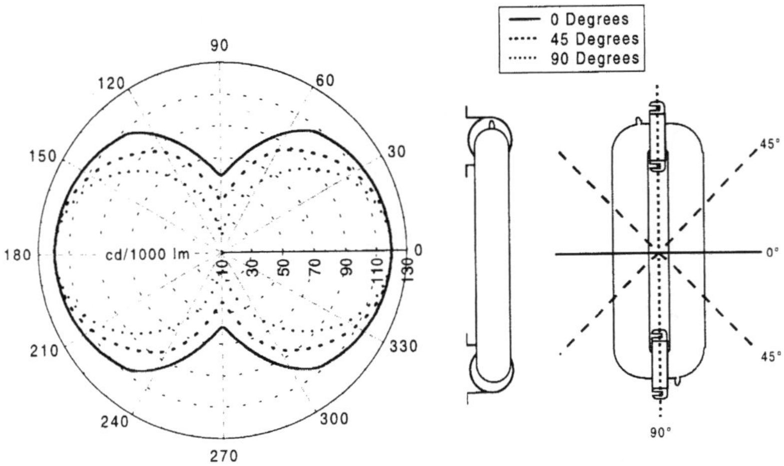

Fig. 5. Luminous intensity distribution for the 250mm lamp.

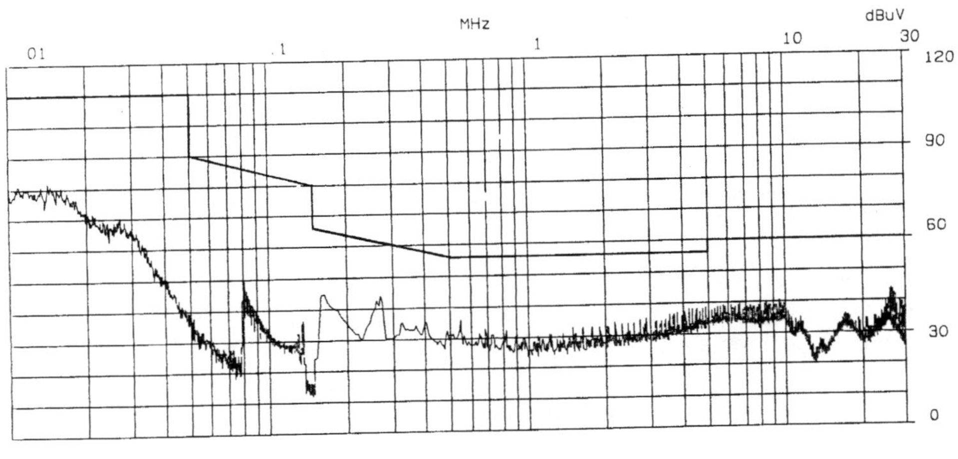

Fig. 6. Conductive noise of the ENDURA system (low trace) and tolerance level EN55015 (upper trace).

The lamp operation at an order of magnitude lower frequency (0.25 MHz instead of 2.65 MHz or 13.56 MHz) than that of the existing electrodeless fluorescent RF lamps in the market significantly reduces EMI. Conducted noise, induced by the discharge itself and by the ballast, is readily brought into FCC compliance without additional external shielding. A small fraction of RF magnetic fields generated by the discharge current and not confined in the ferrite cores (so called the radiated EMI, which is a subject to regulation in Europe) is addressed by proper fixture design. Fig. 6 shows the conductive EMI level (low trace) for a 230 V line voltage European version of ENDURA system together with the EN55015 tolerance limit (upper trace).

CONCLUSIONS

The development of a low profile electrodeless fluorescent lamps having uniquely high lumen output, high system efficacy, and long life has been described. Such lamps are free from the usual limitations imposed by electrodes and therefore provide essentially instant starting and hot re-strike. Lamp life is unaffected by short duty cycles.
The "white" light and long life available from remotely ballasted electrodeless fluorescent lamps makes them ideal for lighting applications where lamp replacement is hazardous or costly, and wherever the lamps would be operated many hours per day. Applications for such lamps would include bridge and tunnel lighting, decorative street lighting, parking garages, internally lighted signs, and low bay or indirect lighting for retailing and manufacturing areas.

Utilization of lower frequency in this lamp significantly reduces problems associated with EMI and ballast complexity and cost comparing to those in RF lighting systems working at 2.65 and 13.56 MHz. On the other hand, decentralized RF deposition in a long discharge path permits to make slim lamps and allows for broad scalability of the light system operating at low frequency. The potential of this concept has already been demonstrated by design of a compact (integrated with ballast) fluorescent lamp (Godyak et al, 1997) and by a superpower 1 kW electrodeless fluorescent lamp demonstrated at Siemens 150 anniversary in Berlin in 1997.

ACKNOWLEDGMENT

The authors wish to thank Benjamin Alexandrovich, Arunava Dutta, Jonathan Grot, Dieter Hofmann, Michael Kling, James Lester, Joseph Lima, John Peterson, Robert Piejak, Eugen Statnic, David Wentzel, and Gregory Zaslavsky for their many contributions to this project.

REFERENCES

Anderson J. M., Electrodeless Gaseous Electric Discharge Device Utilizing Ferrite Cores, US Patent # 3,500,118 (1970).

Bethenod J. et al, Electromagnetic Apparatus, US Patent # 2,030,957 (1936).

Godyak V. A., Low Pressure RF Discharges for Lighting, Invited talk at IEEE Intern. Conf. on Plasma Sci., Vancouver, BC, Canada (1993).

Godyak V. A. et al, Low Frequency Compact Fluorescent Lamp, US Des. Patent # 384,173 (1997).

Godyak V. A. et al, Ion Flux and Ion Power Losses at the Electrode Sheaths in a Symmetrical RF Discharge, J. Appl. Phys. **69**, 3455 (1991).

Hewitt P. C., Induction Vapor or Gas Electric Lamp, US Patent # 843,533 (1907).

Lieberman M. A. and Lichtenberg A. J., Principles of Plasma Discharges and Materials. Processing, John Wiley & Sons, Inc., New York (1994).

Netten A. and Verheij C. M., The Operating Principles of the Philips QL Lamp System, Philips Lighting B. V. (1991).

Proud J.M. and Smith R. K., Compact Fluorescent Light Source Having Metallized Electrodes, US Patent # 4,266,166 (1981).

Shaffer J. W. and Godyak V. A., The Development of Low Frequency, High Output Electrodeless Lamps, IESNA Annual Meeting, San Antonio, TX, Aug. (1998).

Shinomiya M. et al, Development of the Electrodeless Fluorescent Lamp, J. of IES, 44, Winter (1991).

Tesla N., Tesla's Experiments with Alternating Current at High Frequency, Electrical Engineer **7**, 549 (1891).

Wharmby, D.O., Electrodeless Lamps for Lighting: a review, IEE Proceedings-A, **140**, 465 (1993).

Wharmby, D.O., Electrodeless Lamps, in Lamps and Lighting, p. 216, Ed. by J. R. Coaton and A.M. Marsden, Arnold and Contributors Publ., London (1997).

IL 03

MERCURY REDUCTION IN TL FLUORESCENT LAMPS

Manuel Oomen
Philips Lighting Company, Hoult Road, Fairmont WV 26554, USA

1 INTRODUCTION

The strong position of fluorescent lamps (TL, compact fluorescents, QL) on the lighting market is mainly because of the energy efficient light generation and long lifetimes of these products, combined with good lighting properties. Next to the resulting cost efficiency of fluorescent lighting, fluorescent lamps are more and more positioned as 'green lighting products' because of their high energy efficacy and low material waste.

However, there is one paradox in the simultanuous pursuit of cost efficiency and environmental friendliness of fluorescent lighting products. Fluorescents lamps require mercury, a hazardous material, to generate the highly efficient discharge plasma converting electrical power into UV light (which is converted into visible light by the phosphor layer at the inner side of the glass tube). The mercury discharge is the most efficient low pressure discharge known today and highly efficient phosphors are available to transfer the mercury UV discharge radiation into the desired visible radiation. Approximately 90% of the total environmental burden of a fluorescent lamp during its entire lifetime is due to energy usage [1]. Therefore, the presence of mercury in fluorescent lamps is indispensible for environmental friendly (and cost efficient) lighting.

So mercury elimination is not the way to go for improving the environmental friendliness of fluorescent lighting. A better way is to reduce the amount of Hg in lamps. Graph 1 shows this is indeed happening the latest years for TL lamps. TL lamps with the lowest doses are a T8 lamp containing only 3 mg of mercury, introduced in Europe in 1994, and a T12 lamp with less than 5 mg of mercury (USA).

Figure 1: Mercury content (mgr) in 4ft fluorescent lamps

In the USA low Hg dosing is a major topic in the TL business today, because low mercury lamps pass the so-called 'TCLP'[1] test implicating that these TL lamps are not considered to be chemical waste. These low mercury lamps can therefore, at end of life, be disposed of free of additional chemical waste charges. Additional charges for chemical waste in the US may, depending on the state, be as high as $1 per 4 ft TL lamp. So, next to a lamp with the lowest mercury dose, the customer also has an important cost advantage once the lamp is at end-of-life.

Further reduction of mercury levels in the coming years is expected. It should be realized that for optimal operation of a 4ft T8 lamp only 50 microgrammes of mercury are required in the discharge. Mercury levels in the order of milligrammes are in this respect far more than the minimum mercury level required for energy efficient fluorescent lamp operation. The main reason for this overdosing is the use of mercury during lamp operation. This so-called Hg consumption determines the amount of mercury (Hg) required to meet rated lamplife. Another practical issue is the accuracy mercury can be dosed with in the lamp. Hg consumption phenomena and means to reduce Hg consumption will be described in the next section. Also the state of the art of Hg dosing will be shortly described.

2 TECHNICAL ISSUES

2.1 Mercury consumption

Mercury consumption occurs upon operation of the lamp. A typical plot of Hg consumed in a T8 fluorescent lamp as function of lamp operation time is shown in figure 2. The shape of the curve and the amount of Hg consumed depend strongly

Figure 2
Typical Hg consumption - lifetime curve for a 4ft TLD lamp

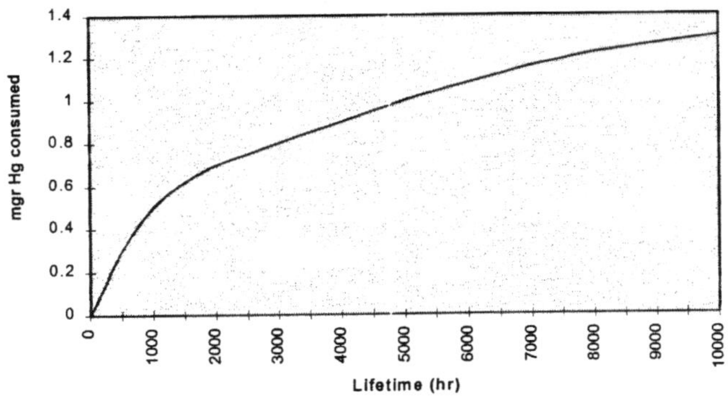

on lamp design and lamp operation, but the Hg consumed per time unit is in the early burning hours always higher than at elevated lifetimes.

The amount of Hg consumed can be measured in various ways.

2.1.1 Diagnostic tools

Several methods have been developed to measure the amount of bound Hg. An overview of available diagnostic tools is presented in table 1.

---

[1] TCLP = Toxic Characteristic Leaching Procedure developed by Environmental Policy Act, see [2]

TABLE 1
Comparison of diagnostic methods for Hg determination in fluorescent lamps

Method	Accuracy	Main advantages	Main disadvantages
Cataphoresis	5% relative not<0.1 mg	Fast, cheap Lamp intact	Not site selective Not for <0.5 mg Hg
Wet-chemical analysis	5% relative	Site selective (every lamppart)	Lamp destructed Laborious
Tracer Hg	2% relative	Site selective (axial) Lamp intact	Limited to 5000 hrs Radio-active

Cataphoresis is a well-known method to determine quickly the amount of free mercury in a lamp [3,4]. It is a relatively fast measurement and lamps do not have to be destructed for this diagnosis, allowing to make mercury consumption profiles as a function of lifetime for the same lamps. The accuracy of the method is about 5% and not less than about 0.1 mg. The method is therefore not suited to study Hg consumption phenomena in the order of tenths of milligrammes or less. Recently, cataphoresis equipment has become commercially available [5].

Wet-chemical analysis is a reliable method to determine bound Hg in the different lampparts with a good accuracy (5% relative). Main disadvantages of this method are that the total procedure is time consuming and that lamps have to be destructed.

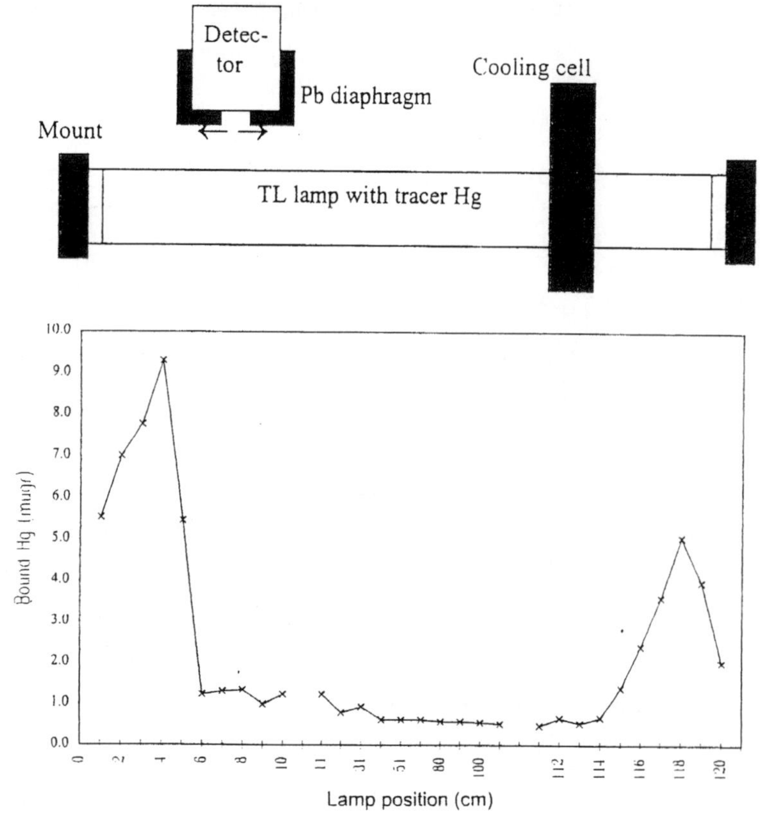

Figure 3 Equipment for tracer Hg analysis (schematic) and a typical result.

The tracer-Hg method (figure 3) is an accurate method (2% relative error) which is site selective (at least along the axis of the lamp) and the lamps remain intact. Disadvantages of this method are the relative short lifetime of the ^{203}Hg isotope used for diagnosis ($t_{1/2}$ = 1100 hrs), as a result this method can only be used to determine mercury consumption with an acceptable accuracy up to a lifetime of about 5000 hrs. Furthermore, this method requires all precautions to be taken for working with radioactive materials during lampmaking, lifetesting, transport, and measuring.

The present level at which Hg analyses can be performed is the result of years and years of devoted work on Hg consumption in fluorescent lamps. With the presented set of analytical tools a complete set of Hg consumption data can be obtained. These analytical tools are the key to understanding Hg consumption and further reduction of Hg contents in fluorescent lamps.

2.1.2 Reducing mercury consumption

Many lampparts contribute to mercury consumption. In table 2 an overview is given of which lampparts contribute to what extend to Hg consumption in TL lamps at end-of-life (EOL), which is typically 15000 hrs for European lamps using a 3 hr cycle (165' on, 15' off). The data in the table are typical data and vary significantly for other materials and quality of the materials used for that lamppart.

TABLE 2
Hg consumption of the different lampparts in TL lamps

Lamppart	Hg consumption at EOL (mg Hg)	Comment
Glass bulb (no glasscoating)	5 mg	Depends on shielding glass by phosphor layer Depending on wall load (W/cm2) Depending on glass type
Glass bulb (with glasscoating)	0.1 mg	Depending on materials, quality and thickness of glasscoating
Glass stems	0.2 mg	
Phosphor layer	0.1 - 2 mg	Depending on phosphors applied, thick-ness phosphor layer, and additives in layer Hg consumption in, on,or in-between the individual phosphor grains
Emitter (on glass bulb and cathode shield)	0.1 - 1 mg	Depending on lamp operation (fig.5) Depending on configuration electrode region and emitter material & quantity
Others	< 0.3 mg	E.g. impurities, metals, Hg capsule

The glass tube is a well-known Hg consumer, and Hg consumption in the glass wall has been studied since many years [6-10]. All studies learn that Na(+) ions in the glass play an important role in Hg consumption. However, Na-free glass (e.g. quartz) also consumes Hg, be it that the Hg consumption is less (about a factor 2) than in case soda lime glass is applied. The effect of a phosphor layer is also substantial. Two effects of the phosphorlayer to Hg consumption have been reported. One is that the phosphor layer simply shields the glass from bombardment of Hg-ions, thereby reducing the amount of Hg in the glass for a better shielding phosphor layer [7,8,10].

The other is that the phosphor layer strongly influences the amount of UV on the glass wall. UV radiation is reported to contribute to Hg consumption in the glass [10,11].

Applying a protective coating to the glass strongly diminishes the amount of mercury consumed by the glass wall. Coatings of alumina or yttria are known to reduce the Hg consumption of the glass bulb to well below 1 mgr Hg during lifetime. However, tinoxide coatings are known not to reduce Hg consumption of the glass bulb and silica coatings protect the glassbulb for Hg consumption to a much lesser extent than alumina or yttria. This is explained by the 'charging tendency' of the coating material [12], or iso-electric point (IEP) of the material. We believe that Hg can only migrate into the glasswall (be 'consumed' by the glass) if Hg(+) is stable at the glass surface. IEP calculations or charging measurements show that alumina and yttria will donate an electron to Hg(+) resulting in Hg(0) , whereas silica does not and leaves Hg(+) stable at the surface.

So our overal model for Hg consumption in glass[2] can simply be described in two steps (fig. 4). At the first step Hg(+) gets at the wall, driven by negative charging of the glass wall during lamp operation. The higher the wall load (lamppower per cm^2 glass) the more Hg(+) per time-unit will reach the glass bulb. A first barrier for Hg(+) to overcome is the phosphorlayer. If Hg(+) is at the glasswall and the glass is coated with yttria or alumina Hg(+) is converted into Hg(0) and is available for the discharge. If Hg(+) remains stable (e.g. at the glass surface or at a silica coating) the second step is that it migrates into the glass. In order for Hg(+) to migrate into the glass another ion needs to migrate from the glass to the inner surface of the glass tube (law of charge neutrality). Na(+) is known to be a very mobile ion in glass, which explains that more Hg(+) is consumed for Na-containing glass. However also other ions can act as countermigrating ion (H(+)). UV can also play an important role here by formation of defects in the glass which facilitate Hg(+) migration.

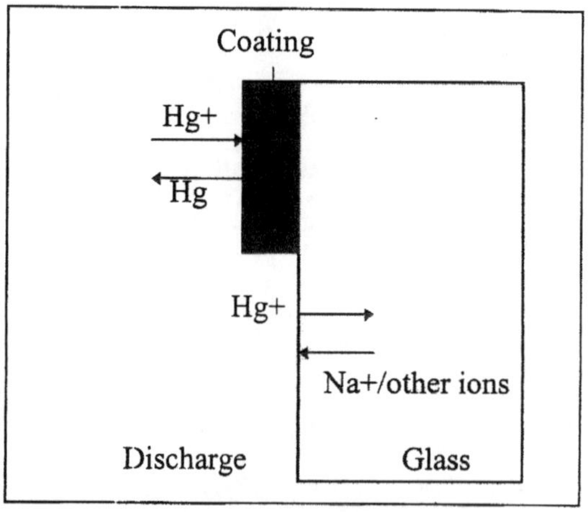

Figure 4. Schematic presentation of Hg consumption into glass

From the presented model it will be clear that quality of the glass coating has a large effect on the Hg consumption of the lamp. A coating which leaves parts of the lamp uncovered, or which has voids or pores (is diffuse for Hg(+)),will still allow Hg(+) to reach the glass surface resulting in Hg consumption.

Stem glass also contributes to Hg consumption, but to a much lesser extent than bulb glass. Of course this is mainly a result of the small surface of the glass stems as compared to the glass bulb. Furthermore Hg consumption mechanisms at the stem are probably different from the glass bulb because the stem is not covered with a phosphor or glass coating, the wall load is different, and charging of the glass is not similar as for the glass bulb.

[2] Model courtesy of C. Ronda and coworkers at Philips Research Labs, Aachen, Germany

Mercury consumption in the phosphorlayer depends strongly on the type of phosphor applied. In terms of Hg consumption of the phosphor layer we can discriminate into the following classifications (Hg consumption is estimated for the total phosphor layer of single component (one phosphor only) 36W TLD lamps at 15000 hrs burning).

TABLE 3
Discrimation of Hg consumption in phosphors

Phosphor classification	Typical Hg consumed EOL	Examples
Strongly Hg consuming	1-3 mg	Willemite, BSP
Medium Hg consuming	0.1 - 1 mg	Halophosphates,LAP,BAM
Low Hg consuming	0.0 - 0.1 mg	YOX,CBT,CAT

Note that silicate phosphors have the highest Hg consumption. This is explained by the charging tendency of silica, as explained for glasscoatings. Hg(+) will generally not be stable at yttria and alumina based phosphors resulting in a low Hg consumption. BAM (an aluminate material) has a higher Hg consumption. A possible explanation lies in the effect of dopants on the IEP of the material (Eu(2+) may provide an electron). Medium and strong Hg consuming phosphors can be coated with, for example, an alumina coating resulting in much less Hg consumption [12,13].

The classification as above can largely be explained by Hg(+) migrating into the phosphor. However, we have also seen significant amounts of Hg being present at the surface of the phosphor grains or in between the phosphor grains. Apparently the phosphor surface, possibly in combination with impurities on that surface, also has a contribution to Hg consumption.

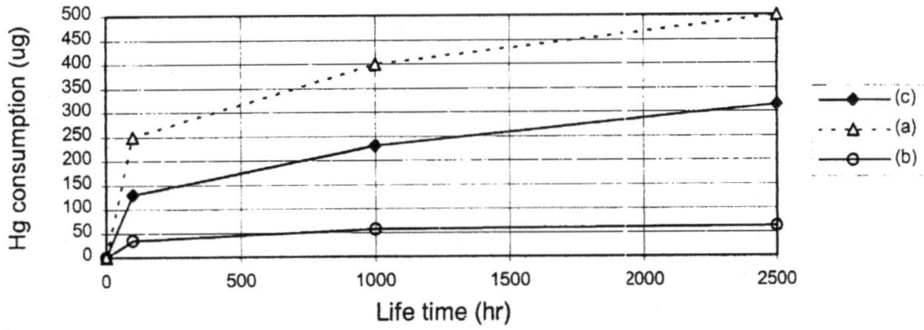

Figure 5:
Hg consumption in electrode regions for TLD 36W lamps as a function of lifetime for three different operation conditions: (a) HF ballast, cold start, 20' cycle (15' on, 5' off) ; (b) HF ballast, warm start, 180' cycle (165' on, 15' off) ; (c) as (b) but at a 10% dimming level and too much filament heating. All lamps have identical Hg dose, glass coating and phosphor layer.

The Hg consumption due to emitter is strongly dependent on the operation conditions of the lamp. Figure 5 shows the Hg-consumption as a function of lifetime for identical lamps operated at different conditions. It should be mentioned here that for dimming we on purpose used a dimming ballast which at the imposed dimming level (10%) heats the electrode too much. Tracer-Hg and wet-chemical analysis learned that the differences in Hg-consumption result from Hg bound in the electrode regions of the lamp.

The pattern of Hg consumption is in good agreement with the pattern one expects for emitter exhaust. Many cold switches and dimming with too much filament heating will result in a large exhaust of emitter materials. The mechanisms of Hg consumption due to emitter exhaust are not resolved yet. Presumably Ba plays an important role because Ba and Hg are known to amalgate [14].

Other major Hg consumers except those mentioned here have not become apparent during our extensive studies. This makes us believe that the Hg consumption due to 'other consumers' (impurities, metals, etc..) during lamplife is less than 0.3 mgr Hg in total.

2.2 Mercury dosing

Low mercury lamps need production equipment for accurately dosing a low amount of mercury in fluorescent lamps. One method which is applied for many years in production is the so-called 'capsule technology' (figure 6). Hg is accurately dosed

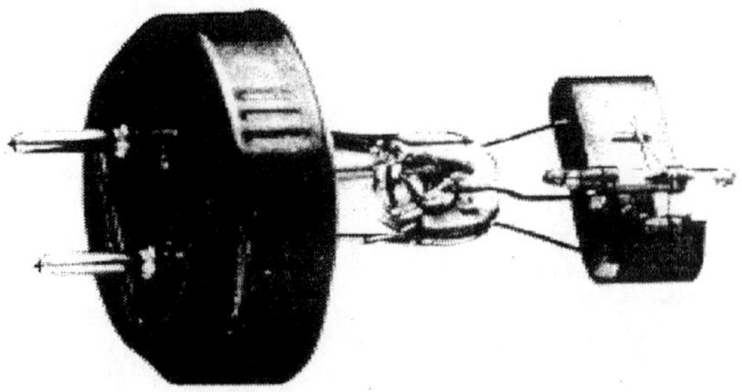

Figure 6 Stem with Hg capsule mounted on cathode guard shield

in a glass capsule which is mounted at the cathode guard with a nickel wire over the capsule. After mounting and pumping of the lamp the ends holding the capsule are put in an HF field, which causes thenickel wire to heat and the Hg pressure in the capsule to increase. This results in bursting of the capsule and the release of an accurately dosed amount of Hg in the lamp.

Other methods are known such as dosing with metal capsules [14], shields with Hg compounds such as Ti_3Hg releasing the mercury at elevated temperatures [15,16], metallic solids releasing a dosed amount of Hg upon exposure to lamp vacuum and temperature [17] and amalgams with similar mercury pressure - temperature curves as metallic mercury [18].

All these dosing methods are known to be able to dose Hg with a rather high accuracy (< 10%) down to a level of a few mgs to 1 mg Hg.

3 CONCLUSIONS

Hg reduction in fluorescent lamps is a discernible environmental issue which is asked for by the market. Further Hg reduction is expected to go on for the years to come. This means that in production of fluorescent lamps measures to reduce Hg consumption have to be included, as well as equipment to dose accurately and reliably the amount of Hg in the lamp.

The most effective measure to reduce Hg consumption is application of a protective coating on the glass tube. Using a good coating the Hg consumption by the glass bulb can be reduced from mgrs of Hg to well below 1 mgr at end-of-life (for a 4ft T8 lamp). Depending on lampdesign (especially phosphor layer and electrode region) and lamp operation the Hg consumption of the total lamp can then be reduced to 1 to 3 mgr at end-of-life. Further reduction of Hg consumption will require modifications at the phosphor, the phosphor layer, the emitter, and/or the electrode region.

Production equipment for Hg dosing is available for Hg doses as low as 1 mg of Hg. Hg consumption during lifetime in present lamps exceeds 1 mg. Assuming that no sacrifices in lamp life are made for lowering the Hg dose, it is therefore expected that the amount of Hg in future fluorescent lamps is determined by the Hg consumption during lifetime.

4 ACKNOWLEDGEMENTS

The major part of this contribution describes the results of many years of work of numerous people working within various Philips departments (Central Development Lamps of Philips Lighting, Centre of Manufacturing Technology, and Philips Research Labs, all located in Eindhoven, The Netherlands; Fluorescent development in Roosendaal, The Netherlands; Fluorescent product development in Fairmont, WV, USA, and Philips Research Labs in Aachen, Germany).

REFERENCES

[1] European Lighting Companies Federation, Discharge lamps and the environment, 4th edition, Brussels, May 1997

[2] The Toxic Characteristic Leaching Procedure os prescribed on pages 26987-26998 of vol.55, nr. 126 of the June 29, 1990 issue of the Federal Register (Washington, USA). The protocol for TL lamps is prescribed in NEMA Standards Publication LL-1-1997, National Electrical Manufacturers Association, Rosslyn,VA, 1997

[3] H.C.G. Verhaar & B..J. Roelevink, Abstract 12:L in Proc. of the 6th Int.Symp. on the Science & Techn. of Light Sources, Budapest, Hungary, Aug.30-Sept.3, 1992

[4] H. Tomioka, T. Higashi & K. Iwama, in Proc. of the 7th Int.Symp. on the Science & Techn. of Light Sources, Kyoto Japan, 1995

[5] Cataphoresis equipment is available through EPS - Augsburg (Germany)

[6] U. Kuemmel, Appl. Phys., 7 (1975) p.93

[7] B.J. Mulder & S. van Heusden, J. Electrochem. Soc., 130 (1983) p.440

[8] P.W.C. Verhees, Abstract 105:L, Proc. of the 5th Int.Symp. on the Science & Techn. of Light Sources, York UK, 10-14 Sept. 1989

[9] E.G. Thaler, R.H. Wilson, D.A. Doughty & W.W. Beers, J. Electrochem. Soc., 142 (1995) p.1968

[10] D.A. Doughty, R.H. Wilson & E.G. Thaler, J. Electrochem. Soc., 142 (1995) p.3542

[11] Japanese patent JP-29547, (1993), (MEC)

[12] US Patents 5289081 (1991) and 5402036 (1994), (Toshiba);
H. Ito, Y. Yuge, A. Taya, M. Tamatani, & K. Terasima, Abstract 11:L in Proc. of the 6th Int.Symp. on the Science & Techn. of Light Sources, Budapest, Hungary, Aug.30-Sept.3, 1992

[13] US Patent 5309069 (1992), (GTE)

[14] G. Bruzzone & F. Merlo, J.Less-Comm.Met., 39 (1975) p.271

[15] Eur.Patent 0788142 (1996) (Osram-Sylvania)

[16] World Patent 97/00532 (1996)

[17] Hg dispensers can be obtained from SAES-getters, Milan, Italy

[18] W. Schuster, in ' Technisch-wissenschaftliche abhandlungen der Osram-Gesellschaft, 12-er band' , G. Haehnel (ed.), Springer-Verlag, Berlin, 1986

[19] T.R. Brumleve, S.C. Hansen & L.G. Kaczorowski, High temperature lamp Chemistry III symposium at the 184th meeting of the Electrochem. Soc., Oct. 10-13, 1993, New Orleans

MODELING THE BREAKDOWN AND GLOW PHASES OF HID LAMP IGNITION

L.C. Pitchford
Centre de Physique de Plasmas et Applications de Toulouse (CPAT)
118 route de Narbonne, 31062 Toulouse, FRANCE

I INTRODUCTION

Over the past ten years, the Discharge Modeling Group at the Centre de Physique de Plasmas et Applications de Toulouse has developed a series of models to describe the electrical characteristics of glow discharges. These models have been applied to a variety of applications such as rf discharges for plasma processing, plasma display panels, glow discharge for mass spectrometry, discharge light sources, and other applications. This talk will focus on our work related to the breakdown and glow phases during HID lamp ignition [1].

Previous studies of starting in HID lamps have concentrated on the breakdown phase and the transition to a hot cathode discharge without identifying explicitly the takeover phase during which a cold cathode, glow discharge must be sustained. In this talk, we will illustrate the effect of the breakdown pulse on the takeover voltage, the minimum voltage needed to sustain the glow after breakdown has occurred. The conditions are limited to those for which breakdown occurs via a Townsend mechanism; that is, the product of the gas pressure and the electrode spacing is less than about 30 torr cm. Thus, streamer breakdown is not considered here.

2 PHYSICAL AND NUMERICAL MODELS

Our approach to the problem of HID lamp ignition is to use a physical model which has been simplified as much as possible in order to reduce the number of parameters and to have a numerically tractable problem. To this end, we use the 'idealized' open circuit voltage shown in fig. 1 to represent high voltage spikes superimposed on a 50 Hz generator voltage. The ignitor is supposed to produce high voltage pulses and the ignitor impedance is represented as a series resistance. After a delay t_d to represent the phase shift between the high voltage spikes and the peak of the 50 Hz voltage, a takeover voltage pulse is applied to the discharge. The ballast is represented by a series resistance and a series inductor. A parallel capacitance is included to account for the capacitance of the leads. These parameters can all be varied in the model.

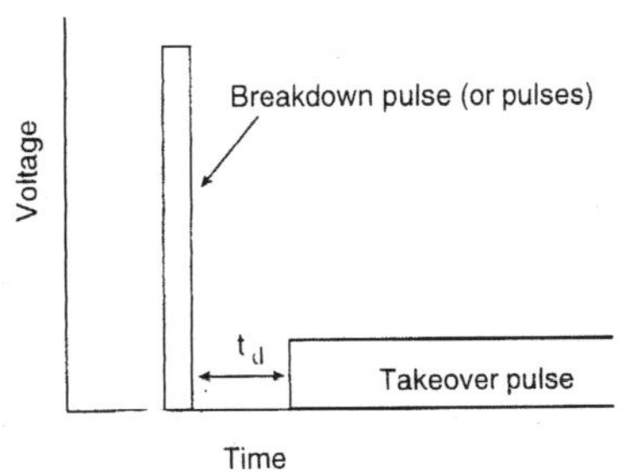

Fig. 1. Idealized open circuit voltage.

In the 'off' state, the lamp fill consists of a rare gas (usually argon) at a pressure of some 10's of torr or more, mercury vapor at a few mtorr (vapor pressure at room temperature), and other solid components which have essentially zero vapor pressure at room temperature. During the breakdown and glow phases of HID ignition, the gas pressure does not change.

We further simplify the problem and consider discharge initiation in a one-dimensional (parallel-plane electrode) geometry. Although lamp geometries are considerably more complicated, these one-dimensional model results provide insight and predict trends which are expected to be generally valid. A few results in two-dimensional geometries will be shown to illustrate the effect of the geometry.

The fundamental variables in our model are the electron density, the ion density and the electric field. These are functions of the axial distance between the electrodes and of time. The model is self-consistent; that is, the space charge field is calculated self-consistently from the charged particle densities at each point in space and time. The equations which are solved are therefore the continuity equations for the electrons and ions and Poisson's equation for the electric field. The electron and ion fluxes are supposed to be the sum of drift (mobility) and diffusion terms. Thus the model yields the electron and ion densities and the potential distribution between the electrodes as a function of time. From these quantities, the discharge current can be deduced.

3 RESULTS

It is convenient to consider three different intervals of time corresponding to (a) breakdown (b) afterglow and (c) glow discharge.

3.1 Breakdown

We suppose that a few initiating electrons are present at the time the breakdown pulse is applied (due, for example, to UV illumination). Breakdown is defined as a 10% decrease in the lamp voltage. The plasma density at breakdown depends on the magnitude of the voltage pulse, pd and on the source impedance. However, the time to breakdown depends only on the voltage and pd, and not on the source impedance The height and width of the breakdown voltage pulse can be reduced if multiple pulses are applied and if the time between successive pulses is less than the ion transit time (if ions remain in the gap between the voltage pulses). Breakdown occurs on a time scale of about 1 μs (for 10's of torr argon and a voltage on the order of 1 kV).

3.2 Afterglow

In the afterglow (between the end of the breakdown pulse and the application of the takeover voltage), the plasma density decays rapidly (by recombination) when the plasma density is high in the early afterglow and then more slowly (by ambipolar diffusion) in the late afterglow. The ambipolar time constant is given by (L^2/D_a) where D_a is approximately $(10^4/p_{Torr})$ $cm^2\ s^{-1}$ in a nonthermal plasma and L is the diffusion length. Thus ambipolar diffusion is slower for higher pressures. The plasma density is 10^{10} cm^{-3} or more for times as long as 400 μs in an afterglow in 20 torr argon for an initial density of 10^{11} to 10^{12} cm^{-3}.

3.3 Glow phase

If the discharge voltage is sufficient, a quasi-steady-state glow discharge is reached after several 100's of μs for argon for pd on the order of 10's of torr cm. This glow discharge must be sustained until the electrodes are heated to temperatures high enough for thermionic emission to become important.

The results from our model calculations show that, after breakdown has occurred, the minimum voltage needed to sustain the glow discharge is a decreasing function of the plasma density present in the gap at the time the takeover pulse is applied. The takeover voltage is shown in fig. 2 as a function of the initial plasma density for argon at 20 torr cm, a 10 pf parallel capacitance and a 10 kW series resistance. By 'takeover' voltage, we mean the lowest voltage which can sustain the glow. If the plasma density is less than about 10^8 cm^{-3}, the takeover voltage is not very different from the breakdown voltage given by the Paschen curve. As the plasma density increases, the takeover voltage decreases until a minimum is reached which is about 10 V about the steadt-state glow voltage. The takeover voltage is higher for higher values of the series resistance.

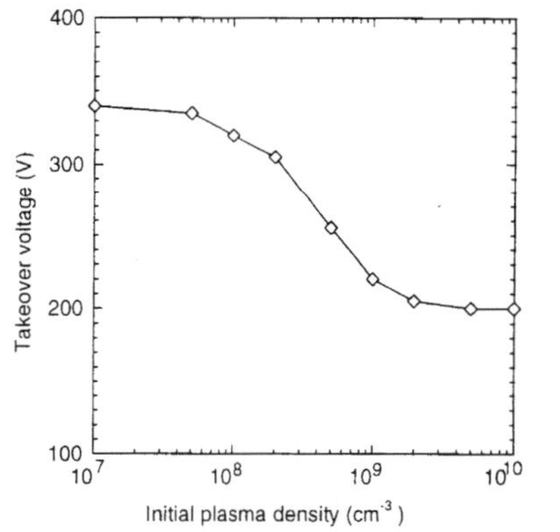

Fig. 2. Takeover voltage as a function of the initial plasma density for conditions in text.

3. CONCLUSIONS

Although our model is too approximate to draw qualitative conclusions, it is sufficient to provide a basis for understanding the interplay of the breakdown and takeover voltages and to deduce trends. The results from the model are consistent with experiment.

Some conclusions from this work are:

(a) Lamp starting characteristics are not very dependent on the high voltage breakdown pulse as long as breakdown has occurred during the high voltage pulse. However, the source impedance is an important parameter.

(b) High voltage spikes superimposed on a 50 Hz generator voltage are effective if they are applied before the peak of the voltage. The phase shift between the high voltage pulse and the peak in the generator voltage is an increasingly more important parameter for lower pressures.

(c) Multiple breakdown pulses can be effective if the time between pulses is less than the ion transit time in the gap (some 10's of μs).

(d) The takeover voltage depends strongly on the series resistance in the takeover circuit.

(e) The presence of HI as an impurity causes a large increase in the takeover voltage.

4. ACKNOWLEDGEMENTS

This work was supported in part by the Central Development Laboratory of Philips Lighting. I would like to acknowledge contributions of colleagues in Philips Lighting (especially Herman Gielen and Pieter Postma) and at the CPAT (J.P. Boeuf, K. Liland, and I. Pérès) to this work.

5. REFERENCES

[1] L.C. Pitchford, I. Pérès, K.B. Liland, J.P. Boeuf and H. Gielen, J. Appl. Phys, **82** 112 (1997).

IL 05

ULTRA HIGH PERFORMANCE DISCHARGE LAMPS FOR PROJECTION TV SYSTEMS

E.Fischer

Philips Research Laboratories, Weisshausstrasse 2, D-52066, Aachen, Germany

1. INTRODUCTION

In the last years TV projection systems using liquid crystal panels to generate the TV picture have become of increasing importance. The rapidly growing market depends strongly on the availability of suitable projection lamps. The usual way to encode the image information into the projecting light beam is to split the light beam into three colors by dichroic filters, then pass the partial beams through three LCD panels, put them together again into a single beam by a suitable filter arrangement, and then project it onto a screen. A schematic sketch of such a light path is shown in fig.1. There are many additional features and variations of this principle to improve the homogeneity and intensity of illumination, but common to all systems is, that the light path in the system is long, the angle of acceptance of the filters and the LCD panels is small and the light losses in the optical components are high. To minimize the losses these projection systems require point-like light sources of extremely high luminance. In addition the light of the lamp should contain equal amounts of red, green and blue radiation, so that all colors can be projected with the same brightness.

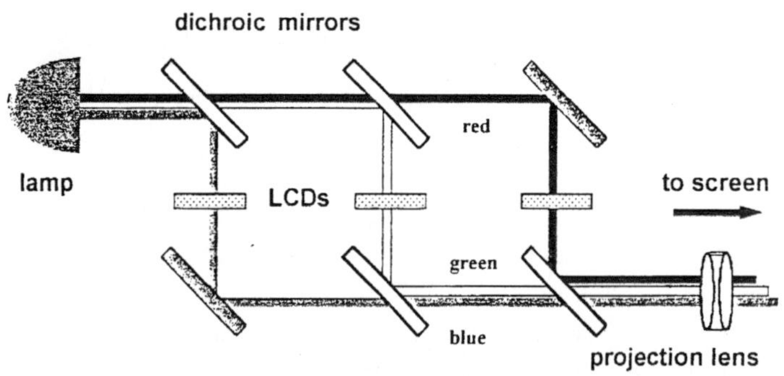

Fig.1: Schematic representation of the light path in a LCD TV-projector.

What set manufacturers actually would like, is a point source of about 8000 K color temperature. As the brightness of a thermal radiator is limited by the black body intensity, maximum brightness would be obtained with a lamp emitting a black body spectrum at this temperature. Of course the lamp should live as long as conventional TV sets at least, and should not change its properties during life.

The development of the UHP (ultra high performance) lamp system was the attempt, to make a step in the direction of this ideal light source. First attempts to use incandescent lamps as light sources for projection TV had failed due to the low radiation level in the blue spectral region. As the surface temperature of solids is limited by melting and evaporation, the brightness of the source is limited by the black body intensity at this temperature. To reach higher brightness gas discharge lamps appeared as the only viable alternative.

Thus most developers have focussed on high pressure short arc lamps with metal halide filling. These lamps emit atomic line radiation from various metals and molecular bands from the corresponding halides. Due to the low ionization energy of the additives, the plasma temperature in these lamps is between 5000 K and 6000 K in the center. Most molecules are not even stable at these temperatures, so that the effective emission temperature of molecular bands is still lower. Even for the strongest emitters the intensity of radiation cannot exceed the black body intensity at these temperatures.

Rising the power density by shorter electrode gaps can bring the radiation intensity closer to the black body limit, but this method is limited by the stability of the electrodes. Though metal halide lamps represent a major improvement compared with incandescent lamps, they are still far away from the ideal point source.

The only way, to reach still higher luminosity with discharge lamps, is to use only such fillings, that allow higher plasma temperatures. That means, the plasma should contain only chemical elements with high ionization energy. By principle it should be possible with every such filling to obtain an intensity close to the black body, if only the gas phase concentration of the radiating atoms is high enough. Unfortunately the periodical system contains only very few elements, which fulfil this condition of high ionization energy and which, in addition, are gaseous or can be brought into the gas phase by suitable chemical processes. A further restriction is, that the filling should not be too aggressive to the tube and electrode material.

With all these restrictions the only remaining candidates are rare gases and mercury. Rare gas plasmas appear more attractive with respect to the higher plasma temperature, but are very poor radiators in the visible. Pressures in the order of several thousand bars would be required to approach black body emission from the center. In mercury vapor the plasma temperature only reaches 7000 K to 8000 K, but the emission properties of mercury vapor are much better, so that an attractive spectrum is obtained already in the pressure range of 200 bar. The occurrence of strong continuous visible emission in mercury vapor discharges at high operating pressures has already been described by Elenbaas [1] fifty years ago, but has never been used as the dominant radiation process in lamps.

2. THE UHP CONCEPT

2.1 Radiation Processes

To create projection lamps with extremely high luminance the path directed by Elenbaas has been followed. To understand the origin of the continuous emission spectrum, we must look somewhat closer at the chemical properties of mercury vapor. Normally mercury vapor is regarded as an atomic gas.

The existence of diatomic molecules is well known, but the molecular ground state is only loosely bound, so that the concentration of Hg_2 is low at normal pressure. The spectrum emitted by mercury plasmas is dominated by atomic lines, molecular band radiation is very weak. But as the molecule concentration rises with the square of pressure, the contribution of molecular band radiation becomes important, when the operating pressure is increased.

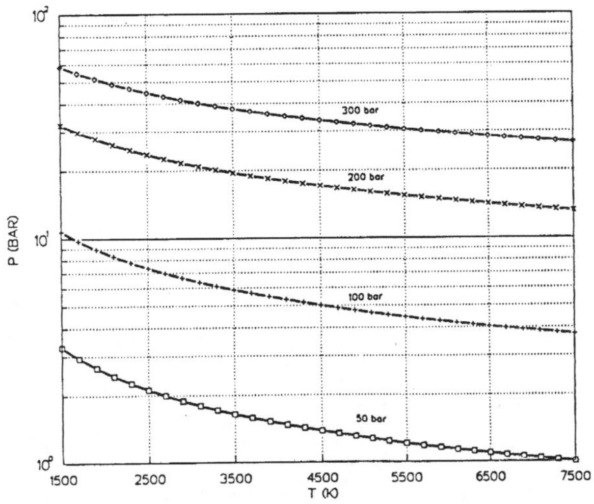

Fig.2: Partial pressure of diatomic mercury molecules dependent on temperature and total mercury pressure.

Fig.2 shows the temperature dependence of the Hg_2 partial pressure for different total mercury pressure. As the enthalpy of formation for the molecules is small, the concentration of molecules decrease very slowly with increasing temperature, so that in a 200 bar plasma the partial pressure in the central region of the arc is still in the order of 10 bar.

In this region with temperatures of 7000 to 8000 K excited molecules exist in various electronic states, so that besides of the resonant molecular bands in the far UV also radiation from transitions between excited states is observed in the visible and near IR. The concentration of ground state molecules is so high, that below 280 nm all radiation from the center is absorbed in the outer layers. Not even the strong 254 nm atomic line can be detected in such lamps.

The advantage of the non-resonant molecular radiation is, that it is not reabsorbed in the outer layers and can thus theoretically could be emitted with an intensity close to the black body intensity at center temperature. To reach the black body limit, a pressure of 200 bar is still too low. Only the atomic lines approximately reach this intensity. They are self absorbed and strongly broadened by collisional interactions. Fig.3 shows spectra of mercury discharges at 100, 200 and 300 bar operating pressure. The spectra are calculated using transition probabilities and broadening parameters derived from spectroscopic measurements. They show, that already at 200 bar more light is emitted as continuum radiation than in the atomic lines.

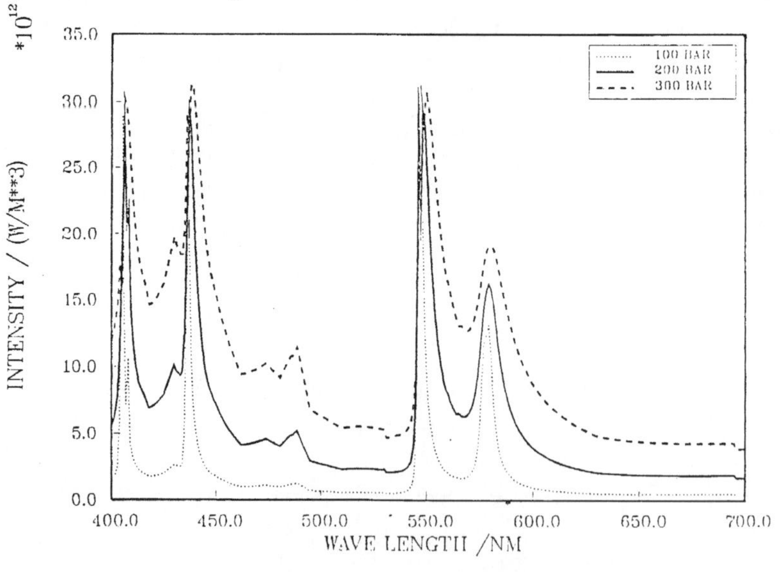

Fig.3: Emitted spectra of mercury arc lamps at 100, 200 and 300 bar operating pressure, calculated for 100W lamps with 1.3 mm arc length.

The spectrum exhibits the broadened lines and the continuous molecular radiation, which extends from the near UV into IR with decreasing intensity. As the spectrum is not very selectively concentrated in the visible, the luminous efficacy is not as high as in some metal halide lamps. 65 to 70 lm/W can be obtained at most. Thus such lamps will never be world record holders in efficacy, but well in luminance.

The spectra show, that the continuous radiation is the only contribution in the red part of the spectrum. Mercury lamps with lower pressure, which radiate preferably the atomic lines, are not usable as TV light sources. To reach an acceptable color balance, an operating pressure of at least 200 bar is necessary.

2.2 Mechanical and Thermal Aspects

Building a lamp, which operates at 200 bar is not a trivial task. Apart from the fact, that the bulb and the electrode feed through must withstand the high pressure, also the thermal requirements are high. To evaporate mercury to a pressure of 200 bar, the temperature of the coldest spot on the inner surface of the bulb must be high. On the other side the hottest point must not exceed the devitrification temperature of the quartz wall.

The fact, that there is only a narrow range of allowed temperatures, limits the size of the bulb. In every lamp the density gradients in the gas lead to an internal convective flow, which transports energy from the lower to the upper side of the bulb. Fig.4 shows a typical flow pattern (from [2]). The flow velocity and with it the energy flux increases with increasing linear dimensions of the bulb and with increasing pressure. To stay within the narrow temperature limits between the upper and lower side, the bulb has to be very small. At an operating pressure of 200 bar the linear dimensions of the inner volume should not exceed 4 to 5 mm. On the other hand this small dimensions are advantageous to reach sufficient pressure resistance of the bulb.

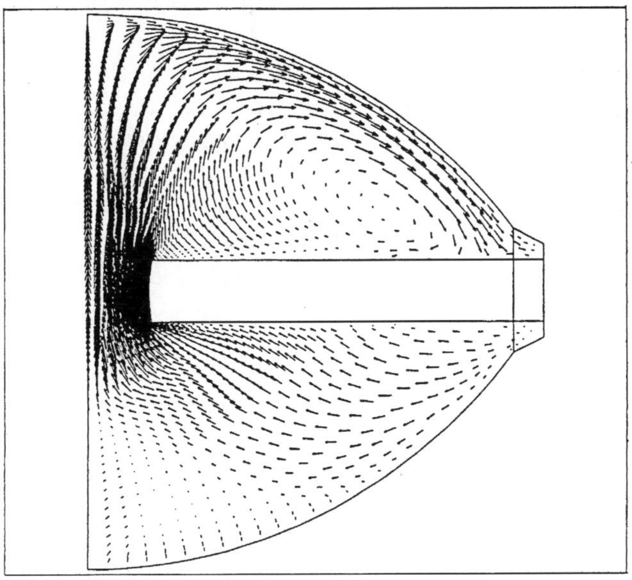

Fig.4: Pattern of the convective flow in a high pressure short arc lamp.

Another feature of high pressure mercury concept is, that the required electrical field strength in the arc column increases with pressure. At 200 bar the field strength is in the order of 60 V/mm. High burning voltage means low arc current and thus low electrode losses. The low thermal load of the electrodes makes it easier, to keep the arc gap constant over long periods of time. Thus the concept is favorable preferably for short arc lamps, just for the point sources, which are necessary for projection TV.

A typical 100W lamp of this type with an electrode gap of 1.3 mm has a burning voltage of 85V. It has been proven, that the arc gap and with it the burning voltage can be kept constant over thousands of hours. But this constancy is not only a consequence of the low current, but also of the small chemical transport rates, as can be seen from the considerations in the next section.

2.3 Halogen Transport Cycle

In the last section we mentioned, that due to the low arc current the thermal load of the electrodes is low and consequently also the evaporation of the tungsten electrodes. But due to the small inner surface even this rate would lead to severe blackening of the tube wall. The absorption of radiation would strongly influence the thermal balance of the wall.

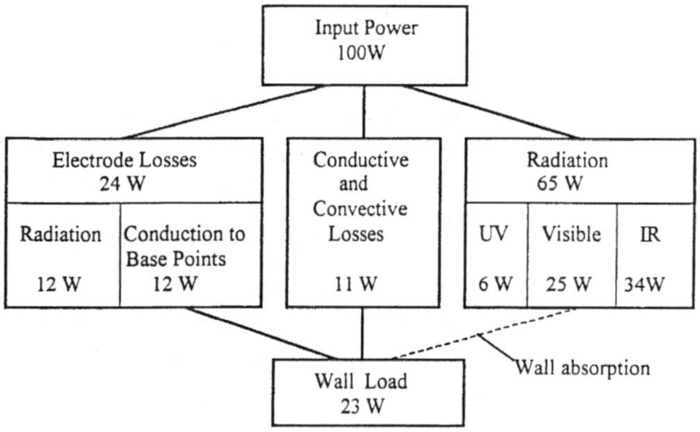

Fig.5: Schematic representation of the energy flow in an UHP lamp.

The energy balance of the lamp is dominated by radiation processes. Fig.5 shows a schematic energy flow chart for a 100W lamp. Only 11 % of the input power are carried directly from the plasma to the wall by conduction and convection, another 12 % reach the wall from the electrode base points, while the rest of 77 % leaves the lamp as radiation from the plasma and from the electrode. If only a small amount of this radiation is reabsorbed, the heat balance of the wall is strongly disturbed.

For a reasonable life time of this type of lamps the wall should not blacken at all. A concept to avoid wall blackening is known for incandescent lamps for long time. Adding a certain amount of halogen and oxygen to the lamp atmosphere prevents the tungsten evaporated from the incandescent filament from condensing on the walls, as in the colder regions the tungsten atoms react chemically to form oxihalide molecules. The vapor pressure of the oxihalides is higher than that of tungsten by many orders of magnitude, so that no tungsten can condense on the walls as long as the tungsten vapor pressure at the filament temperature remains below the saturation pressure of the oxihalides at the wall.

This principle works in HID lamps as well, as long as it is possible to establish a stable gas phase, containing sufficient amounts of oxygen and halogen. Most lamps with metal halide fillings do not allow a halogen cycle, as many additive metals are strong oxygen getters and lower the oxygen level so far, that no tungsten-oxihalides can be formed. In pure mercury lamps the most stable oxide that can be formed is WO_2 and the vapor pressure of this oxide determines the oxygen content of the plasma.

The topic has been discussed in detail in many papers (see e.g. [3-5]). The qualitative description of the transport processes in these papers is based on the "solubility model". The tungsten solubility in a given gas atmosphere is defined as the sum of partial pressures of tungsten containing components (multiplied with the respective number of tungsten atoms), when the gas is in contact with a solid tungsten surface. Fig.6 shows the typical equilibrium composition and the tungsten solubility for the situation in a UHP lamp containing a bromine filling. The curves are calculated under the condition, that the oxygen content of the gas phase is limited by the formation of solid tungsten oxide at the cold spot.

Tungsten transport occurs from the temperature regions of high solubility to those with low solubility. While at the electrode tips the solubility is primarily determined by the tungsten vapor pressure, at wall temperature WO_2Br_2 molecules dominate. By choosing an appropriate bromine level, the solubility at the wall can be fixed on a level slightly above the value at the electrode tip, even at the melting point of tungsten. In this case no tungsten can be deposited at the wall.

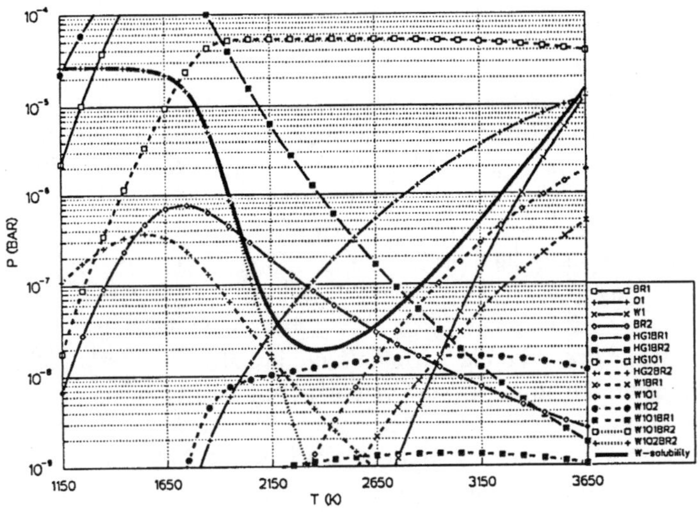

Fig.6: Chemical composition and tungsten solubility.

It is important to rise the solubility not too much, because the transport mechanism, which keeps the wall clean, removes tungsten also from the cold ends of the electrodes (see [4]). This corrosion at the electrode rods is a major problem in metal halide lamps, where halogen comes free from the salts during life and thus enhances the solubility. In a pure mercury lamp the halogen level remains constant during lamp life.

3. THE UHP LAMP

Based on the concept described in the last section, a new 100 W lamp for projection TV applications has been developed. The lamp has an arc length of 1.3mm and operates at a pressure of about 200 bar. The quartz bulb has a spherical shape, to avoid optical distortion of the emitted light bundle.

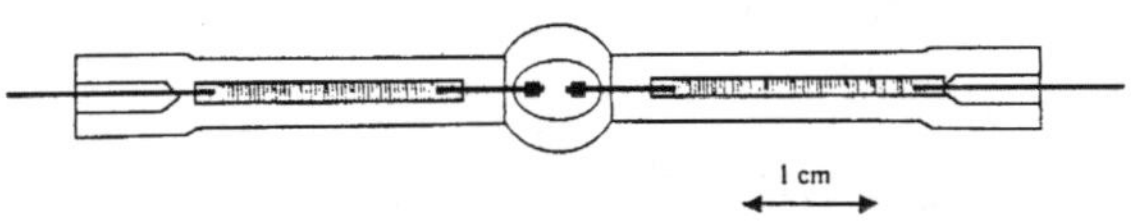

Fig.7: Schematic representation of the UHP lamp geometry.

Fig. 7 shows a sketch of the burner geometry. The cylindrical parts extending from the spherical bulb have been chosen very long to avoid thermal corrosion of the current lead-in wires. This is necessary, because the lamp is designed to burn in air without an outer bulb, to keep as much as possible of the high luminosity in the projected light beam. Any additional glass surface would lower the brightness due to reflection or scattering.

The nominal operating voltage of the lamp is 85 V, the arc length is 1.3 mm A critical point of lamps with such small arc gaps is, that a small deviation from the nominal value leads to a large change of the arc voltage. A deviation by only 0.1 mm leads to a voltage change of about 6 V. When operating the lamp with a conventional inductive ballast, this would lead to a change of the consumed power by 15 % . This again would disturb the thermal balance of the lamp tube and thus lead to reduced lamp life or light quality. Therefore a special electronic ballast with power stabilization has been developed, to keep the lamp in optimum burning conditions.

The spectrum of the lamp is close to the theoretical data shown in fig. 3. The color temperature is 8500 K with a color point close to the day light line ($x = 0.289 \pm 0.015$, $y = 0.302 \pm 0.015$) As there is no wall blackening and no chemical reactions, which could change the plasma composition, the color rendering properties are practically constant during life.

The same holds for the total emitted light. The efficacy remains at 100%. Tests of up to 8000 hours switched operation (3.5h on /0.5h off) were carried out to prove this fact.

The maintenance of collectable light of lamps mounted in parabolic reflectors – which is the standard type of lamps sold to customers – is less than 100 %, as in spite of the sophisticated thermal balance over long periods of time the high wall load in many lamps leads to devitrification of the quartz wall and thus to scattering of light on the upper side of an horizontally burning lamp. Additional light losses may result from degradation of the parabolic mirror surface. But also for the bundle maintenance (the light in a beam of 3° aperture) after 8000 h values of 80 to 85 % have been obtained.

One problem with UHP lamps is that there is no clear life determining process. Metal halide lamps get blackened or they do ignite no more, when the content of free halogen rises to much. UHP lamps are extremely stable with respect to their optical and electrical properties. So the end of life is determined in most cases by material degradation or by the formation of thermal stress in the quartz tube. Thus explosion is a possible end of life.

For this reason all lamps, which are given to customers, are mounted in reflectors with front glass. The glass parts are so thick, that they cannot be broken by the lamp fragments. In spite of the very high operating pressure the explosive energy is not higher than in conventional metal halide lamps, as the inner lamp volume is only 55 mm^3. The reflectors are not hermetically sealed, but allow pressure equalization with the environment. The apertures are so small, however, that no quartz fragments or mercury can escape and cause damage in the TV set.

4. CONCLUSIONS AND OUTLOOK

It has been proven, that the high pressure mercury concept with halogen cycle allows the realization of point-like light sources with by now unknown brightness and life time. The stability of color rendering properties and light level is superior to metal halide lamps, used in the past in projection systems.

Due to its unique properties the UHP concept will open a new application field, as it allows the development of small scale portable TV projection systems. The advantages of increasing brightness for the efficiency of projection systems will be addressed in a separate paper of this conference [6].

Of course we have not yet reached the ideal point source mentioned at the beginning of this paper. Still better spectra and still shorter arcs would be desirable. But even at this time the UHP lamp can help to make LCD projection the TV system of the future.

5. REFERENCES:

[1] Elenbaas, W., „The High Pressure Mercury Vapour Discharge“, North Holland, Amsterdam, 1951
[2] Born, M., Giese, H., Krücken, T., paper of this conference
[3] Dittmer, G., Klopfer, A., Schröder, J., Philips Res. Repts., 32,341 (1977)
[4] Fischer, E., Proc. Symp. High Temp. Lamp Chem.II, p 111, Toronto, Canada (1985)
[5] Fischer, E., Schnedler E., Proc. 5th Int Symp. on the Sci. & Techn of Light Sources, York, England (1989)
[6] Derra, G., Mönch, H., paper of this conference

DIAGNOSTICS AND MODELING OF MICRO-DISCHARGE FOR PDP'S AND OTHER APPLICATIONS

Kunihide Tachibana

Department of Electronic Science and Engineering, Kyoto University,
Yoshia-Honmachi, Sakyo-ku, Kyoto 606-8501, Japan

1 INTRODUCTION

In recent plasma application technologies, various kinds of micro-discharges have become of great concern in the sense of one to three dimensionally small scales. For examples, a narrow gap planer discharges may be categorized in the one-dimensional micro-scale discharge, a streamer-like discharge or a discharge confined along a thin wire may be in the two-dimensional one, and a discharge confined in a small volume may be in the three-dimensional one. The first one is interested in the application to a large-area flat light source as well as in the application to surface plasma-treatment of various materials. The second one is also going to be used in the material processing such as patterning of solar cells, slicing Si ingot etc. The major application of the third one is to plasma display panels of which cell size is 0.2 to 0.5 mm, but it may also be useful as a tool of micro-machining. In this invited talk, I would like to explain the present status of diagnostics and simulations of such micro-discharges by taking typical examples from our recent analyses of discharge and light-emission phenomena in PDP cells.

2 DISCHARGE SCHEMES USED IN PDP CELLS

There have been developed two types of discharge schemes for PDP's, i.e, DC-type and AC-type. In a typical DC-type PDP cell a pair of naked anode and cathode are located on the opposite sides of glass plates with a certain gas gap as shown in Fig.1(a). A series resistor is tends to be connected for the discharge current restriction purpose. On

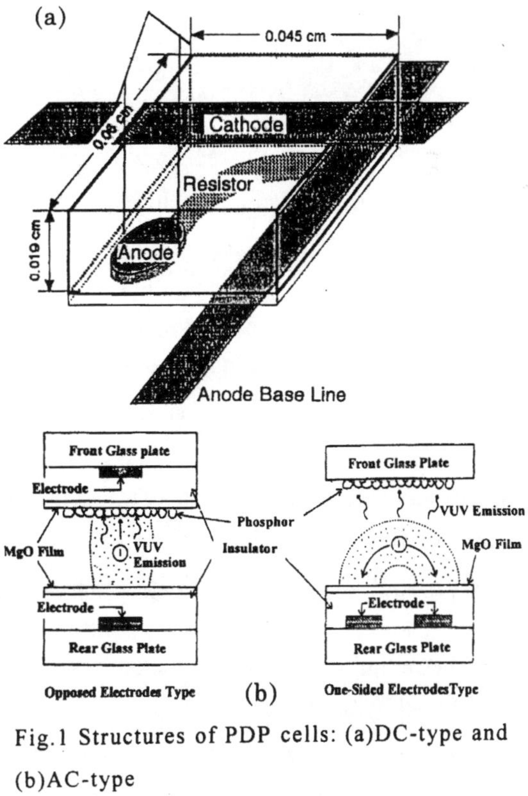

Fig.1 Structures of PDP cells: (a)DC-type and (b)AC-type

model since the filling gas pressure is kept relatively high. We have employed a one-dimensional fluid model under the local field approximation (LFA) in which transport coefficients, such as electron drift velocity, ionization and excitation coefficients etc., are specified by the local electric field strength.[1]

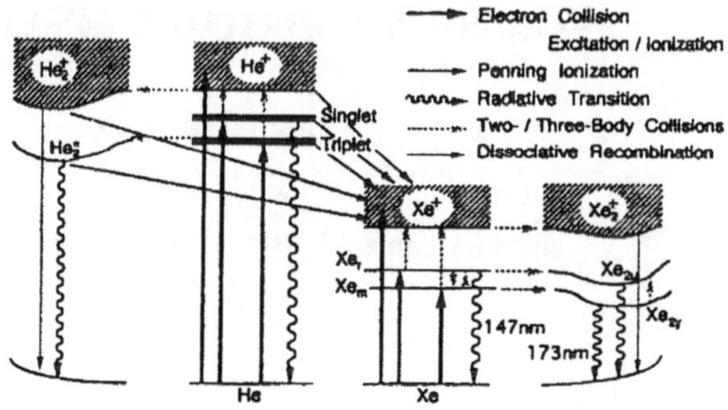

Fig.3 Atomic and molecular processes in a discharge in He-Xe mixture

The basic set of equations are composed of the continuity equation for a particle i, the Poisson's equation and the circuit equation as follows;

$$\frac{\partial n_i}{\partial t} = -\frac{\partial J_i}{\partial x} + G_i(x,t) - L_i(x,t), \quad (1)$$

$$\frac{\partial^2 V}{\partial x^2} = -\frac{\rho}{\varepsilon}, \quad (2)$$

$$V_g(t) = V_a(t) - ZI(t), \quad (3)$$

where n_i is the density of the species, G_i and L_i are the production and loss terms, respectively, J_i is the flux given by the drift velocity v_i and the diffusion coefficient D_i as $J_i = v_i n_i - \partial(D_i n_i)/\partial x$, V is the potential related to the electric field E as $E = -\partial V/\partial x$, ρ is the space charge density, V_g is the gap voltage, V_a is the applied voltage, Z is the external resistance and I is the discharge current. The boundary condition for n_i is set zero at the electrode surfaces except for the secondary electrons produced at the cathode due to collisions of ions and excited species. We have measured the excitation coefficient of 1s levels of Xe in a He-Xe mixture as a function of the reduced field strength E/N, where N is the total gas density.[2] Transport coefficients for electrons, ions and neutrals and the rate coefficients for collisions between heavy particles have been compiled from numerous references.[1] The coupled equations have been solved numerically by a difference method, where the Scharfetter-Gummel's scheme has been used. This scheme is based on the implicit difference method which is less sensitive to the setting of time steps. The radiation transport of the Xe resonance line at 147 nm has been treated by the method developed by Sahni *et al.*[3]

An example of the calculated results is show in Fig.4 for the electric field, Xe ion density (major ionic species), electron density, and densities of metastable Xe($1s_5$) atoms, resonance the contrary, in an AC-type PDP cell both anode and cathode are covered by dielectrics as shown in Fig.1(b), so that the discharge self-terminates due to the accumulation of charges on the surface. However, in the next reversed cycle the electric field by the accumulated charge is added to the external applied voltage, and the discharge is going to be ignited successively. Two types of the electrode configurations have been employed in the AC-type: the opposed electrodes type and the one-sided parallel electrodes type. In the latter type an address electrode is mounted on the other side of the cell to form the xy-matrix. The dielectric surface is usually covered with MgO thin film for higher ion-bombardment resistivity and higher secondary electron emission capability.

Gas mixtures of Xe with other rare gases have been used, but He-Xe mixture and Ne-Xe mixtures are commonly used for the DC-type and the AC-type, respectively, based on a long history of experimental try-and-errors. Typical range of the gas pressure p is 200 to 400 Torr with a mixing ratio of Xe at 3 to 10%. Under this condition the similarity parameter pd lies nearly at the minimum of the Paschen curve with the electrode gap d of the order of 0.1 to 0.2 mm. However, in order to decrease the sputtering effect and then increasing the lifetime, a little bit higher pressure is preferred. In a PDP the VUV emissions from Xe($1s_4$) atoms at 147 nm and from Xe_2^* excited dimers 153 nm are converted to RGB dots by phospors.

Among several problems remaining in the development of commercially available PDP's, the improvement of electrical efficiency is the key issue. At the moment the efficiency lies around 1 lm/W at most. However, it should be enhanced up to 5 lm/W for panels of 40 to 50 inches scale in order to keep the power consumption on the same level of CRT. To get to this target, every possible efforts must be done from various sides of approach, say, diagnostics and simulation of micro-discharge phenomena, developments of electrode and/or dielectric materials and fluorescent materials, and efforts for searching new discharge schemes. Here, we will concentrate on the mechanisms of discharge and VUV light-emission phenomena.

3 MODELLING OF A DC-PDP DISCHARGE CELL

For the optimization of the cell design and the operating discharge parameters such as gas composition, pressure, applied voltage etc., a reliable simulation model is desirable as a CAD tool of the PDP panel. Typical dimensions of a discharge cell for the DC-type PDP is shown in Fig.2. Simplified atomic and molecular levels in a He-Xe system are shown in Fig.3 together with the relevant collisional-radiative processes. Molecular states must be considered in the

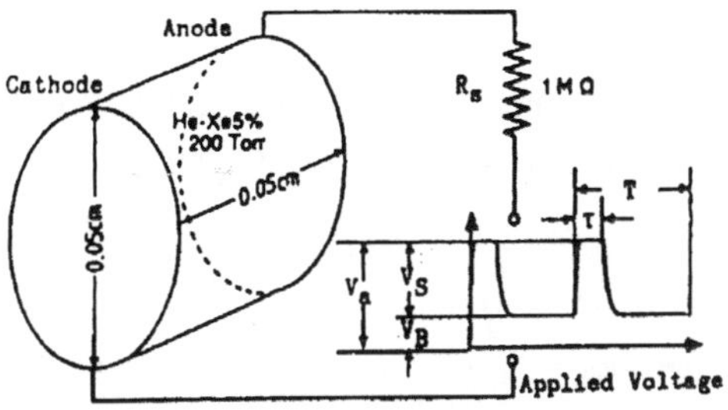

Fig.2 Boundary conditions of DC-type cell for one-dimensional model

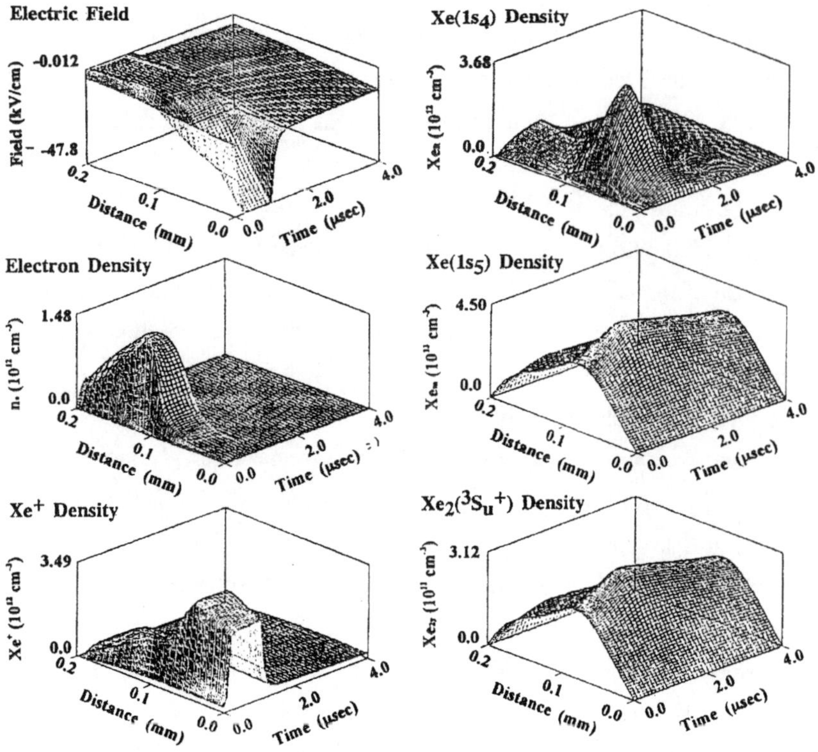

Fig.4 Example of simulation results in a DC-type cell with 0.2mm gap at 1 μs pulse duration in 4μs period

$Xe(1s_4)$ atoms and excited Xe_2* molecules (denoted as Xe_m, Xe_r and Xe_{2y}, respectively). Characteristic features of the results are as follows. As the the discharge current rises, the densities of electrons and ions increase rapidly, and then ions are attracted toward the cathode side and electrons are repelled there forming the cathode fall region. The secondary electrons emitted from the cathode by ion bombardments gain energy in the region, and the excitation and ionization processes become dominant between the plasma bulk and the cathode fall region. Since $Xe(1s_4)$ atoms have an effective lifetime of $\tau = 0.5$ to 1 μs, the density of $Xe(1s_4)$ atoms is accumulated and reaches to a quasi-steady state before the fall of the pulsed voltage, and then decays with the time constant τ. However, metastable $Xe(1s_5)$, atoms have much longer lifetime, so that the density decays quite slowly. Therefore, it remains considerably even at the end of one cycle. This helps in lowering the sustain voltage for successive pulses and enabling the so-called pulsed memory operation. Since the excited Xe_2* molecules are formed from $Xe(1s_5)$ atoms by three-body collisions with 2Xe or Xe + He, the Xe_2* density follows that of $Xe(1s_5)$ atoms.

The validity of the local field approximation has been argued in some cases, especially at higher field regions such as the cathode fall region. Therefore, we have developed another approximation method.[4)] In the model a beam component of electrons has been taken into account, of which energy is calculated from the energy balance. The calculated results in this model differ in the spatial distributions of electric field and charged particles, but no appreciable difference has been noticed on the densities of excited Xe atoms and molecules.

The LFA model can also be applied easily to the AC-type PDP discharge. Up to now there reported one- and two-dimensional simulations from the same group.[5,6)] We are also extending the DC-type model to two dimensional scheme and getting some preliminary results.[7)]

4 DIAGNOSTICS OF MICRO-DISCHARGES

In order to verify simulation results and improve the model, it is necessary to make comparisons with experimental data. In the first step, we have performed the measurement of VUV emissions by using a specially designed PDP panel with sapphire front glass which can transmit up to about 140 nm.[8)] From the measurement we have succeeded in the estimation of the relative contribution of the 147 nm atomic line to the 153 nm molecular band. However, it is not possible by this

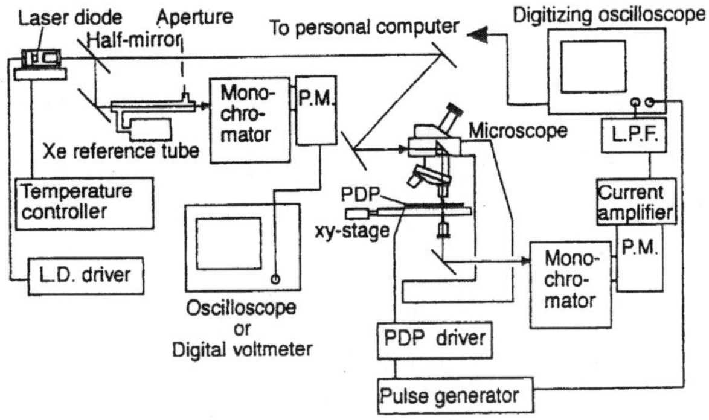

Fig.5 Schematic of experimental setup

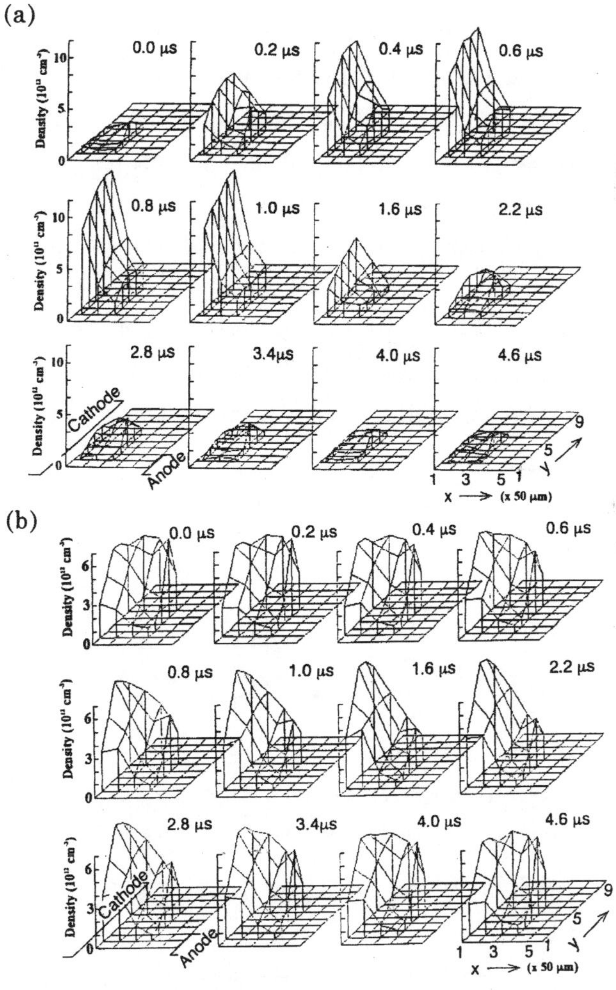

Fig.6 Spatio-temporal behavior of (a) Xe($1s_4$) and (b) Xe($1s_5$) densities measured by laser spectroscopic microscopy

method to measure the spatio-temporal behaviors of those emissions, so that we have developed a new technique by combining a microscope with a laser spectroscopic technique.[9)] The principle of the technique is to measure directly the densities of Xe($1s_4$) and Xe($1s_5$) atoms by absorption using a diode laser tunable to the espective transitions from those levels to appropriate upper levels in the 2p configuration in the near-IR (around 800 nm) wavelength region. The schematic setup is shown in Fig.5. By this method a spatial resolution better than 20 μm and a temporal resolution better than 5 ns have been attained, and the two-dimensional analysis has become possible for dynamic behaviors of VUV emitting atomic and molecular species.

Examples of the measured results by this technique in a cell filled with He-Xe(10%) mixture at 200 Torr are shown in Figs.6 (a) and (b) for the Xe($1s_4$) and Xe($1s_5$) levels, respectively. The structure of the cell (0.5 x 0.5 mm square with 0.2 mm thickness) has been specially designed for this measurement with electrodes arranged on the left-hand and right-hand sides with a 0.3 mm gap. When these results are compared with the results shown in Fig.4, we can notice fairly good qualitative agreements are obtained on the behaviors of Xe($1s_4$) and Xe($1s_5$) atoms. Agreements of their densities in the orders are also in satisfaction. However, for more quantitative comparisons our one-dimensional model is not good enough, so that the need for a reliable two-dimensional model is urgent.

In Fig.7 shown is the dependence of spatial distribution of Xe($1s_4$) atom density measured along the center line in several cells with different electrode gaps from 0.25 to 0.45 mm. It can be seen that the second peak appears in the mid gap other than the negative glow in front of the cathode. This suggests that the positive column is formed in longer gap cells. The sustain voltage does not increase so much as the gap becomes longer, so that the enhancement of luminosity and efficiency can be expected.

Spatio-temporal behavior of visible to near-IR emissions from the plasma has also been measured by using an image-intensified CCD camera at 100ns resolution. A typical result is shown in Fig.8, from which a dynamical behavior of electrons in their density and energy is suggested. If we can measure the changes in the

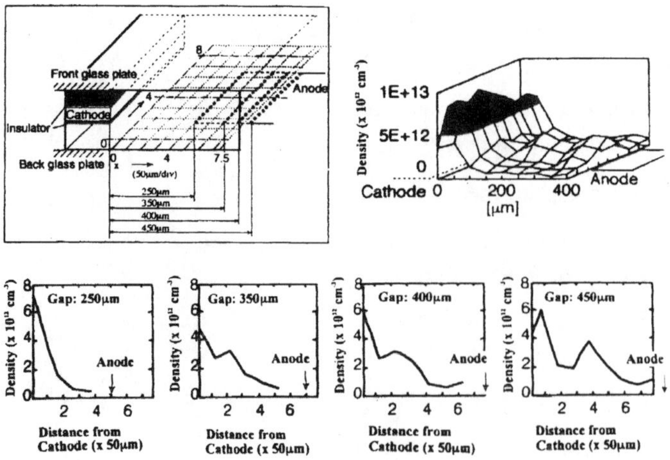

Fig.7 Variation of spatial distribution of Xe($1s_4$) atoms measured at several different electrode gaps

wavelength resolved emissions from several levels with different excitation energies, the mechanisms of production, acceleration and energy relaxation of electrons will be clarified in more detail.

Similar diagnostics are also going on in AC-type cells. Those results will be shown at the time of the conference in comparison with the results in DC-type cells.

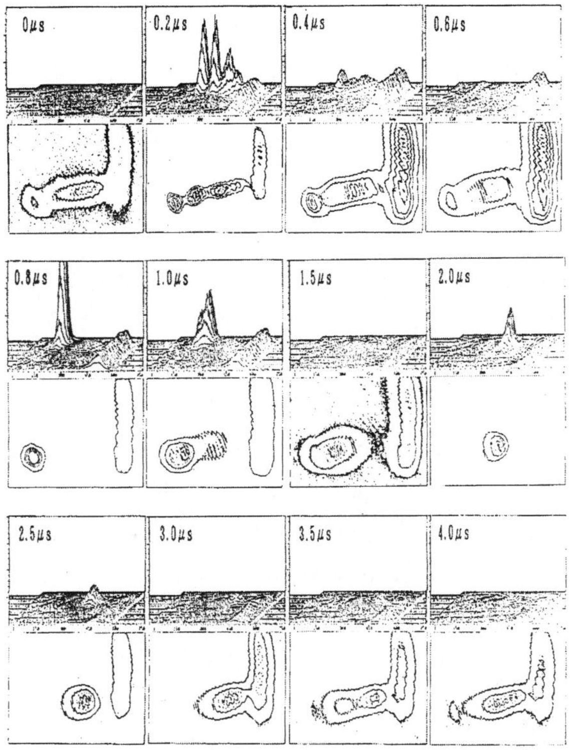

Fig.8 Spatio-temporal behavior of the total emission measured in a DC-type PDP cell with 100 nm resolution

5 TOWARDS THE IMPROVEMENT OF PDP'S EFFICIENCY

A recent large-step success in the efficiency improvement of AC-type PDP is due to the adoption of one-sided electrodes configuration and the change in the coating manner of fluorescent materials from the transmission type to the reflection type. This configuration has become realized by arranging transparent ITO films for the display electrodes on the front panel and the address electrode on the back panel. In this configuration, the discharge may take a u-shaped longer path in the gas gap, although it may depends on the value of pd. If it is the case, there may appear a positive column-like region even in an AC-type cells. As mentioned above, this contributes to the enhancement of efficiency in a DC-type cell.

Search for the electrode materials with higher γ values for the secondary electron emission and ion-bombardment resistively are crucial for DC-type cells, and it is also true for the dielectric materials for AC-type cells. As for the optimization of the operating parameters such as the gas compositions, the driving voltage waveforms etc. predictions by simulations may be effective. Therefore, the construction of reliable models are strongly expected in two- or three-dimensional scheme.

Even though the efforts mentioned above must be continued along the line of improving present technology, completely new ideas are wanted to find out breakthroughs for the final target of the 5 lm/W efficiency. We have recently organized a society for fundamental research on micro-discharge phenomena within University level researchers in Japan. The society is going to collaborate with the "PDP developments consortium" composed of about 30 manufacturing companies in the following aspects: (a) developments of fast and reliable simulation codes in multi-dimensions, (b) developments of new microscopic diagnostic-techniques, (c) developments of new materials for electrodes and dielectrics, and (d) developments of completely new discharge schemes for PDP's.

REFERENCES

1) K. Takahasi, S. Hashiguchi, Y. Murakami, M. Takei, K. Itoh, K. Tachibana and T. Sakai: Jpn. J. Appl. Phys. **35**, 251 (1996).

2) K. Takahasi and K. Tachibana: Trans. IEE Japan, **111-A**, 182 (1991) in Japanese.

3) O. Sahni, C. Lanza and W.E. Howard: J. Appl. Phys. 49, 2365 (1978).

4) S. Hashiguchi: Jpn. J Appl. Phys. **32**, 2865 (1993).

5) J. Meunier, P. Belenguer and J.P. Boeuf: J. Appl. Phys. **78**, 731 (1995).

6) J.P. Boeuf and L.C. Pitchford: IEEE Trans. Plasma Sci. **24**, 95 (1996).

7) Y. Murakami and K. Tachibana: Bull. Am. Phys. Soc. 42, No.8, 1735 (1997), (Abs. 50th Annual Gaseous Electronics Conference, #JTP5-4).

8) T. Ssakai, Y. Murakami, K. Takahashi and K. Tachibana: ITEJ Tech. Rep. 14, No.67, 49 (1990) in Japanese.

9) K. Tachibana, N. Kosigi and T. Sakai: Appl. Phys. Lett. **65**, 935 (1994).

DIELECTRIC BARRIER DISCHARGE

Frank Vollkommer and Lothar Hitzschke
OSRAM GmbH, Hellabrunner Straße 1, D-81536 München

1 INTRODUCTION

Dielectric barrier discharges (DBD's) have already been used by W. Siemens in 1857 [1] for the purpose of ozonizing air. Nevertheless only about a decade ago barrier discharge activities were directed to develop UV radiation sources, especially with rare gas excimer systems like Xe, Kr, Ar, XeCl, KrCl or XeCl [2]. Discharges within these systems allow to generate spectrally broad UV transitions at specific wavelengths between a bound upper level and a dissociatively unstable ground state. Important industrial applications [3-4] of excimer lamps on the basis of DBD's are material deposition, surface modification, pollution control and excitation of phosphors in plasma displays. Usually, DBD's are generated using AC power sources with frequencies between a few tens up to a few hundreds of kHz and driving voltages with amplitudes of typically 10 kV. For technically relevant power densities corresponding efficiencies are about or less than 20 % for xenon [5], 15% for KrCl [6], and less than 12 % for XeCl [7-8].
Last investigations have revealed now that much higher efficiencies can be achieved by exploring a different excitation method. Using excitation pulses which are interrupted by idle times and matching the time constants to the specific gas kinetic properties, it is possible to increase the efficiencies dramatically. For xenon, there was reached an efficiency of more than 60 % [9].
In this paper, we will point out the main differences and properties between discharges which are excited by traditionally AC voltages and the new pulsed power input.

2 FUNDAMENTALS AND AC EXCITATION

Common dielectric barrier discharge configurations are shown in Fig. 1. The DBD's are characterized by the presence of a dielectric layer disposed between at least one electrode and the discharge gas. Normally, the electrodes of these discharges are extended over large areas. The dimension of the discharge gap is small compared by at least one dimension of the electrode surface [3]:

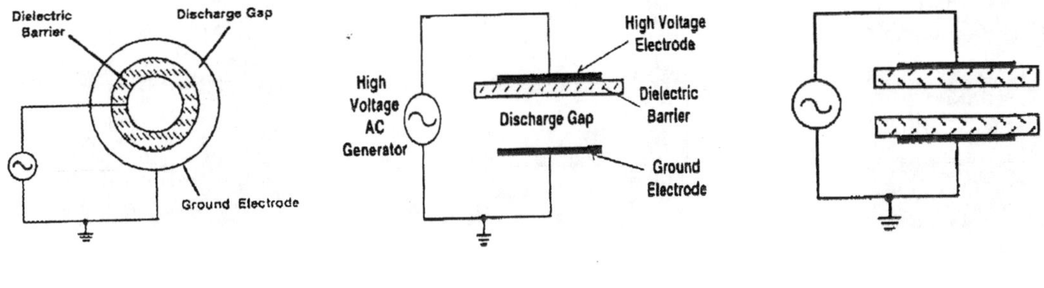

Fig. 1: Common dielectric barrier discharge configurations

The purpose of the dielectric barrier is to limit the electron current between the electrodes:
A microdischarge is initiated once the breakdown field is reached at a certain location in the gap. An electron flux to the barriered anode leads to an accumulation of electrons at the dielectric surface and is generating a space charge in front of the anode. The electric field of this space charge has the opposite direction of the external electric field (Fig. 2). Within a very short time, the compensation of both fields leads to a collapse of the microdischarge at this location. Therefore, the barrier can be characterized by a very fast intrinsic switch which suppresses the generation of a LTE-electron gas. If the external voltage is rising, additional microdischarges will occur at new positions because the presence of residual charges on the dielectric has reduced the electric fields at positions where microdischarges have already occurred. The current density in a microdischarge channel can reach 100 to 1000 A cm^{-2} and the filament radii about 0.1 mm [2]. Fig. 3a and Fig. 3b show snapshots of microdischarges, typical for a configuration with a 5 mm xenon gap and AC excitation. Fig. 4 is displaying the increasing number of microdischarges with rising frequency and power input [11]. The common understanding of the microdischarge formation is stressing an extremely fast streamer propagation toward both electrodes when an overvoltage is applied to a discharge gap. Investigations in dried atmospheric air with the aid of a Streak-camera record have confirmed this idea [10]. Usually with AC excitation the electron densities are reaching values between 10^{14} and 10^{15} cm^{-3} [3]. Detailed investigations on the geometry of microdischarges have been performed [11]:

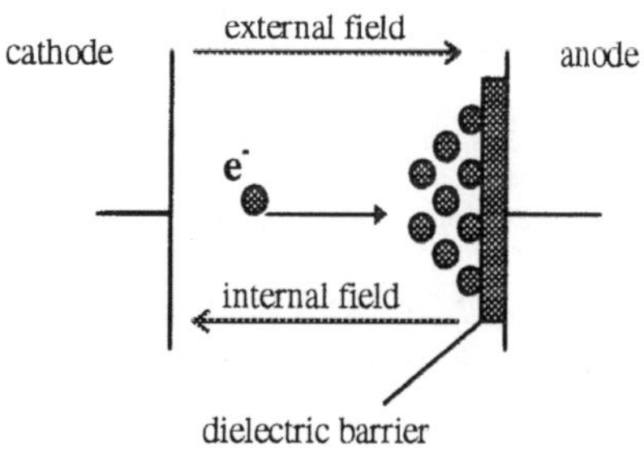

Fig. 2: Function of the dielectric barrier

Fig. 3a: End-on view of micro-discharge

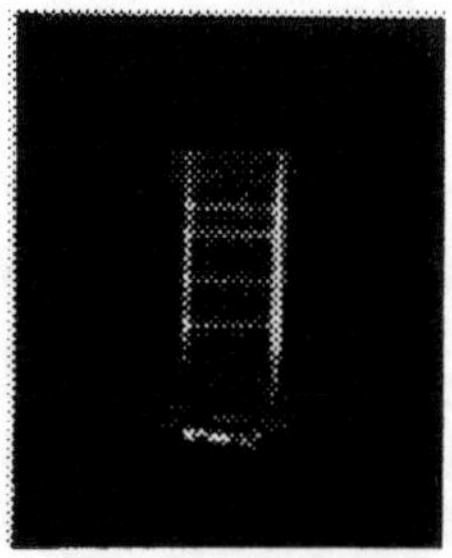

Fig.3b: Side-on view of microdischarges

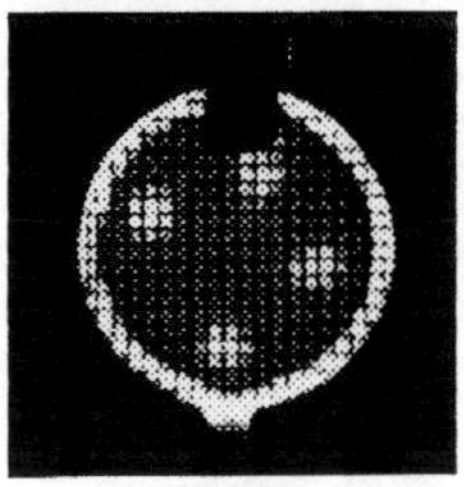

17kHz

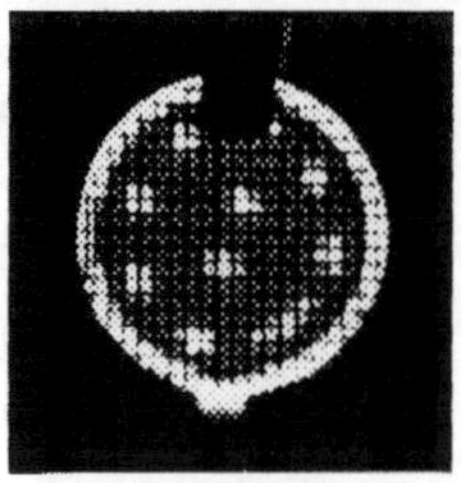

32kHz

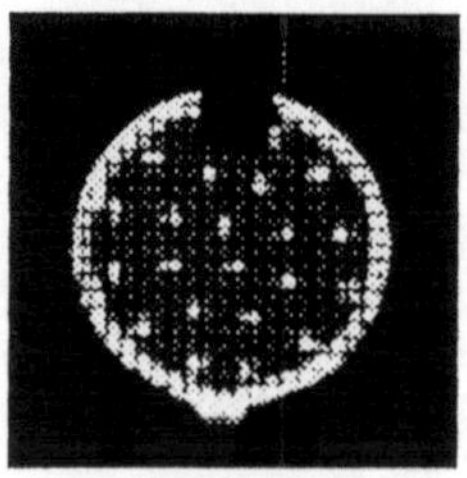

60kHz

Fig. 4: End-on view of microdischarges with increasing frequency and increasing power input

In most cases a lot of microdischarges with narrow filament structures and diameters of about 100-300 µm are emerging. Only at very low voltages and specific frequencies single ignitions of local discharges are appearing with much wider structures and external efficiencies of 20 % [5]. Because an increase of the power density can only be realized by increasing the external voltage, this efficiency declines below 10% for technically relevant power densities [5].
Fig. 5 is demonstrating the main aspects of the radiation generation at the example of xenon: Electrons from gas discharge are exciting neutral xenon atoms to a metastable state (Xe*). After three-body collisions of Xe* with two ground-state Xe atoms a Xe^*_2 excimer is generated. This molecule is dissociating by emitting a high energy photon. A typical spectrum of a Xenon excimer system at 140 Torr is shown in Fig. 6.

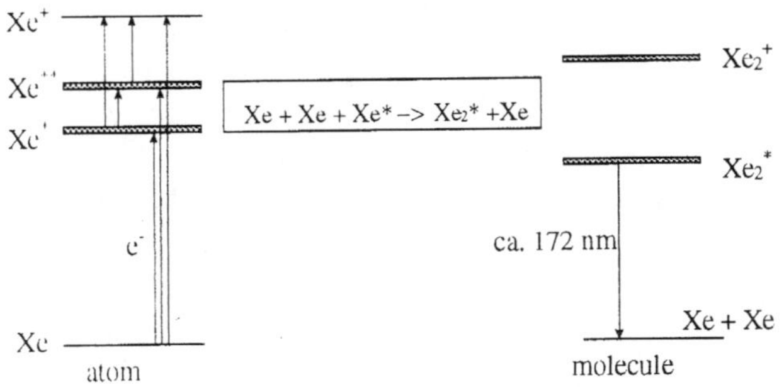

Fig. 5: Simplified picture of the Xe excimer radiation generation

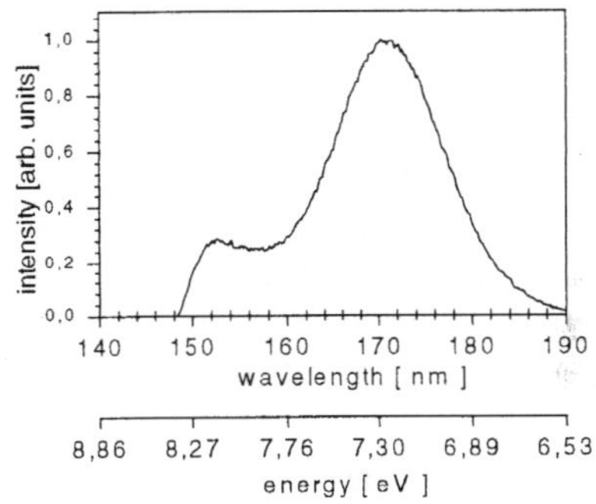

Fig. 6: Spectrum of a Xe excimer discharge (140 Torr)

3 THE NEW OSRAM SYSTEMS

To increase the efficiency of UV generation conditions had to be found where a maximum of the power input is used to create metastable Xe^* atoms. In other words, among all different inelastic collisions we had to increase that portion of collisions in the process of which Xe^* were produced. Simultaneously, the energy loss due to electron quenching collisions had to be kept as small as possible. In this sense, the generation of higher excited atoms had to be reduced. Generally, in discharges quenching processes can not be suppressed totally. However, a reduction of these processes can be achieved when the electron energy distribution is allowing only a minor part of electrons to quench Xe*. In conjunction with the first condition these demands are guiding to operate DBD's within a regime of extreme non equilibrium. Further, we had to avoid a fast relaxation of the electron gas to a Maxwellian distribution due to frequent electron-electron collisions. For this reason, the electron density in our discharge had to be low. And finally, we had to ensure, that all of these conditions were satisfied at technically relevant power densities. The complete set of conditions can be summarized as follows:
The gas volume between the electrodes has to be filled tightly in space and in time with discharges, which are characterized by low electron densities and electron distribution functions, far away from equilibrium. A simple AC excitation at technically relevant power densities has the consequence that thin filamentary discharges are appearing stochastically in space and time. The conditions for getting a maximal efficiency are violated. Further, AC driven discharges unavoidably are running into

stepwise ionization and increased loss processes. Therefore, for an optimization of the efficiency of DBD's it is necessary to swich to a complete different type of excitation: Depending on the heavy particle system, well defined short excitation pulses have to be applied which are followed by especially chosen idle times. The latter have to be selected in such a manner, that during this time period the whole plasma is able to relax into a state, which regarding the space pattern is equivalent to that state, the whole system was existing in, before the excitation had started. Note: In doing so, the development of the electron subsystem into the direction of thermodynamic equilibrium due to a stepwise ionization is efficiently reduced. A correct choice of the idle time duration has been made when the remaining space charges are able to support the ignition process such that periodically discharge patterns in space and time are promoted to appear.

In other words: **The pulsed excitation of DBD's is the necessary condition for a high efficient generation of excimer radiation.**

The explicite external voltage shape, which has to be applied depends on the specific filling gas, the pressure and the electrode configuration. In more detail, Fig. 7 is showing a longitudinal view of a discharge lamp. A circular-cylindrical tubular glass discharge vessel (2) is provided on its outer wall with two diametrically opposed, axially located striplike external electrodes (4a, 4b). A metal rod like internal electrode (3) acts as a cathode and is located centrally axially inside the discharge vessel. The discharge is dielectrically impaired on one end. In the case of unipolar voltage pulses (Fig. 8) from the surface of the internal electrode, various triangle-like discharge pattern arrise where each widens toward one of the external electrodes (Fig. 9). The filling (5) is xenon at 100 Torr. It can clearly be seen, that the patterns illuminate essentially uniformly diffusely. Only at their narrow cathode-side end points do they each have a somewhat brighter-illuminating region, which percentage wise is quite small. Moreover, the high uniformity is notable, both with respect to the distance between the various patterns and with respect to the shape and light density distribution of the various patterns in comparison with one another.

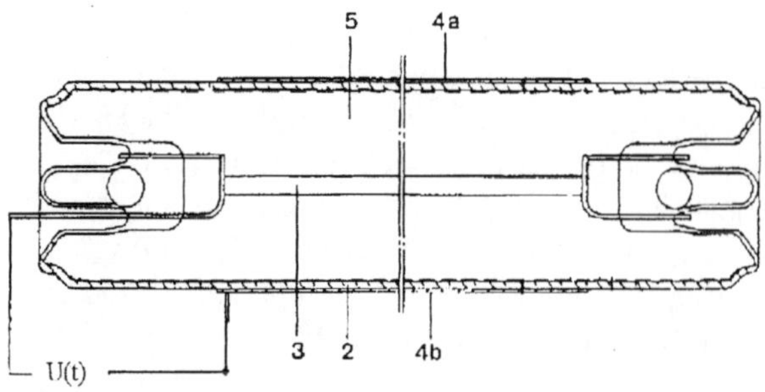

Fig. 7: Longitudinal view of a discharge lamp

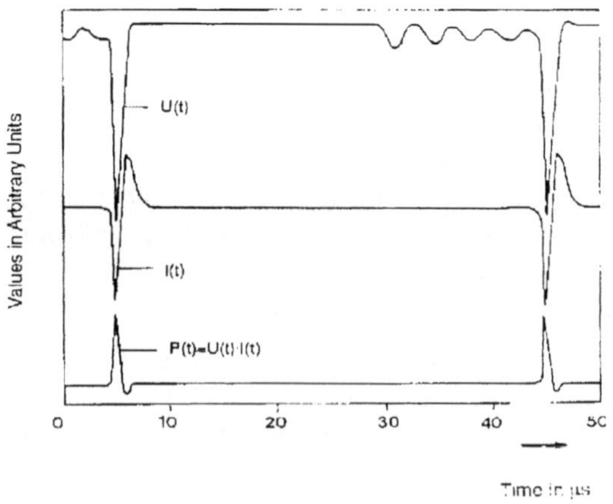

Fig. 8: Shapes of excitation voltage, current and power

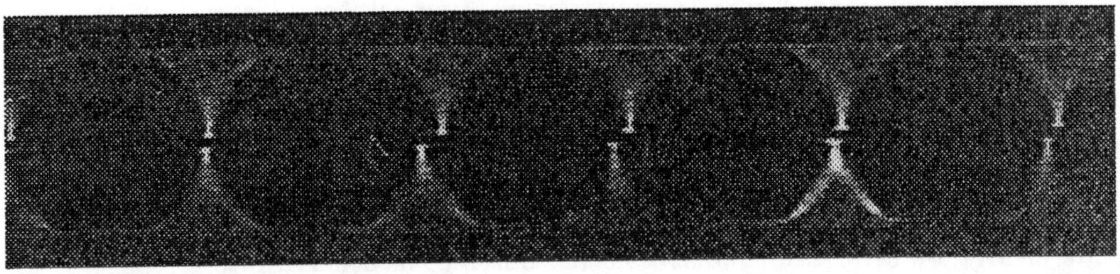

Fig. 9: Discharge pattern

The development of the microdischarge formation has been investigated. Besides a very strong VUV, the spectrum of xenon has lines in the IR. The corresponding transitions from high excited (6p) terms to the resonant (6s) levels have been used as a monitor of how the plasma is developing in space and time. The space and time resolved framing pictures of a series of microdischarges demonstrate that the microdischarges are extremely reproducible. A correlation of the external excitation voltage and the development of the microdischarges is shown in Fig. 10-11.

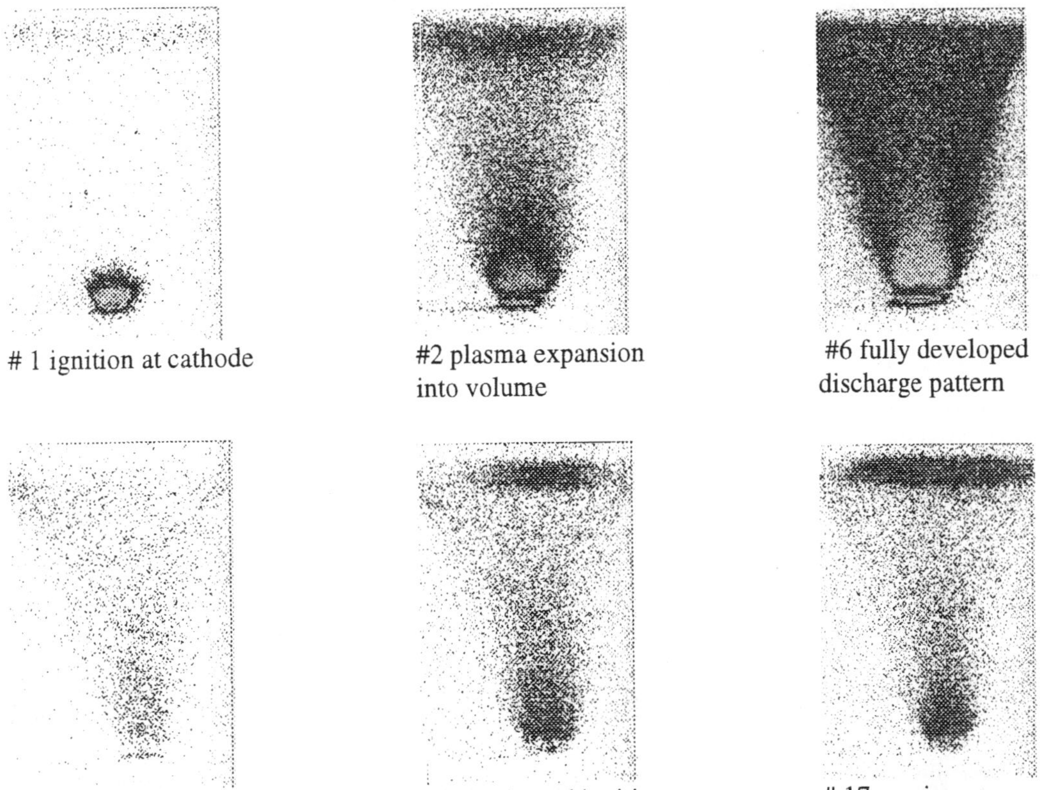

1 ignition at cathode

#2 plasma expansion into volume

#6 fully developed discharge pattern

12 state near current zero

#14 backward ignition

17 maximum backward current

Fig. 10: Framing sequence: dielectrically impeded anode at the top, metal cathode at the bottom; time correlation from Fig. 11

In contrast to the previous physical understanding, we identified our DBD processes as typically glow discharges. No doubt, space charges play a very important role but if a maximum of UV efficiency is to be reached they must not arrive at charge densities which are typical for streamers. The efficiency of the generation of VUV radiation is depending on the gap distance, the pressure and the electrode configuration. With a gap of 1.2 cm and a xenon pressure of 100 Torr we have transformed about 60% of the input energy into the typical Xe excimer VUV radiation [9].
A lot of the above findings can be verified by using a suitable discharge modeling. However, before starting the theoretical description the question has to be answered which the essentials for an adequate modeling of our typical DBD are.
During the last decade a large number of papers appeared which were dealing with the modeling of DBD's. Two important of these are [12] and [13]. As mentioned, the common understanding of DBD's had been closely related to streamer formations. Preconditions of the latter consist in high charge densities and high locally limited electric fields. Practically all previous DBD modeling activities were relying on these basic assumptions by assuming that there is no need to consider the time dependent exact outer voltage shape once the discharges had been started. Furthermore, the resulting high electron densities due to the streamer assumption forced all succeeding thinking into the opposite direction.

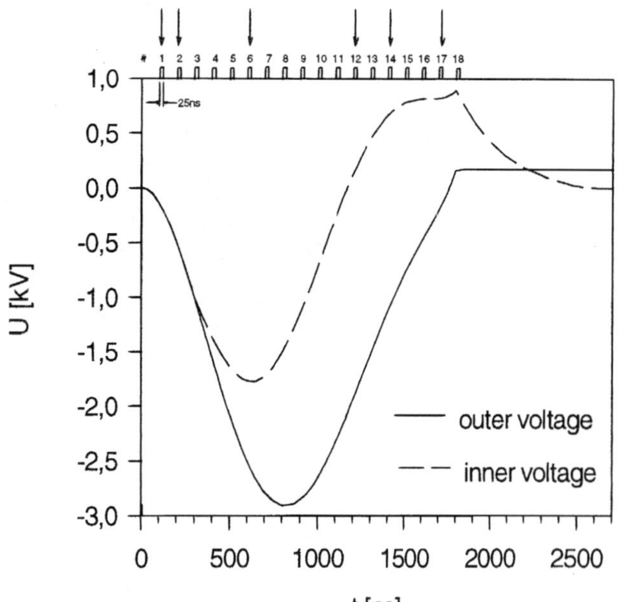

Fig. 11: Voltage shapes and framing sequence

At this place we do not want to discuss the problem whether the adequate theoretical fundamentals for the simulation of AC driven DBD's really streamer formation processes are. Nonetheless, for a better understanding of how the efficiency of DBD can be increased to such a level that the latter are competitive on market, the streamer approach premise was rather contraproductive. On the other hand, independent from the general approach, the previous modeling activities established valuable roots also for our activities. So, the treatment of the electron and heavy particle kinetics as well as the general description of geometry was possible to be adopted without introducing important changes.

The main physical ideas how to increase the efficiency of DBD's have already been outlined above. Of course, a corresponding modeling has to incorporate these principles. Therefore, our theoretical description is relying on the following essentials:

1. The discharge processes can be considered as glow discharges.
2. Streamer conditions do not arise. The time development of the external electrical field is significant. The electrical field strength at each moment forcing the electrons to move is the result of a superposition of the outer and inner fields, see Fig. 2. In other words, the time dependent outer voltage shape has to be treated explicitly within the model.
3. The discharge processes are periodically in space and time. This means that corresponding starting conditions have to be formulated. Further, due to the periodicity of the processes the modeling has to be extended at least over one complete cycle.

The quality of both model and data can be checked by comparing the densities after a complete cycle with the corresponding values at the starting time. A further sensitive quality check is given by comparing the calculated current with corresponding measured values. In fact, the current reflects not only intrinsic parameters which are governing the collision dynamics of the discharge as electron energy distribution function, different excitation and ionization processes, electron loss processes and the like, but it reflects also all the screening dynamics which is mainly depending on geometry.

On the basis of the above essentials and in dependence on what the targets of simulation are, different kinds of geometrical approaches have to be used. Regarding the plasma development, a lot of valuable information can already be acquired by applying the correct time dependence of processes to an everywhere homogeneous plasma column. On the first impression this fact could seem to be somewhat surprising. However, referring to Fig. 10 this approach is suggesting itself. Really, excluding the space in front of the electrodes, i. e. considering only the plasma column, at each moment a quite uniform plasma distribution is revealing. As an example we discuss a discharge very similar to that from which the framing camera depositions had been taken. Fig. 11 is displaying that part of the corresponding experimental outer voltage shape which is different from zero. Of course, the latter with the corresponding U = 0 continuation has been used as the starting condition for the simulation. Note that Fig. 11 is indicating also the calculated inner voltage shape. The chosen frequency was 33 kHz. For a judgment of the reliability of the simulations Fig. 12 delivers a comparison between measured and calculated current shapes which demonstrate a rather good agreement.

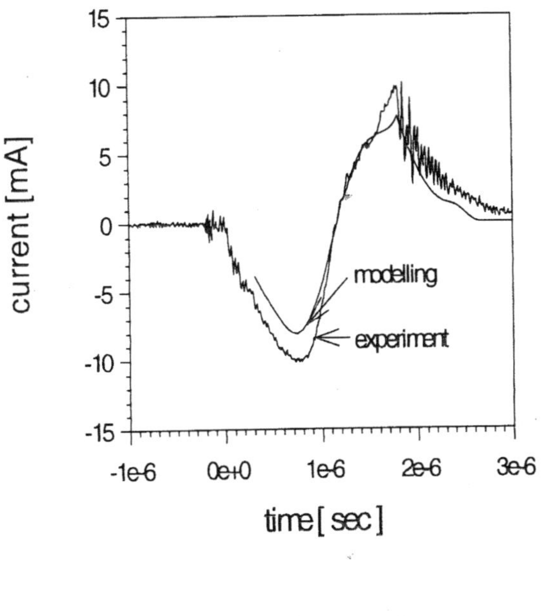

Fig. 12: Comparison of current shapes

As one important result, the modeling is delivering the discharge efficiency: 78 % of the input power is transferred into excimer radiation. For making this result more transparent, Fig. 13 presents further insides into the energy dissipation processes. Mention that as good as all input power is used to excite the xenon atom and that only a very minor portion of collisions leads to ionization. That is why the resulting electron densities are in the order of 10^{11} cm^{-3}. In other words: Electron densities are at least three orders of magnitude lower than those usually reported for AC discharges.

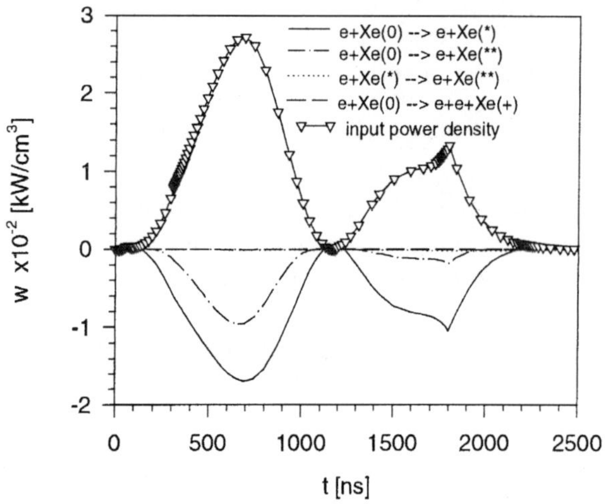

Fig. 13: Energy balance

The calculated efficiency seems to be very reasonable. In fact, by the experiment we got a value of about 60 %. However, the latter result was containing electrode losses which are impossible to be considered within a uniform plasma distribution approach. Really, for treating electrode losses a much more refined geometrical model has to be applied. Our preferred approaches here are two dimensional.

3 FIRST COMMERCIAL APPLICATION FROM OSRAM: OSRAM PLANON - A FLAT MERCURY FREE FLUORESCENT LAMP

At Hannover Fair 1998 OSRAM has introduced a flat Hg-free fluorescent lamp for backlighting applications. The lamp is operated on the basis of DBD's and is able to generate a high luminance over large areas in a very homogeneous manner. The new light source has a number of unique product characteristics which are:

- extremely flat (approx. 10 mm)
- no restriction on size (currently 5 inches to 21 inches)
- lamp and ECG available as a system
- immediate light
- luminous flux is independent on temperature (-30°C and + 85°C)
- luminance is above 5000 cd/m² (without prism foils which limit viewing angle)
- luminance is constant over the entire area
- no ramp-on/ ramp-off effects (energy saving mode)
- no damage of susceptible electrodes
- reduced darkening of phosphor

The OSRAM PLANON system is well suited for a large number of different LCD backlighting types. The specific advantages have been demonstrated in flat panel displays for desktop monitors as well as in industrial applications, medical technology or transportation systems (automotive, plane, train) which require highest technological standards.

Technical Data:

Lamp reference	PLANON 15/24/50
Supplied voltage	24V DC
Luminance behind diffuser	5000 cd/m^2
Lamp power	48W
Dimension L * W * H	324mm * 258mm * 10mm
Color temperature	6500K

5 CONCLUSIONS

Dielelectric barrier discharges have overcome a long existing disadvantage, their relatively low radiation efficiencies. Using a well defined excitation procedure it became possible to achieve radiation efficiencies of more than 60 % in xenon. To come up with these results, during the excitation processes the electrons have to be kept away, as far as possible, from any thermalization. For realizing this, the influencing measures had to go beyond the well known current interruption effect of the dielectric barrier. Moreover, the spatial stochastics of the microdischarges had to be eliminated. A distinct precondition for both demands is to hold the electron densities as low as possible, but as high as necessary for suppressing the stochastic appearance of filamentary microdischarges. On the strength of these ideas definite power pulses are used. The power pulses are separated by idle times. The duration of these idle times has to be chosen such, that the system is able to relax into the state in which it was existing before the excitation had been started.

On the grounds of these findings a wide field of additional applications has been opened up. Due to the high VUV efficiency, for the first time it became possible to take the chance for commercializing DBD's with the target of visible light generation. The introduced flat lamp is delivering a convincing example.

6 ACKNOWLEDGMENTS

The authors are indebted to S. Jerebic for his important contributions and J. Mücke and R. Siebauer who did a lot of work in preparing and developing lamps.

7 REFERENCES

[1] W. Siemens , Poggendorfs Ann. Phys. Chem.,**102,** (1857), 66

[2] U. Kogelschatz, Pure & Appl. Chem., **62,** No. 9, (1990), 1667

[3] U. Kogelschatz, B. Eliasson, W. Egli, Invited Plenary Lecture, ICPIG XXIII, 1997, Toulose, France, C4-47

[4] F. Vollkommer, L. Hitzschke, Phys. Bl., **53,** (1997), 887

[5] K. Stockwald, M. Neiger, Contrib. Plasma Phys., **35**, (1995), 15
[6] J.-Y. Zhang, I. W. Boyd, H. Esrom, Applied Surface Science, **109/110**, (1997), 482
[7] Y. Aiura, H. Matsuno, T. Igarashi, T. Hiramoto, 7th International Symposium on the Science & Technology of Light Sources, 1995, Kyoto, Japan, 169
[8] J.-Y. Zhang, I. W. Boyd, J. Appl. Phys., **80**, (1996), 633
[9] F. Vollkommer, L. Hitzschke, IEEE23rd ICOPS, 1996, Boston, U.S.A., 270
[10] C. Heuser, Dissertation, Aachen, 1984, Zur Ozonerzeugung in elektrischen Gasentladungen
[11] K. Stockwald, Dissertation, Karlsruhe, 1991, Neuartige Xenon-und Xenon/Quecksilber-Lampen im VUV/UV-Spektralbereich
[12] E. Eliasson, U. Kogelschatz, Appl. Phys., **B 46**, (1988), 299
[13] H. Müller, Dissertation, Karlsruhe, 1991, Modellierung von Excimer-Gasentladungen zur Erzeugung spektral selektiver Strahlung

ADVANCES IN FIBER OPTICS: FIBER APPLICATIONS MOVE INTO THE MAINSTREAM

J. M. Davenport[1], W. J. Cassarly[2]

[1]GE Lighting, 5138 Graham Drive, Lyndhurst, Ohio, 44124, USA
[2]Optical Research Associates, 3601 Green Road, Suite. 104, Beachwood, OH, 44122, USA

1 INTRODUCTION

High efficacy distributed lighting systems for general lighting applications are now feasible due to the development of high brightness light sources and the advent of new low loss, low cost light pipes. Most conventional lighting systems are now possible with one or two such sources in combination with a distribution harness and delivery optics.

Major changes in fiber based systems have occurred as fiber system efficacy has increased. When delivered task illumination efficacy first approached 10 lumens-per-watt (LPW) in the early 90's, fiber approaches became the system of choice for many specialty lighting applications. Present state of the art delivered illumination efficacy is approaching 40 LPW. At this efficacy, halogen as well as some discharge applications are possible with fiber optics. As we approach the 21st Century, the delivered efficacy of fiber systems is expected to cross current conventional discharge efficacy. When that happens, fiber optics will have come, in less than a generation, from a novel way of lighting toys to a mainstream lighting technology.

In achieving practical systems, a number of significant technical problems have been overcome. In this paper we shall review some of these solutions as well as indicate our view of the direction of future advances.

2 BACKGROUND

In 1881, William Wheeler [1, 2] was granted a patent for a fiber optic lighting system. While not practical at the time, Wheeler's use of an arc source illuminator, "light pipes" (literally pipes with reflective coatings), and output diffusing optics (Figure 1) outlined the fundamental elements necessary for a viable system. Driven by the needs of a few pioneering applications, automotive in particular [3,4], Wheeler's vision is nearing realization. High efficacy, long life metal halide lamps are now available as practical high brightness light sources. "Light pipes" made from total internal reflecting (TIR) plastic fibers provide light transmission at more than an order of magnitude higher transmission than Wheeler's reflective pipes. Simple plastic optics, designed by computer using Monte Carlo ray-tracing techniques, can now provide lighting systems with both high efficiency and controlled light distributions.

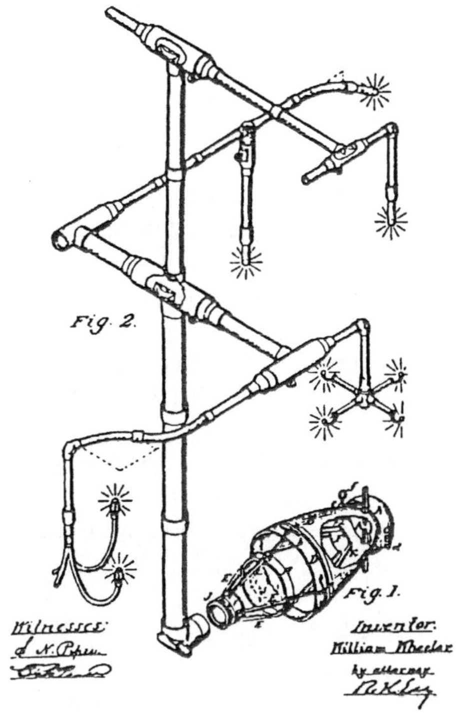

Figure 1: Wheeler's System.

3 SOURCE TECHNOLOGY

Fiber illuminators in the early 80's were typically a light source placed within either an elliptical or a parabolic reflector. With an elliptical reflector, fibers were placed at the second focus of the ellipse. With a parabolic reflector, a lens was added to focus the light onto the fibers. For either reflector, the numerical aperture (NA) of the collected flux was adjusted to match the fiber NA, which was typically about 0.5.

In the early 80's, most long life fiber illuminators were based upon halogen source technology. To minimize the size of the 'spot' at the second focus and hence the size of the fiber, the Tungsten coil needed to be compact. This usually meant the use of low voltages to drive the filament (typically 12 volts and below). The collected brightness[3] was in the range of 5 to 20 lm/mm^2, source efficacy was about 20 lumens per watt (LPW), and source life was typically in the range of 50 to 3000 hours. With Tungsten source technology, the high end of the brightness range was correlated with the short end of the life range. Tungsten source technology was not able to provide both high brightness and long life.

Xenon sources were also used as fiber illuminators. The most prevalent Xenon fiber illuminators were composed of a ceramic parabolic reflector combined with a condensing lens. More recently, ceramic elliptical reflectors were produced. The collected brightness was typically in the 50 to 100 lm/mm^2 range, and source life was in the 250 to 1000 hour range. Similar to halogen sources, the source efficacy is in the 15 to 25 LPW range. Xenon source illuminators did not provide the combination of high efficacy and long life that was needed for widespread use of fiber optic systems.

By the mid to late 80's, 150W, 5000 hour life metal halide sources began to be used as fiber illuminators. The efficacy of these sources was in the 60 to 80 LPW range Used in combination with the plastic large core optical fibers (LCPOFs) developed around the same time [5], these systems showed a marked increase in delivered luminous efficacy. For the first time, delivered fiber optic system efficacy approached 10 LPW making them competitive ***on an efficacy basis*** with incandescent and even some conventional halogen systems. The limitation of these 150W metal halide sources was the brightness, which was in the 5 to 15 lm/mm^2 range. The brightness was limited by the electrode gap size (> 5mm) and because condensed metal halides on the wall of the arc tube scattered a significant fraction of the emitted light. Nevertheless, the combination of metal halide source and LCPOF spawned significant interest in fiber optic systems both for specialty applications as well as demonstrating their potential as a serious lighting alternative.

As lamp companies recognized the market for fiber illuminators, development of short gap lamps with relatively clear walls began. Short arc metal halide based fiber illuminators with 5 to 40 lm/mm^2 collected brightness, lives of >4000 hours, and wattages in the 30 to 75W range came to market in the early to mid 90's. These new sources produced a startling 3-4 times improvement in delivered luminous efficacy. Fiber optic systems could now compete, on an efficacy basis, with halogen and even some discharge systems. Besides spawning a whole new class of applications, this development also encouraged light source investigators to again stretch the limits of fiber optic systems.

The electrodes in conventional metal halide lamps limit both the life and the efficacy of metal halide lamps. Electrodeless lamps [6, 7, 8] offer the possibility of providing 50 to 100 lm/mm^2 brightness with long life (>5000 hours), and high efficacy (>120 LPW). Fiber optic

---

[3] Throughout most of this paper, brightness and illuminance are equivalent. In some communities, brightness and luminance are equivalent. The difference is minor here because the 0.5 NA gives ~1 steradian.

systems, powered with such sources, are expected to reach the performance of conventional fluorescent and high efficacy discharge systems.

A summary of the progress in source technologies is shown in Figure 2.

	Low Voltage Halogen	Xenon	Short Arc Metal Halide	**Electrodeless Sources**
Power Supply	Transformer	Ballast	Ballast	**Ballast**
Collected Brightness	5 to 20 lm/mm^2	50 to 100 lm/mm^2	20 to 40 lm/mm^2	**50 to 100 lm/mm^2**
Life	50 to 3000 hour	250 to 1000 hours	2000 to 10,000 hours	**>5,000 hours**
Color Rendering Index	~100	~95	75 to 90	**>90**
Source Efficiency	20 LPW	15-25 LPW	60 to 80 LPW	**>120**

Figure 2: Fiber Optic Illuminator Source Technology Comparison.

4 FIBER TECHNOLOGY

In Wheeler's system, the light distribution optics were comprised of hollow coated tubes. To collect a reasonable amount of light, a tube with a diameter of at least 12mm was needed. Even with high mirror reflectivities of ~97%, the lumen loss/foot with a 12mm tube was about 25%. At about 2.4 feet of tube, 50% of the starting flux was lost. This was too short to allow widespread fiber applications.

Mainly spurred by the telecommunications industry, glass multimode fibers with ~3% loss per foot in the visible wavelength range became available. To provide flexibility, glass fibers must have small diameters. To create a 12mm fiber, numerous fibers were bundled together to form cables (see [9] for example). The combination of cladding on the individual fibers, interstitial spaces between fibers (packing fraction loss), and the Fresnel reflection losses at the input and output end the fibers limited the transmission of such cables. A cable length of about 7 feet provided 50% of the flux launched into the cable. While this is long enough to attract interest, it limited placement of the illuminator and therefore widespread attention.

Polymethylmethacrylate (PMMA) based plastic fibers provided flexibility up to about a 3mm diameter. Bundles of these larger PMMA fibers were used to provide an improvement over glass fiber cables because the thickness of the cladding relative to the core was lower. This lower clad/core ratio minimized the cladding loss; however, bundles of PMMA fibers still had the same packing fraction loss as glass fiber cables. With the improved cable transmission, a cable length of about 13 feet provided 50% of the flux launched into the cable. Thirteen feet was long enough to provide designers with reasonable control of the source placement and began to attract significant attention.

The major technical advance in fibers came when *flexible* 12mm larger core plastic optical fibers (LCPOF) were produced [5]. The important change introduced by these fibers was that they remove the packing fraction and cladding losses. The typical NA of LCPOFs was >0.6 whereas the typical glass NA is ~0.5. Glass cables with NA's of 0.65 were fabricated, but the absorption was higher than 0.5 NA glass fibers. The higher LCPOF NA was part of the reason why LCPOFs were shown to provide single fiber losses of <1.5% per foot using a 0.5 NA illuminator. At a cable length of about 40 feet, more than 50% of the starting flux was preserved. Forty feet was long enough to place the source and fiber output in separate rooms, thus providing significant flexibility to designers.

The transmission of fibers has been investigated with scattering [10] and bend losses [11, 12] being the most prevalent loss mechanisms. Once scattering losses are controlled, the fundamental limitation of LCPOF transmission is the bulk absorption. It is not unreasonable to expect that fibers with <0.2% loss per foot will be produced at some point in the future. In this case, the 50% preserved flux distance is >300 feet. Three hundred feet is long enough to allow the source to be placed at one side of a building and the output at the other.

Progress in fiber technologies is summarized in Figure 3.

Type of Light Pipe	Flexibility	Loss / ft	Packing Fraction Loss	Cladding Loss	Fresnel Surface Reflection Loss	Distance to 50% Transmission
Hollow Coated Tube	Poor	0.250	0.00	0.00	0.00	2.4
Bundle of Glass Fibers	Excellent	0.030	0.21	0.15	0.08	6.6
Bundle of Plastic Fibers	Excellent	0.025	0.21	0.04	0.08	13
Large Core Plastic Fiber	Good	0.015	0.00	0.00	0.08	40
New Fibers	**Good**	**0.002**	**0.00**	**0.00**	**0.08**	**305**

Figure 3: Fiber Technology Comparison.

5 SOURCE PLUS FIBER OUTPUT

In the early 80's, a 150W halogen fiber illuminator combined with a 6-foot-long-12mm-diameter bundle of glass fibers provided roughly 250 lumens at the output end of the cable. By the mid 90's, a 60W short arc metal halide illuminator with a 6-foot-long-12mm-diameter LCPOF could provide more than 2000 lumens. For comparison, a 60W A-line lamp (the typical incandescent bulb purchased for home use) provides about 900 lumens. This means that a 12mm diameter fiber illumination system has progressed from delivering the equivalent of <1/3 of an A-line to >2 A-lines. Perhaps even more important, the system LPW has improved from <2 to >30. The system LPW is now greater than the typical A-line by a factor of 2 and continues to improve.

Delivered LPW and lumens for a 6-foot-long-12mm-diameter fiber system are summarized chronologically in Figure 4. Projected performance of an electrodeless lamp used with a high transmission fiber is also shown. As discussed earlier, the higher efficacy of electrodeless lamps provides the means to make the next major improvement in fiber system performance.

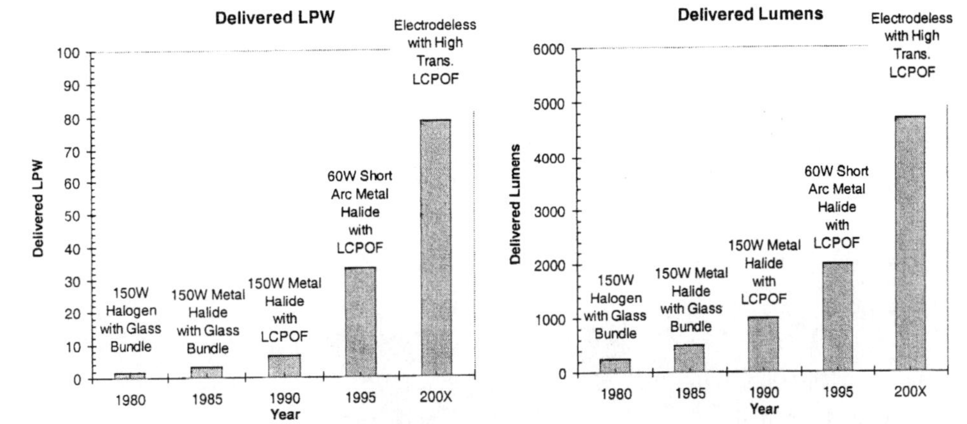

Figure 4: Delivered Lumens and LPW. For comparison, a 60W A-line lamp provides about 900 lumens at 15LPW.

6 SPLITTING

Now that fiber systems are being developed that will be capable of simultaneously achieving both high efficacy and high flux, the application of efficient light splitting methods are essential in preserving that efficacy at the task level. LCPOFs enable transporting significant flux from one location to another with relatively low loss by eliminating the packing fraction loss. However, if a bundle of LCPOFs is used to provide flux to multiple output locations, the total delivered flux is once again limited by the packing faction loss and the fiber cladding. An important difference, however, is that we are only dealing with a few macroscopic fibers—from two to twenty or so—instead of thousands of tiny glass fibers. This time the problem yields to careful optical design which can be modeled effectively using commercial software tools such as LightTools®. An important example is the use of fiber illuminators which produce multiple output ports [13, 14, 15]. Improvements in optical modeling software has enabled the design of multiport illuminators that collect as much, if not more, flux than is possible from a single port illuminator. Improvements in glass molding technology have enabled the fabrication of theses multiport illuminators.

There are many examples where the need for efficient light splitting is evident. Although, one could envision an automobile with only a single headlamp (See Figure 5), there are both performance as well as styling limitations to the design. "Real" cars have two headlamps. Thus, if fiber systems are to be applied to vehicles, both high efficacy and high brightness multiple port fiber illuminators are required. One example of a solution to the problem is the dual port illuminator [16] shown in Figure 6. Reflective optics, refractive optics, and combinations of reflective/refractive optics have been used to produce multiple port illuminator systems. An example four port fiber illuminator system is shown in Figure 7 along with a typical application.

Figure 5: A Single Port Automotive Application (1955 FUJICABIN from the Toyota Auto Museum).

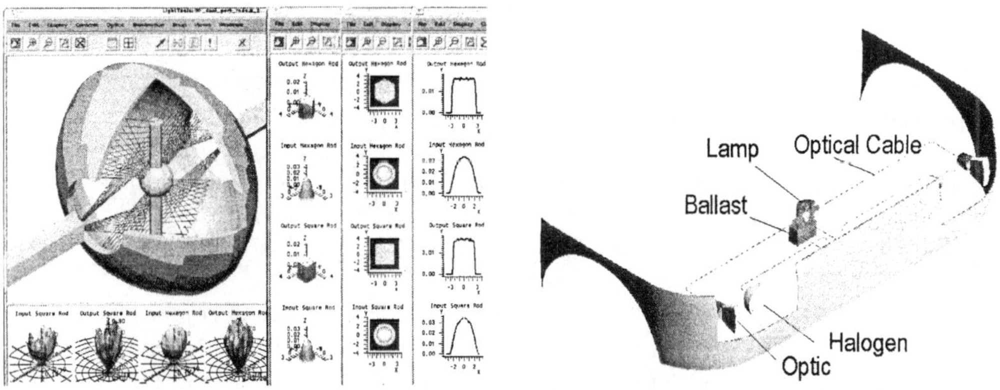

Figure 6: LightTools® Model of a Dual Port Illuminator (left) and Front End of an Automobile (right).

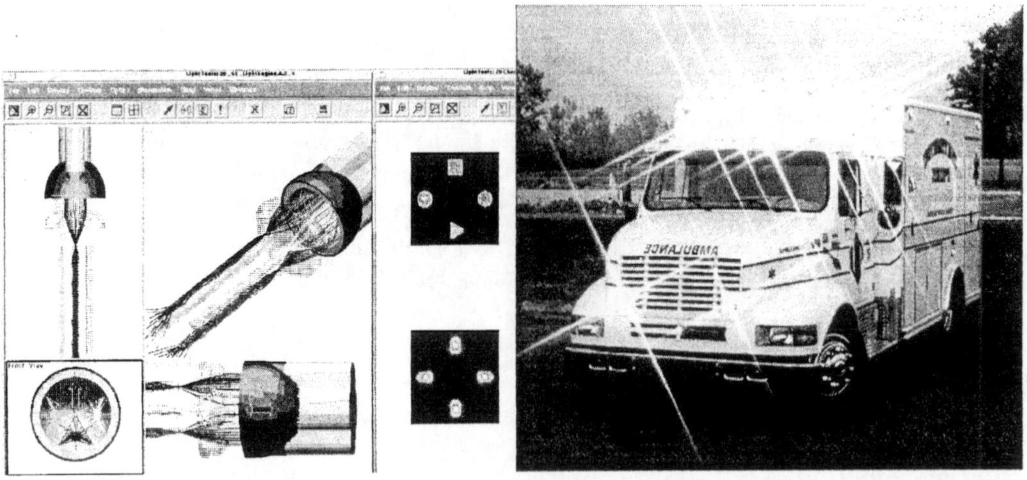

Figure 7: LightTools® Model of a Four Port Illuminator (left) and an Emergency Vehicle with Multiple Output Ports (right).

In conventional lighting applications, a lamp is placed into a luminaire to control the directionality of the light. With a lamp that radiates into a full sphere, the process of providing directionality in a compact space means that some of the light is invariably lost. Typically this loss is between 20 and 40%, with even higher losses found in systems with very demanding beam pattern requirements (e.g. conventional headlamps). With a fiber system, this loss is incurred back in the illumination optics. Thus, the delivered LPW of systems with directionality constraints can be as much as ***25 to 60% higher*** than those that might be inferred from Figure 5.

7 EXPECTATIONS FOR THE FUTURE

The output from a single high efficacy source can be used to illuminate multiple locations. As the transmission of fibers improves and the average luminance of the collected flux increases, the number of locations which can be illuminated with a single source will increase. As the number of output locations increases, fiber system architectures will evolve towards those which split the light at the source and then split the light again at some intermediate distance. In essence, the output of a fiber becomes the input to a second multiple output port fiber system. The approaches described for splitting at the source can be adapted for fiber-to-fiber splitting, with the added advantage that plastic optics may be used instead of glass.

As more and more outputs become available, efficient optical switching will become increasingly important. Switching systems are beginning to appear. The novel approach shown in Fig. 8 uses a matrix of high-speed plastic piezoelectric shutters to create a large area, full color, video fiber optic display. A typical display might include several hundred thousand switches, each addressable individually [17]. Uniform fiber illumination, essential for producing a high quality video image, is achieved by illuminating fibers through rectangular TIR mixing rods [18].

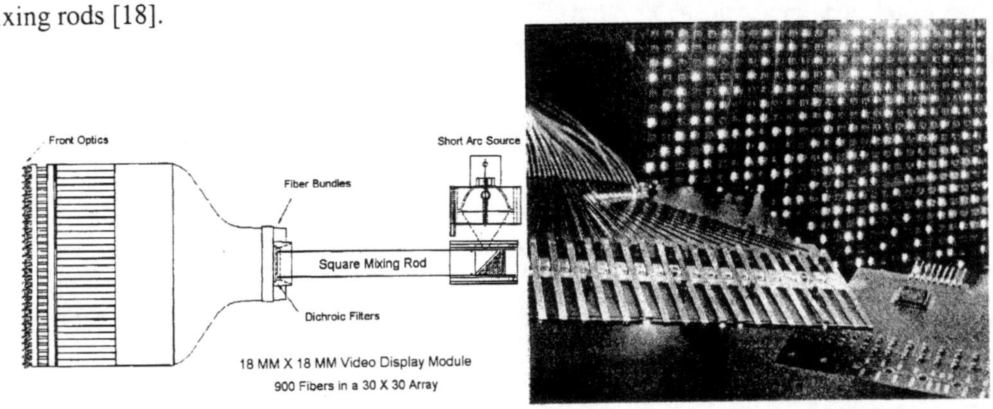

Figure 8: Large Area Fiber Optic Display Using Piezoelectric Shutters.

Since the LPW of fiber systems is already higher than halogen and some discharge systems, fiber systems are already becoming competitive with conventional systems on a performance basis. The high efficacy of electrodeless sources will enable a new generation of fiber systems to exceed the delivered LPW of Fluorescent and conventional HID lighting systems. An example of an electrodeless lamp driving a bundle of fiber optics is shown in Figure 9. Given the high light output power typically available from these new sources, the cost of a fiber system on a per light point basis is expected to eventually be lower that that of conventional systems. Thus, fiber systems are expected to competitive with conventional high performance lighting systems on an efficacy, life, and cost basis.

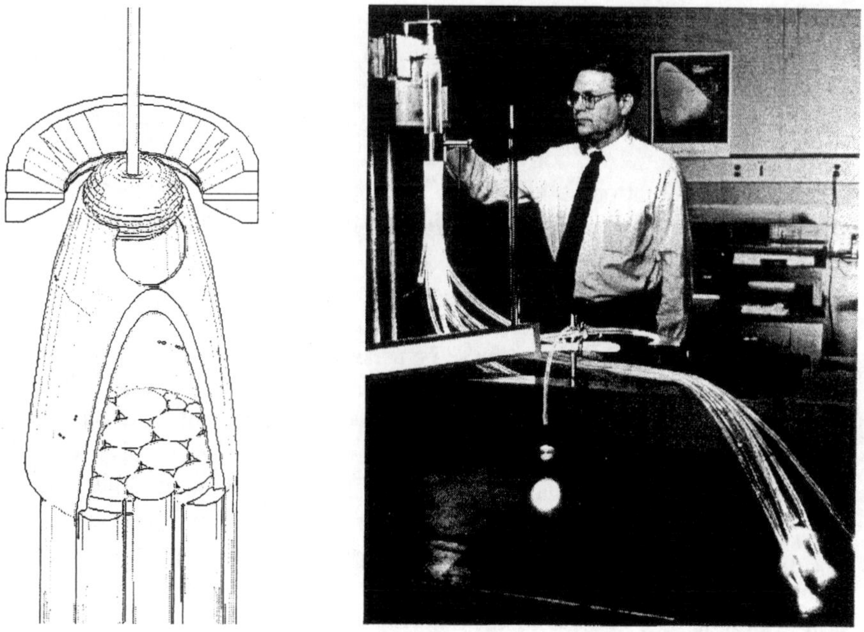

Figure 9: LightTools® Model of an Electrodeless Fiber Optic Lighting System (left) and a Proof of Concept Demonstration (right).

8 CONCLUSIONS

Higher performance fiber optic systems have been developed in the past 15 years. They are the result of significant advances in light source, fiber and delivery optics technology. The systems are driven by a new generation of light sources which combine efficacy, brightness and long life. They utilize new low cost, low loss, large core fibers to distribute light. Novel optical techniques have been devised both for efficient splitting as well as efficient delivery of the light to a task. In short, fiber optic systems have been developed which realize Wheeler's original vision. With the application of a new generation of electrodeless lamps together with fibers that are an order of magnitude more efficient that those in use now, they will go well beyond it.

9 ACKNOWLEDGEMENTS

The authors wish to thank Tom, Mary Sue, and Eggbert for their help in proofreading this paper.

10 REFERENCES

[1] W. Wheeler, 'Apparatus for Lighting Dwelling or Other Structures,' US Patent 247,229, 1881.

[2] J. Hecht, Understanding Fiber Optics, Howard W. Sams and Company, 1989.

[3] J. M. Davenport 'Fiber Optics for Automotive Lighting,' SAE, Oral Presentation (Invited), 1991.

[4] H. Tiesler-Wittig and J. T. Bindels, 'Discharge Light Source - The basis for fiber optics automotive lighting,' SAE, Feb. 1995.

[5] J. J. Zarian and J. A. Robbins, 'High Temperature Plastic Light Conduit and Composition of Matter Therefor,' US Patent 5,298,327, 1994.

[6] J. M. Anderson, 'High Intensity Discharge Lamp Geometries,' US Patent 4,180,763, Dec. 1979.

[7] J. T. Dolan, M. G. Ury, and C. H. Wood, 'Sulfur/Selenium Lamp,' US Patent 5,404,076, Apr. 1995.

[8] C. N. Stewart, M.E.Duffy, J. T. Dakin, V. D. Roberts, S. El-Hamamsy, H. L. Witting, A. Inouye, K. Shimizu, K. Araki, 'Inductively Coupled HID Lighting System,' Paper 78:L, The 6th International Symposium on the Science and Technology of Light Sources (LS-6), pp. 319-320, Sept. 1992.

[9] N. S. Kapany, Fiber Optics, Academic Press, 1967.

[10] J. T. Remillard, M. P. Everson, and W. H. Weber, 'Loss mechanisms in optical light pipes,' Applied Optics, Vol. 31, No. 34, pp. 7232-7241, Dec 1992.

[11] T. L. Davenport, R. L. Hansler, T. E. Stenger, W. J. Cassarly, G. R. Allen, and R. F. Buelow 'Changes in Angular and Spatial Distribution Introduced into Fiber Optic Headlamp Systems by the Fiber Optic Cables,' SAE, Paper No. 981197, 1998.

[12] R. M. Saraiji, R. G. Mistrick, and M. F. Modest, 'Modeling Light Transfer through Optical Fibers for Illumination Applications,' Jou. IES, pp. 128-139, Summer 1996.

[13] W. J. Cassarly, J. M.. Davenport, and R. L. Hansler, 'Uniform Light Delivery Systems,' SAE paper 960490, Feb. 1996.

[14] R. F. Mathis, 'Efficient Coupling of Noncoherent Light to Fiber Optic Light Guides,' SPIE Vol. 3139, pp. 165-169, 1997.

[15] D. Feuermann, J. M.Gordon, and H. Ries, "Nonimaging optical designs for maximum-power-density remote irradiation," Applied Optics, Vol. 37, No. 10, pp. 1835-1844, Apr. 1998.

[16] W. J. Cassarly, T. L. Davenport, R. L. Hansler, and G. R. Allen, 'Comparison of Dual Focus Collector Schemes for Fiber Systems,' SAE paper 970254, Feb. 1997.

[17] J. P. Vanderreydt, "Large-Area Fibre Optic Display Using Piezoelectric Shutters," SID SYM 95, Vol. 26, Paper 20.3, pp. 311-314, 1995.

[18] W. J. Cassarly, J. M. Davenport, and R. L. Hansler, "Distributed Lighting Systems: Uniform Light Delivery Systems," SAE Paper 950904, Feb. 1995.

IL 09

Basic Aspects of the Plasma Modelling of Excimer Lamps

D. Loffhagen and R. Winkler
Institut für Niedertemperatur-Plasmaphysik,
Robert-Blum-Str. 8-10, 17489 Greifswald, Germany

1 Introduction

In recent years UV and VUV light sources using excimer mediums have been developed and their potential applications have experimentally been evaluated. Excimers (*exc*ited di*mers*/tri*mers*) are short-lived excited states of molecules that, under normal conditions, do not possess a stable ground state [1]. Typical examples are the rare gas excimer Xe_2^* and the rare-gas halide excimers KrF^* and $XeCl^*$. Their spontaneous emission lifetime is of the order of nanoseconds and during their decay the excitation energy is released in form of UV or VUV radiation.

As sources of intense pulsed coherent UV and VUV radiation discharge-pumped rare-gas halide excimer lasers are used in a large number of scientific researchs and in industrial fields, such as photochemistry and material processing. As a high-power incoherent light source UV and VUV excimer lamps pumped by a dielectric-barrier discharge have become available. In the development and optimization of these excimer light sources the experiments have often been accompanied and supported by more or less sophisticated numerical model calculations. The efficient generation of UV and VUV excimer radiation can also be used for fluorescent lamps and plasma display panels. Plasma display panels and flourescent lamps operate on the same principle converting the UV or VUV radiation of excimers into visible light by the use of appropriate phosphors. Both applications that are based on the generation of excimer radiation are pumped by dielectric-barrier discharges. In the present paper main aspects of the modelling of excimer discharge plasmas are reported.

2 Modelling of Gas Discharges

A complete model of a gas discharge has to include an electrical (or plasma) model, a plasma chemistry model and a surface model [2, 3]. The different components of the gas discharge model and their interrelationship are shown in Fig. 1. Each of these components involves equations that determine the spatial and temporal variation of important quantities. The electrical model describes the behaviour of the charged particles in the discharge and comprises the charge particles transport and kinetic equations as well as Maxwell's equations. The plasma chemistry model treats the transport and kinetics of the different atomic and molecular species and photons. In the surface model the interaction of the various species in the discharge with the walls and electrodes is described.

As each of these components is coupled to the other two, this classification is in a certain sense arbitrary and a simplification. To date, a consistent multi-dimensional model including all these components does not exist. Especially, the ability to model the plasma-surface chemical and physical processes is quite limited. Thus, most models concentrate on the coupled treatment of the electrical and the plasma chemistry model, where certain

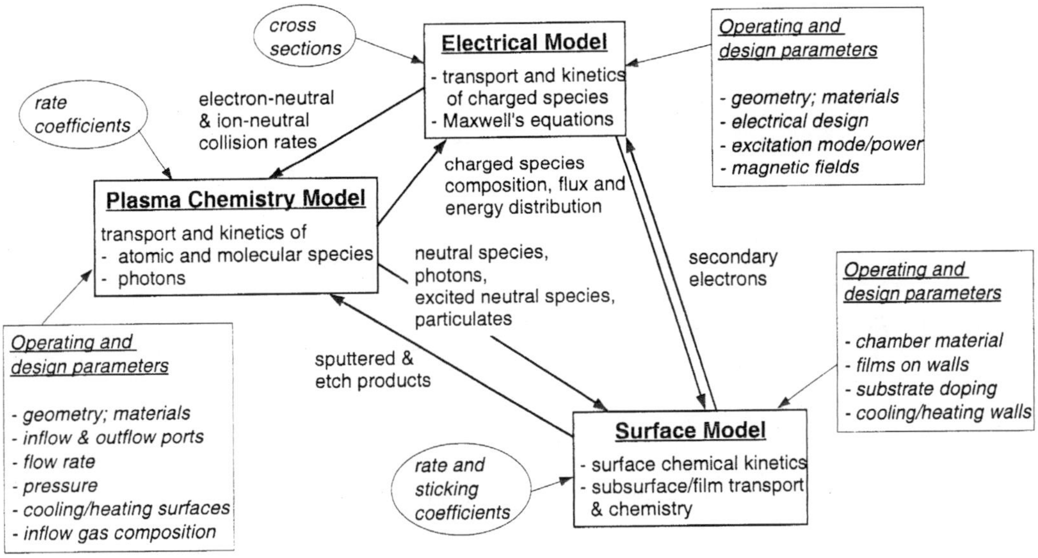

Figure 1: Components of a gas discharge model (following [3]).

aspects of the surface are introduced, if possible, to illustrate their influence.

3 Plasma Modelling of Excimer Discharges

The discharges which are utilized to pump the excimer light sources are weakly-ionized, non-equilibrium discharges at pressure around or above atmospheric pressure. In excimer lasers a transient high-pressure glow discharge between two metal electrodes initiates and sustains the pre-ionized discharge plasma. For the generation of incoherent UV and VUV radiation dielectric-barrier discharges are used to pump the excimer discharge plasma.

Dielectric-barrier discharges are characterized by the presence of a dielectric layer between the discharge gap and at least one of the electrodes. At about atmospheric pressure the electrical breakdown occurs in a multitude of independent microdischarges. The discharge plasma of each individual short-lived microdischarge has characteristics similar to those of transient high-pressure glow discharges, i.e. to those in pulsed excimer lasers. The principle of dielectric-barrier discharges and a review of its applications is presented in [4].

For the theoretical investigations of discharge-pumped excimer discharge plasmas several numerical models have been developed. Many self-consistent models have been used to study discharge-pumped excimer laser discharges [5–33], while only a few models for discharge-pumped excimer lamps, in particular, for the plasma in a microdischarge channel of the dielectric-barrier discharge have been developed [34–37]. In recent years models of the discharge and plasma formation in plasma display panel cells have become available and have improved the understanding of the physical mechanisms with respect to the light emission from a picture element [38–43]. A review of the physics and modelling of plasma display panels can be found in [44].

In many cases spatial gradients in the discharge plasma of excimer lamps have been disregarded and a spatially homogeneous treatment of the plasma has been employed to determine the temporal characteristics of the excimer discharge [5–20,34–36]. These so-called

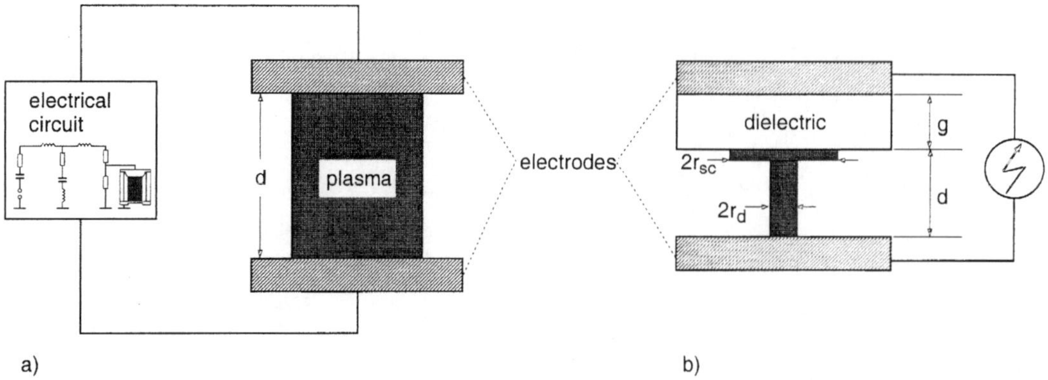

Figure 2: Arrangement of the laser discharge (a) and the dielectric-barrier discharge (b) with the gap width d, the thickness g of the dielectric, the radius r_d of the microdischarge channel and the radius r_{sc} of the surface discharge occurring on the dielectric surface.

zero-dimensional models usually include the steady-state or the time-dependent electron Boltzmann equation for the determination of the electron kinetics, a system of equations for the electrical discharge circuit and an extensive reaction kinetics for the various atomic, molecular and ionic species and the photons occurring in the plasma. On this basis the temporal behaviour of the discharge voltage and current and of the various plasma components can be obtained and a physical understanding of the kinetics of the electrons, photons and the various heavy particles is provided. Once the reliability of the model is checked comparing its predictions with experimental data, zero-dimensional models are suitable to optimize the discharge with respect to the electrical power input, the light output and the efficiency. However, the impact of the discharge conditions on the spatial profile of the discharge plasma and the radiation and on the development of discharge instabilities cannot be explained by means of these models.

Current one-dimensional [21–31, 37–39, 43] and two-dimensional [32, 33, 35, 36, 40–42] models take into account spatial features of the discharge plasma. These models often use a simplified reaction kinetics, which is based on the results obtained by zero-dimensional model calculations. Instead of solving the multi-dimensional Boltzmann equation the temporal behaviour of the electrons is described by velocity-averaged "fluid-like" quantities such as particle density, particle current density and mean energy. In these fluid models the local field approximation is commonly used, assuming that the electron velocity distribution is in equilibrium with the electric field, locally in space and time, and that the transport and rate coefficients in the system of equations can be determined as sole functions of the mean electron energy or the reduced electric field. These coefficients are usually calculated in advance from the steady-state version of the electron Boltzmann equation and are tabulated in look-up tables. At present, one- and two-dimensional models can be used to determine certain spatial characteristics such as the structure of microdischarges and the spatial profile of the electron density and the radiation.

In the following main aspects of the different approaches will be represented. For the discussion the discharge plasma of an excimer laser between two plane-parallel electrodes and a plane dielectric-barrier discharge configuration of a microdischarge in an excimer lamp are used. The corresponding geometrical arrangements are shown in Fig. 2.

4 Zero-Dimensional Models

For a self-consistent zero-dimensional treatment of the discharge-pumped excimer plasma the coupled solution of the time-dependent electron Boltzmann equation, of the equation system for the electrical discharge circuit and of the system of rate equations for the various atomic, molecular and ionic species and photons has to be performed. The coupling of the electron kinetic equation with the electrical circuit and the rate equation system is caused by the electric field E and the various heavy particle and photon densities N_j involved in electron-heavy particle collision processes and in further collision processes leading to the generation of electrons. On the other hand, the electron kinetics determines the plasma conductivity σ for the solution of the electrical circuit equations and the electron density and the rate coefficients for the electron-heavy particle collision processes in the solution of the rate equation system.

4.1 Electron Boltzmann Equation

The temporal behaviour of the electron density n_e and velocity distribution $f(\vec{v}, t)$ is determined by the time-dependent spatially homogeneous Boltzmann equation

$$\frac{\partial f}{\partial t} - \frac{e_0}{m}\vec{E}(t) \cdot \nabla_{\vec{v}} f = \left.\frac{\partial f}{\partial t}\right|_c , \tag{1}$$

where t is the time, e_0 the elementary charge, m the electron mass and $\vec{v}$ the velocity. The collision term $\partial f/\partial t|_c$ describes the change of f due to binary collisions of electrons with various heavy particles. The electron kinetics has to include all relevant collision processes of the excimer plasma such as elastic collisions, excitation, deexcitation, dissociation, attachment, ionization and recombination. Furthermore, the electron-electron interaction can become important and it can be taken into account by introducing a Fokker-Planck-term in the kinetic equation.

The time-dependent treatment of this equation is based on the conventional two-term expansion of the velocity distribution function in Legendre polynomials. A justification of the validity of this two-term approximation for excimer laser plasmas is given in [45]. Methods for solving the kinetic equation are described in [17, 19, 20, 46, 47]. It should be mentioned, that recently a numerical method for solving the time-dependent electron kinetic equation in a multi-term approximation has been represented [48].

Models for excimer laser plasmas developed prior to 1991 and models for the microdischarges in dielectric-barrier discharge plasmas generally make use of a simplified stationary treatment of the Boltzmann equation. A review of the former models is given in [45].

4.2 Electrical Circuit

In the high-pressure glow discharge of excimer lasers the discharge current of the spatially homogeneous plasma is determined by the external electrical circuit. The discharge is described as a time-dependent ohmic resistance $R_d(t) = d/\left(\sigma(t)A\right)$, where d is the electrode gap and A is the cross-section area of the plasma. Thus, according to Ohm's law the voltage U_d of the discharge is given by its resistance and the imprinted current. Neglecting the cathode fall the electric field between the parallel plane electrodes is given by $E = U_d/d$.

Applying Kirchhoff's laws to the electrical circuit, a system of equations for the voltages and currents of the different branches in the circuit can be derived. The equations for the electrical circuit can be described by a system of coupled first-order ordinary differential equations

$$\frac{dY}{dt} = F(t, Y) \tag{2}$$

with the initial values $Y(t_0) = Y_0$, where Y is the vector of the voltages and currents of the different branches in the electrical circuit. This system is solved e.g. by classical Runge-Kutta technique or it is appended to the set of chemical kinetics equations and solved using the corresponding algorithm.

In case of the dielectric-barrier discharge the discharge current is not imprinted and the intensity of the electric field in the gap is determined by the space charge density in the plasma and the external voltage applied to the electrodes. Since the plasmas in dielectric-barrier discharges are strongly inhomogeneous, spatially homogeneous models can only give a rough description of the real discharge development. To approximately treat the plasma by a spatially homogeneous model, a linearly increasing electric field in the gap which reaches a maximum value during a certain time and decays exponentially with a decay time constant has been employed in [34, 35]. The maximum field value and the decay time constant are chosen in such a way that the current density pulse which is generated by the field pulse corresponds to measured current density pulses in the gap. Using this temporal course of the electric field, the coupled system of the electron Boltzmann equation and the rate equation system for the various atomic, molecular and ionic species and photons has been solved.

4.3 Plasma Chemistry Model

Studies of excimer light sources have been performed for most of the known excimers. The best understood systems are the rare gas excimers He_2^*, Ne_2^*, Ar_2^*, Kr_2^* and Xe_2^* and the rare-gas halide excimers ArF^*, KrF^*, $XeCl^*$ and XeF^*. The latter are mainly used in excimer lasers.

Excimer plasmas generally involve complex branches of energy flow from initial excitation of the background gas to the generation of UV and VUV radiation. These branches include collision processes of electrons leading to excitation, deexcitation, dissociation, ionization, recombination and attachment, various collisional mechanisms including neutralization, harpooning reactions and ion-ion recombination leading to the formation of the excimers and various quenching processes of the excimers before the emission occurs. Furthermore, in lasers the extraction of stimulated emission, including absorptions by various processes in the plasma and, if there is a lower laser level, reabsorption by this level, cavity oscillations of the laser and the coupling through output mirrors, has to be taken into account. In spatially homogeneous models of excimer discharge plasmas usually the temporal behaviour of various gas components – typically about ten in pure rare gas plasmas and between twenty and forty in rare-gas halide plasmas – is considered, accounting for atoms and molecules in the ground state and in several excited states, atomic and molecular ions, photons and electrons.

The formation of excimers is favoured by high collision rates which requires a high gas pressure and an efficient excitation or ionization of the precursor species. Typical examples are the formation of Xe_2^* and of XeCl* in a Xe/HCl mixture. The corresponding main formation processes are shown in Fig. 3. Xe_2^* is essentially formed by the three-body collision processes between the lowest excited state in atomic xenon, a ground state xenon atom and a buffer gas or a further ground state xenon atom, while the XeCl* formation occurs mainly via ion-ion recombination.

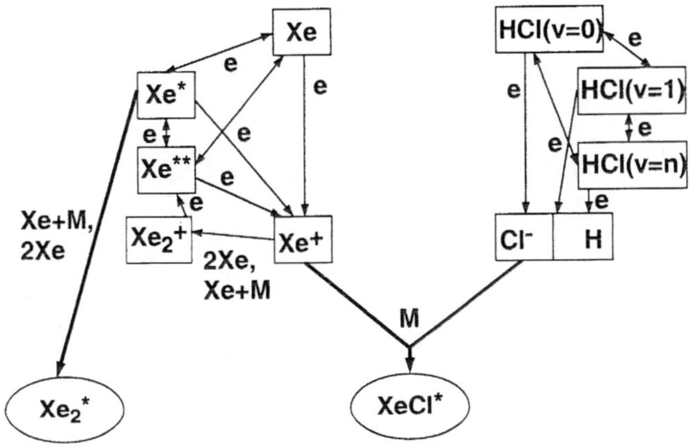

Figure 3: Scheme of the main reactions of the Xe_2^* and the XeCl* excimer formation, where e denotes the electron and M the buffer gas atom.

The temporal evolution of the various components of the plasma can generally be described by a system of first-order ordinary differential equations (2) – one for each component – including the various rates of change of the species densities. The solution of such a rate equation system requires a suitable choice of the initial conditions and some precautions in order to avoid numerical instabilities because of the stiffness of those equations. The stiffness is due to the very different magnitudes of the rate coefficients. Different stiff integration schemes for the chemical kinetics equations have been used successfully.

5 One-Dimensional Models

The spatially one-dimensional models can be divided into two groups: models describing the spatial behaviour of the discharge plasma transverse to the direction of the electric field and models for the study of the spatial discharge characteristics longitudinal to the field direction. One-dimensional and two-dimensional models have in common that the multi-dimensional solution of the Boltzmann equation for the electrons and ions is generally not taken into consideration. This is caused not only by the enormous computing times expected, but also by the lack of appropriate and efficient numerical techniques.

Instead of solving the multi-dimensional Boltzmann equation the time and space behaviour of the charged particles is described by fluid or hydrodynamic models. In these models the properties of the charged particles are characterized by macroscopic quantities such as density, current density and mean energy of the particle. These quantities are solutions of the first three moment equations deduced from the Boltzmann equation. The fluid models commonly use the local field approximation to determine the rate and transport coefficients of the electrons in the coupled system of equations. This approximation assumes that the velocity distribution of the electrons at any position and time is given by the solution of the steady-state, spatially homogeneous Boltzmann equation obtained for the local (in space and time) reduced electric field. This approximation is sufficient as a first approach. However, it becomes less appropriate, if the relative changes of the electric field are not negligible over lengths of the order of the energy relaxation length or during times of the order of the energy relaxation time of the electrons. The cathode region of glow discharges is a typical example of the failure of the local field approximation [49].

5.1 One-Dimensional Transversal Models

Several one-dimensional models have been developed to investigated in excimer laser discharge plasmas the effects of an inhomogeneous preionization, of local fluctuations of heavy-particle densities and of the profile of the electrodes on the discharge characteristics [21–26, 29–31]. To model the behaviour in time and in space in the direction transverse to the electric field, a parallel-resistor description of the discharge plasma has been employed. The total discharge region has been divided into a number of spatially homogeneous discharges which are assumed to ignite parallel to each other between the electrodes. Each discharge region is characterized by the cross-section area A_i determining the current density j_i, the gap length d_i determining the electric field $E_i = U_d/d_i$ and the initial electron density according to the spatial distribution of the preionization. The gap lengths and cross section areas are matched to the profile of the electrodes. The total discharge current is given by the sum of the currents $I_i = A_i j_i$ of the individual discharges and the time-dependent resistance of the total discharge is determined according to $R_d(t) = (\sum_i A_i \sigma_i / d_i)^{-1}$.

A self-consistent parallel-resistor model taking into account the relevant electrical circuit equations as well as, for each discharge region, the time-dependent electron Boltzmann equation including electron-electron interaction for the determination of the electron kinetics and an extensive reaction kinetics for the various heavy particles and photons has become available recently [31]. Earlier models have accounted for simpler reaction kinetics, and a rate equation for the electrons has been employed for each discharge region of the plasma. The local field approximation has generally been used in these models to calculate the transport and rate coefficients.

Due to the basic assumptions the coupling between the various spatially homogeneous discharge regions occurs via the discharge voltage U_d and no spatial derivative terms are considered in the system of equations. Thus, parallel-resistor models are justified only if space charge generated fields homogenize the electric field in longitudinal direction. Furthermore, these models cannot describe any smoothing or homogenizing of spatial fluctuations, e.g. of the initial electron density due to inhomogeneous preionization, in the further temporal course of the discharge evolution.

5.2 One-Dimensional Longitudinal Models

For the one-dimensional longitudinal description of the evolution of excimer discharges fluid models have been used. These models typically consist of a set of the continuity equations

$$\frac{\partial N_j}{\partial t} + \nabla_{\vec{x}} \cdot \vec{\Gamma}_j = S_j \qquad (3)$$

for all important species, i.e. for the electrons, for positive and negative ions and for the neutral species, with the particle density N_j, the particle current density $\vec{\Gamma}_j$ and the net production rate S_j for the species j. The charged particle current density, usually deduced from the momentum transport equation in the drift-diffusion approximation, is given by $\vec{\Gamma}_j = -D_j \nabla_{\vec{x}} N_j + N_j \mu_j \vec{E}$, while for the neutral species only the diffusive current density

$\vec{\Gamma}_j = -D_j \nabla_{\vec{x}} N_j$ is taken into account. Here D_j is the diffusion coefficient and μ_j is the mobility for the species j. The rate S_j includes all relevant reactions leading to the formation or the depletion of the species. Each individual reaction rate is determined by the product of the densities of the interacting species and the rate coefficient of the reaction. As mentioned above the local field approximation is used to determine the transport and rate coefficients in these equations. With respect to the electrons these coefficients are tabulated as functions of the reduced electric field or of the mean electron energy. In the latter case additionally an energy conservation equation for the electrons [28,37] has been taken into account assuming that the electron energy distribution is Maxwellian. The secondary electron emission from the cathode and from the phosphor surface, respectively, has been considered in the respective models for the excimer lasers [27, 28] and for the plasma display panel cell [38,39].

The fluid equations are coupled to Poisson's equation

$$\nabla_{\vec{x}} \cdot \vec{E} = \frac{\rho}{\varepsilon_0} \qquad \text{with} \qquad \vec{E} = -\nabla_{\vec{x}} V \tag{4}$$

for the determination of the electric field and the plasma potential V, where ρ is the net charge due to ions and electrons in the plasma and ε_0 is the vacuum permittivity. The voltage across the gap is deduced from the applied sustaining voltage between the electrodes and, in dielectric-barrier discharges, from the total voltage across the dielectric. The sustaining voltage is either given in accordance with experimental conditions or is determined by solving an appropriate system of circuit equations. In order to obtain the solution of the system of fluid equations and Poisson's equation with appropriate initial and boundary conditions different numerical techniques have been employed.

The excimer laser discharge model developed by [27] differs from the other one-dimensional models. This model deals with the cathode sheath formation in high-pressure glow discharges and consists of two submodels: a zero-dimensional model of the spatially homogeneous bulk plasma and a one-dimensional longitudinal fluid model for the description of the cathode fall region.

6 Two-Dimensional Models

Spatially two-dimensional models can reveal spatial features of the discharge plasma transverse and longitudinal to the direction of the electric field at the same time. Thus they can provide useful informations e.g. with regard to the impact of different electrode and barrier geometries and spacings as well as of the orientation of phosphors relative to the plasma. Expect for the laser model presented in [32], time-dependent two-dimensional fluid models have been developed for describing the evolution of the discharge plasma. In part, these models are extensions of corresponding one-dimensional longitudinal models. In the frame of the two-dimensional models the charged particle transport equations (3) coupled with Poisson's equation (4) are solved. Often the solution in arbitrary two-dimensional geometries is possible. A coupling of these models to the kinetics of the neutral species can easily be done, but has only been presented in [40, 41]. Instead of including an external circuit in these models, an appropriate voltage has usually been applied to the electrodes.

The models presented in [35,36] are similar to that one of the dielectric-barrier discharge in air detailed in [50]. However, these models of excimer discharge plasmas use the further simplification that the ionic drift can be ignored.

The two-dimensional excimer laser model described in [32] has been developed to study the impact of the discharge electrode profile on the laser performance. Besides a simple reaction kinetics the space charge generated electric field has been neglected. By solving the current conservation equation the electric potential and the electric field at each position between arbitrarily curved electrodes has been obtained. Neglecting diffusion of the species the rate equation systems determining the temporal behaviour of the plasma components at each mesh point are coupled only through the electric field equations.

7 Conclusion

Models of different degrees of complexity have been developed and are used in research laboratories and in industry to determine the plasma properties of excimer discharges. Although a complete description of the complex system of an excimer light source is not possible to date, current models have provided useful details for its development and optimization. A self-consistent multi-dimensional model is still missing. Main reasons for that are the general lack of needed data like cross sections, transport and rate coefficients, secondary electron emission coefficients etc. as well as the shortage of appropriate and efficient numerical techniques to overcome the enormous computing times. Further efforts are necessary to surmount these limitations.

References

[1] Ch. K. Rhodes, editor, *Excimer Lasers*, Topics Applied Physics, Vol. 30 (Springer, Berlin, Heidelberg, 1984).

[2] G. G. Lister, J. Phys. D: Appl. Phys. **25**, 1649 (1992).

[3] D. B. Graves, IEEE Trans. Plasma Sci. **22**, 31 (1994).

[4] U. Kogelschatz, B. Eliasson and W. Egli, J. Phys. IV, Coll. C4, Suppl. J. Phys. III **7**, C4-47 (1997).

[5] A. E. Greene and Ch. A. Brau, IEEE J. QE-**14**, 951 (1978).

[6] T. H. Johnson, L. J. Palumbo and A. M. Hunter, IEEE J. QE-bf 15, 289 (1979).

[7] M. Maeda, A. Takahashi, T. Mizunami and Y. Miyozoe, Jpn. J. Appl. Phys. **21**, 1161 (1982).

[8] H. Hokazono, K. Midorikawa, M. Obara and T. Fujioka, J. Appl. Phys. **56**, 680 (1984).

[9] A. V. Dem'yanov, V. S. Egorov, I. V. Kochetov, A. P. Narpatovich, A. A. Pastor, N. P. Penkin, P. Y. Serdobintsev and N. N. Shubin, Sov. J. Quant. Electron. **16**, 817 (1986).

[10] M. Ohwa and M. Obara, J. Appl. Phys. **59**, 32 (1986).

[11] G. Stielow, Th. Hammer and W. Bötticher, Appl. Phys. B **47**, 333 (1988); Th. Hammer and W. Bötticher, Appl. Phys. B **48**, 73 (1989).

[12] V. Mihkelsoo, P. Miidla, V. Peet, A. Sherman, R. Sorkina, E. Tamme and A. Treshchalov, J. Phys. B **22**, 1489 (1989).

[13] M. Ohwa and M. J. Kushner, J. Appl. Phys. **65**, 4138 (1989).

[14] S. Longo, C. Gorse and M. Capitelli, IEEE Trans. Plasma Sci. **19**, 379 (1991); C. Gorse, M. Capitelli and A. Dipace, J. Appl. Phys. **67**, 1118 (1991).

[15] U. Krause and J. Kleinschmidt, Contrib. Plasma Phys. **31**, 101 (1991).

[16] E. Estocq, G. Delouya and J. Bretagne, Appl. Phys. B **56**, 209 (1993).

[17] D. Loffhagen and R. Winkler, J. Comput. Phys. **112** 91 (1994).
[18] A. Riva, M. Legentil, S. Pasquires and V. Puech, J. Phys D: Appl. Phys. **28**, 856 (1995).
[19] O. Lamrous, A. Gaouar and M. Yousfi, J. Appl. Phys. **79**, 6775 (1996).
[20] S. Gortchakov, D. Loffhagen and R. Winkler, Contrib. Plasma Phys. **37**, 539 (1997).
[21] M. J. Kushner, A. L. Pindroh, C. H. Fischer, T. A. Znotins and J. J. Ewing, J. Appl. Phys. **57**, 2406 (1985).
[22] I. O. Blinov, A. V. Dem'yanov, I. V. Kochetov, A. P. Napartovich, A. A. Pastor, P. Yu. Serdobintsev and N. N. Shubin, Sov. J. Quantum Electron. **18** 1531 (1988).
[23] R. Sorkina, J. Phys. D: Appl. Phys. **23**, 806 (1990).
[24] M. Bähr, W. Bötticher and S. Choroba, IEEE Trans. Plasma Sci. **19**, 369 (1991).
[25] U. Krause and J. Kleinschmidt, J. Appl. Phys. **72**, 1237 (1992).
[26] A. Belasri, S. Longo, M. Capitelli, J. P. Boeuf and L. C. Pitchford, XX ICPIG, Barga, Italy, Contr. Papers **6**, 1170 (1991).
[27] A. Belasri, J. P. Boeuf and L. C. Pitchford, J. Appl. Phys. **74**, 1553 (1993).
[28] H. Akashi, Y. Sakai and H. Tagashira, J. Phys. D: Appl. Phys. **27**, 1097 (1994).
[29] K. Kasuya, K. Horioka, N. Hikida, M. Watanabe, Y. Kawakita, S. Kato and H. Okuda: Laser and Particle Beams **12**, 647 (1994).
[30] A. V. Dem'yanov, I. V. Kochetov, A. P. Napartovich, M. Capitelli and S. Longo: Quantum Electron. **25**, 645 (1995).
[31] S. Gortchakov, D. Loffhagen and R. Winkler, Appl. Phys. B **66**, 313 (1998).
[32] M. M. Turner, J. Appl. Phys. **71**, 2113 (1992).
[33] G. Simon and W. Bötticher, J. Appl. Phys. **76**, 5036 (1994).
[34] B. Eliasson and U. Kogelschatz, Appl. Phys. B **46**, 299 (1988).
[35] B. Eliasson and U. Kogelschatz, IEEE Trans. Plasma Sci. **19**, 309 (1991).
[36] H. Müller, "Modellierung von Excimer-Gasentladungen zur Erzeugung spektral selektiver Strahlung", Ph.D. thesis, LTI Karlsruhe, 1991.
[37] A. Oda, Y. Sakai, H. Sugawara and P Ventzek, XXIII ICPIG, Toulouse, France, Contr. Papers **V**, 74 (1997).
[38] J. Meunier, Ph. Belenguer and J.P. Boeuf, J. Appl. Phys. **78**, 731 (1995).
[39] R. Veerasingam, R. B. Campbell and R. T. McGrath, IEEE Trans. Plasma Sci. **23**, 688 (1995).
[40] R. B. Campbell, R. Veerasingam and R. T. McGrath, IEEE Trans. Plasma Sci. **23**, 698 (1995).
[41] K. C. Choi and K.-W. Whang, IEEE Trans. Plasma Sci. **23**, 399 (1995).
[42] J. P. Boeuf and L. C. Pitchford, IEEE Trans. Plasma Sci. **24**, 95 (1996).
[43] K. Takahashi, S. Hashiguchi, Y. Murakami, M. Takei, K. Itoh, K. Tachibana and T. Sakai, Jpn. J. Appl. Phys. **35**, 251 (1996).
[44] J. P. Bœuf, C. Punset, A. Hirech and H. Doyeux, J. Phys. IV, Coll. C4, Suppl. J. Phys. III **7**, C4-3 (1997).
[45] R. Winkler and M. W. Wuttke, Appl. Phys. B **54**, 1 (1992).
[46] S. D. Rockwood, Phys. Rev. A **8**, 2348 (1972).
[47] W. L. Morgan and B. M. Penetrante, Comput. Phys. Commun. **58**, 2348 (1990).
[48] D. Loffhagen and R. Winkler, J. Phys. D: Appl. Phys. **29**, 618 (1996).
[49] R. Winkler, G. Petrov, F. Sigeneger and D. Uhrlandt, Plasma Scources Sci. Technol. **6**, 118 (1997).
[50] D. Braun, V. Gibalov and G. Pietsch, Plasma Scources Sci. Technol. **1**, 166 (1992).

IL 10

Performance and applications of high-brightness InGaN LED

Kanji Bando
Opto-electronics division, Nichia Chemical Industries, Ltd.
491 Oka, Kaminaka, Anan, Tokushima 774, Japan

1. Introduction

LED (Light Emitting Diode) is p-n junction diode with the structure combining n-type and p-type semiconductor. When n-side of LED is connected to (-) and p-side to (+) by using an electric power supply, the careers inject into the p-n junction and the LED lights on.

The first LED in the world was developed in 1968 using the material of GaAsP which is the compound semiconductor. After that, the LED business has grown over this thirty years. Recent days, LEDs is widespread among the general public as moving information signs installed along highways, in the transport stations, and in commercial advertising. There are several reasons for such variety of applications. First, LEDs have good visibility for high luminous intensity and pure color light. Second, power consumption is low for the high luminous efficiency. Third, reliability is excellent and lifetime is very long (at least ten times of common incandescent lamps). In addition, LEDs have the good usefulness such as of simple drive circuit or low current driving directly by IC.

Because of such superior characteristics, much research has been done on high-brightness blue LEDs and green LEDs for use in full-color displays, full-color indicators and light sources for lamps with the characteristics of high efficiency, high reliability and high speed.

In the past, Ⅱ-Ⅵ materials such as ZnSe, SiC and Ⅲ-Ⅴ materials such as GaN have been researched eagerly. However, it was impossible to obtain high-brightness blue LEDs with a brightness over 1cd and a reliability over 10000hr. The short lifetimes prevent Ⅱ-Ⅵ based devices from commercialization. It is considered that the short life time of these Ⅱ-Ⅵ based devices is caused by crystal defects at a density $10^3/cm^2$, because one crystal defect would cause the propagation of other defects leading to a failure of the devices. SiC is another wide band-gap material for blue LEDs. The brightness of SiC blue LEDs is only between 10mcd and 20mcd because of indirect bandgap of this material. Despite this poor performance, 6H-SiC blue LEDs have been commercialized for a long time because there has been no competition for blue light emitting devices.

On green LEDs, the wavelength of conventional materials such as GaP, GaP:N and AlInGaP is in the region of 555-575nm. The color of the region is yellowish green, not pure green which is 510-545nm. The external quantum efficiency of green GaP LEDs is only 0.1% due to the indirect band-gap structure of this material and the peak wavelength is 555nm. Green AlInGaP LEDs is an emission wavelength of 572nm and maximum external quantum efficiency of 1%. When the emission wavelength of AlInGaP is reduced to the pure green region, the external quantum efficiency drops sharply, because the band structure becomes nearly indirect.

Recently, considerable progress has been made in MOCVD (metal organic chemical vapor deposition) technology with the growth of thin compound semiconductor film from vapor phase. LEDs made by MOCVD method has dramatically improved the luminous intensity and increased the color variation. When only warm-color LEDs, red (GaAlAs) or yellow green (GaP), were commercialized several years ago, color reproducible area of moving signs was narrow. But after the development of blue InGaN LEDs and pure green InGaN LEDs by Nichia in 1993, the reproducible area was much wider and fulfilled the most part of color diagram[1][2][3].

The achievement of 3 primary colors (red (R), green (G), blue (B)) in LED technology means that if three colors of "RGB" are mixed in a suitable ratio, a full color display same as CRT can be achieved. High luminance full color LED displays for outdoor are already installed on walls of buildings in many cities. Additionally, new applications for smaller and longer-life lighting sources such as automobile indicators, traffic signals and color scanners are considered with interest.

In this paper, we report high-brightness blue and pure green LEDs based on InGaN materials. Furthermore, we described new emitting devices such as white and UV LEDs which have been developed in the process of studying InGaN LEDs (table 1)[4][5].

Table1 Characteristics of LEDs and Other lamps

Classification	Material	Color	Chromaticity Diagram		Peak Wavelength	Half Width	Output		External Quantum Efficiency	Luminous Efficiency
			x	y	nm	nm	mW	lm	%	lm/W
LED	InGaN/YAG	white:6500K	0.31	0.32	460/555	30/150	2.8	0.70	7.0	10
	InGaN	ultraviolet	-	-	371	11	1.7	-	2.5	-
	InGaN	blue	0.13	0.08	465	30	4.2	0.35	7.8	5
	InGaN	blue-green	0.08	0.40	495	35	3.5	0.77	7.0	11
	InGaN	green	0.10	0.55	505	35	3.2	0.98	6.6	14
	InGaN	green	0.17	0.70	520	40	2.7	1.19	5.6	17
	GaP	yellow-green	0.37	0.63	555	30	0.04	0.03	0.1	0.6
	GaP:N	yellow-green	0.45	0.55	565	30	0.2	0.10	0.4	2.4
	AlInGaP	yellow-green	0.46	0.54	572	12	0.4	0.28	1	6
	AlInGaP	yellow	0.57	0.43	590	15	2.1	0.84	5	20
	AlInGaP	red	0.70	0.30	635	18	2.3	0.76	6	20
	GaAlAs	red	0.72	0.28	655	25	5.7	0.24	15	6.6
Incandescent Lamp	W-filament	white:2856K	0.45	0.41	—	—	—	485	—	12.1
Fluorescence Lamp	phosphor	white:6500K	0.31	0.33	—	—	—	2700	—	67.5

All data for LEDs were measured at a forward current of 20mA,and temperature of 25 C
The package type except UV LED were normal 5mm diameter epoxy lamp.
For UV LED, TO-18 with glass cap and silver coated stem was used.
InGaN LEDs are now commercialized by Nichia, and others are extracted from reference papers.
Incandescent bulb is LW100V-40W and fluorescent lamp is FL40SD-100V-40W.

2. High-brightness blue and green InGaN LEDs

GaN and related materials such as AlN, InGaN and AlGaN are Ⅲ-Ⅴ nitride semiconductors with the wurtzite crystal structure and a direct energy band structure which is suitable for light emitting devices. The band gap energy (Eg) of InGaN varies between 3.40eV and 1.95eV depending on its to indium mole fraction at room temperature. The energy corresponds to ultraviolet (365nm) - red (635nm). Therefore, these Ⅲ-Ⅴ nitride semiconductors are useful for visible light emitting devices especially in short wavelength regions.

IL 10

Sapphire (Al_2O_3) substrates have been used for the films growth process of GaN LEDs. But sapphire has a big lattice mismatch (13%) in comparison with GaN. Recent research on Ⅲ-V nitrides have paved the way for realization of high-quality crystals of GaN, AlGaN and InGaN and p-type conduction in GaN and AlGaN. The mechanism of the acceptor-compensation which prevents obtaining low-resistivity p-type GaN and AlGaN has been elucidated. In Mg-doped p-type GaN, Mg acceptors are deactivated by atomic hydrogen which is produced from NH_3 gas used as the N source during GaN growth. High high-brightness single-quantum-well structure (SQW) blue and green InGaN LED have been fabricated and commercialized on the basis of these results, and luminous intensities over 1cd have been achieved. .

The output power, the external quantum efficiency and luminous efficacy of the blue SQW LEDs were 4.2mW, 7.8% and 5.0lm/W, respectively at a forward current of 20mA and at a room temperature. Those of the green LEDs were 2.7mW, 5.6% and 17lm/W, respectively. A typical on-axis luminous intensity of the blue and green LEDs with 15° cone viewing angle was 4.5cd and 9cd. By combining these high-power and high-brightness blue InGaN SQW LED, green InGaN SQW LED and red AlInGaP LED, many kinds of applications, such as full-color displays, signals, automotive lighting and LED white lamps for use in place of light bulbs or fluorescent lamps, are now expected with characteristics of high reliability, high durability and low energy consumption.

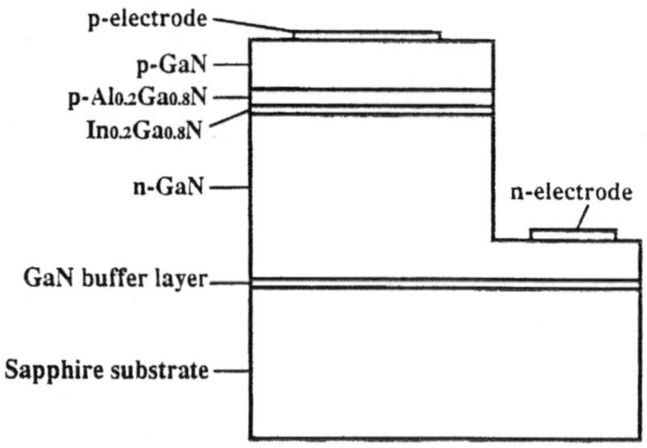

Figure 1 Structure of blue InGaN LED

The structure of InGaN blue die is shown in Figure 1. This structure is called single-quantum-well, due to having a very thin active layer of 30-Å-thick. If the active layer becomes thin, quantum size effects comes out and radiation efficiency is improved. Such high performance InGaN LEDs was developed by using the new film growth method that S.Nakamura designed and named Two-Flow MOCVD. The blue InGaN SQW LED device structures consist of 300-Å-thick GaN buffer layer grown at a low temperature (550℃), a 4-μm-thick layer of n-type GaN:Si, a 30-Å-thick active layer of undoped $In_{0.2}Ga_{0.8}N$, a 1000-Å-thick layer of p-type $Al_{0.2}Ga_{0.8}N$:Mg, and a 0.5-μm-thick layer of p-type GaN:Mg. The active layer of the green SQW is consisting of a 30-Å-thick $In_{0.45}Ga_{0.55}N$.

The active layer (well layer) is sandwiched by n-type GaN layer and p-type $Al_{0.2}Ga_{0.8}N$:Mg barrier layer which have wider band-gap than the active layer. The light of LEDs is emitted by recombination of donor and acceptor inside the active layer.

P-type GaN:Mg layer was used as the contact layer for the p-type electrode in order to improve the ohmic contact. After the growth, N_2 ambient thermal annealing at a temperature of 700℃ was performed to obtain highly p-type GaN and AlGaN layers.

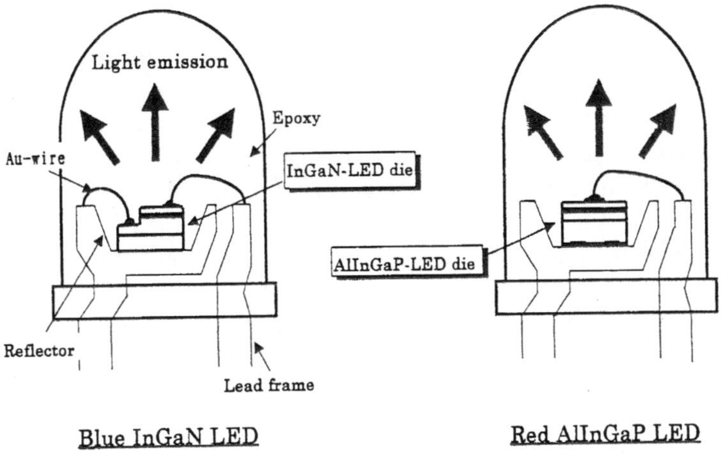

Figure 2 Schematic diagram of blue InGaN LED

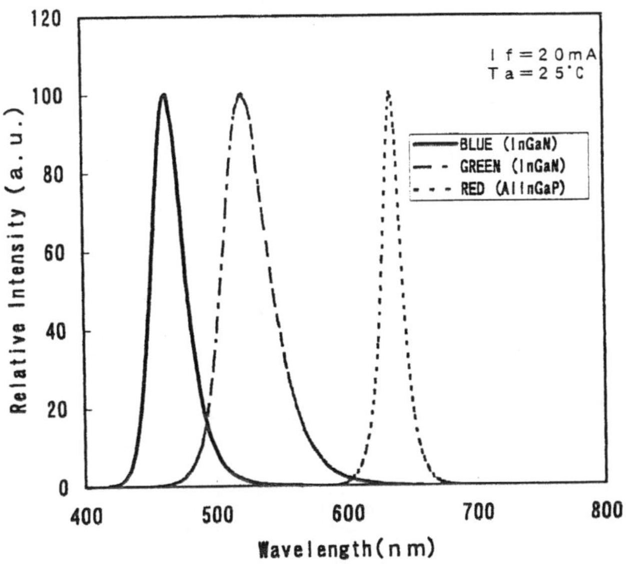

Figure 3 Spectra of InGaN and AlInGaP LEDs

IL 10

Fabrication of LED chips was accomplished as follows: the surface of p-type GaN layer was partially eched until the n-type GaN layer was exposed. Next, a Ni/Au contact was evaporated onto the p-type GaN layer and a Ti/Al contact on to the n-type GaN layer. The wafer was cut into a rectangular shape (350 μm × 350 μm). These chips were set on the lead frame, and were then molded. Figure 2 shows schematic diagram of InGaN blue/green LEDs. That of AlInGaP LEDs is just a reference and a representative of all conventional LEDs except InGaN LEDs.

Figure 3 shows the typical spectra of the blue and green InGaN SQW LEDs with different indium mole fractions of the InGaN well layer. As reference, red AlInGaP is shown in the same area. The peak wavelength and the full-width at half maximum (FWHM) of the typical blue SQW LEDs are 470nm and 30nm, respectively, and those of the green SQW LEDs are 525nm and 40nm. When the peak wavelength becomes longer with the indium mole fraction increased, the FWHM of the spectra increases and the output power decreases. It is due to the crystal quality degradation of the InGaN active layer or the strain between the well and the barrier layers, which is caused by mismatch of the lattice and the thermal expansion coefficients between well and barrier layers. In figure 3, the peak wavelength and the FWHM of the typical AlInGaP LEDs is 625nm and 18nm.

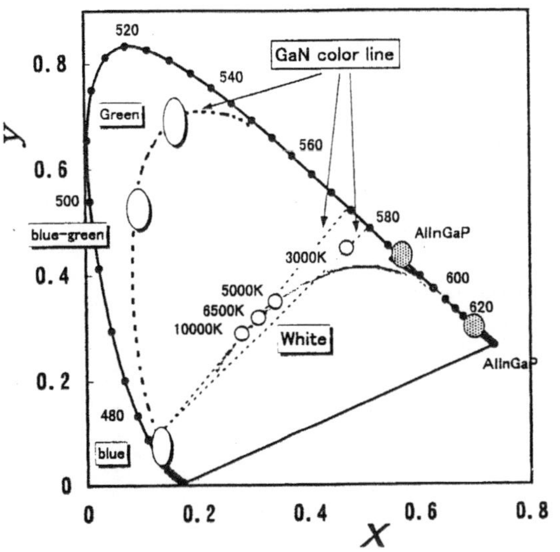

Figure 4 Chromaticity of InGaN LEDs

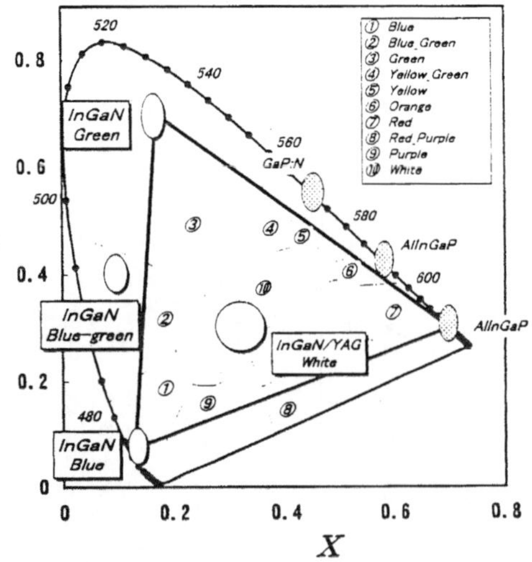

Figure 5 Color reproducible area of LEDs

Figure 4 is a chromaticity diagram in which the blue, blue-green and green InGaN SQW LEDs are shown. Commercially available green GaP LEDs, amber AlInGaP LEDs, red AlInGaP LEDs and white InGaN/YAG LEDs are also shown. The color range of InGaN SQW LEDs obtained by changing the indium contents is shown as dotted line.

Using the high-brightness blue and green InGaN SQW LEDs and AlInGaP red LEDs full-color LED displays can be fabricated especially for outdoor use at present. The color reproducible range by these LED lamps in the chromaticity diagram is shown as the region inside a triangle which is drawn by connecting the positions of three primary color LED lamps, in figure 5. The color reproducible ranges (triangle) are quite different only in the green LED (green InGaN, green GaP:N and green AlInGaP LEDs). From this figure, It is found that the color range shown by using a green InGaN SQW LED is the widest. This means that the InGaN green SQW LEDs show much better color and color purity in comparison with other green LEDs. Using these LEDs, the full color LED display which can express much more beautiful and a nature color have been demonstrated in Tokyo Japan, 1996. After that, Such new type of display are installed increasingly in the world wide..

Figure 4 shows also blue-green InGaN LEDs with other color LEDs. This blue-green LEDs are within the applicable regions for traffic signals and railway signals. Traffic siganls may prove to a great application for the blue-green LEDs. Total power consumption by traffic signals reaches the 100 megawatt range in Japan. Blue-green InGaN LED traffic signals, with an electrical power consumption of only 10% of present incandescent bulb, promise to save vast amounts of energy and avoid carbon dioxide (CO_2) above 100kton. With its extremely of long life time several tens of thousands of hours, the replacement cost of burn-out bulbs will be dramatically reduced. Using these high-brightness blue-green LEDs, safe and energy-efficient signals can be achieved[6].

3.White LEDs

This new technology was achieved by combining high-brightness blue InGaN LED die and YAG fluorescent powder (phosphor)[7]. White LEDs are the fourth color made for commercial use following blue, green and red. The white LEDs typically have the following characteristics, the luminous intensity of 6cd, the efficacy of 10lm/W, the color temperature of 6500K, the xy chromaticity coordinates of (0.31,0.32) and the general color rendering index of Ra=85 at a forward current of 20mA, an ambient temperature of 25℃ and a viewing cone angle of 15° . In addition to these characteristics, the white LEDs have the high-speed responsivity of 120nsec. This is because the afterglow of YAG phosphor is very short. Table 1 summarizes the comparison of commercially available visible LEDs. The white LEDs have achieved high efficiencies equivalent to other visible LEDs. At present, White LEDs of the color temperature of 3000K-10000K and color variations from cool white to warm white are commercially available. Accordingly, the color tones of white LEDs can be tuned for each application.

Figure 6 shows a schematic diagram of white LEDs. InGaN LED die is mounted in the cup on the lead. YAG phosphor described by $(Y, Gd)_3 (Al, Ga)_5O_{12}$ (a kind of Yttrium Aluminum Garnet) is coated on the die. The phosphor layer is approximately 100 μm thick. The die and the phosphor are packaged by epoxy resin formed by the shape of optical lens.

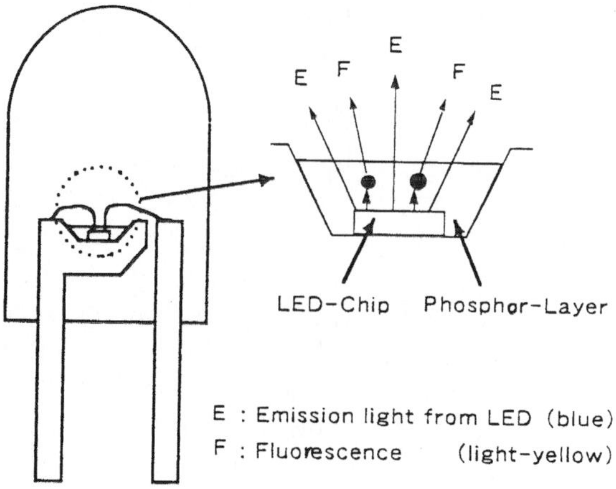

Figure 6 Structure of white InGaN/YAG LED

The phosphor used in this study was prepared by a chemical reaction carried out at around 1400℃ in a furnace. The raw material is a fine chemical powder that Y2O3, Al2O3, Gd2O3 ,Ga2O3 and Ce2O3 are mixed by proper amount. After baking this mixture, the phosphor is got through the serial production process of handled flush, desiccation and classification.

The mechanism of white LED is represented in figure 6. The blue light from the LED die is scattered many times in the phosphor layer and partially absorbed by the phosphor. The phosphor is excited by the blue light, and emits the yellow fluorescence. The white light is made by mixing these blue and yellow lights. Such mixed light is felt white for human eyes in according to the principle of the additive mixture of colors. The light is focused by the lens of epoxy package and taken to outside from LED lamp.

Figure 7 shows the typical spectrum of the white LEDs at a forward current of 20mA. For the white LEDs, the spectrum consists of a narrow band and a broad one. The peak wavelength and the FWHM of the narrow band is 460nm and 30nm, respectively, while those of the broad one is 555nm and 150nm. The narrow band is blue emission from InGaN LED and the broad one is yellow fluorescence from YAG phosphor. Comparing the spectrum of white LEDs with that of incandescent bulbs, it is much broader and most of the radiated energy is in the infra-red.

In the chromaticity diagram of Figure 4, white LEDs is described together with other visible LEDs. As predicted from the additive mixture of colors, the color of white LED locates on the dotted line which connects the coordinates of blue (LED) and yellow (phosphor). It is interesting to note that white LEDs can be freely controlled the own color on the color line. Bigger coating amount comes in yellowish color and smaller amount more blue. In addition to this, various colors of phosphor can be made by changing the composition ratio of (Ya,Gd1-

a)$_3$(Alb,Ga1-b)$_5$O$_{12}$:Ce^{3+}.

Figure 8 shows the color reproducible area (the fan-shaped area in the center) of the white LEDs as a function of coating amount and composition ratios. In the result, very wide color variation is able to come true. That is cool white, daylight, warm white, incandescent white, purplish white, etc.

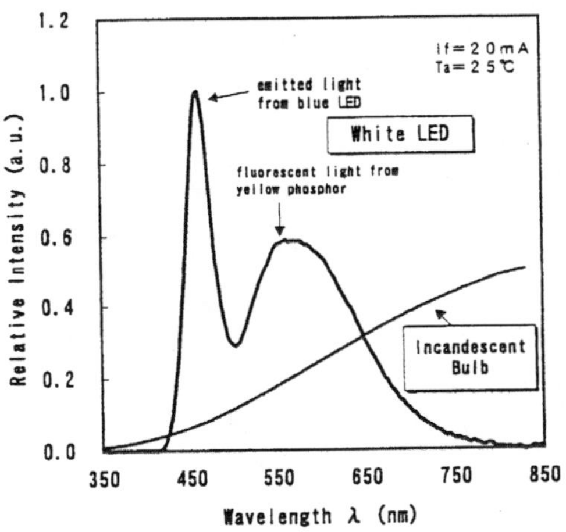

Figure 7 Spectrum of white InGaN/YAG LED

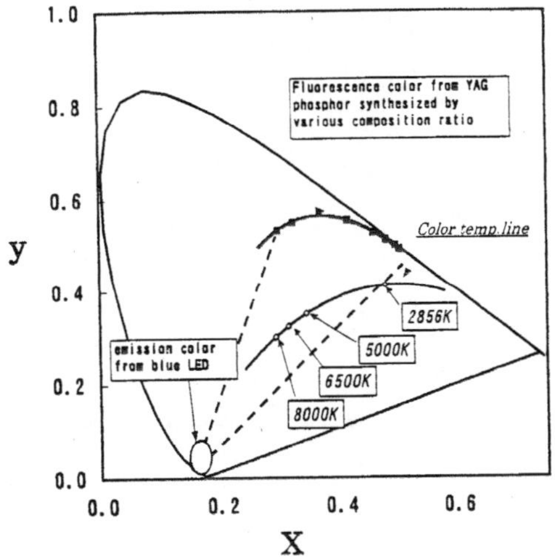

Figure 8 Color reproducible area of white LEDs

The white LEDs achieved small and long-life lighting sources. White color is the most friendly of all colors and is used in many applications. Compared the white LEDs with the common light sources such as fluorescent lamps or incandescent bulbs, white LEDs hold several advantages:

*variety of color temperature including luxury cool white
*highly stability of luminosity for temperature drift
*flexibility in lamp shape and light distribution
*prompt light up and quick response
*shock and vibration resistance
*downsizing of lighting system
*long life beyond 10000hr

For these reasons, white LEDs hold great potential for replacing conventional lighting sources. In the near future, white LED lighting will open new markets for automotive interior lighting, moving signs, LCD back-lights and pedestrian signals.

IL 10

4.UV (ultraviolet) LEDs

High-power UV (ultraviolet) LEDs is very new devices which was developed recently. The structure of the UV InGaN SQW LEDs is similar to the blue InGaN SQW LEDs shown in Figure 1. Only different point is a presence of a 300-A-thick cladding layer of n-type $Al_{0.1}Ga_{0.9}N$:Si between n-type GaN layer and InGaN active layer. InGaN active layer has In composition of nearly zero. High-power UV LEDs can be obtained only using InGaN active layer involving very small amount of indium in stead of GaN of no Indium.

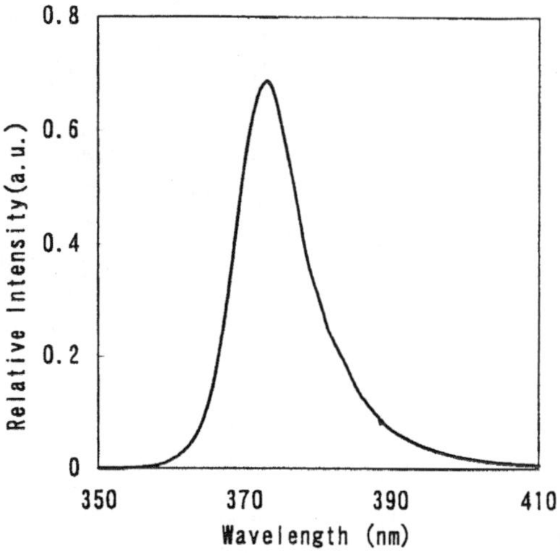

Figure 9 Spectrum of UV LEDs

Figure 9 shows the typical spectrum of the UV SQW LEDs. The output power was 1.7mW and the external quantum efficiency was 2.5%. The measurements of these values were based on TO-18 package with a flat glass window and silver coated stem at a forward current of 20mA and at a room temperature. The peak wavelength and the FWHM were 371nm and 11nm, respectively.

The UV LEDs can be used for the application of excitation light sources of phosphors and other materials, sensors and a forgery bill finder.

5.Summary

Blue, green and UV (ultraviolet) light emitting diodes (LEDs) were fabricated using InGaN material of single-quantum-well (SQW) structure. Furthermore, white LED was made by combining a blue InGaN SQW LED and yttrium aluminum garnet (YAG) phosphor. At 20mA, the output power, the external quantum efficiency and luminous efficacy of the blue SQW LEDs were 4.2mW, 7.8% and 5.0lm/W, respectively. Those of the green/UV/white LEDs were (2.7mW, 5.6%, 17lm/W) /(1.7mW, 2.5%)/(2.8mW, 7.0%, 10lm/W), respectively. A typical on-axis luminous intensity of the blue/green/white LEDs with 15° cone viewing angle was 4.5cd/9cd/6cd at 20mA. LEDs based on these materials are expected to many applications including full-color displays, signals, automotive lighting and substitute of incandescent bulb. UV LEDs is expected to the specialized use for the excitation light sources.

References

[1] H.Amano,M.Kito,K.Hiramatsu and I.Akasaki:
"P-type conduction in Mg-doped GaN
treated with low energy electron beam
irradiation(LEEBI)",
Jpn.J.Appl.Phys.,Vol.28(1989)p.2112-2114.

[2] S.Nakamura, T.Mukai and M.Senoh: "Candela-class high-brightness
InGaN/AlGaN double-heterostructure blue-light-emitting diodes",
Appl.Phys.Lett.,Vol.64, No.13, (3.1994) pp.1687-1689.

[3] S. Nakamura, M.Senoh, N.Iwasa, S.Nagahama, T.Yamada and T.Mukai:
"Superbright green InGaN single-quantum-well-structure light-emitting
diodes",
Jpn.J.Appl.Phys., Vol.34, No.10B, (10.1995) pp.L1332-L1335

[4] M.G.Craford:"LEDs challenge the
incandescents",Circuit and Devices,
(Sept.1992)p.24-29.

[5] M.G.Craford: Commercial light emitting diode technology,ed.J.Rarity
(Kluer Academic Publishers,1996) pp.323-331.

[6] M.W.Mayer: "Affordable green LED traffic signal", Traffic Technology
International,
(Oct./Nov.1996) pp.95-98.

[7] K.Bando, Y.Noguchi, K.Sakano, Y.Shimizu: "Development and applications
highbright white LED lamps", Proc.264th Keikoh-tai Doh-gakkai,Tokyo,
(11.1996), pp.5-14

IL 12

HIGH-POWER EXCIMER SOURCES

Erich Arnold, Ralf Dreiskemper, and Silke Reber
Heraeus Noblelight GmbH, Reinhard-Heraeus-Ring 7, Postfach 63797, 63801 Kleinostheim

1 INTRODUCTION

With the advent of excimer lamps with their incoherent narrow-band radiation (band half-width <15nm) a lot of experiments were performed in the last years to prove the range of applications. To mention only a few XeCl-lamps (308nm) are used to cure coatings and dry inks on the surface of paper, foils and glass fibres without increasing the surface temperature significantly or damaging the substrate. KrCl- and XeBr-excimer radiation at 222nm and 282nm has germicidal effects in water treatment and on surfaces for the packing industry. Xe_2^*-excimer radiation at 172nm has been proved to be efficient for surface modifications and in CVD processes. Most of these application tests were done on a laboratory scale. To fit the requirements for industrial applications on a production scale more intensive excimer sources are requested. High power excimer lamps with reasonable lifetime and reliable operation were developed by Heraeus Noblelight to serve the market requests. A good understanding of the underlying electrical, physical and chemical processes in generating excimer radiation is essential for the development of high-power excimer sources in the UV and VUV.

2 GENERAL CONSIDERATIONS

In the past, descriptions of various kinds of excimer sources with different geometry are found in literature [1]. Coaxial quartz tubes are often used because of its easy manufacturing and high mechanical robustness, even if in general the geometry of these sources can be adapted to special requirements. Such a lamp consists of an inner and outer quartz tube connected together at both ends. After carefully cleaning and annealing the quartz body the lamp is filled with an appropriate gas composition for the desired excimer generation. The gas can be excited dielectrically with a high frequency, high voltage power supply (200-400 kHz U<10kV) connected via two electrodes to the outer surface of the lamp body. The electrodes and quartz surface can be cooled easily with deionized water. Beside the cooling effect water helps to couple the high frequency electrical power into the discharge chamber due to its high permittivity of $\varepsilon = 80$ (at 20°C) compared to the quartz body ($\varepsilon=3.7$). The excitation of the gas through the dielectric quartz barriers forms self-extinguishing discharge channels (Δt = 5-10ns) in which excimer formation takes place [2]. These channels and their related surface discharges at the inner quartz boundary are the sources of radiation and have to be optimized. The number of discharge channels created simultaneously and distributed randomly across the inner quartz surface depend on the form and height of the applied external voltage. The energy conversion efficencies defined as UV radiation flux / electrical power can be as high as 15%. In case of rare-gas halide mixtures the most efficient systems are found combining light halides with heavy rare gas atoms. Due to the molecular levels formed in the excitation process ArF (192nm), KrF (248nm), XeF (351nm), XeBr (282nm) and XeCl (308nm), KrCl (222nm) are the most promising candidates. Formation and quenching reactions are described elsewhere in literature [5, 6, 7]. In pure noble gas fillings the emission of radiation at 126nm, 146nm and 172nm using Ar, Kr and Xe respectively are observed. As far as closed quartz tubes are used to confine the gases, fluorine compounds cannot be used because of its high reactivity to quartz. On the other hand even quartz of the Suprasil type is not able to transmit wavelength at 126nm and 146nm. Windows made out of MgF_2 or LiF are required for those wavelength. In spite of this severe limitations XeCl, KrCl, XeBr and Xe filled sources remain the most promising candidates for large-scale industrial applications.

3 ELECTRICAL EXCITATION

To optimize the conditions for forming excimers one has to be aware of the fact that the complete system of lamp, power supply including the transmission cable has to be considered. The task is to optimize the whole system for proper operation.

Fig. 1 shows a simple equivalent electrical circuit for an excimer lamp. The external voltage delivered from the power supply is in its simplest form of sinusoidal type

$$U(t) = U_0 \sin(\omega t)$$

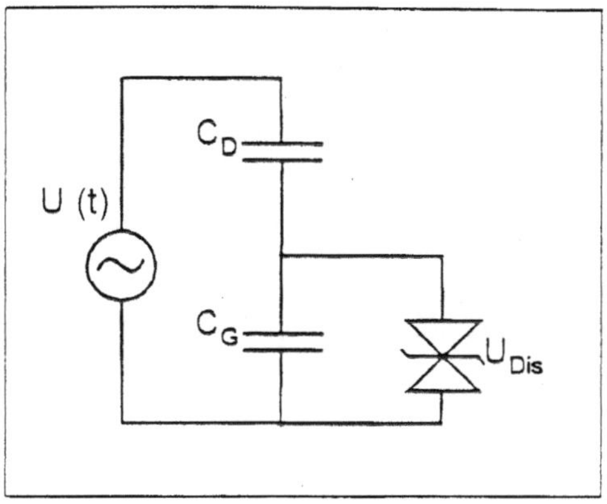

Fig. 1: Equivalent circuit (C_D: dielectric capacitance, C_G: gap capacitance, U_{Dis}: discharge voltage)

The lamp forms a capacitive voltage divider consisting of the capacitance of the dielectric wall C_D and the gap C_G. As long as the gap voltage $|U_G| < |U_{DIS}|$, where U_{DIS} is the average discharge voltage, no discharge ignites. Trying to increase $|U_G|$ above $|U_{DIS}|$ is counteracted by the formation of microdischarges limiting the gap voltage to U_{DIS} . This is represented by two antiparallel Zener diodes in the equivalent circuit in Fig. 1. The introduction of an average and constant discharge voltage neglects the fact that discrete microdischarges are formed but is a reasonable approximation towards reality, especially at high power levels.

Fig. 2 [2] shows the applied external voltage U, the gap voltage U_G and current I as a function of time.

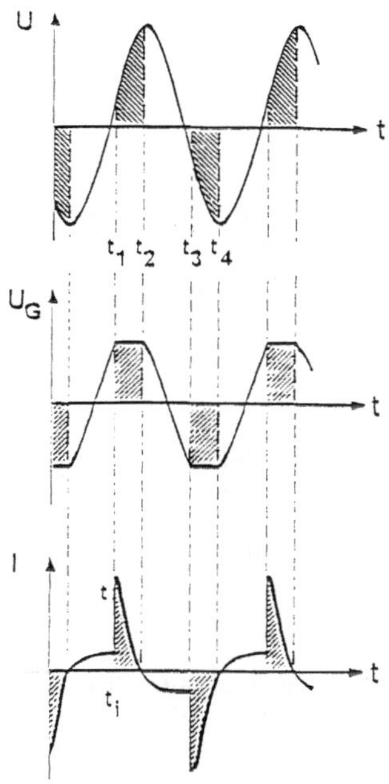

Fig. 2: External sinusoidal voltage U(t), gap voltage U_G and current I . The marked areas indicate the present of silent discharges.

Discharge activities start at

$$t_1 = \frac{1}{\omega}\arcsin\left(\frac{U_{DIS}}{U_0}(1+\beta)\right) \qquad \text{with } \beta = \frac{C_G}{C_D}$$

and end at $t_2 = \frac{\pi}{2\omega}$ where dU/dt = 0.
During the active phase the effective capacitance is C_D and the voltage across the lamp becomes

$$U_2(t) = U_0\sin(\omega t) - U_{DIS} \qquad \text{for } t_1 < t < t_2$$

and the current reads

$$I_2(t) = C_D\frac{dU_2}{dt} = \omega U_0 C_D\cos(\omega t)$$

During the passive phase the capacitance is $C_{TOT} = \frac{C_D C_G}{C_D+C_G}$ and the voltage and current becomes

$$U_1(t) = U_o\sin(\omega t)$$

$$I_1(t) = C_{TOT}\frac{dU_1}{dt} = \omega U_0 C_{TOT}\cos(\omega t)$$

Using the common definitions of the effective values for voltage and current yield

$$U_{eff}^2 = \frac{1}{T}\int_0^T U(t)^2 dt = \frac{U_0^2}{2}$$

$$I_{eff}^2 = \frac{1}{T}\int_0^T I^2 dt = \frac{1}{4\pi}\omega^2 U_0^2 C_D^2\left(\pi\left(1+\frac{C_D^2}{(C_D+C_G)^2}\right) - \left(1-\frac{C_D^2}{(C_D+C_G)^2}\right)(2\omega t_1 + \sin(\omega t_1))\right)$$

The power dissipated in the discharge can be calculated according to

$$P = \frac{1}{T}\int_0^T U(t)I(t)dt = \frac{1}{\Delta T}U_{DIS}\int_{\Delta T} I_2(t)dt = \frac{2}{\pi}\omega C_D U_{DIS}(U_0 - (1+\beta)U_{DIS})$$

Where ΔT is the active time interval for the discharge. This is the well-known power formula for ozonizers [3]. By introducing the voltage U_{MIN} as the external applied voltage at which the discharge starts, we can express P by measurable quantities using the relation

$$U_{MIN} = (1+\beta)U_{DIS}$$

The maximal power is dissipated at $U_{MIN} = \frac{U_0}{2}$. For a fixed lamp geometry U_{MIN} can be adjusted by changing fill-pressure or gas composition. The power consumption rises linearly with increasing U_0 and frequency. Increasing C_D by changing the permittivity ε, lowering the wall thickness or enlargement of the electrode area can further optimize the power dissipated by the discharge.
Another important quantity for optimizing the system performance is the power factor

$$\cos\varphi = \frac{P}{U_{eff}I_{eff}}$$

For a pure sinusoidal voltage from the relations derived above one yields

$$\cos\varphi = \sqrt{\frac{2}{\pi}}\,\frac{4}{1+\beta}\,\frac{\frac{U_{MIN}}{U_0}-\left(\frac{U_{MIN}}{U_0}\right)^2}{\sqrt{\pi\left(1+\frac{C_G^2}{(C_G+C_D)^2}\right)-\left(1-\frac{C_G^2}{(C_G+C_D)^2}\right)(2\omega t_1+\sin\omega t_1)}}$$

The power factor depends on the capacitance ratio β. Lower β-values yield larger power factors, but even with values as low as $\beta = 0.16$ the power factor hardly approches 0.7 Fig.3. The power factor rises sharply after ignition reaching a maximum early between $U_{MIN} < U_0 < 2U_{MIN}$ and declines slowly for higher voltages.

Fig. 3: Power factor for different β-values as a function of U_0 / U_{Min}.

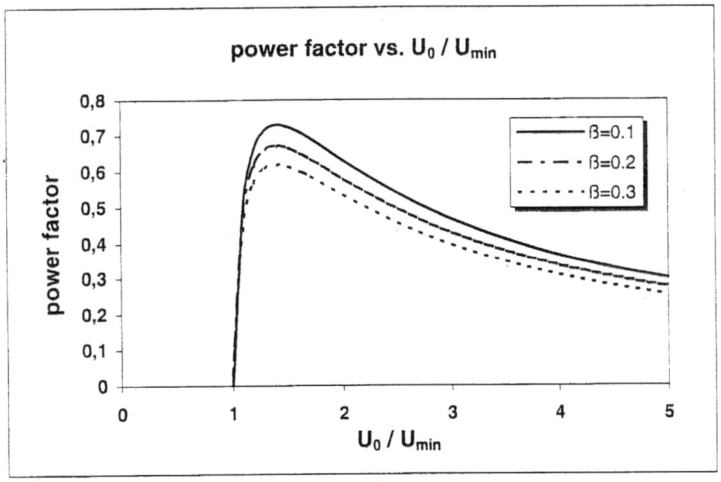

Higher power factors are possible with square-wave currents. In addition to a simple sinusoidal external voltage, other voltage forms for pulsed excitations are presently under investigations [4]. For most industrial applications beside high lamp effiencies, high irradiation densities combined with a long lifetime are important requirements and will be discussed later.

4 RARE GAS / HALIDE SYSTEMS

RgX* excimer molecules are formed in reactions between metastable rare gas atoms produced by electron impact and halide molecules via [5]

$$Rg^* + X_2 \rightarrow RgX^* + X \qquad \text{(harpoon reaction)}$$

and

$$Rg^+ + X^- + M \rightarrow RgX^* + M \qquad \text{(recombination).}$$

(Rg) represents the rare gas and (X) the halide, (M) an additional collision partner. The excimer molecule decays into a repulsive or only slightly bounded ground-state under photon emission

$$RgX^* \rightarrow Rg + X + h\nu.$$

Due to the extremely short lifetime of the excimer ground-state no self-absorption of the emitted photon occurs.

Concerning the properties mentioned above we have based the actual system design of a high power excimer lamp on two coaxial quartz tubes with outer diameter of 30mm, a gap spacing of 5.5mm and a wall thickness of 1.5mm for the inner and outer tube respectively. This leads to ß=0.16, reasonable low to reach a reasonable power factor. The external surface of the inner and outer tube is cooled with deionized water (ρ<2μS). Spiral electrodes, water couplings and the lamp body are mounted on a single PTFE-block. The length of the lamp can be varied between 30-200cm. A stainless steel housing including an aluminium reflector and an exit window capsulates the whole arrangement, ensuring electromagnetic interference compatibility. A shielded high voltage power cable links the lamp electrodes to the external power supply. The power source can deliver up to >50 W/cm into the lamp. Power level can be regulated by changing frequency (200-400kHz) or using pulse-width modulation. Special care was taken to optimize the system for compactness. To ensure a long operational time U_0 is limited to <7kV. This is no principal limitation, but a consequence of the compactness of transformer, flexibility of cable and size of the lamp housing.

With an electrical power density of >1W/cm^3 the discharge belongs to the high loaded type. These mode of operation is characterized by microdischarges with visual thin constricted discharge channels and flat surface discharges at the quartz boundary [7, 8].

Fig. 4 shows the UV-C irradiance for a Kr:Cl filled excimer lamp at 222nm measured at a distance of 1m from the lamp surface as a function of rare gas concentration for three different values of U_0. The concentration of chlorine atoms was kept constant at N[Cl]=3x10^{16} 1/cm³ while the krypton density was varied from 2 to 10x10^{18} 1/cm³. No additional buffer gas was used. Highest irradiance could be obtained at a krypton density of 7x10^{18} 1/cm³ for all chosen U_0. A larger or smaller rare gas density results in a lower UV-C production. The gain in UV-C output at 222nm increasing U_0 from 5 to 6 kV is larger compared to an further increase from 6 to 7 kV. This indicates the onset of saturation for UV-C photon production at values U_0>6kV.

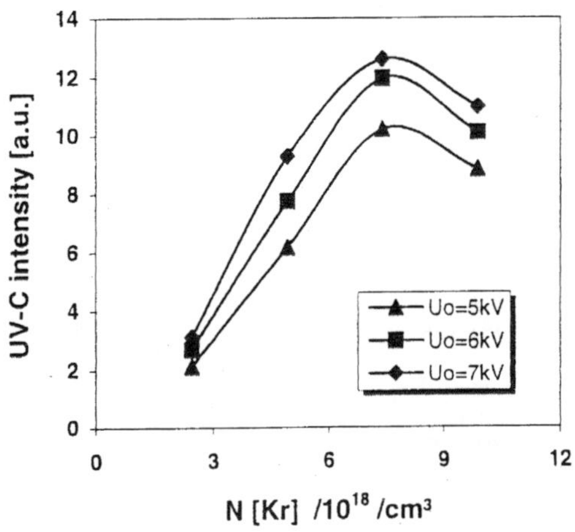

Fig.4: UV-C output of a KrCl lamp as a function of krypton density at fixed chlorine density of N[Cl]=3x10^{16} 1/cm³.

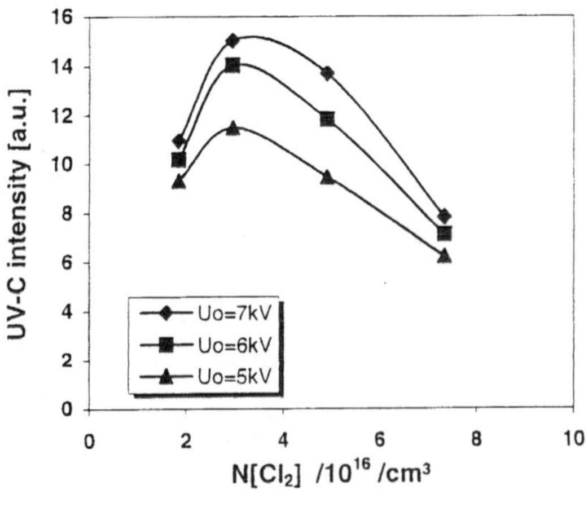

Fig.5: UV-C output of a KrCl lamp versus chlorine density at fixed krypton density of N[Kr]=7x10^{18} 1/cm³.

Fig. 5 shows the influence of changing chlorine concentration on UV-C radiation at constant krypton density of 7x10^{18} 1/cm³ corresponding to a partial pressure of 300mbar.

While the power dissipated into the lamp is smoothly rising with increasing chlorine concentration an pronounced maximum in UV generation is observed. An optimal UV-C flux is obtained with a chlorine concentration of $N[Cl_2]\sim 3x10^{16}$ cm^{-3}. Analogous to KrCl-sources an optimal mixing ratio between halide and noble gas and an optimal total fill-pressure for UV generation in XeCl and XeBr [9] lamps can be found. High chlorine concentration leads to stronger quenching reactions due to a reduced microchannel diameter caused by the electronegativity of chlorine. A high fill pressure has a similar effect on the channel diameter. A high fill pressure changes the electron energy distribution and enhances the quenching processes during the formation of KrCl*.

Table 1 summerize the typical features of a 60cm long excimer lamp system under optimized conditions concerning fill-pressure and gas mixture.

Higher UV-C fluxes can be obtained using lamps with larger diameter or increasing the operational frequency. With an outer diameter of 40 mm and the same gap width and equal length a flux of >300W was achieved. Lamps with even larger output diameters are currently under investigations.

active length	**60**	**cm**
outer diameter	**30**	**mm**
gap width	**5.5**	**mm**
electrical input power	**3**	**kW**
U_0	**<6**	**kV**
frequency	**280**	**kHz**
surface load	**5.9**	**W/cm^2**
specific power	**50**	**W/cm**
power density	**13.5**	**W/cm^3**
photon flux	**>150**	**W**
UV-density	**0.7**	**W/cm^3**
	$7.5x10^{17}$	**$photons/s/cm^3$**
irradiance at 1m	**>2000**	**$\mu W/cm^2$**
irradiance at quartz surface	**>250**	**mW/cm^2**
efficiency	**>6**	**%**

Table 1: Characteristic values of an optimized high power excimer system operating at 3kW.
surface load = electrical power / inner surface area
specific power = electrical power / lamp length
power density = electrical power / discharge volume
UV-density = photons/s/discharge volume
efficiency = photon flux / electrical power

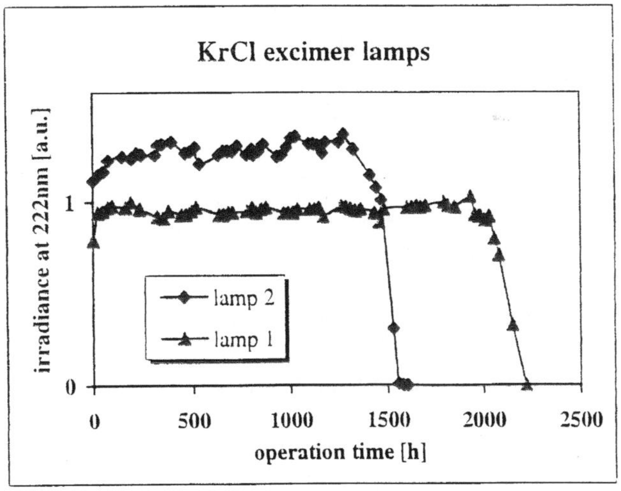

Fig. 6: UV-C irradiance as function of time for two lamps with different chlorine concentration

In DC-driven low pressure excimer glow-discharge lamps a similar radiation flux (>100W) was produced on a laboratory scale [10, 11, 12]. This kind of lamp shows a remarkable efficiency of up to 15%, but suffers from the fact that the electrodes are in direct contact with the reactive discharge gas. Lamp lifetime of 100h limits the practical application of this lamp type. In competition with hot medium pressure mercury lamps an operational time of >1000h is essential. During operation a loss of free chlorine occurs due to reactions with SiO_2. Even though quartz is not chemically attacked by chlorine molecules, lifetime of dielectric driven excimer lamps with external electrodes is limited by the reaction of atomic chlorine and their onic spezies with quartz. If no surface inertisation is realized, the only way to counteract for loss of chlorine is to use a larger chlorine concentration from the beginning. Under these conditions lifetimes of >30000h are possible. Fig.6 shows the lifetime behaviour of two KrCl lamps with different Cl_2 content. As long as no surface inertisation is realized one pays for longer lifetimes with lower UV-efficiencies according to the curves in Fig.6. Investigations concerning surface inertisation are currently under progress.

5 XE-EXCIMER LAMPS

In xenon filled excimer lamps the formation of excimer molecules occur via

$$Xe^* + 2Xe \rightarrow Xe^*_2 + Xe.$$

Metastable Xe^* are formed via direct electron impact on xenon atoms in its ground state or in recombination processes of Xe^+. A detailed description can be found elsewhere [2, 5, 14]. Radiation at 172nm (Fig.7) with spectral half-width of 15nm is emitted in the decay of Xe^*_2 molecules with an expected efficiency of 10%. These photons can be used to crack organic molecules, to modify and clean surfaces or to initiate photochemical reactions [13, 14, 15, 16]. The actual design is similar to the one used for rare gas / halide systems. Because photons at 172nm penetrate water barely 10µm only the inner tube of a coaxial lamp is cooled by water. The inner electrode is of spiral type while the outer electrode is a conducting mesh directly in contact with the outer quartz surface. The outer quartz surface can reach temperatures of 200-300°C during operation at a specific power load of 15W/cm. The gap width is fixed to 3.5mm. The outer quartz tube is made from synthetic quartz of Suprasil type with a transmission of >80% at 172nm. The inner tube is made out of Suprasil or Heralux. To optimize the UV-generation the fill pressure was varied between 100-800mbar. Fig.8 shows the obtained UV at 172nm as a function of fill pressure. The highest UV-flux is generated at highest fill pressure but the gain in irradiance increasing the fill pressure

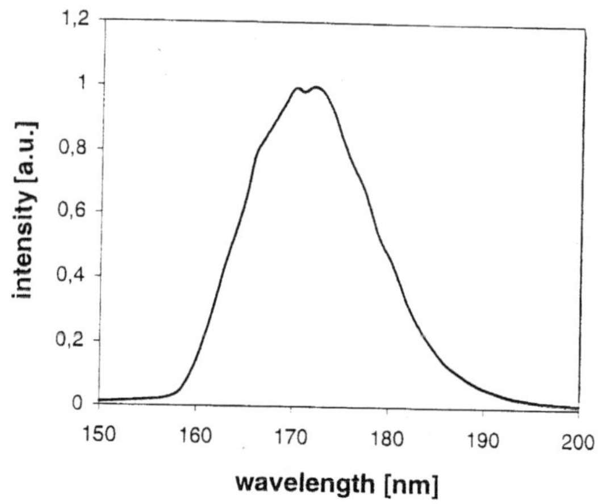

Fig. 7: Spectrum of a xenon filled excimer lamp.

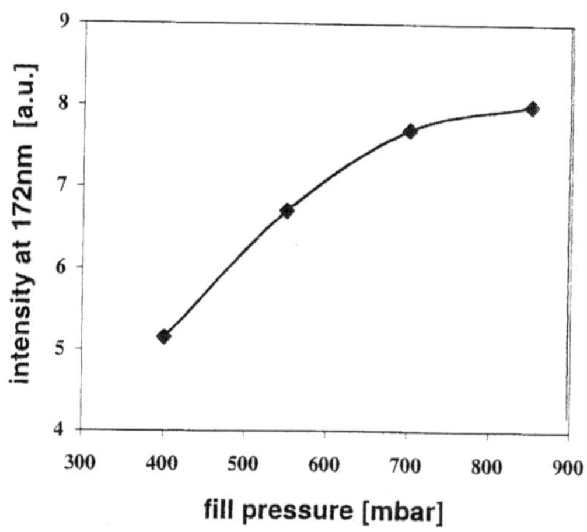

Fig. 8: VUV-signal (172nm) at different xenon fill pressures (U_o = const.)

above 700mbar is small.

The lifetime of xenon filled excimer lamps is not limited by any reaction between quartz and fill gas, but due to the formation of colour centers in the quartz bulk created by the radiation at 172nm. Beside the loss of transmission in quartz tubes the generation of non-bridging oxygen hole centers (NBOH-bounds, absorption at 165nm) and E'-centers (absorption around 215nm) as dominating defects create internal stress in the quartz bulk during operation [18]. Especially in the inner quartz tube these stress generation together with a high temperature gradient between the cooled surface and the surface directed toward the hot plasma (dT/dx>200K/mm) creates cracks during lifetime. With specific power loads of 15W/cm the lifetime is limited to 1000-2000h by the absorption of radiation at 172nm in the created colour centers of the outer tube and finally by the mechanical destruction of the inner tube due to the build-up of internal stress leading to cracks [18]. For many applications the power load and correlated life-time performance is sufficient, but quartz materials with higher resistivity against radiation in the VUV are strongly requested.

To overcome these limitations for future market demands, an increase in diameter is recommended. Life-time tests with power loads of 25-50 W/cm are currently in progress.

6 CONCLUSIONS

In this paper design parameters for high loaded excimer lamps were discaussed. The emitted radiant power, electrical power consumption and life-time behaviour of silent electrical discharges have been optimized for halogen / rare gas and pure xenon filled lamps as function of partial pressures. Changing the density of halogene or rare gas in $KrCl_2$ filled lamps clearly revealed the optimal gas composition for maximal photon production in the UV. For the first time life-time measurements of high loaded excimer lamps were presented as function of chlorine concentration.

In contrast to halogene / rare gas systems pure xenon filled excimer lamps show an enhanced VUV output at 172nm with increasing fill pressure in the range 400 - 900 mbar. The importance of color center formation on life-time behaviour was emphasized.

7 ACKNOWLEDGEMENTS

The authors would like to thank all coworkers from Heraeus Noblelight who made the studies possible, especially Mr. F. Schilling for the fabrication of the quartz lamps and Mr. K. Ernesti and Mr. M. Kessler for the development of power supplies.

8 REFERENCES

[1] G.A. Volkova, J. Opt. Technol. **64**, 637 (1997)

[2] B. Eliasson and U. Kogelschatz, IEEE Trans Plasma Sci. **19**, 309 (1991)

[3] T.C. Manley, Trans. Electrochem. Soc. **84,** 83 (1943)

[4] F. Vollkommer, L. Hitzschke, Phys. Bl. **53**, 887 (1997)

[5] I.W. Boyd, J.Y. Zhang , Nucl. Instr. and Meth. in Phys. Rev. B **121**, 349 (1997)

[6] Jun-Ying Zhang and Ian W. Boyd, J. Appl. Phys. **80**, 633 (1996)

[7] G.J. Piersch and V.I. Gibalov, Proc. 12th Int. Conf. on Gas Discharges & Their Applications, Greifswald 750 (1997)

[8] R. Kling, Dissertation Karlsruhe (1997)

[9] Z. Falkenstein and J.J. Coogan, J. Phys. D: Appl. Phys. **30**, 2704 (1997)

[10] V.F. Tarasenko, M.I. Lomaev, A.N. Panchenko, E.A. Sosnin, Proc. SPIE Int. Soc. Opt. Eng. **2992**, 24 (1997)

[11] A. Schwabedissen and W. Bötticher, Contrib. Plasma Phys. **35**, 517 (1995)

[12] K. Tiedke, A. Schwabedissen, G. Schröder and W. Bötticher, Contrib. Plasma Phys. **35**, 537 (1995)

[13] U. Kogelschatz, B. Eliasson and W. Egli, Int. Conf. on Phenomena in Ionized Gases ICPIG XXIII Toulose (1997)

[14] B. Eliasson and U. Kogelschatz, Appl. Phys. B **46**, 299 (1988)

[15] J. Wieser and D.E. Murnick, Rev. Sci. Instrum. **68**, 1360 (1997)

[16] W. Sasaki, S. Kubodera and J. Kawanaka, SPIE **3092**, 78 (1997)

[17] Jun-Ying Zhang and I.W. Boyd, Electronic Letters **33**, 911 (1997)

[18] Y. Morimoto, T. Igarashi, T. Okanuma, J. Non-Cryst. Solids **179**, 260 (1994)

Inductively Coupled Electrode-less HID Lamp System

Akihiro Inouye

Toshiba Lighting & Technology Corporation

1-201-1,Funakoshi-cho, Yokosuka-city, 〒237-8510, Japan

1. Introduction

1.1 Background

There are such kinds of electrode-less discharge operating systems, as a capacitively coupled discharge, a microwave discharge and an inductively coupled discharge [1].

The capacitively coupled discharge lamp has been used as light sources for facsimile, automobile panel indicator and LCD back lighting, etc. The microwave discharge lamps have been used as photochemical lamps and are now started being used as public lighting. The inductively coupled discharge lamp filled with low-pressure gas is also expanding the application as a fluorescent lamp for public lighting [2,3].

Inductively coupled HID lamp is now under development and has been actively studying [3-6]. In the arc tube, some ten thousands Pascal gas is filled as a starting gas and metal-halides are dosed. Some turns coil is wound around the arc tube and high frequency current is applied to the coil. In the arc tube, an arc ring is induced and the arc tube is warmed up and emits dosed metal spectra. There are some tasks to operate the electrode-less HID lamp system stable.

1.2 Tasks for successful system operation

(A) Starting

To start the inductively coupled HID lamp system, at first high voltage is applied to an arc tube and an open loop corona discharge occurs. If the induction coil current is high enough, the open loop corona turns to a ring arc.

Normally the high voltage is obtained by resonance of the induction coil with a capacitor. The resonant state is very sensitive to the coil inductance and the resonant capacitance. If these values sift a little, the resonant circuit output voltage decreases extremely and the lamp fails to start.

(B) Transition from corona to arc

Lamp impedance changes very extremely before and after arc transition. If the circuit is adjusted to no discharge state in an arc tube, a corona discharge will occur but an arc transition will not happen due to the lack of coil current. If the circuit is adjusted to a corona discharge state, the arc transition will occur but enough power will not be consumed in the arc tube. It is necessary to match the circuit impedance to the lamp impedance before and after lamp starting. Also during a lamp warming up, the arc tube plasma impedance varies rapidly and widely.

(C) System efficiency and cost

It is a key point that the system operates with high efficiency and it can be made cheap, for widely use in the market. For making a circuit cheap, a simple circuit is desirable. For example, a dc power supply composed with only a rectifying bridge, not with a switching regulator, operates with high efficiency and is very cheap.

(D) System element characteristic deviation during operation

Under a long operation of the fixture, the circuit component characteristics gradually deviate. In such case, the resonant condition shifts and lamp performances, especially lamp power and light output shift widely, and sometimes, switching devices of DC AC converter fail. In such cases, the resonant circuit has to provide enough voltage and power, and has to keep the converter current constant not to fail switching devices.

The arc tube plasma is induced by the induction coil and is attracted toward the coil. It means that the arc plasma exists very close to the tube wall and the wall temperature rises up high. The tube wall reacts with dosing materials under such high temperature and the arc tube life will be short.

This paper describes on the tasks and ways for the stable operation of the electrode-less HID lamp system.

2. System configuration and operation

2.1 System configuration

The inductively coupled HID lamp system is composed with a low pass filter, a supplied voltage rectifier and regulator, a DC AC converter, a matching circuit, a

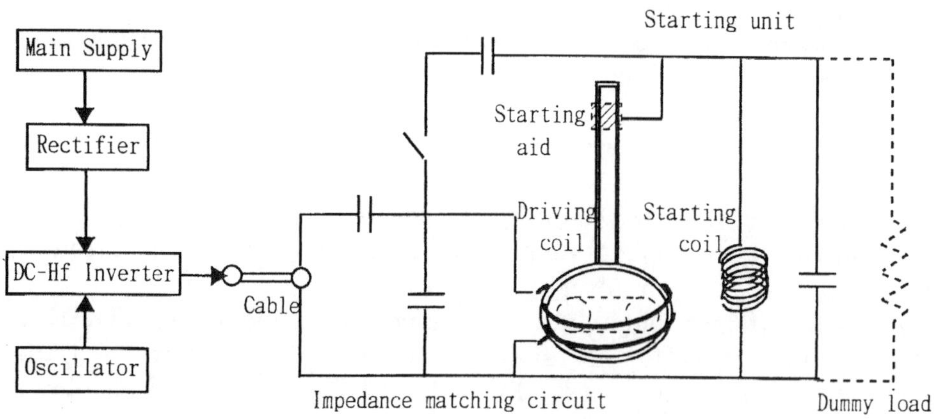

Fig.1 Inductively coupled HID lamp system

resonant circuit composed with a resonant capacitor and an induction coil and an arc tube. To ignite a corona discharge into the arc tube, there is a small gas discharge chamber on the arc tube, which acts as an auxiliary conductor. These are shown in figure 1.

It is convenient to choose ISM band frequency for the operation of an inductively coupled arc tube. In this case, 13.56MHz is chosen. Supplied power through AC line is about 380w and input power of matching circuit is almost 300w.

2.2 Starting

(A) Damping resistor in starting circuit

To get a high voltage for igniting a corona discharge into the arc tube, it is used a resonant circuit in the matching circuit and the starting circuit. But in the resonant circuit, a small shift of an element of resonant components such as coil or capacitor makes the output voltage extremely low.

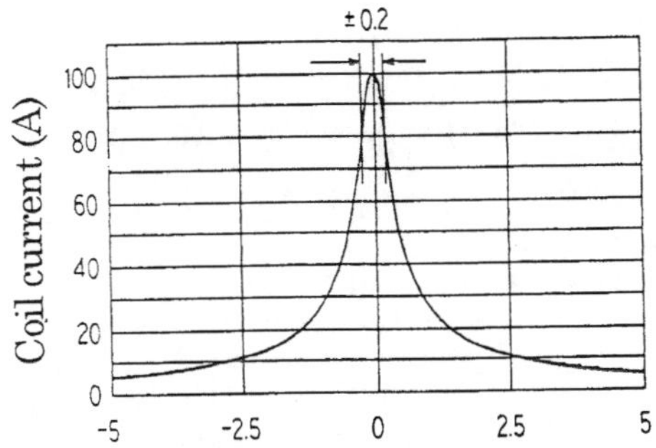

Fig.2 Induction coil current variation by inductance deviation.

In figure 2 , the induction coil current variation by inductance deviation is shown. The coil current deviation has very high peak. If the coil inductance varies only 2.5%, the coil current decreases to 1/10. In such case it is impossible to start lamp and also to have an arc transition. It is the same situation on the deviation of a resonant capacitor. To keep the coil current decrease within 20%, the resonant circuit part deviation should be suppressed within ±0.2%. In actual use, it is impossible to keep the parts deviation in such narrow range during a fixture life.

To overcome this problem, there are some ideas. One of them is adding resister as a dummy load in the auxiliary starting circuit. As shown in figure 1 by dotted line, some thousand-ohm resister is connected in the starting resonant circuit. By adjusting the circuit parameters, it is possible to widen the resonant Q curve with not so much decrease of the peak voltage and the arc tube starting is quit well. After the arc transition, the auxiliary circuit is opened.

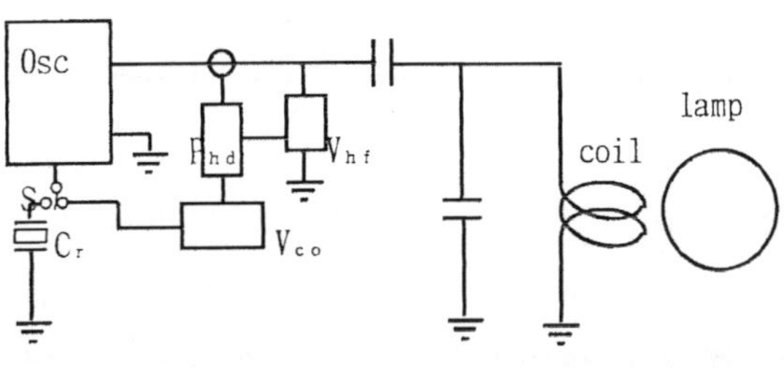

Fig.3 Resonant frequency following up system at starting.

Osc: oscillator Cr: crystal oscillator

Vhf: volt meter Phd: phase detector

Vco: voltage controlled oscillator S : switch

(B)Coil driving frequency shift during start

The other method is shifting the coil driving frequency. A block diagram of the shifting system of the coil driving frequency is shown in figure 3. The phase difference between voltage and current of the converter output is detected, amplified and the output frequency is followed up to keep the phase difference constant. Consequently, even if the coil or resonant capacitor deviates, the impedance matching state is kept quit well, and the coil current is kept constant. As a result the arc tube starts very easily. This is shown in figure 4. White marks are data in case of the controlled driving frequency and black ones are data in case of constant frequency. Lines show calculation. Starting probabilities are shown in table 1 [6].

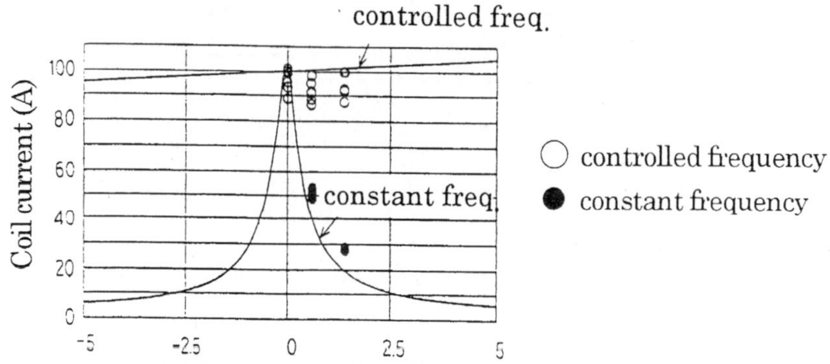

Fig.4 Induction coil current variation under controlled drive frequency by coil inductance deviation.

Table 1 Starting probabilities under controlled riving frequency and constant frequency.

	Part deviation percent	Starting probability percent
Constant frequency	0	100
	0.76	0
	1.67	0
Controlled frequency	0	100
	0.76	100
	1.67	100

2.3 Arc discharge operation

It has been found that it is better using a co-axial cable as one of impedance matching circuit elements to operate the system without failure, even in the resonant circuit elements deviating. As described in chapter 1, a stable operation of the arc tube is obtained by keeping the resonant circuit part deviation low. But the shape or turn pitch of the inductively coupled coil is very easy to be deformed and as a result the reactance deviates.

The Plasma resistance, Plasma inductance, induction coil inductance, mutual inductance and coupling coefficient between coil and plasma are denoted as R_p, L_p, L_t, M, and k respectively. F parameters A, B, C and D of the circuit shown in figure 5 are described as follows,

$$F = \begin{pmatrix} A & B \\ C & D \end{pmatrix} = \begin{pmatrix} \frac{tX + bY}{bd} & \frac{t(1-k^2)X + bY}{bd} pZ_0 \\ \frac{tX_0 + bY_0}{bdgZ_0} & \frac{t(1-k^2)X_0 + Y_0}{bdg} p \end{pmatrix}, \tag{1}$$

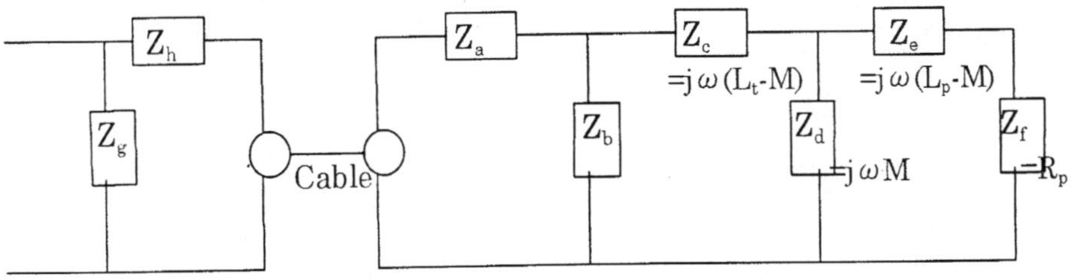

Fig.5 Matching circuit system

where $a=Z_a/Z_0$, $b=Z_b/Z_0$, $c=Z_c/Z_0$, $d=Z_d/Z_0$, $e=Z_e/Z_0$, $f=Z_f/Z_0$, $g=Z_g/Z_0$, $h=Z_h/Z_0$, $t=j\omega L_t$, $p=j\omega L_p$, $Z_c=j\omega(L_t\text{-}M)$, $Z_d=j\omega M$, $Z_e=j\omega(L_p\text{-}M)$, $Z_f=R_p$, and Z_0 is the co-axial cable characteristic impedance, γ is propagating constant and l is cable length.

$$X_0 = X + [(a+b)x + y]g = [x + (a+b)y] + [(a+b)x + y](g+h), \tag{2-1}$$

$$Y_0 = Y + (ax+y)g = x + ay + (ax+y)(g+h), \tag{2-2}$$

$$\eta = \frac{x}{y} = \frac{\sinh(\gamma l)}{\cosh(\gamma l)} = \tanh(\gamma l), \tag{2-3}$$

And also the following is valid,

$$M = k\sqrt{L_p L_t} \tag{3}$$

The arc tube characteristics α_p and Q_p are defined as follows,

$$\alpha_p = \frac{1}{Q_p} = \frac{R_p}{\omega L_p} \tag{4}$$

Under any small value of deviation, $\delta t/t$, to keep the arc tube power constant, i.e. $\delta P/P=0$, the following equation is derived from $\delta B/B=0$,

$$\xi = \frac{tX}{bY} = \frac{1}{1-k^2}. \qquad (5)$$

For switching devices in an inverter, it is better to minimize the devices current deviation. A condition for the minimum current deviation is obtained at a point that the following equation is satisfied.

$$\xi_0^2 = \left(\frac{tX_0}{bY_0}\right)^2 = \frac{1+Q_p^2}{1+Q_p^2(1-k^2)^2}, \qquad (6)$$

To supply the higher voltage on the arc tube, it is selected the near resonant state, i.e. between Z_b and $j\omega L_t$,

$$Z_b = \frac{1}{j\omega C} = -j\omega L_t(1+\theta)$$
$$b = -t(1+\theta) \qquad (7)$$

where θ is the difference from the resonant value $1/j\omega C = -j\omega L_t$. From equation (5),

$$a = b\left[1 - \frac{b}{b - t(1-k^2)}\right] - \frac{\eta + h}{\eta h + 1} \qquad (8)$$

By the way, the input impedance of the matching circuit should be set to the desirable value, i.e. the inverter internal impedance.

$$Z_{in} = R_{in} + jX_{in} \qquad (9\text{-}1)$$

$$R_{in} = \frac{R_p}{D^2 - C^2R_p^2} \qquad (9\text{-}2)$$

$$jX_{in} = \frac{DB - ACR_p^2}{D^2 - C^2R_p^2} \qquad (9\text{-}3)$$

From these equations, a, b, g, h and Z_0, γl are determined.

The experimental results are as follows.

In figure 6, the lamp power and the switching device loss variations due to the resonant circuit part deviation are shown [7]. The lamp power decreases according to the shift of the reactant element and the lamp is on stable operation without failure of the switching devices.

(1) The design center of the circuit is determined by the equations from (5) to (9).

(2) The switching device failure has been extremely reduced by using the current limitation of the co-axial cable.

3. Lamp characteristics

3.1 Modulated high frequency operation

It is desirable to keep the high frequency coil current amplitude constant. In such

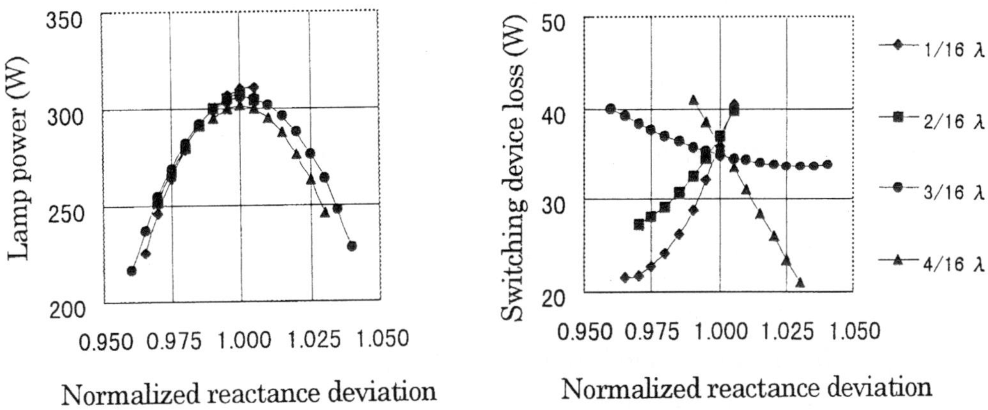

Fig.6 Lamp power and switching device loss variations by resonant reactance deviation using a cable as a impedance matching element.

case it is necessary to use a switching regulated and completely smoothed dc current power supply. But by using it, the system efficiency decreases about 10%, and also system cost goes up.

It has been investigated to operate the system with only rectified, non-smoothed dc power supply. In this case, the coil is driven by a modulated 13.56MHz current. It has been found that the arc ring pulsates, i.e. expands and shrinks in accordance with change of the power supply voltage, and in case of the arc ring pulsation, the lamp efficacy is low. The higher the filled gas pressure of arc tube and the higher the modulating frequency are, the higher the efficacy of lamp goes up, as shown in

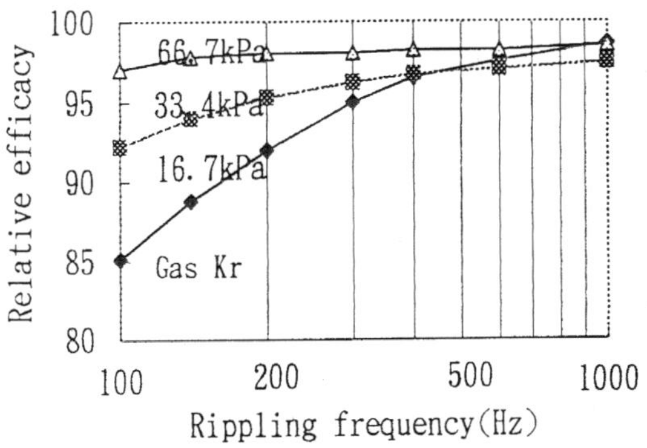

Fig.7 (a) Relative lamp efficacy by modulated high frequency operation.

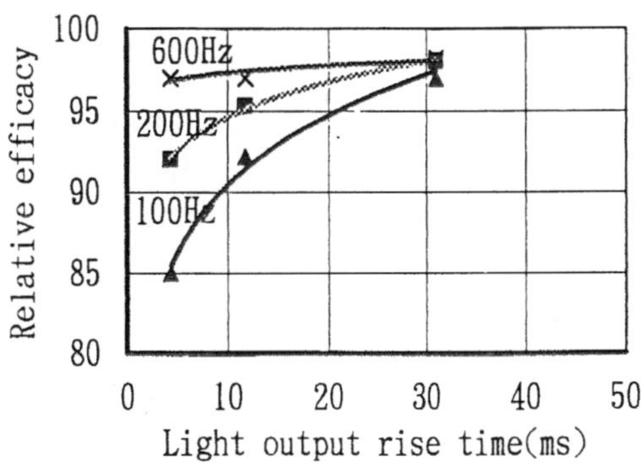

Fig.7 (b) The relation between light output rise time and lamp efficacy.

figure 7(a). At higher filled gas pressure, the light output rise time or decay time is longer than that of the lower filled gas pressure case and also the lamp efficiency has a close relation to a light output rise time, as shown in figure 7(b). The reason of these phenomena is the gas temperature of the arc column is swinging up and down. At low pressure, the arc gas temperature varies extremely, but at higher pressure, the gas heat capacity becomes big and the gas temperature changes very slow.

It is desirable to fill the gas pressure high, or to operate the lamp with higher modulating frequency, for getting a higher lamp efficacy even under a rippling dc power supply.

3.2 Arc tube chemical reaction and performance

As an arc ring induced by a coil is attracted toward the coil, the tube temperature goes up high. In case of the scandium and sodium iodides dosed into the arc tube, by chemical reaction between tube wall and dosed materials, free iodine is formed in the arc tube soon after the lamp burning. Free iodine makes the lamp flickering and finally the lamp drops out. So the lamp life becomes short. In a higher loaded arc tube, the free iodine formation has been accelerated. It is said that in case of a conventional metal halide lamp doped with Zn, Pb, Sn or such metals or halides, formation of the free iodine is suppressed.

Also, in case of the electrode-less HID lamp, it has been found that a proper dose of such metal halides suppress the formation of free iodine. This suppressing effect is found in case of an arc tube dosed with not such metals but only such metal halides. And also the lamp efficacy is kept high even such metal halides exist. In figure 8, the effect of free iodide suppression by adding Sn metal halide is shown. The detail will be presented in the poster session at the LS:8 [8].

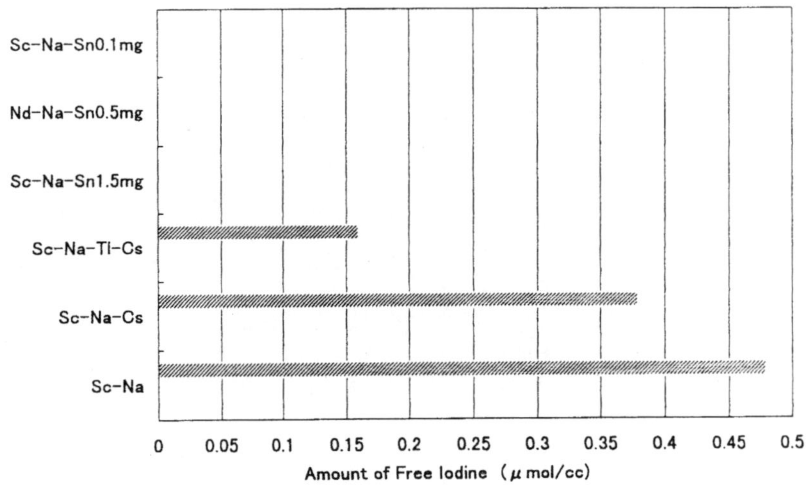

Fig.8 Amount of the free iodine in the arc tubes. (1,000 hours operation)

4. Conclusions

It has been studied to operate the inductively coupled electrode-less HID lamp system stable. There had been some technical items to be overcome before commercial use.

The obtained technical conclusions are as follows.

(1) Good starting has been realized by using a resistor as a dummy load of auxiliary starting circuit. The dummy load resistor makes the resonant circuit Q curve wide.

(2) It is very effective to adopt the system that the frequency of the driving coil follows up the circuit resonant frequency. At starting the circuit impedance is automatically matched by the following up frequency system. But the shift out from the admitted ISM band frequency range is new problem. Electro magnetic noise under the shift out condition should be suppressed.

(3) By using co-axial cable as a part of a matching circuit, the extreme lamp power deviation suppressed and the system operates stable without failure of switching devices. The resonant circuit parts deviation can be permitted up to $\pm 2.5\%$ if 10% of lamp power decrease is admitted.

(4) A 3/16 or so wave length co-axial cable shows best performance, because the current shift of the switching device is suppressed in minimum.

(5) The lamp efficacy is low when the system is operated by the only rectified non-smoothed dc power supply. The higher the modulating frequency or the higher the filled gas pressure is, the higher the lamp efficacy is.

(6) Light out put rise or decay time is fast when the filled gas pressure is low, because the arc column gas temperature goes up and down rapidly. At higher filled gas pressure, the gas temperature is kept almost constant, the light output variation is slow and the efficacy is high.

(7) At low filled gas pressure, the arc ring diameter changes accordance with modulating voltage.

(8) To get the high efficacy lamp system, it is necessary to operate the lamp with high filled gas pressure or higher modulating frequency.

(9) Sn or such metal halide suppresses the chemical reaction between the tube wall material and the dosed conventional metal halides such as Sc/Na halide.

(10) The complete system for the inductively coupled HID lamp is obtained to use the following system.

1. A rectifying bridge to produce non-smoothed dc power.
2. A switch to change the matching state from a starting state to a nominal arc discharge state.
3. A co-axial cable as a matching circuit part.
4. The arc tube filled the moderately high-pressure gas.
5. The arc tube dosed with Sn or such metal halide with conventional metal halides.

Acknowledgements

The cooperation of General Electric Company Lighting Division engineers and

IL 13

Toshiba Lighting & Technology Corporation engineers has been performed on the development of the inductively coupled HID lamp system. The author thanks to those colleagues such as Messrs. K. Shimizu, I. Yokozeki, K. Uemura, A. Itoh, T. Kakitani and et al.

References

[1] C. N. Stewart, M. E. Duffy, J. T. Dakin, V. D. Roberts, S. A. El-Hamamsy, H. L. Witting, A. Inouye, K. Shimizu and K. Araki, *"Inductively Coupled HID Lighting System"* Proceedings of the Sixth International Symposium on the Science & Technology of Light Sources (LS:6) Budapest (1992) 78:L

[2] M. Shinomiya, *"Matching Network of the Inductively Coupled Discharge Lamps"* J. Illuminating Engineering Institute of Japan, Vol.78 No.2 (1994) pp101-106 (In Japanese)

[3] H. Hanada, *"Lighting Industries In Japan"* Proceedings of the Seventh International Symposium on the Science & Technology of Light Sources (Proceedings of the Seventh International Symposium on the Science & Technology of Light Sources (LS:7) Kyoto (1995) 1:I

[4] D. O. Wharmby & S. A. El-Hamamsy, *"Low power compact electrode-less lamps"* Proceedings of the Seventh International Symposium on the Science & Technology of Light Sources (LS:7) Kyoto (1995) 2:I

[5] K. Shimizu, I. Yokozeki, K. Uemura, A. Itoh and A. Inouye, *"Pulsating Operation of Electrode-less HID Lamp"* Proceedings of the Seventh International Symposium on the Science & Technology of Light Sources (LS:7) Kyoto (1995) 37:P

[6] I. Yokozeki, K. Shimizu, K. Uemura, M. Matsuda and A. Inouye, *"Study of an Electrode-less HID Lamp System with Frequency Control"* J. Illuminating Engineering Institute of Japan, Vol.80 No.5 (1996) pp325-330 (In Japanese)

[7] I. Yokozeki, T. Kakitani, K. Shimizu and A. Inouye, *"Matching circuit of electrode-less lamp system using distributed constant line"* to be presented at the Annual conference of IEI-Japan (1998) (In Japanese)

[8] A. Itoh, K. Uemura, T. Ishigami and A. Inouye, *"Life Performance of Inductively Coupled Electrode-less Metal Halide Lamp"* to be presented at the Proceedings of the Eighth International Symposium on the Science & Technology of Light Sources (LS:8) Greifswald (1998)

IMPROVED ENVELOPE MATERIALS IMPACT ENVIRONMENTAL AND LIGHTING PERFORMANCE

William H. Rhodes

OSRAM SYLVANIA Development Inc., Beverly, Mass. 01915, U.S.A.

1 INTRODUCTION

Performance and environmental issues are currently driving materials research and development for lighting applications. Materials properties are often a critical factor in limiting efficacy of current light sources, consequently researchers seek improved properties with a goal of improved performance. Cost-performance interrelationships have guided materials decisions for generations. Now, environmental manufacturing and disposal issues have equal footing in materials decisions. Not that the environment was ignored by our predecessors, but the weight given to environmental impact has risen to be equal to performance and cost.

Translucent polycrystalline alumina, PCA, has been the most important crystalline lamp envelope material for over 30 years. Its discovery in 1959 by Coble [1] made possible a practical high-pressure sodium (HPS) lamp. More recently, Philips introduced a metal halide lamp employing a PCA envelope [2]. Consequently, the importance and worldwide production of PCA has grown to an estimated 50 x 10^6 per year. Happily, there are no environmental issues with the production and disposal of the PCA envelope once the starting powder has been converted to Al_2O_3. Depending on the synthesis route there may be environmental considerations during decomposition of the precursor salts.

Sapphire, the single crystal form of PCA, was fabricated by the edge defined growth method, and used for HPS lamps in the first decade of HPS lamp production. Gradually PCA optical properties improved to the point where there was no clear advantage to sapphire. Recently a new method of forming sapphire by PCA conversion was discovered [3] thus the benefits of sapphire will be revisited.

In the last 30 years many programs were aimed at producing an improved crystalline lamp envelope. Initially, $MgAl_2O_4$ was developed and tested [4]. The author worked on Y_2O_3 [5]. Others tested include BeO [6] and $Y_3Al_5O_{12}$ [7]. Many other oxides have been sintered to transparency for window, infrared dome, and laser applications. In general these have much higher optical quality than PCA. Not only are the physical properties of potential lamp envelopes important, but the thermodynamic stability relative to the useful arc radiating species is crucial.

The lighting industry blossomed and matured using various glasses as lamp envelopes. These range from relatively low melting high lead glass to the most refractory, fused quartz, which is melted at 2100°C. The lead glasses date back to the 1800s, an era where environmental issues were less important than today. Recently, several lead-free glasses have been introduced with properties closely mimicking their leaded cousins. These glasses are expected to have improved environmental performance.

2 CERAMICS
2.1 PCA
2.1.1 Processing

The starting powder and processing of PCA has improved since the early days of manufacturing. All of the early arc tubes were manufactured from alum derived Al_2O_3. Today competitive grades are synthesized from alkoxides, hydroxycarborates, chlorides, and hydroxides. All of these powders have different purity levels, phase and particle size assemblages. Further, the degree of agglomeration differs with precursor, and post calcination milling.

Forming processes have not changed significantly in the last 20 years. Some manufacturers use dry bag isostatic pressing while others use extrusion. The binders and plasticizers may have improved to give a green body free of density variations and hard agglomerates. Surface finish has improved, particularly for isostatic pressers who have instituted green finishing steps. Extrusion manufacturers now pay more attention to the surface finish of the die.

Sintering has also improved through better understanding of the sintering process. The temperature-time cycles are more carefully controlled. Sintering aids have been optimized, and play a significant role in the optimum sintering cycle as well as in performance (see Section 2.12). The sintering atmosphere has evolved from pure H_2 to N_2-H_2 mixes with control of dew points for the various parts of the cycle [8].

These changes have resulted in a gradual improvement in optical properties over the last 30 years as shown in Figure 1. Total transmittance was measured by the light source within the tube within an integrating sphere technique. In-line transmittance, shown in Figure 2, was measured by passing a light beam through two walls. This latter technique has been found to correlate with the 2-angle in-line and single wall techniques. Because it passes through two walls it may be more sensitive to surface finishes. The gradual improvement in properties is impressive, but obviously can't continue indefinitely as theoretical limits are approached.

High pressure sodium, HPS, lamp performance was shown to correlate with optical properties by Ingold and Taylor [9] in the following:

$$LPW = aT^* + b \tag{1}$$

where a and b are numerical constants and T^* is

$$T^* = (T_t \times T_f/[\eta T_t + (1-\eta)T_f] \tag{2}$$

where T_t is total transmittance, and T_f is in-line transmittance. The referenced work employed a single wall transmittance technique.

Data collected over several years at OSRAM SYLVANIA was examined for possible correlation with equations (1) and (2). The present study used the two wall technique thus to obtain equivalence, T_f is taken to the ½ power, and using $\eta = 0.6$ as employed for arc absorptance by Ingold and Taylor,

$$T^* = (T_t x T_f^{0.5})/[0.6T_t + 0.4T_f^{0.5}] \tag{3}$$

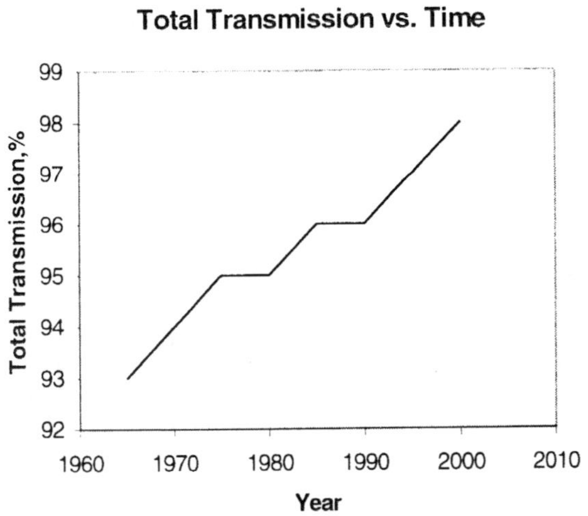

Figure 1. Qualitative depiction of improvements worldwide in PCA total transmittance over the years since HPS lamps were introduced.

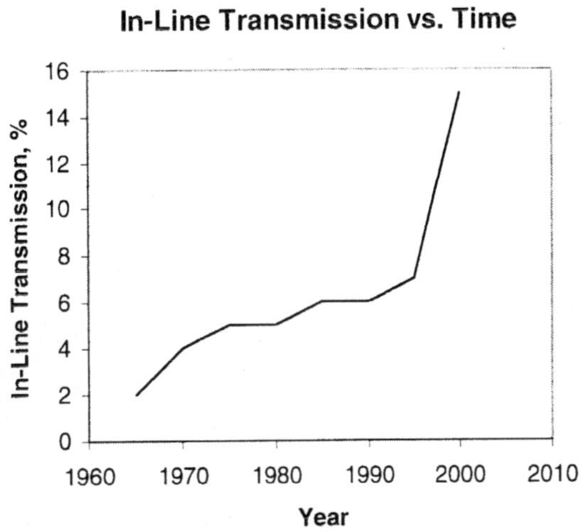

Figure 2. Qualitative depiction of progress in improving in-line transmittance worldwide over the years since HPS was introduced.

Using this relationship, eleven sets of lamp data involving 156 lamps were reviewed. The data set with the best correlation coefficient is shown in Figure 3. All of the data showed an increase in LPW with T_t , and four sets of the data showed a good correlation when T_f was included at the predicted dependence. The reason for the lack of correlation in the other sets is thought to be due to arc tube dimensional variations, and lamp design and/or construction parameters such as emitter, electrodes, Xe pressure, CRI, etc.

A comparison of these results with other literature studies is shown in Table 1. The role of T_t is always important, and the effect of T_f is clearly demonstrated in cases A, B, and C. Case D had a very weak dependence, which could have been caused by uncontrolled parameters similar to those, discussed above. Taking an average of all the eleven data sets in the present study gave the dependence in Case E while averaging only the data sets where a strong correlation with T* was demonstrated gave the results of Case F. This is the dependence expected if arc tube and lamp fabrication are closely controlled. It is noteworthy that the effect of T_t is stronger than the other studies while the effect of T_f is intermediate. It is clear that improved optical properties have resulted in improved lamp performance.

2.1.2 The Effect of PCA Doping on Na Resistance

2.1.2.1 Reactions

Maekawa reviewed recent advances in improving the Na resistance of PCA at LS-7 [12]. He stated that improvements were realized when using doping systems consisting of $MgO + ZrO_2$ or $MgO + ZrO_2 + Y_2O_3$. He stated that the reasons for the improvement were not clear, but that the following reaction was proposed:

$$2Na + 11\ Al_2O_3 \rightarrow Na_2O.11Al_2O_3 + 2/3\ Al \qquad (4)$$

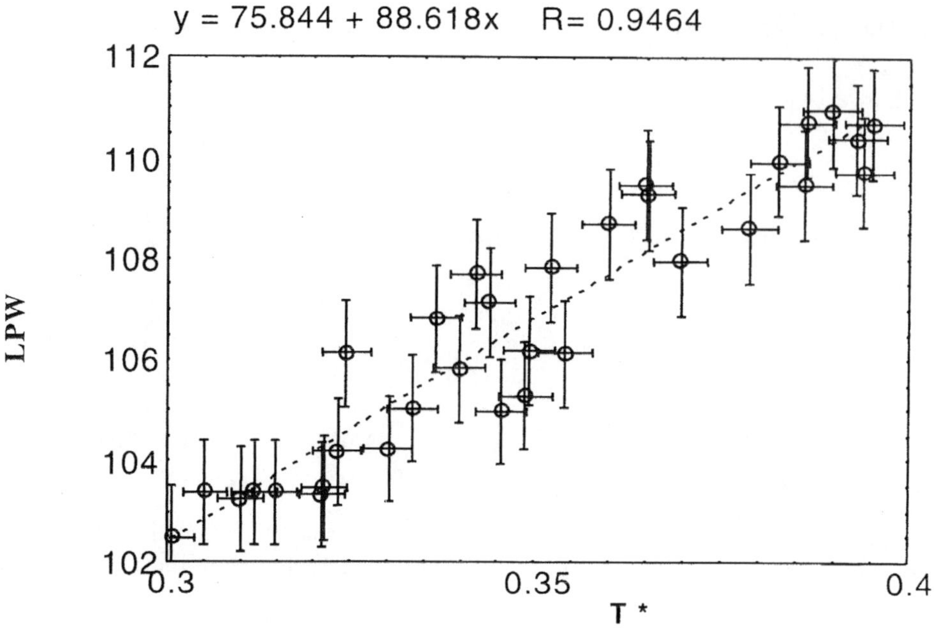

Figure 3. Efficacy of 200W HPS lamps plotted according to equations (1) and (3) showing excellent correlation with both total and in-line transmittance.

Table 1. LPW vs. Total and In-line Transmittance.

Case	Reference	Range of LPW	Range of T_{total} Value (%)	Range of $T_{in\text{-}line}$ Value (%)*	Lamps Number	% Increase in LPW per 1% Increase in the $T_{in\text{-}line}$ Value	% Increase in LPW per 1% Increase in the T_{total} Value
A	[9] (Ingold and Taylor of GE)	110-140 (400W)	75-96	5-12	25	2.9	1.0
B	[10] (Otani and Suzuki of Mitsubishi)	120-140 (400W)	92.5-94.0	3-20	4**	0.8	***
C	[11] (Kaneno and Oda of NGK)	113-132 (400W)	90-94	1-6	41	0.8	2.2
D	[11] (Kaneno and Oda of NGK)	125-130 (400W)	90-94	6-20	6	0.02	2.2
E	OSI Present	78-124	94.4-97.5	4.3-11.7	156	1.1	1.0
F	OSI Good Correlation	92-114	94.4-97.4	4.3-11.7	69	1.53	3.4

*$T_{in\text{-}line}$ values were the OSI in-line transmittance values estimated from the reported single-wall in-inline transmittance values.
**The exact lamp number was not known. There were 4 types of lamps representing arc tubes of various transmittance values.
***Not reported in Reference 10.

resulting in free Al which diffused via grain boundary diffusion to the outer PCA wall and evaporated onto the outer glass causing darkening and lumen loss. Recent work at OSRAM

SYLVANIA verified the general thesis proposed by Maekawa, but amplifies on understanding the controlling mechanisms.

A more general form of reaction (4) is:

$$2Na + (x + 1/3)\ Al_2O_3 \rightarrow Na_2O \bullet x\ Al_2O_3 + 2/3\ Al \quad (5)$$

where x = 1, 5, and 11 for sodium mono-aluminate, sodium β"-alumina, and β-alumina. In addition, it is believed that spinel second phase is a major factor for sodium resistance, since it is preferentially attacked by Na to form β''' alumina by the reaction:

$$6Na + 12\ MgO \bullet Al_2O_3 + 34\ Al_2O_3 \rightarrow 3\ (Na_2O \bullet 4MgO \bullet 15\ Al_2O_3) + 2Al \quad (6)$$

In the literature sodium β"-alumina was sometimes reported to be stabilized with the addition of MgO [13]. For example, $Na_2O \bullet 0.6MgO \bullet 5\ Al_2O_3$ was reported by Togwa et al., [14], $Na_2O \bullet 0.8MgO \bullet 6.2\ Al_2O_3$ reported by Verweij et al., [13], and $Na_2O \bullet 0.2MgO \bullet 6.3\ Al_2O_3$ reported by Tan et al., [15].

2.1.2.2 Experimental Verification

Arc tubes (7.39 mm OD by 5.87 mm ID by 89 mm cavity length) were sealed with 100 torr of Xe. The arc tubes were the standard 250W size except for cavity length. Hg-Na amalgam dose was (21.6 wt% or 70 atom% Na) 13.5 mg (3 mg Na, 10.5 mg Hg). The lamp was designed such that it would have a normal cold spot temperature of 1000K (D-line of 120Å) when operated at 400W. Typically five lamps were tested per composition group with lumen output measured every 500h. The transmittance of the outer jacket was found to correlate with lumen loss, and was measured more often. Wall temperature was measured by the d-line pryometry technique, and found to be 1350 ± 20°C, some 170°C above normal.

Sapphire as well as many PCA compositions were tested. Only the most illustrative tests are shown in Figure 4. Sapphire clearly has the best performance. This was followed by a MgO + ZrO_2 + Y_2O_3 composition having no spinel phase, but residual YAG phase. The doping system of MgO + Tb_2O_3 + Y_2O_3 exhibited the next level of improved behavior, which is slightly better than straight MgO + ZrO_2. This is followed by what used to be the standard composition of straight MgO. It is thought that all of these relationships can be explained based on second phase and grain boundary absorption phenomena.

2.1.2.3 Mechanisms

The Al product of reactions (4, 5, or 6) diffuses through the wall to deposit on the glass outer jacket, causing darkening of the jacket. Al metal is the predominant component in the dark deposit on the outer glass [12]. Other species such as Al_2O_3 and Na were also found in the deposit sometimes, but to a much lesser degree. Thus, it appears that Al diffuses faster than Na, which is consistent with the diffusion data reported in the literature [16]. This is reasonable, as the size of Al^{+3} ions is much smaller than Na^{+1} ions. Since the diffusion rate of Al is faster than Na along grain boundaries, reaction (2) should be controlled by the arrival of the reactant, i.e., Na at the reaction front.

McVey [17] reported that sodium diffused through the PCA wall, and β-alumina served as a fast-diffusion path for Na transport. Since the structure of β-alumina consists of blocks of spinel

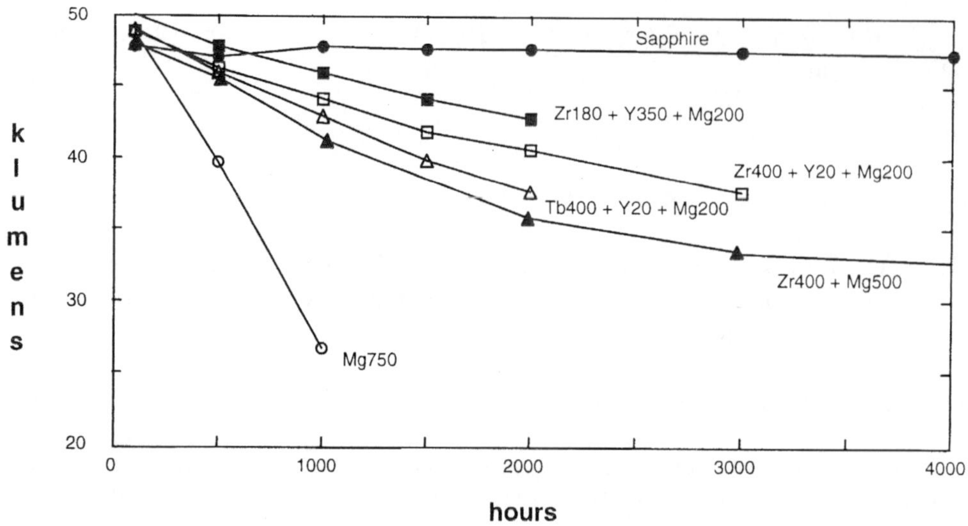

Figure 4. Effect of PCA dopants on lumen maintenance for force test running at 1350°C wall temperature.

cells [18], the spinel phase provides ideal sites for nucleation of the β-alumina by reaction (6). It was thus argued that the presence of spinel phase accelerated the Na attack in PCA. While this holds true if the MgO level is above the solid solubility, the nature of the adsorbed Mg level may be important in low MgO level compositions. Zr^{+4} doping ties Mg^{+2}, as the positively charged Zr at Al sites compensate pairs with the negatively charged Mg at Al sites. When this pairing happens, less Mg becomes available to reside at grain boundaries or form spinel.

Many studies in the literature [19, 20] showed Ca impurity enhanced the loss of Na, and increased the rate of β-alumina formation. The fact that YAG second phase can absorb Ca and Si impurities [21], explains the beneficial effect of the Y_2O_3 addition using MgO plus Y_2O_3 dopants as compared to the MgO sole dopant [22]. The solubility of Ca in YAG was 500 ppm, the solubility of Zr was >800 ppm. The solubility of Mg in YAG was believed to be very low [23]. As further confirmation of the proposed mechanism, Y_2O_3-doped alumina was found to exhibit a lower creep rate [24] than pure alumina due to segregation of Y to grain boundaries.

Tb aluminate second phase was found to absorb MgO, which means that not only is spinel eliminated, but also the grain-boundary second phase can act as a sink for MgO throughout lamp life. This dopant combined with Y_2O_3 for the reasons explained above effectively inhibit Na from reaching the reaction front thereby slowing the release of Al to the outer jacket.

It should be clear from the discussion that straight MgO-doped PCA would have the highest lumen loss rate due reaction (6) and the increased supply of Na to the reaction front. Vacuum or hydrogen anneals can reduce the residual spinel and MgO level at surfaces or grain boundaries. This can improve performance, but not to the level of the special doping compositions.

Sapphire had the best performance in this study. The explanation is that there are no high diffusion rate grain boundaries present to speed the reactants or products from the reaction fronts. Lattice diffusion is orders of magnitude slower than grain boundary diffusion. Maekawa [12] found that sapphire performed equal to an MgO + ZrO_2 composition which is in conflict with these results and the proposed mechanism.

2.2 Sapphire

2.2.1 Edge Defined, Film Feed Growth (EFG)

Growing sapphire or other single crystal materials to any cross sectional shape by EFG was invented by H. LaBalle [25]. The work was initiated to find a method of growing sapphire fibers for the US Air Force composite program. The final process is the result of three successive inventions: (1) the use of a floating orifice through which to pull the fiber, (2) the use of capillary action to bring liquid to the solid interface from a continuously fed reservoir, and (3) obtaining consistent geometric cross section tolerances by allowing the liquid to flow to a flat edge of the die which would define the outer diameter for example.

Sapphire tubing can be obtained from several companies under license from the original invention. A third company began an ambitious pilot program to supply arc tubes to the lighting industry. Early field installations of HPS lamps using this tubing suffered a high incidence of failures due to fatigue crack growth. Another problem was that the sapphire tubing only had about 95% total transmission due to a fine layer of Mo precipitates (from the die). The lamps reported in Figure 4 were made from EFG grown sapphire. They performed well with no failures even under the high temperature stress conditions. As expected, they exhibited superior sodium resistance due to the lack of grain boundaries for Na or Al diffusion.

2.2.2 Solid State Crystal Conversion (SSCC)

This process for producing sapphire tubing was recently invented by General Electric [3]. PCA doped with MgO sintering aid is first sintered to translucency, then heat treated for extended times to reduce the MgO level to <60ppm, a level that would be free of spinel second phase [26]. Further heat treatment can result in the nucleation and growth at near intrinsic rates of a single crystal. The growth rates are so fast that the advancing boundary sweeps past any remaining porosity, which becomes entrapped within the grains. Occasionally, entire islands of polycrystalline material are trapped. Figure 5 shows PCA tubing which is partially converted exhibiting a high degree of transparency in the single crystal region. Heretofore, a complete understanding of the nucleation and growth process is not in hand. Clearly, the potential for a new method of achieving sapphire tubing has been demonstrated which are not only interesting, but also a marvelous technological feat. The remaining issues appear to be the ability to reproducibly control the nucleation process, demonstrate the advantages in lamps, and conduct a cost/benefit analysis.

3 GLASS

3.1 Soft Glass

Two soft glasses, SG10 and SG12, are generally employed in lamp manufacturing to seal electrical wires into lamps. SG10 is commonly referred to as the "low lead" glass [21% PbO] and SG12 the "high lead" glass [28% PbO]. SG10 has excellent sealing properties to both soda lime glass envelopes, and niron and dumet wires used in the manufacture of incandescent andfluorescent lamps. For high wattage incandescent lamps, the higher electrical resistivity glass, SG12, is required.

SG10 is the higher volume glass. Several companies have had active programs to develop substitute glass compositions to eliminate lead. Typically, Li_2O, BaO and SrO are the major chemicals substituted for PbO. Such a lead-free glass would ease environmental issues throughout their life cycle. The properties of two new glasses, SG64 by OSRAM SYLVANIA INC. and PH360 by Philips are compared with SG10 and SG12 in Table 2.

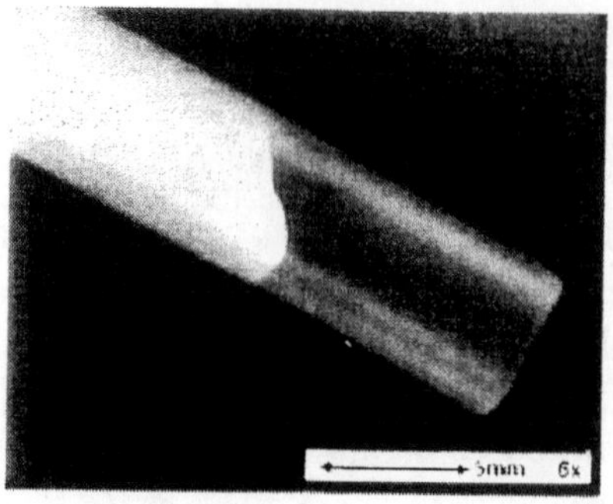

Figure 5. Large sapphire crystal growing on the hot (1880°C) end of a sintered tube of MgO-doped alumina placed in a longitudinal temperature gradient of 100°C. Image taken in polarized light.

Table 2. Properties of Soft Glass Compositions.

	SG10	SG64	PH360	SG12
Thermal Expansion [ppm/°C] 23-300°C	9.20	9.17	9.03	8.90
Viscosity [Poise]				
Working Point [Log = 4.0]	1002	978	1015	985
Softening Point [Log = 7.6]	634	667	676	630
Working Range	368	311	338	355
Annealing Point	435	487	485	435
Electrical Reistivity [Log]				
250°C	8.5	8.8	8.6	9.6
350°C	6.7	7.0	6.9	7.7

The properties of both lead-free glasses are strikingly similar to SG10. They have very similar viscosity curves, given the long working range required for dimensional control in lamp making. Thermal expansion and electrical properties are also similar to SG10, resulting in similar seal strains and the necessary resistivity required to prevent lead-wire electrolysis during lamp life. Properties are, however, not exactly identical, such that lamp plants must typically make minor flame adjustments for lamp sealing and annealing. PH360 has been successfully implemented and is now being used by Philips and OSRAM.

3.2 Hard Glass

Ever since the introduction of the high intensity mercury discharge lamp in the 1930s, a lead borosilicate glass [SG772, GE772], commonly referred to as Nonex, has been used to manufacture HID lamps in the U.S.A. In Europe lead-free Schott 8486 bulbs and 8487 tubing are also used to make HID lamps, however UV absorption of these bulbs is not sufficient for many HID lamps manufactured in the US. With all the environmental and regulatory concerns associated with melting a lead and arsenic containing composition, a new glass [SG773] has been developed which has all the glass physical properties required of a tungsten sealing glass, as well as sufficient UVB protection to be a direct replacement for SG772. Because the new glass does not contain lead oxide, the associated lamp manufacturing problems with blackened, reduced lead do not occur, and a corresponding improvement in lamp sealing has been achieved. Melting improvements have also been realized. The new glass has a higher resistance to surface "scumming" and phase separation, problems typical of most borosilicate glasses. The properties of these glasses are compared in Table 3.

Table 3. Properties of Several Hard Glasses Compared to the New Lead-Free SG773.

	SG772	SG773	Schott 8486	Schott 8487
Thermal Expansion [ppm/°C]				
23-300°C	3.60	3.90	4.10	3.90
Seal to Tungsten Wire, Radial [psi]	745 tension	495 tension	NA	730 tension
Viscosity [Poise]				
Working Point [Log = 4.0]	1159	1181	1220	1135
Softening Point [Log = 7.6]	756	773	820	774
Working Range [C°]	403	408	400	361
Annealing Point [°C]	523	542	555	542
Electrical Resistivity [Log 10]				
250°C	8.8	8.8	7.5	8.3
350°C	7.1	7.1	6.1	6.9

Ultraviolet light absorption is an important property for a HID bulb. The principal role of lead in SG772 is to protect the observer from UV exposure. There are several additives that could be employed to impart the necessary UV absorption, but solarization during lamp life is a phenomenon that must be guarded against. The solution to the problem in SG773 development was to stabilize the oxidation state of iron to primarily ferric [Fe^{3+}] iron by the addition of from 100-300 ppm ceria. This 10X reduction in ceria from that traditionally used for UVB absorption had two major impacts: (1) iron could be reduced to a concentration which would not color the glass green, and (2) lumens were not reduced in life through solarization. Glass manufacturing was greatly assisted by having a composition that was resistant to the normal fluctuations in glass oxidation state which occur due to changes in cullet ratio, furnace atmosphere, tramp contaminants, etc.

The physical properties of SG773 were extremely close to SG772, such that only minor differences were noted in lamp sealing. Since lead-free glass will not darken in the sealing fires, improvements were realized in seal uniformity and quality. This glass was introduced in production in the spring of 1998.

4 THERMODYNAMIC COMPATIBILITY OF ENVELOPE MATERIALS

Chemical compatibility between the lamp components and fill components is essential for the long lives expected of lamps. For example, it wasn't until the development of PCA that the HPS lamp became practical. As discussed in Section 2.12 some reactions occur even in this lamp, so complete thermodynamic compatibility is not required. As long as the kinetics of some favorable corroding reactions are slow, practical lamps can be produced. The new metal halide lamps in PCA envelopes present an interesting case. In addition, there is evidence [27] manufacturers are considering metal halide lamps in other envelopes such as $Y_3Al_5O_{12}$. The author considered Y_2O_3 envelopes for a ScI_3 based metal halide chemistry in the late 1970s.

In considering whether or not an arc tube material will be stable with a metal halide fill, one must have knowledge of the free energy of formation for both the oxides and the iodides. When one considers just the oxides, one might choose a very stable oxide like Y_2O_3. However, this is only half of the problem. The calculation must compare the stability of the oxide relative to the iodide as shown in Figure 6.

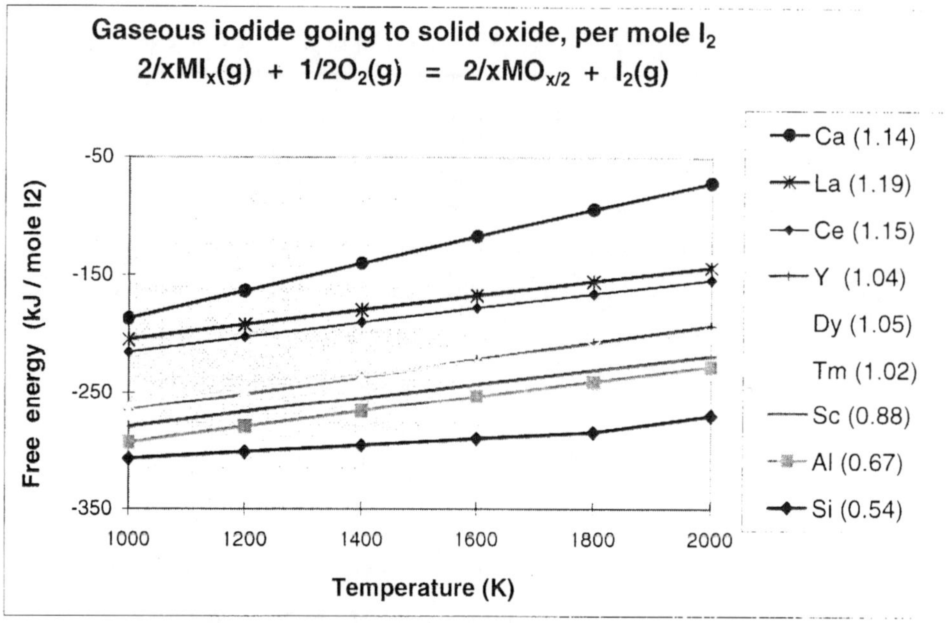

Figure 6. Relative thermodynamic stability of potential envelope materials in the presence of typical iodides employed in metal halide lamps.

When the calculation is completed in this manner it is apparent that SiO_2 is the most stable oxide while CaI_2 is the most stable iodide shown. Al_2O_3 or PCA is somewhat less stable than SiO_2 (assumed to be the same thermodynamically as quartz). PCA does appear to be stable relative to the rare earth halides, but the reason the lamp based on PCA envelopes and rare earth halide fills is so interesting has more to do with the higher use temperature and adaptability to precision fabrication than to its thermodynamic properties.

Y_2O_3 turns out not to be a very good envelope for rare earth halide fills, and this diagram explains the author's experience with Y_2O_3 envelopes and a ScI_3 based halide fill. In this case the rare earth iodides are more stable than the oxide, so the expected Sc emission disappeared within a matter of minutes as scandium oxide formed, and Y lines entered the spectrum. Thus, the intended emission changed and performance degraded radically. Similar behavior is expected for YAG envelopes. These envelopes might perform well for CaI_2 lamps, however, based on this diagram.

5 CONCLUSIONS

Polycrystalline alumina lamp envelopes have undergone evolutionary improvements over the last 30 years resulting in improved performance for HPS lamps. This material is proving to be quite versatile as a new generation of metal halide lamps has been introduced using this material. Thermodynamic calculations suggest this may be the best polycrystalline oxide for rare earth halide fills. Fused quartz, the envelope now in general use is more stable than alumina, but doesn't have as high a use temperature.

Environmental considerations have resulted in several new glasses free of lead for lighting applications. Glass scientists have done a remarkable job in mimicking the physical properties of the lead glass cousins yet achieving glasses with improved processing behavior. These glasses will benefit the glass plant, the lighting manufacturing factories, and lamp disposal considerations.

6 ACKNOWLEDGEMENTS

The author thanks G. Wei, R. Snellgrove, and R. Marlor for their and their group's work, which is condensed for this paper. He thanks C. Greskovich from the General Electric R&D Laboratory for sharing the sapphire conversion contribution. Finally, he thanks B. Loschi for preparing the manuscript.

7 REFERENCES

1. R. Coble, Transparent Alumina and Method of Preparation, U.S. Patent 3,026,210.
2. P. A. Seinen, 7th Int. Sym. Sci. & Tech. of Light Sources, Kyoto (1995).
3. C. Scott, L. Levinson, R. Maxwell, and M. Kaliszewski, U.S.Patent 5,451,553.
4. P. Hing, J. Matl. Sci. **11**, 1919 (1976).
5. W. Rhodes, 2nd Int. Sym. On Incoherent Light Sources, Enschede, 103 (1979.
6. J. Richter, ONR Contract No. N00014-71-C-0384.
7. G. DeWith and J. Parren, Matl. Res. Bull., **19**, 13 (1984).
8. W. Rhodes, G. Wei and G. Fryburg, U.S. Patent No. 4,762,655.
9. J. H. Ingold and W. L. Taylor, J. Illu. Eng. Soc., 223 (1982).
10. K Otani and R. Suzuki, J. Light & Vis. Env. **3**, 18 (1979).
11. M. Kaneno and I. Oda, *Energy and Ceramics, Materials Science Mono. 6*, ed. P. Vincenzini, (Elsevier, New York 1980).
12. K. Maekawa, 7th Int. Sym. Sci. & Tech. of Light Sources, Kyoto (1995).
13. H. Verweij and C.M.P.M. Saris, J. Am. Cer. Soc., 69, 2, 94-98 (1986).
14. H. Togwa, K. Tsuchida and A. Kato, J. Mat. Sci. Lett. 10, 96 (1990).
15. A. Tan, C. K. Kuo and P. Nicholson, Solid State Ionics, 67, 131-136 (1993).

16a. K. Kitazawa and R. L. Coble, J. Am. Cer. Soc., 57, 6, 250-253 (1974).

16b. G. H. Frischat, paper presented at the Technical Ceramic Congress of the French Society of Ceramics, Paris (1972).

17a. C. I. McVey, U.S. Patent No. 4,567,396.

17b. C. E. Scott and C. I. McVey, European Patent Applicaton No. 93319948.9.

18. J. D. Hodge, in High Temperature Lamp Chemistry, ed. E. G. Zubler, The Electrochem. Soc., 261 (1985).
19. P. Hing, J. Illum. Eng. Soc. 10, 194-204 (1981).
20. G. de With, P. J. Vrugt and A. J. C. Van de Ven, J. Mater. Sci., 20, 1215-1221 (1985).
21. G. C. Wei, S. Jean, C. Sung and W. H. Rhodes, paper presented at Ceramic Microstructure Conferences, Berkeley, CA, June (1996).
22. R. J. Tiernan and D. B. Shinn, in High Temperature Lamp Chemistry, ed. E. G. Zubler, The Electrochem. Soc. (1985).
23. L. H. Schuh, thesis Eindhoven Univ. (1989).
24. S. Lartrigue, C. Carry and L. Priester, Coll. De Phys. C1-985 (1990).
25. H. E. LaBelle, J. Cry. Gr. 50, 9 (1980).
26. C. Greskovich, J. Brewer and C. Scott, U.S. Air Force Contract No. F33615-94-D-5801.
27. KYOCERA Corp., Japanese Patent No. 09045287.

, 15

Lighting and seeing, the difference between night and day

Mark S. Rea, Ph.D, FIES
Director, Lighting Research Center
Rensselaer Polytechnic Institute
Troy, NY, USA 12180-3590
April 27, 1998

STRACT

ninous efficacy and color rendering are two important properties for lamp design. Both properties change with lighting application. As light levels are reduced, the relative luminous efficacy ratings for different lamps can rse and color rendering can disappear. In general, a variety of spectral power distributions with energy ributed throughout the visible spectrum can retain high efficacies and can retain good color rendering in most iting applications. The purpose of this presentation is to show how luminous efficacy and color rendering are cted by light level.

RODUCTION

ht sources can be evaluated by two important criteria, luminous efficacy and color rendering. Luminous cacy is the effectiveness of a light source in converting electrical power into visible radiation; color rendering is effectiveness of a light source in making objects appear like those seen under a reference source, like daylight ure 1 shows a scatter plot of some commercially available lamps with luminous efficacy (photopic lumens per) and color rendering index (CRI) as the two axes. The "perfect" light source, with high efficacy and good r rendering ability, would plot in the upper right corner of this graph, but, generally, lamp designers must npromise between the two criteria. A lamp with higher luminous efficacy usually has a lower color rendering vice versa.

ure 1. CRI plotted as a function of photopic (2°) luminous efficacy for several lamp types.

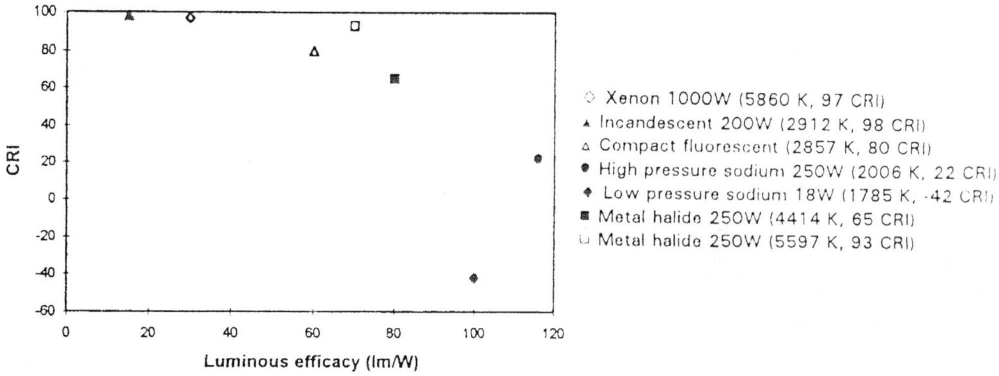

rescent and metal halide (MH) lamps are widely used in commercial and industrial applications because they mize luminous efficacy and color rendering. Xenon lamps have relatively low luminous efficacy, but provide ellent color rendering for specialty applications in the field of photography. Low and high pressure sodium S and HPS) lamps are used in outdoor applications where luminous efficacy is paramount and color rendering nimportant. There is no perfect lamp for every application and lamp manufacturers continue to play the "lamp gn game" offering different lamps for different applications, featuring luminous efficacy, color rendering or ı.

The "lamp design game" has an added dimension which has been ignored until recently. It is still not widely appreciated among lamp manufacturers that both luminous efficacy and color rendering can change with radiance, or light level. The spectral sensitivity of the human visual system shifts to shorter wavelengths and our ability to see colors is reduced as light level is reduced. Depending upon the reflectances of the objects being seen, below about 100 lx (10 fc) the spectral sensitivity of the entire visual system begins to shift from a peak of about 555 nm to shorter wavelengths; under very low light levels peak sensitivity shifts to about 507 nm. The ability of light sources to reveal object colors also begins to change at lower radiances; yellows and blues become less distinct hues and color perception becomes more dominated by reds and greens (Kaiser and Boynton, 1997). At very low radiances, the visual system is totally blind to hue.

For many applications where light levels must remain relatively high these radiance-dependent changes to the operation of the visual system are unimportant. For example, illuminances in most commercial, industrial and retail lighting applications are usually well above 100 lx [Illuminating Engineering Society of North America (IESNA), 1993]. For applications above this level the cone photoreceptors dominate visual response. The lumen used by all lamp manufacturers for luminous efficacy measurements is usually based upon a luminous efficiency function which closely mimics the cone photoreceptor spectral sensitivity. It is referred to as the *photopic* luminous efficiency function, $V(\lambda)$. CRI has been used extensively since the 1960s (Nickerson, 1960) to quantify the ability of a light source to render object colors. CRI is based upon the calculated color shift of several (eight or fourteen) color chips produced by the test light source relative to a blackbody radiator of the same color temperature (Wyszecki and Stiles, 1982). At high light levels CRI orders light sources in terms of their ability to render object colors. Photopic luminous efficacy and CRI have, therefore, been widely used by lamp manufacturers and specifiers to characterize light sources.

For many night-time applications, however, light levels are low enough for the visual system to shift its peak spectral sensitivity to shorter wavelengths and to experience some impairment of color perception. What is largely unappreciated in the lamp industry is the magnitude of these changes as light level is reduced. This paper provides an overview of the changes in spectral sensitivity and color perception as light level is reduced and suggests a new "strategy" in the "game" of lamp design and specification for low light level applications. Specifically, recent measurements of lamp efficacy and color rendering at low radiances suggest that different lamp spectral power distributions should be used at low light levels for maximum efficacy and color rendering.

Luminous efficacy

Photometry is the measurement of light. There are several accepted luminous efficiency functions published by the Commission Internationale de l'Eclairage (CIE) that can be used in photometry. All are based upon heterochromatic brightness matching experiments using human subjects, but each is based upon a different set of experimental conditions. The proper choice of function for photometric measurements depends mainly upon the light level and the part of the retina being stimulated in a particular application. As previously described, the spectral sensitivity of the visual system changes with light level. At very low light levels, only the highly sensitive rods are active. This class of photoreceptors is found throughout the retina except in the fovea, which contains a high density of cone photoreceptors. To measure light at very low light levels, below approximately 0.001 cd/m^2, the fovea does not function and the scotopic luminous efficiency function (CIE, 1951) should be used. Above approximately 3 cd/m^2 (corresponding to an illuminance of 100 lx on a surface of reflectance 10%) one of two luminous efficiency functions may be used, depending upon which part of the retina is being stimulated. For visual targets limited exclusively to the fovea, the 2-degree function, $V(\lambda)$, (CIE, 1926) should be used. [Recent data by He *et al.* (1997) show that the 2-degree luminous efficiency function is appropriate for the fovea at any light level because it contains only cones.] For larger visual targets the 10-degree function $V_{10}(\lambda)$ (CIE, 1964) should be used. Figure 2 shows these three luminous efficiency functions currently sanctioned by the CIE.

Table 1 shows luminous efficacy for different commercially available light sources using these three luminous efficiency functions. It will be noted that there is very little difference between the efficacy rankings for the 2-degree and the 10-degree calculations. Probably for this reason, most photometrists and lamp manufacturers ignore the 10-degree function even though there are many applications where this luminous efficiency function would be more appropriate (e.g., playing basketball requires both foveal and peripheral vision at relatively high light levels.) The lamp efficacy ranking using the scotopic luminous efficiency function is very different than the ranks obtained with the two photopic functions. The scotopic function is never used, however, because few if any applications using electric lighting would produce light levels below 0.001 cd/m^2.

Table 1. *Luminous efficacy for several light sources using different luminous efficiency functions.*

Source	Photopic 2° lm/W	Photopic 10° lm/W	Mesopic			Scotopic lm/W
			0.3 cd/m²	0.1 cd/m²	0.03 cd/m²	
HPS (100 W)	95 (1.00)	98 (1.00)	86 (1.00)	74 (1.00)	66 (1.00)	61 (1.00)
MH (100 W)	78 (0.82)	83 (0.85)	99 (1.15)	115 (1.54)	125 (1.89)	130 (2.12)
2815 K incandescent	15 (0.15)	16 (0.16)	17 (0.20)	19 (0.25)	20 (0.30)	20 (0.33)
Clear mercury (400 W)	52 (0.55)	54 (0.54)	58 (0.67)	62 (0.84)	65 (0.98)	67 (1.09)
Xe (1000 W)	30 (0.32)	32 (0.33)	45 (0.52)	56 (0.75)	64 (0.96)	68 (1.10)
LPS (180 W)	180 (1.89)	180 (1.83)	136 (1.58)	89 (1.20)	58 (0.88)	41 (0.66)
4150 K cool-white fluorescent (F40)	77 (0.81)	82 (0.83)	93 (1.08)	105 (1.42)	113 (1.70)	117 (1.91)
3500 K T8 (F32)	85 (0.89)	89 (0.90)	98 (1.14)	107 (1.44)	113 (1.70)	116 (1.89)
5500 K sulfur (1375 W) Solar 1000™	94 (0.99)	101 (1.03)	140 (1.62)	176 (2.37)	200 (3.02)	213 (3.47)

Figure 2. *Luminous efficiency functions: a. 2-degree photopic function; b. 10-degree photopic function; c: scotopic function.*

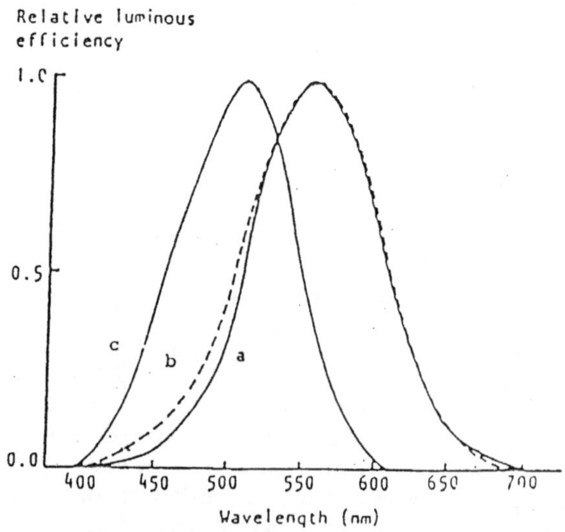

There are many night-time application where the light level is below 3 cd/m^2 (approximately 100 lx). But presently there are no accepted luminous efficiency functions between scotopic and photopic light levels (CIE, 1989). Recently, Rea and his colleagues have presented a series of papers (He *et al.*, 1997; He *et al.*, 1998; Rea *et al.*, 1997) which attempt to develop a foundation for unifying photometric measurements under scotopic, photopic and mesopic light levels. They used unique psychophysical procedures to determine a set of mesopic luminous efficiency functions and have presented a series of equations to calculate the luminous efficacy of different light sources at any light level (He *et al.*, 1998). Table 1 also presents the results of those calculations for several commercially available light sources; Appendix 1 presents their calculation procedure for determining mesopic luminance. Details of our experiments can be obtained from the published papers.

Color rendering

Under scotopic conditions there is no color perception. As light levels increase through the mesopic region color perception begins. Red and green hues dominate under low, mesopic conditions. As light levels reach photopic levels, yellow and blue hues begin to be important. At very high light levels, these yellow and blue hues dominate color perception. Interestingly enough, monochromatic stimuli, which are often believed to have fixed colors, will also change apparent hue as radiance changes. The same phenomenon is observed with object colors illuminated by the same spectral power distribution at different radiances. This shift in apparent hue with radiance, without changing spectral power distribution, is known as the Bezold-Brücke effect (Kaiser and Boynton, 1997).

CRI is commonly used to characterize the color rendering properties of a light source. It is sometimes referred to as a system of measuring "the effects of a light source on color appearance" (IESNA, 1993). Since CRI is constant for a particular light source at any radiance, this cannot be strictly true. However, at photopic light levels commonly experienced in electrically illuminated interior environments (between approximately 200 and 1000 lx), CRI is probably a useful measure of color rendering. Generally, the more continuous the spectral power distribution of the light source, the higher will be the CRI and the better will be color rendering for the most object colors.

Since CRI is not strictly related to color appearance, it is not really appropriate to relate this metric to the accuracy of naming colors under different light sources. Color naming accuracy is important when a person is describing both manufactured and natural colored objects. Further, since color appearance changes with light level, CRI will be unable to reveal any radiance-dependent changes in color appearance. Our laboratory recently undertook a study of color naming accuracy under different light sources and light levels. Subjects were required to describe Munsell (IESNA, 1993) matte, color chips in terms of their primary and secondary hues. For example, the chip designated as 7.5PB 8/4 in the Munsell system appears "lilac" under bright daylight and would be described by most people as having a blue primary hue and a red secondary hue. Two sets of color chips were chosen for study. One set was composed of seven chips with yellow or blue primary hues and the other was composed of seven chips with red or green primary hues. All chips in each set had the same value (reflectance) under daylight. Seven light sources (those listed in Table 2) provided four light levels (0.01, 0.1, 1.0 and 10 cd/m^2). The accuracy of color naming was based upon correctly identifying both the primary and secondary hues. Maximum color naming accuracy is 100%; "blind" chance was associated with a probability of 25%.

Table 2. *Color naming accuracy of several light sources at several luminances.*

Light source	Luminance (cd/m²)			
	0.01	0.1	1.0	10.0
Xenon 1000 W (5860 K, 97 CRI)	54%	82%	97%	100%
Incandescent 200 W (2912 K, 98 CRI)	53%	78%	91%	98%
Compact fluorescent (2857 K, 80 CRI)	61%	81%	93%	97%
High pressure sodium 250 W (2006 K, 22 CRI)	48%	61%	81%	84%
Low pressure sodium 18 W (1785 K, -42 CRI)	29%	33%	38%	40%
Metal halide 250 W (4414 K, 65 CRI)	51%	78%	96%	100%
Metal halide 250 W (5597 K, 93 CRI)	56%	85%	97%	100%

Figure 3 shows the combined results for the yellow-blue and the red-green primary sets. In general, color naming accuracy deteriorates with light level, but is good or excellent for many light sources at photopic (10 cd/m^2) and high mesopic (1 cd/m^2) levels. The HPS lamp was, as expected, poorer in color naming accuracy than most of the other light sources at every light level. Only performance for LPS was worse; indeed color accuracy was near "blind" chance at all light levels. Of some significance, the lamps with the highest CRI were not always associated with the highest color naming accuracy. Although not significantly different, the incandescent lamp did not result in the highest color naming accuracy; several lamps with lower CRI values produced higher color naming accuracy.

These data lead to two interpretations about light sources. First, which was known before the study began, CRI cannot be used to predict the effects of light level on color naming accuracy and another metric is needed to predict radiance-dependent changes in color perception. Color naming accuracy may be a good candidate. Second, CRI may be a hypersensitive or even misleading measure of color appearance. All lamps tested with a CRI greater than 65 performed almost equally well at any light level. While this may mean that color naming accuracy is a *hypo*sensitive measure of color appearance, it does reveal the fact that many commercially available light sources are suitable for accurate color judgments.

Figure 3. *Color naming accuracy with several light sources as a function of luminance.*

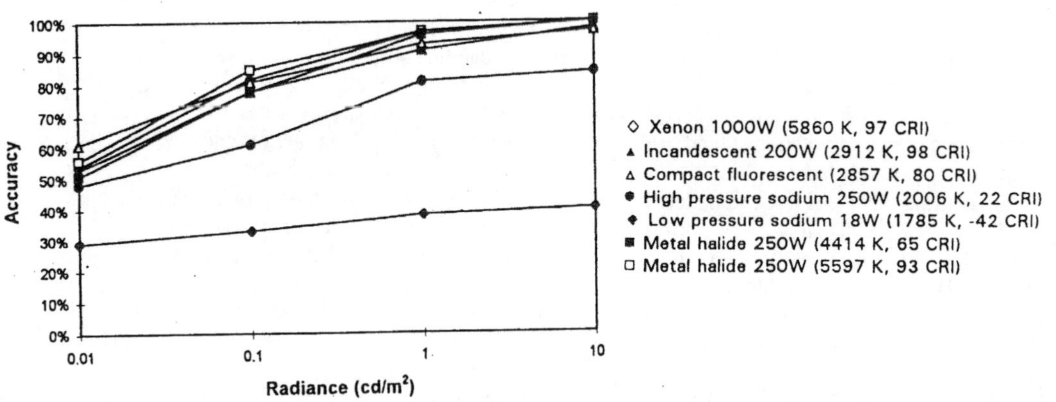

Radiance as an important dimension in lamp design

Two light sources from Table 2, MH and HPS, will be used to illustrate the changes that occur to luminous efficacy and color rendering with radiance. By definition, CRI does not change with radiance. Since color perception *does* change with radiance, color naming accuracy was chosen as a criterion for evaluating the color rendering properties of a light source. Figure 4 shows the luminous efficacies for MH and HPS lamps plotted with color naming accuracy at 10 cd/m^2. Figure 4 is comparable to Figure 1 except color naming accuracy is used as the measure of color rendering rather than CRI. The relative positions of MH and HPS in Figures 1 and 4 are the same and, therefore, using color naming accuracy as a measure of color rendering should not distort the illustration.

Figure 4. *Color naming accuracy plotted as a function of luminous efficacy at a luminance of 10 cd/m².*

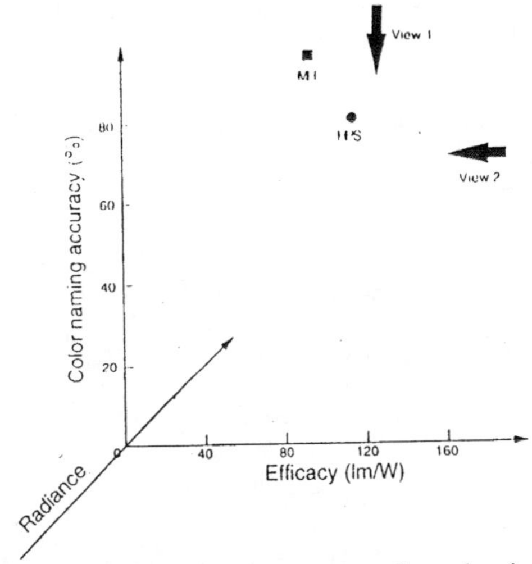

Figure 4 also introduces another dimension, radiance. The three-dimensional space can be viewed from two other perspectives. View 1 can show the change in luminous efficacy with radiance, and View 2 can show the change in color naming accuracy with radiance. These two views are shown in Figures 5 and 6, respectively.

Luminous efficacies of MH and HPS are shown at three radiances, 10, 0.1 and 0.001 cd/m^2 in Figure 5. Luminous efficacies were determined by three different luminous efficiency functions. At 10 cd/m^2 the conventional photopic luminous efficiency function was employed, at 0.001 cd/m^2 the scotopic function was used, and at 0.1 cd/m^2, the corresponding luminous efficiency function based upon the He *et al.* (1998) equations were used. Figure 5 shows that the energy effectiveness of the two lamps changes significantly with radiance. It should be noted again that few if any lighting applications can be found at 0.001 cd/m^2 so the scotopic efficacy value is only of academic interest. However, roadways in North America are commonly at luminances of approximately 0.1 cd/m^2 or lower, implying that HPS lamps typically used on roadways today are not the most energy efficient light source. It should be stressed, however, that the full story is not known. Studies are presently underway in our laboratory to ascertain the effectiveness of different spectral power distributions for driving speed and accuracy; preliminary results confirm the implication, however, that HPS is not as effective as MH for driving automobiles.

Figure 5. *The luminous efficacy of MH and HPS light sources at three luminances: 10 cd/m² (photopic); 0.1 cd/m² (mesopic); and 0.001 cd/m² (scotopic).*

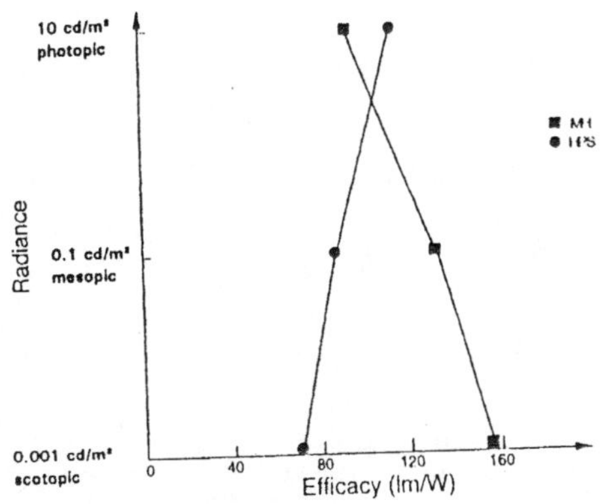

View 1

Figure 6 shows color naming accuracy for the two light sources at 10, 0.1 and 0.001 cd/m^2. Unlike luminous efficacy there is no reversal in the relative positions of MH and HPS as light level is reduced. Rather, the difference between MH and HPS become nonexistent as light level is reduced to scotopic light levels (0.001 cd/m^2) where the visual system becomes completely color blind.

Figure 7 shows luminous efficacy and color naming accuracy for these two light sources at different radiances.

Figure 6. *Color naming accuracy of MH and HPS at three luminances: 0.001, 0.1 and 10 cd/m^2.*

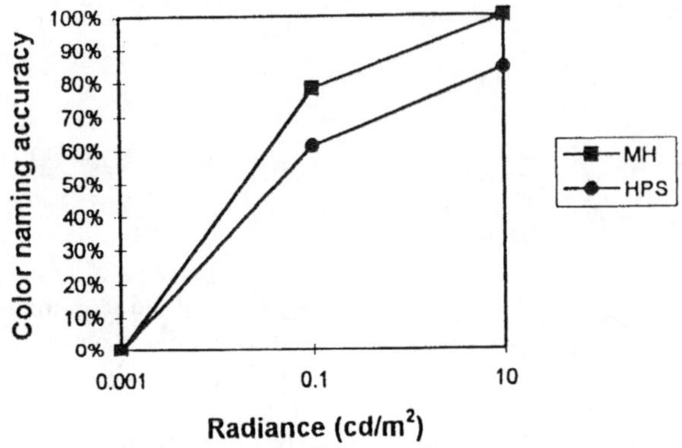

Figure 7. *Luminous efficacy and color naming accuracy for MH and HPS at three radiances: 0.001, 0.1 and 10 cd/m^2.*

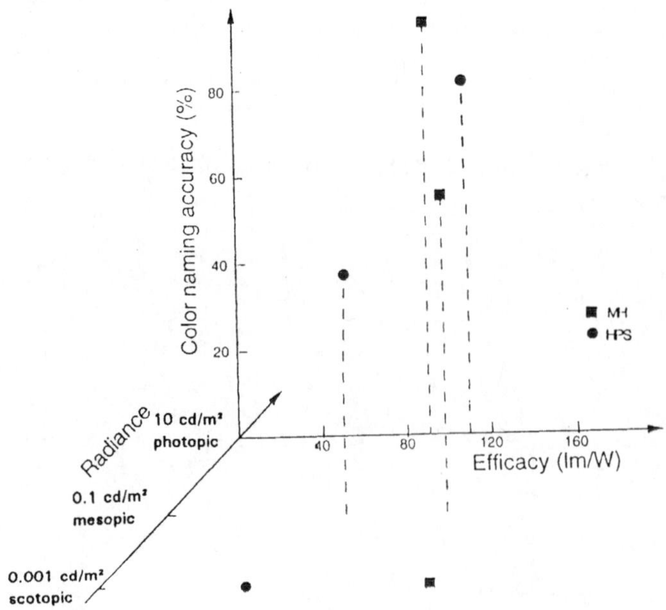

DISCUSSION

Light sources are designed and specified to meet different application requirements. Luminous efficacy and color rendering are very important to lamp design and specification, but these measures of lamp performance are not constants. Both luminous efficacy and color rendering change with radiance.

At photopic levels, above approximately 100 lx, the conventional criterion of luminous efficacy based upon the 2-degree luminous efficiency function is suitable for most applications. The same cannot be said of CRI, however. Considering color naming accuracy, CRI is slightly hypersensitive, and perhaps a misleading measure of color rendering. Many broadband sources provide good color rendering for accurate color naming, and some sources with relatively low CRI may result in more accurate color naming than sources with higher CRI.

Many outdoor night-time activities are performed at mesopic light levels. Light sources for night-time applications are often specified on the basis of their luminous efficacy, ignoring color rendering altogether. These recent data suggest, however, that the conventional photopic luminous efficiency function may not be appropriate for characterizing luminous efficacy for applications such as driving automobiles or flying aircraft at night. These applications require lamps with spectral power distributions tuned to shorter wavelengths, and since color naming accuracy remains high for many light sources at many mesopic levels it may not be necessary to surrender good color naming accuracy in favor of luminous efficacy. In other words, LPS and HPS may not be the best light sources for night-time applications.

Some important caveats should be identified, however, in the strategy of lamp design for night-time applications. First, luminous efficacy and color rendering are not the only criteria important to lamp design. Lamp life, maintenance and cost are also important considerations in lamp specification. A lamp that lasts only 100 hours will not be specified no matter how high its luminous efficacy. Second, the measures proposed here have not reached an orthodox status that can only be provided by sanctioning bodies such as the CIE or the IESNA. These bodies are, quite rightly, very conservative and new ideas require many years of debate before they are "officially" accepted. The new metrics presented here may never be "officially" accepted, but it seems safe to predict that some new measure of luminous efficacy and, perhaps, of color rendering will be accepted by these sanctioning bodies at some future date. Certainly there is a great deal of discussion on the subject and discussion is often the harbinger of orthodoxy. Despite the conservative nature of sanctioning bodies, the marketplace seems to be demanding greater energy efficiency and better color rendering (Rea, 1996). Our laboratory hopes to continue to respond to this market demand by providing more information on the role of light spectrum and light level for visual performance under mesopic levels.

ACKNOWLEDGMENTS

The author gratefully acknowledges the following lamp manufacturers and government agencies who provided funding for the research on which this paper is based: Advanced Lighting Technologies, GE Lighting, the New York State Energy Research and Development Authority, OSRAM SYLVANIA Inc., Philips Lighting and the United States Department of Energy. Also acknowledged are my colleagues and students at the Lighting Research Center who were important to this research: Andrew Bierman, John Bullough, Linsheng Chen and Yunjian He.

REFERENCES

Commission Internationale de l'Eclairage (CIE). *CIE Proceedings 1924*, Cambridge University Press, Cambridge, UK, 1926.

Commission Internationale de l'Eclairage (CIE). *CIE Proceedings 1951*, Bureau Central de la CIE, Paris, 1951.

Commission Internationale de l'Eclairage (CIE). *CIE Proceedings 1963*, Bureau Central de la CIE, Paris, 1964.

IL 15

Commission Internationale de l'Eclairage (CIE). *Mesopic Photometry: History, Special Problems and Practical Solutions*, Publication No. CIE 81, 1989.

He, Y., Bierman, A. and Rea, M.S. A system of mesopic photometry, submitted for publication to *Lighting Research and Technology,* 1998.

He, Y., Rea, M.S., Bierman, A. and Bullough, J. Evaluating light source efficacy under mesopic conditions using reaction times, *Journal of the Illuminating Engineering Society*, **26**(1), 125-138, 1997.

Illuminating Engineering Society of North America (IESNA). *Lighting Handbook*, M.S. Rea (ed.), IESNA, New York, 1993.

Kaiser, P. and Boynton, R.M. *Human Color Vision*, 2nd ed., Optical Society of America, Washington, 1997.

Nickerson, D. Light sources and color rendering, *Journal of the Optical Society of America*, **50**(1), 57-69, 1960.

Rea, M.S. Essay by invitation, *Lighting Design and Application,* (October), 15-16, 1996.

Rea, M.S., He, Y. and Bierman, A. Toward a system of mesopic photometry based upon M-channel response. *Proceedings of the NPL/CIE-UK Conference: Visual Scales: Photometric and Colorimetric Aspects,* 24-26 March, 47-51, 1997.

Rea, M.S. and Ouellette, M.J. Relative visual performance: A basis for application, *Lighting Research and Technology,* **23**(3), 135, 1991.

Trezona, P.W. Luminance level conversions to assist lighting engineers to use fundamental visual data, *Lighting Research and Technology,* **15**(2), 83-88, 1983.

Wyszecki, G., and Stiles, W.S. *Color Science*, John Wiley & Sons, New York, 1982.

APPENDIX 1: CALCULATING MESOPIC LUMINANCE (from He *et al.*, 1998)

The following computational procedure illustrates the approach by which mesopic light level can be evaluated for any light source spectra.

1. Measure the photopic luminance **L** (in cd/m²) for the spectral radiance distribution $L_e(\lambda)$ (in W/m²/sr).

2. Calculate the pupil size **A** (in mm²) of an average observer under **L** using the following equation adapted from Trezona (1983):

   $$A = \{5 - 3\tanh[0.41 \log L]\}^2 / 4\pi$$

   or from Rea and Ouellette (1991):

   $$A = \{4.77 - 2.44\tanh[0.3 \log L]\}^2 / 4\pi$$

3. Calculate the retinal illuminance T_{10} (in photopic Td). Use T_{10} as an initial value for T_{mes}. $V_{10}(\lambda)$ is the 10 degree photopic luminous efficiency function.

   $$T_{10} = 683\, A \int L_e(\lambda) V_{10}(\lambda)\, d\lambda$$

4. Calculate the value of $x(T_{mes})$ using the following equation:

   $$x(T_{mes}) = 0.115\,(T_{mes} + 0.006)^{0.71} \quad \text{for } T_{mes} < 21;$$
   $$x(T_{mes}) = 1 \quad \text{for } T_{mes} \geq 21$$

5. Use the value of $x(T_{mes})$ in the following equation to determine a mesopic luminous efficiency function $V_{mes}(\lambda, T_{mes})$:

   $$V_{mes}(\lambda, T_{mes}) = \{V_{10}(\lambda)x(T_{mes}) + V'(\lambda)[1 - x(T_{mes})]\}$$

6. Calculate the retinal illuminance, T_{mes} (in mesopic Td):

   $$T_{mes} = A\, k_{mes} \int L_e(\lambda) V_{mes}(\lambda, T_{mes})\, d\lambda$$

   where k_{mes} is a scaling factor, equal to $683/V_{mes}(\lambda{=}555\text{ nm})$.

7. Use the resulting value of T_{mes} in step 4, and repeat steps 4 through 7 until changes in T_{mes} are negligible.

8. Calculate the mesopic luminance L_{mes}:

   $$L_{mes} = T_{mes} / A$$

Fluorescent Lamps

A 01

MODELING OF 2-MM ID FLUORESCENT LAMPS

John F. Waymouth
16 Bennett Rd., Marblehead, MA 01945 USA

1 INTRODUCTION

The use of low-power fluorescent lamps for backlighting LCD screens in portable computers has sparked interest in optimization of these lamp types. Patent literature suggests that much higher fill pressures and mercury pressures than optimum for "standard" lamps are appropriate [1]. To explore as rapidly as possible a wide parameter space, I have performed calculations for a 2-mm ID, 40-cm long cold-cathode lamp, using a modified version of the Waymouth-Bitter fluorescent lamp model [2]. The results are reported herein.

2 MODIFICATION OF WAYMOUTH-BITTER MODEL

The Waymouth-Bitter model calculates the net emission rate of 254-nm resonance radiation by assuming that the total net excitation (excitation less quenching) of all the 3P states of mercury eventually escapes as 254-nm radiation. This assumption was first justified by Kenty [3]. Because of the large excitation cross-section for $^3P_2$, and the tripling of the number of excitation channels reaching the $^3P_1$ state, the involvement of the metastables approximately triples the net excitation rate. In small diameter tubes, especially at low fill pressures appropriate to larger-diameter tubes, there is an appreciable probability that metastable atoms in the $^3P_0$ and $^3P_2$ state will diffuse to the wall before a second electron collision further excites them or quenches them to the $^3P_0$ radiating state. The loss of a metastable atom to the wall is damaging on two counts: first, it reduces the net effective excitation rate of the $^3P_1$ state; second, the energy required to create the metastable atom is lost to the plasma. Consequently, the efficiency is doubly degraded.

I have modified the W-B model by introducing a diffusion loss term for metastables, plus an equation of continuity to determine the fraction of metastables that succeed in being transferred by a second electron collision to the $^3P_1$ state to radiate. The diffusion coefficient is unknown; I have adopted a value for argon that produces results consistent with Ref. [1]; for other rare gases, I have estimated values based on the square-root of the mass ratio to argon. No claim is made as to the absolute validity of the results; however, I believe the trends are correctly predicted, and have only attempted to draw conclusions that are not too sensitive to numerical discrepancies.

3 RESULTS

3.1 Optimum mercury vapor pressure.

The effect of imprisonment in the W-B model enters through a parameter $\tau\, n_e$, imprisonment time times electron density. Optimum efficiency is found for a value of this product about 1.2×10^6 per-cm-cubed second. For 2 mm ID, much higher mercury pressures are required for a given imprisonment time; hence higher cold-spot temperatures are optimum. The increase of electron density with increasing current forces optimum mercury pressure to lower levels as current increases.

3.2 Optimum fill pressure

In small diameter tubes, fill pressure must be higher to reduce ambipolar diffusion loss rates, to keep electron temperature low enough to prevent excitation and ionization of argon, relatively useless energy losses. On this ground alone, optimum fill pressure should be higher. In addition, higher-than-normal fill pressures reduce the diffusion loss of metastables, reducing this energy loss. Against this is balanced the increase in elastic collision loss. One of the findings of these calculations is that higher mercury vapor pressures increase energy expended in excitation in proportion; thus higher fill pressures are permitted at higher mercury pressures without increasing the proportion of electron energy dissipation by elastic collision loss. The ratio of fill pressure to mercury pressure is the key parameter.

3.3 Effect of electron density (lamp current)

Higher electron densities force lower mercury pressures for maximum efficiency. Lower mercury pressures dictate lower fill pressures to avoid excessive elastic collision loss. Thus both optimum mercury pressure and optimum fill gas pressure decrease in coupled fashion with increasing lamp current.

3.4 Sample Result

Figure 1 is a 3-D plot of luminous efficacy as a function of argon fill pressure and condensed mercury temperature at 2.0 ma discharge current. Optimum cold-spot temperature is approximately 70C, and optimum argon pressure is approximately 28 torr. An unexpected result is that near-optimum efficacy is obtained between 60C and 100C cold-spot temperature, and between 15 and 50 torr fill pressure. Also surprising is the level of efficacy achieved, nearly 70 lm/watt. Other results will be presented in the poster.

4 ACKNOWLEDGEMENT

It is a pleasure to acknowledge that this work was supported by Matsushita Electric Works R&D Laboratory, Woburn, MA, USA, and to express my thanks to them for permission to present these results.

5 REFERENCES

[1] F. Funada, et al; Eur Pat Appl O 178 646 "Small size Discharge Lamp" April 23 (1986)
[2] J. F. Waymouth & F. Bitter; J Appl Phys **27**, 122 (1956)
[3] C. Kenty; J Appl Phys **21**, 1309 (1950)

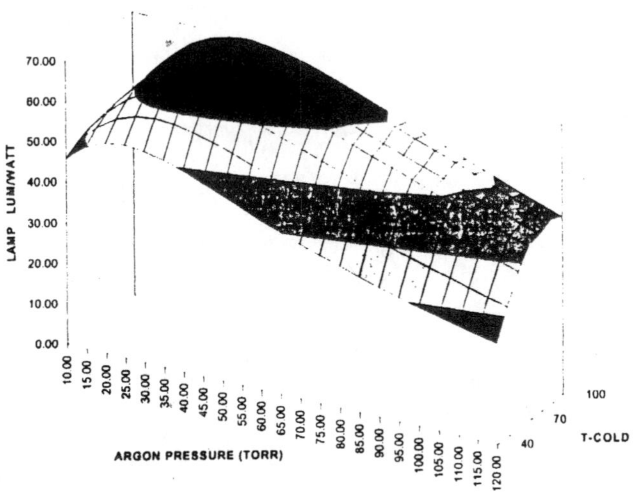

Fig 1; Luminous efficacy of 2-mm ID tube at 2.0 ma current versus argon fill pressure and mercury cold-spot temperature.

A 02

PREHEATING CHARACTERISTICS OF A FLUORESCENT LAMP CATHODE

Minoru Myojo and Atsuo Waki
Technology Center, Lighting Group, Matsushita Electronics Corporation
1-1 Saiwai-cho, Takatsuki, Osaka 569-1143, Japan

1 INTRODUCTION

This study concerns the requirements for proper preheating. A few methods for measuring a thermionic emission time have been presented. In this study the difference in the emission time between the lamp ignition method[1)] and the lamp voltage monitoring method[2,3)] is investigated in the current-controlled preheating mode. A new measuring criterion of thermionic emission times for the lamp voltage monitoring method is examined for the purpose of more practical use.

2 EXPERMENTAL

The lamp voltage monitoring method having a very small lamp current is based on detecting the decrease in lamp voltage, depending on the decrease in cathode fall from cold to hot. The measuring criterion is considered to be that the curvature of the first derivative of lamp voltage after the passage of the negative peak crosses over the defined lines as shown in Figure 1(a). The defined lines and the first derivative of lamp voltage are as follows:

Defined line(1) : $y = 10t - 30$ $(0 < t <= 2.5)$

Defined line(2) : $y = -5$ $(2.5 < t)$

$dv_L/dt = \{v_L(t+\Delta t) - v_L(t)\} / \Delta t$ $(\Delta t = 0.3s)$

where t is the preheating time and v_L is the lamp voltage. These equations have the same y-axis.

Previous criterion[3,4)] using the second derivative of lamp voltage almost corresponds to the defined line(2) combined with the curvature of the first derivative of lamp voltage from the beginning of the preheating time. The defined line(1)[*1] is introduced so that the emission times by both methods fit to each other as closely as possible, referring to the results of examining some fluorescent lamp(FL) cathodes. This means that when the emission time is short, the time by the lamp voltage monitoring method is likely to be longer than that by the lamp ignition method according to the previous criterion. The criterion in this study, however, is under consideration.

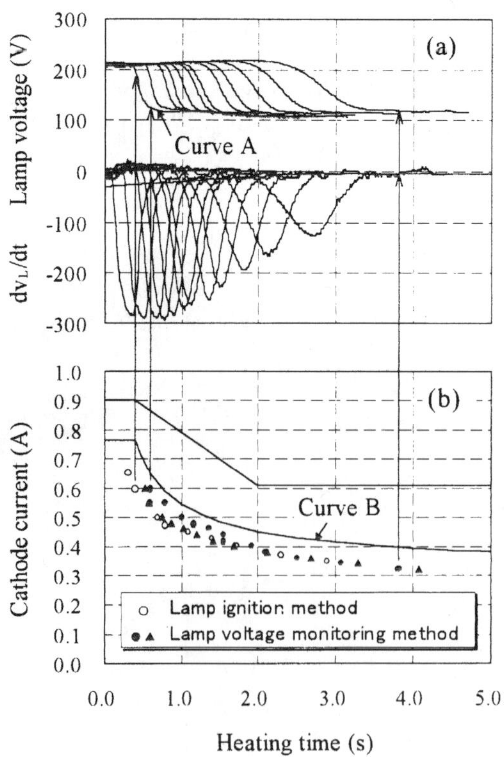

Fig. 1: A measuring criterion for lamp voltage monitoring method (a) and thermionic emission curves by lamp voltage monitoring and lamp ignition methods (b).

---

*1 The defined lines in the voltage-controlled preheating mode may be provided as follows:

Defined line(1) : $y = 25t - 30$ $(0 < t <= 1.0)$

Defined line(2) : $y = -5$ $(1.0 < t)$

3 RESULTS AND DISCUSSION

Figure 1(b) shows the difference between the emission times by the two methods in the case of a T5 diameter, 2 ft length FL (high-output type). When the emission time is short, at least the time in one of the emission curves of both cathodes in the lamp, obtained under the new criterion for the lamp voltage monitoring, is longer than that by the lamp ignition method. Curve A in Figure 1(a) is the curvature of the lamp voltage function according to the lamp voltage monitoring method when the cathode current is 0.6 A. The emission time by the lamp ignition method is obtained before the lamp voltage reaches its minimum value. It follows from this that the shorter emission time by the lamp ignition method may be understood in terms of the relaxation of the limited thermal conductivity of the emissive material from the uncoated part due to a round amount of initial flow of this lamp current through the coated part under the existence of an accelerating field at the cathode surface. Nevertheless, it is not believed that the lamp ignition method provides a premature thermionic emission. From the point of view of the requirements for proper preheating before ignition, however, the lamp voltage monitoring method is more available at short emission times.

On the other hand the emission time by the lamp voltage monitoring method seems to be shorter than that by the lamp ignition method when the emission time is long. That may be why heat losses through radiation and conduction against the smaller energy input make the curvature end of the first derivative function more gentle and will make it difficult to measure long emission times correctly by the criterion of the defined line(2) combined with the fixed interval of $\Delta t = 0.3$s.

Figure 2 shows the curvatures of the average resistance of a cathode together with the thermionic emission points. The average resistance over the period up to the standardized minimum emission time (Curve B in Figure 1(b)) should relate to a substitution resistor in the IEC (International Electrotechnical Commission).

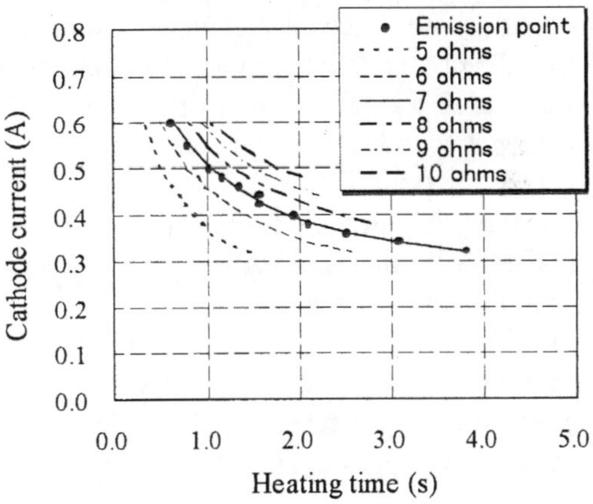

Fig. 2: Curvatures of the average resistance of a cathode having each constant value and its thermionic emission points in the case of a high-output typed T5-2 ft FL.

4 CONCLUSIONS

1) As for the requirements, it is preferable that the preheating period is not exceed the time of around 3.0 s in the current-controlled preheating mode.
2) The reason why the heating energy, until the emission time is reached, by the lamp voltage monitoring method takes a minimum value[2,3] can be understood in terms of the limited thermal conductivity of the emissive material[4].
3) It may be desirable to apply the above-mentioned average resistance of a cathode to a new substitution resistor during the preheating test for electronic drivers.

5 REFERENCES

[1] A. Heidemann, W. Denz and W. Roche, Journal of the IES **23**(1), 115-120 (1994)
[2] M. Myojo and I. Okuno, 7th Int. Symp. Sci. Technol. Light Sources (LS-7), **6:P**, 43-44 (Kyoto,1995)
[3] M. Myojo and I. Okuno, J. Light & Vis. Env. **20**(2), 7-14 (1996)
[4] M. Myojo and A. Waki, Lighting Res. Technol. **30**(1), 12-18 (1998)

A 03

ON THE CHEMICAL PROCESSES IN Zr-CONTAINING FLUORESCENT TUBE EMISSION MIX.

Ferenc Nagel [1,2], Chris Flury [1]
[1] IMATEX S.A. CH-1290 Versoix, Switzerland, P.O. box 316
[2] PASCAL Ltd, H-1025 Budapest, Őzgida u 3. Hungary.

1 INTRODUCTION

The zirconia-doped Ba-Sr-Ca oxide emissive material has been used by the fluorescent lamp industry almost exclusively from the 50-ies. It is still unexplained, in spite of a high number of papers about this subject, why a few Mol % additive (which increases the work function) causes a reduction of the Ba loss by more than an order of magnitude. A significant part of the papers are based on equilibrium thermodynamics; so misleading, since the operating electrode is clearly far from equilibrium. From some of the papers containing reliable data, and direct investigations, the following can be stated (as a continuation of [1]):

a. The direct thermal reduction-rate of the oxide, by the tungsten, is low to explain the loss.
b. Ba (and Sr) loss is current-dependent, and follows the current-time curve at 50-60 Hz, (we have not found reliable data about electronically ballasted lamps). It is near-zero at the current-zero value point.
c. The lamp-independent investigation of Ar-ion (and Ar-Hg ion) bombardment of oxide-surfaces, in the 10eV range, shows, that sputtering (which has been used to explain current-dependence), produces insignificant losses only.
d. In [1] it is stated (proved by the presence of reduction products, i.a. W crystals, which can not be produced by thermal processes), that, parallel to the n type conduction, electrolytic conduction is present in the electrode material. This statement explains the reduction during the cathodic half-period, but (in the original form) can not explain the significant losses during the anodic half-cycle. Based mainly on a theoretical investigation of former experiments, and partly on new experiments, this paper connects the cathode- and anode phenomena producing the decomposition-evaporation of the emissive material.

2 TECHNICAL DETAILS

2.1 Electrolytic decomposition.

To confirm the electrolytic decomposition, and to find a method for optimisation of the emissive material production, the decomposition-activation process (the mix, being a semiconductor, can not be described by the chemical composition only, the way we produce it is decisive). To measure the electrolysis-threshold (critical current density, field strength) at working temperature, we tested emissive material samples between indifferent electrodes (Au, Pt) in an Ar flushed oven. Polarisation, etc. pointed qualitatively to electrolysis, but as an optimisation tool the system could not be used quantitatively. The results depended on the contamination of the Ar also under 1 vpm.. Repeating the experiment in an UHV system we found the same problem, the residual gases influenced the readings both in pumped, and gettered devices. The redox ratios: $CH_4/CO/CO_2$; H_2/H_2O influenced the reading more, than the pressure (in the 10^{-8} - 10^{-11} Pascal range). This result is in agreement with the old electron tube experience: the composition of the residual gases correlate with the cathode emissive properties (work function, conductivity, saturation current). Later we have learnt, that other groups have done similar experiments in an Ar-oven (and had similar troubles) [2], after the New Orleans Meeting, 1993, presentation of [1].

2.2 Information from practical fluorescent lamp results.

The fluorescent lamp industry sometimes uses metallic anode extensions; wires (so called horns) or flattened cylindrical plates around the electrodes. It is well known, that these extensions are the main current carriers during the anodic half period, and it is known, that the extensions do not influence significantly the Ba loss.

2.3 Direct Ba-loss measurements by laser induced fluorescence of Ba atoms and ions.

Ashok K. Bhattacharya [3] developed a high-speed, sensitive method to measure the Ba and Sr loss, so was able to follow them during a 60 Hz period. His results show, that a significant part of the loss (sum of atoms and ions), appears during the anodic half-cycle. We have no doubt of Bhattacharya's method and data (and can not fully agree with his interpretation). His results seem to be in contradiction with 2.2, so something must be wrong or missing for the understanding the chemical process(es) of the emissive material.

2.4 Electron tube experience produced evidence of the migration of the electron-supplying defects in Ba-Sr-Ca oxide cathodes.

There is an interface between the oxide cathode and the metal, usually a doped (Si or Mg) nickel in indirectly heated electron tubes. In tubes being continuously under current, this causes no problem, but in tubes with heated cathodes, for a long time in zero current state (closed grid) the resistance of the interface increases. From this phenomenon the electron tube experts concluded (during the 40-s, before having the mental framework of semiconductor physics) that the unknown donors which are responsible for electron-conduction and emission, move in the oxide towards the interface (activating it) under current. These exactly unknown defects must be positive and must have some mobility (perhaps in a high electrical field only) at working temperature. In a burning fluorescent lamp this effect produces a transport to the Tungsten wire, to the tungstate-zirconate interface, where the decisive reductive reactions happen (and from the surface) during the cathodic half-cycle and the opposite during the anodic half-cycle.

There is no convincing evidence to decide, whether oxygen vacancies are mobile only, or Ba^{++} ions are transported also at the working .temperature (1350 - 1450 K) of the spot. At lower temperatures the O-vacancies are mobile only.

3 CONCLUSIONS

The effect described in 2.4 explains, why the loss is lower than expected during the cathodic half cycle (caused by migration towards the wire), and higher than expected during the anodic half cycle. If the current density is the same (or similar) on the cathode- and the anode-side, electrolytic decomposition is possible on both sides, independently of current direction, above the critical current density. Sr loss goes parallel with Ba loss, from the same reasons, but since Sr is less volatile, the Sr-Ba (loss) ratio is lower than the ratio in the mix.

4 MISCELLENOUS PRACTICAL REMARKS

During our experimental works we found the following (all of them in agreement with old electron tube experience, so they should have been obvious for us):

Cr compounds (usually coming from stainless steel) are strong poisons for the emission mix. Cr-oxide is volatile in the presence of water-vapour traces.

Halogen compounds (cleaning) are also poisons. Stainless steel and Tungsten reacts with CCl_4 and perchloreltylene, it is hard to remove the products from the surfaces and are a source of halogen.

5 REFERENCES

[1] F. Nagel, M. Farkas-Jahnke, J. Mészáros, Proc. of the third Int. Symp. on High Temp. Lamp Chem. 226-234. Ed: J.M. Ranish, C. W. Struck. The Electrochemical Soc. Proc. Vol 93-16.

[2] Imre Cinkota, private communication., Chemistry Dept. Agricult. Univ. Gödöllő, Hungary.

[3] A. K. Bhattacharya, LS-5 Fifth International Symposium on the Science and Technology of Light Sources, York, 1989, + GE report no 92-ATD-2176 + private communication.

See also A.K. Bhattacharya, J. Appl. Phys., 65, 4595 and 4603 (1989).

A 04

NEON (MERCURY FREE) FLUORESCENT LAMP WITH AMBER COLOUR

C. Roozekrans[1], F. Ligthart[1†] and J. Geboers[2]
[1]Philips Lighting B.V., P.O. Box 80020, 5600 JM Eindhoven, The Netherlands
[2]Philips Automotive Lighting, B.P. 319, 28006 Chartres Cedex, France
[†]deceased March 29, 1998

1 INTRODUCTION.

One of the features of red neon lamps is the instant and temperature independent light output. Some years ago a narrow diameter neon lamp for automotive red signalling applications has been introduced [1]. This very robust neon lamp is powered by a direct current electronic circuit. Using direct current offers substantial benefits for fulfilling the EMC requirements. Moreover, a higher lamp luminance is obtained compared to alternating current operation. Recent developments show that also high luminance amber coloured fluorescent lamps can be made in this technology.

2 THE LAMP CONCEPT.

It has been known since long that the visual light output of the neon discharge is by far the most efficient of all noble gases. Besides the visual light, noble gas discharges produce vacuum u.v. radiation too. Beese [1] showed already in 1946 that when the v.u.v. radiation is converted via a zinc silicate phosphor into visible light, neon is also the most efficient noble gas in the 3-30 mbar range. In figure 1 the visible light output of this phosphor for different gases is presented (exclusive the light output of the gas itself).

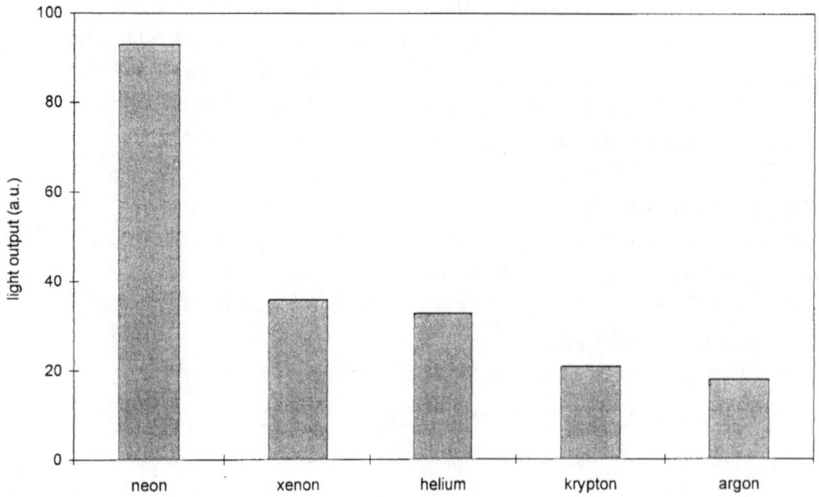

Figure 1: The relative light output of a zinc silicate phosphor for the different noble gases at 5 mbar pressure in T12 lamps [2].

YAG:Ce fluorescent phosphor that converts the 74 nm radiation into green visible light. The green phosphor light plus the red visible neon light makes an amber lamp colour that fulfils the automotive E.C.E. and S.A.E. colour requirements for a direction indicator. In contrary to other phosphors like Willemite, the YAG:Ce phosphor proved to be very stable during lamp operation for many thousands of hours. An example of the visible spectrum of such an amber fluorescent lamp is shown in figure 2.

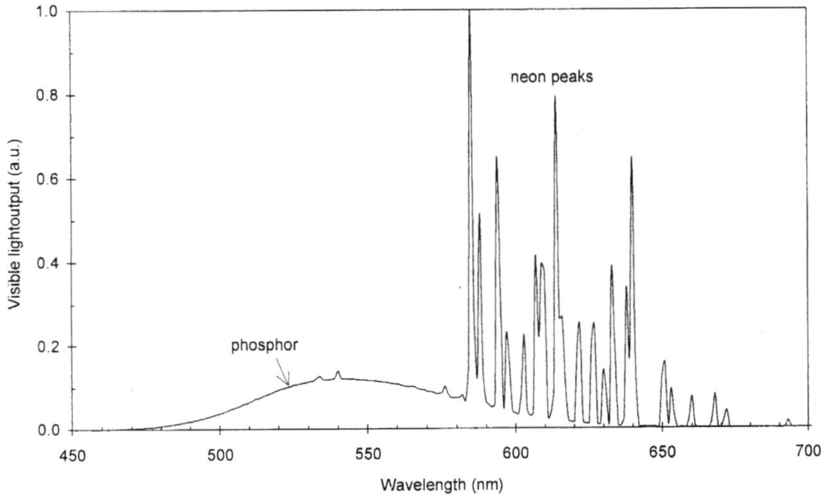

Figure 2: The relative spectrum (corrected for the eye) of a YAG:Ce phosphor coated neon lamp.

With direct current, the vacuum u.v. radiation output and thus the phosphor light output is substantially higher as compared to HF lamp current operation. This makes an amber lamp with a very high luminance (average >20,000 cd/m^2) possible.

3 REFERENCES

[1] The 7th International Symposium on the Science & Technology of Light Sources, Kyoto-Japan, 27th-31 August 1995

[2] N.C. Beese, The fluorescence of phosphors in rare gases, Journal of the Opt. Soc. of America, Sept. 1946

A 05

Technical Issues for Designing the Electronic Compact Fluorescent Lamp with a Diffusive Glass Globe

Takeo Yasuda, Toshiya Tanaka, Kiyoshi Nishio, and Masahiro Izumi
Fluorescent Lamp Engineering Section, Lamp Division, Toshiba Lighting & Technology
1-201-1, Funakoshi, Yokosuka, 237, Japan

1. Introduction

A great number of various electronic compact fluorescent lamps (CFLs) have been developed in the whole world. Although most of them are globeless type, general consumers in Japan have not yet completely got used to the naked light tubes, and they also have a tendency to prefer brighter lamps. Thus, we have developed the 23W electronic CFL with a diffusive glass globe, which generates 1370 lm on 100V main supply and can be substituted for a 100W ball shaped incandescent lamp.

In this paper, some important technical issues for designing the electronic CFL are discussed.

2. The Lamp Structure

Fig.1 schematically shows the structure of the lamp. The lamp has a multi-folded light tube with the outer diameter of 12 mm. The discharge length is about 39 cm, the discharge current and the voltage are 330 mA and 61 V respectively, at the operating frequency of 55 kHz. Therefore, the discharge tube consumes about 20W, which is 87 % of the whole lamp wattage including the loss by the electronic circuit.

The inside of the glass globe is coated with milk-white diffusive layer which consequently gives the soft light, although about 5-7 % of the light is lost for concealing the folded tube.

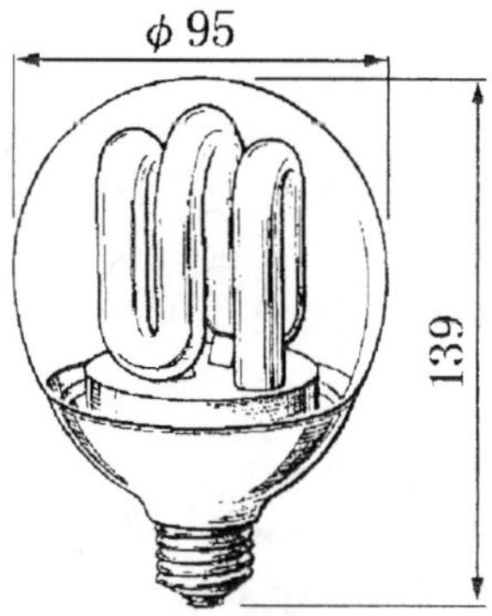

Fig.1 The structure of the lamp. The globe is drawn as if it were transparent, but milk-white in fact.

3. Mercury Vapor Control

The mercury vapor pressure of the light tube is controlled by the amalgam composed of bismuth, indium and mercury, which is located in the exhaust tube at the end of the tube. Its temperature reaches about 120 ℃ when the lamp is operated at an ambient temperature of 25 ℃ in the base-up position, which gives the highest efficiency of luminous output.

For the purpose of improving the run-up characteristic, two auxiliary amalgams are placed near the hot cathode electrodes. The auxiliary amalgam is the indium plated stainless foil, of which the dimension is 7 × 2 mm. The role of the auxiliary amalgam is absorbing mercury while the lamp is turned off, and releasing it immediately after the ignition. The indium plate must be carefully sited as mentioned below to guarantee this function until the end of the life, otherwise the indium on the foil flows to the inner wire of nickel, and makes an alloy that loses the function to absorb mercury[1,2]. In these references, the mesh formed auxiliary amalgam is introduced, but we have confirmed that the partially indium plated stainless plane foil is enough to prevent the indium flowing to the nickel inner wire.

Fig.2 shows the run-up characteristics of the lamps with the partially indium plated auxiliary amalgams and the overall plated ones. The measurements were done after that the both types of the lamps had been aged 3000 hours on normal 3 hours cycle and had been tuned off for 24 hours. The lamps having the partially indium plated auxiliary amalgams show the same run-up behaviour as the initial, on the other hand, it is obvious that overall indium plated auxiliary amalgams lost the function to absorb mercury. This deterioration occurs early within several hundred hours of aging, and 27-indium and 10-nickel alloy is identified by X-ray diffraction analysis.

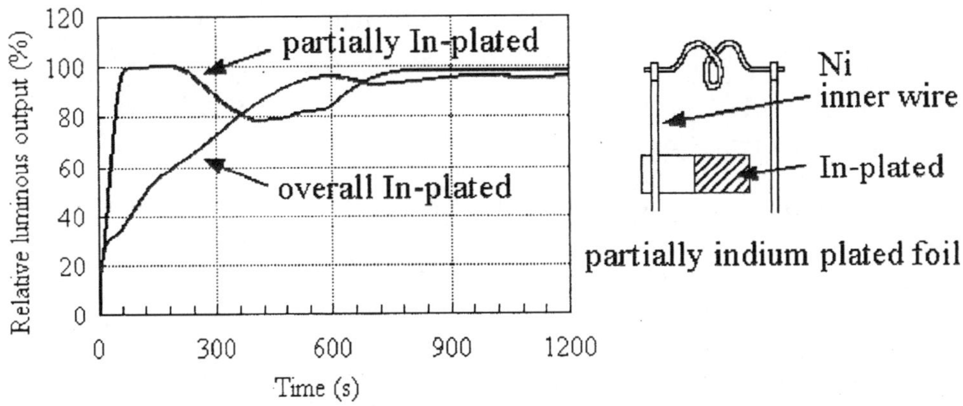

Fig.2 The run-up characteristics of the lamps after 3000 hours burning

4. Life End Phenomena

When the electronic CFL is designed to be put on the market, it must be certified that the lamp comes to the life end safely. Our testing method to investigate the life end phenomena in short term is using the three kinds of the light tubes specially prepared; (1)neither of the cathodes are coated with emitter, (2)only one side is coated with emitter, (3)both of the cathodes are coated with little emitter. Each tube is assembled and operated with 10 % overrated supply voltage, in the tight fixture, at an ambient temperature of 35 ℃. The third lamp which loses emitter within several hundred hours of aging simulates the actual phenomena best. Our lamp adopts the minimum preheat circuit[3] which sustains the glow discharge for a few minutes, even after the emitter is depleted, but since it needs much higher power than normal operation some electronic components get broken and the lamp is finally dead without any apparent damage. More details will be reported in the presentation.

5. References

[1] A.W.Serres and W.Taelman, IESNA Annual Conf. Proc., pp.25-48 (1992)

[2] Japanese Patent Kokai(Laid-open)No.61-121252 (1984)

[3] T.Yasuda and H.Ito, Proc. 7th International Symp. on the Science & Technology of Light Sources, p.41 (1995)

A 06

A LONG LIFE INDUCTION LAMP WITH HIGH LUMEN OUTPUT

John Gielen [1], Piet Antonis [2] and Harry Verhaar [2]
[1] Philips Lighting B.V., P.O. Box 80020, 5600 JM Eindhoven, The Netherlands
[2] Philips Lighting, Turnhout, Steenweg Op Gierle 417, B2300 Turnhout, Belgium

1 THE QL INDUCTION LAMP

In the last three decades the efforts spent in developing electrodeless discharge lamps have increased enormously [1]. As one result the QL lamp was introduced by Philips in 1991. In this lamp the basic principles of induction and gas discharge are combined. In a conventional gas discharge lamp, like TL, energy is coupled into the discharge vessel via electrodes, which are the life limiting components. In the QL lamp energy transfer does not occur via electrodes but via induction; a current through the power coupler coil induces a magnetic field which in turn induces a current in the discharge vessel. This implies that no elements are present in the discharge vessel which reduce the lamp life due to plasma erosion. Other lamp elements, like the power supply, now become the life limiting factor. Currently, the economic life of a QL lamp system is estimated to be over 15 years. In Figure 1 a picture of the QL lamp configuration is given.

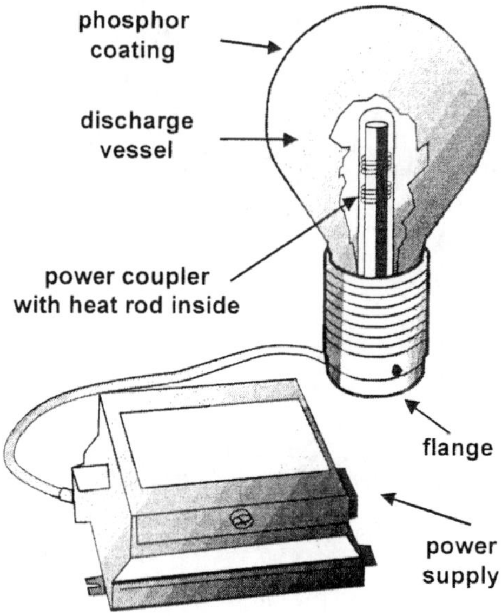

Figure 1: The QL lamp system with its main components.

Presently, the QL lamp is available in two powers: 55 W (nominal luminous flux: 3,5 klm) and 85 W (6 klm) with applications in e.g. public buildings, shops and tunnels. To further increase the field of application of the QL lamp (higher ceilings), a higher luminous flux would be desirable (up to 12 klm). Therefore, currently a QL lamp is being developed operating at 165 W system power, while providing a nominal luminous flux of 12 klm. In this paper several development issues and lamp properties of this lamp will be addressed.

2 DEVELOPMENT ISSUES

In a QL lamp operating at 165W much more heat is generated in a small volume compared to lower power versions. One major problem is the transport of this heat towards its surroundings. One part of the heat is transferred via the outer part of the light bulb towards the surrounding air. The other part is transported via a so-called heat rod. This heat rod is a metallic bar filling up the internal free space inside the power coupler. It allows an effective transport of the heat falling on to the power coupler towards the bottom flange with which the lamp is mounted on the luminaire. The transport capacity of the type of heat rod applied in the 55W and

85W lamp proved not to be sufficient for application in the 165W lamp. Therefore, the development of a new heat rod geometry was started. The main objective was to increase the body mass of the heat rod which results in more heat transport capacity. One way to increase the body of the heat rod is by increasing its diameter. In the lower power QL lamps two exhaust tubes are present in the discharge vessel which limit a substantial increase in diameter of the power coupler, and thus heat rod. In order to allow an increase in diameter without increasing the discharge vessel dimensions dramatically the two exhaust tube concept has been changed into a configuration with one central exhaust tube. With this concept the heat rod is able to transport all necessary heat towards the luminaire [2].

3 LAMP PROPERTIES

In the QL 165W lamp light is generated with ultraviolet (UV) radiation appearing from mercury particles. With this UV radiation a phosphor coating on the inside of the discharge vessel is excited resulting in visible light emission. The mercury pressure depends on the temperature of the mercury amalgam which is controlled via the temperature of the flange (see Figure 1). Proper lamp operation is guaranteed when the flange temperature is between 60 °C and 100 °C. In Figure 2 a schematic representation of the luminous flux of the lamp is given as function of the flange temperature. The luminous flux is given relative to the luminous flux at optimal flange temperature which represents the optimal mercury pressure in the lamp. The level of 100% is between 12 and 13 klm.

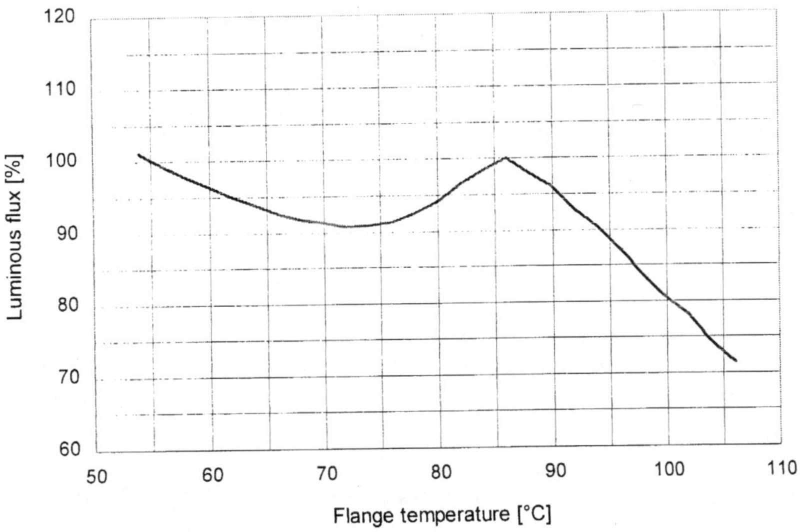

Figure 2: The luminous flux of the QL 165W as a function of the flange temperature.

REFERENCES

[1] D.O. Wharmby, *Proc. of the 5th Int. Symp. on the Science & Technology of Light Sources*, edited by R. Devonshire and J. Meads (University of Sheffield, 1989).

[2] European patent: EP-A 811.240.

A 07

Mercury consumption reduction in fluorescent lamps

Kazuhiro Matsuo, Tomoko Atagi, and Yasuhiro Ikai

Electronics Research Laboratory, Matsushita Electronics Corp.
1-1,Saiwai-cho,Takatsuki,Osaka 569-1193,Japan

1. Introduction

Fluorescent lamps, after a half century of their practical use, are still the major light sources because of their high efficiency and long life. However, they contain a small amount of mercury that adversely affects environment. For environmental protection, the mercury in fluorescent lamps must be reduced. In this paper, we demonstrate a new technology that reduces the mercury consumption by forming a protective film on the phosphor layer.

2. Analysis of Mercury Consumption in Fluorescent Lamps

Mercury consumption was measured by using the device shown in Fig.1. The useful mercury is exhausted by heating the sample lamp at 40℃ (the temperature of the fluorescent lamp in operation) under pressure of 10^{-4}Pa.Then the amount of mercury consumed was measured by chemical analysis. As shown in Fig.2, almost 60% of the consumed mercury was found in the phosphor layer (2ft fluorescent lamps after burning time at 5000h). The consumed mercury compound in the phosphor layer consists of pure mercury of physisorption, Na-amalgam and mercury oxide. This means main cause of mercury consumption can be found in phosphor layer.

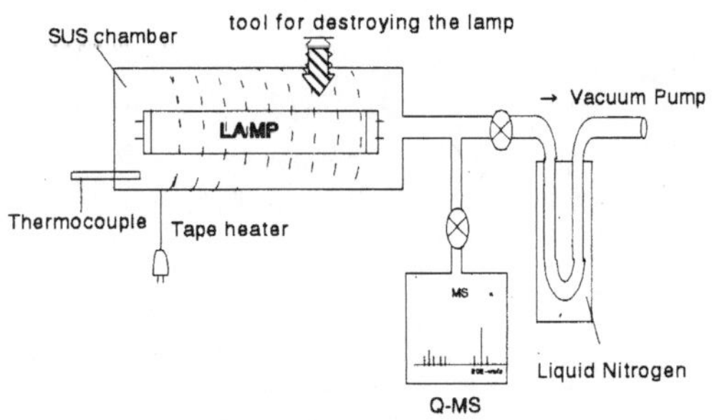

Fig.1 A diagram of useful mercury removal system using vacuum exhaust method in fluorescent lamp

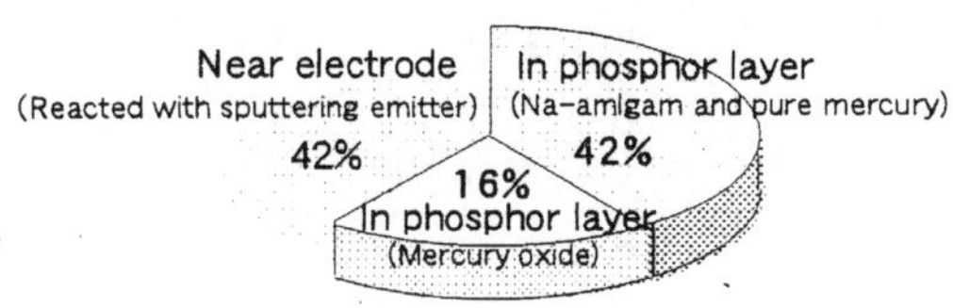

Fig.2 The proportion of the consumed mercury compounds in the 2ft fluorescent lamp after 5000h burning

3. A New Technology for Reducing Mercury Consumption with Y_2O_3 Protective Film

Based on chemical analysis, we found that protective film formed on the phosphor layer could remarkably reduce the consumption of mercury. Therefore, we must find a protective film that has high transmittance at 253.7nm (UV mercury emission) as well as low mercury diffusivity in the low pressure Hg-Ar plasma. Fig.3 shows the measurement results of transmittance of various metal oxide films on fused quartz substrate (film thickness is about 3000Å constant). Y_2O_3 and Al_2O_3 were the materials which showed 80% or more transmittance at 253.7nm. We conducted further experiments to select a steady material not to be diffused by mercury in the low pressure Hg-Ar plasma. We experimentally produced clear lamps, which were formed various kinds of metal oxide thin film inside each glass tube. After 1000h burning, the diffusivity of mercury to each metal oxide film on the glass tube was estimated by mercury depth profiles of SIMS (Secondary ion mass spectrometry). As shown in Fig.4, Y_2O_3 film was diffused by the least amount of mercury. Based on the above results, Y_2O_3 was selected as a material for protective film because of its high transmittance at 253.7nm and low diffusivity of mercury.

Then, we produced the fluorescent lamps having the amorphous continuation film of Y_2O_3 on the phosper layer, and checked the mercury consumption control effect of this film. For this experiment, the 4ft-glass tube with three-band phosphor and 1mg mercury were used. Also, protective layer was formed by the sol-gel dipping method using by the butyl acetate solution of organoyttrium compound.

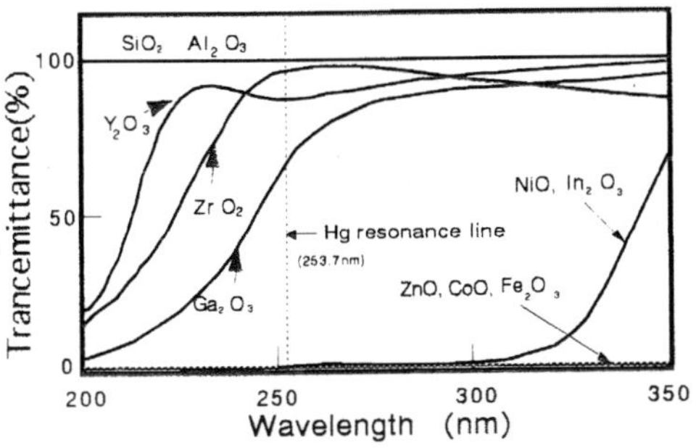

Fig.3 Transmittance of various metal oxide films on fused quartz substrate

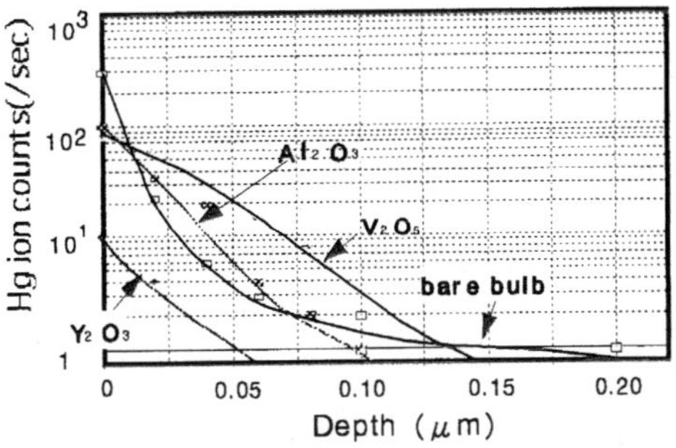

Fig .4 Analysis of amount of diffused mercury into various metal oxide films by SIMS

The relationship between the mercury consumption volume and burning time is shown in Fig. 5. At burning time 12000h, the estimated mercury consumption volume in the lamp with Y_2O_3 protective film was about 1.3mg, while about 2.6mg in the lamp without protective film. It is reasonably estimated that Y_2O_3 protective film reduces the mercury consumption by half. Also, the relationship between lumen maintenance and burning time is shown in Fig. 6. At burning time 12000h, the lamp with protective film achieved 91% estimated lumen maintenance as compared to 85% estimated lumen maintenance without protective film. This resulted in 6 points increase in estimated lumen maintenance. However, the initial total luminous flux value decreased by 4% as the protective film was formed on the phosper layer. After analyzing the structure of the fluorescent lamp with this protective film, we found Y_2O_3 amorphous existed on the phosphor surface and among phosphor particles. The structure helps to reduce the absorption of mercury in phosphor and the diffusion to glass tube. Furthermore, the film prevents the generation of the color center in phosphor, because this protective film hardly transmits at 185nm (VUV mercury emission). Lumen maintenance is expected to be improved as the film controls the ion bombardment to phosphor.

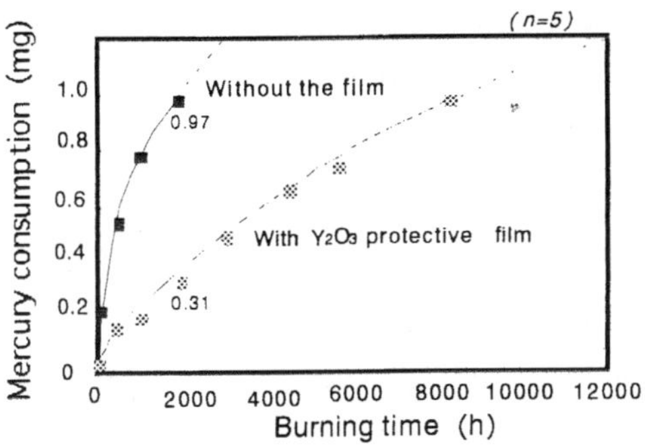

Fig.5 The relationship between the burning time of prototype 4ft fluorescent lamps and the amount of mercury consumption

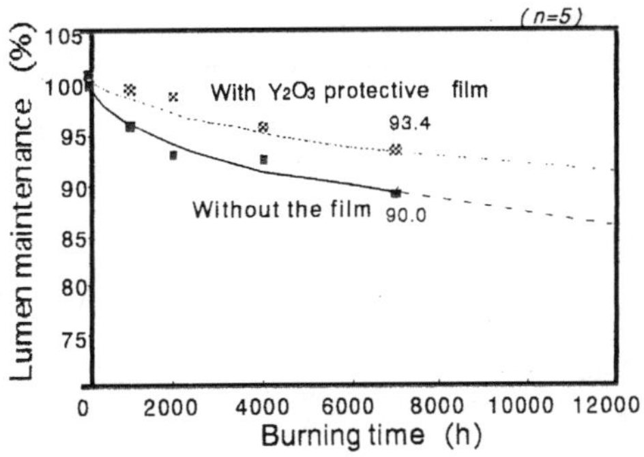

Fig.6 The relationship between the burning time of prototype 4ft fluorescent lamps and the lumen maintenance

4. Conclusion

About 60% of mercury consumption in the fluorescent lamp was made in the phosphor layer. Selection of a material for protective film on the phosphor layer was the key issue to reduced mercury consumption. Our experiment shows that Y_2O_3 is the best material for protective film. Our studies demonstrate that the forming protective film made of amorphous Y_2O_3 among particles and on the surface of phosphor in fluorescent lamps can remarkably reduce the mercury consumption.

5. References

1)P.W.C Verhees.: Mercury consumption in fuluorescent lamps. 5th Int Symp. Light Sources.,1989,p.327-328

2)H.Tomioka,et.al .:A study of mercury consumption in fluorescent lamps.7th Int Symp. Light Sources., 1995,p.323-324

3)S.Van Heusden et.al .: Interaction of a low pressure mercury discharge with discharge with tube wall. Apply phys.23.p355(1980)

A 08

THE INFLUENCE OF THE INERT GAS ON 253.7nm UV EFFICIENCY IN CFL

Shaolong Zhu, Yaoquan Liu, Zipang Sun
Institute for Electric Light Sources, Fudan University, Shanghai 200433, China

1 INTRODUCTION

In recent years various Compact Fluorescent Lamps (CFL) were developed and used widely for interior lighting because of their great energy savings. Although many research works on the CFL have been done[1,2,3], the influence of the various inert gas on the lamp efficiency is still not well investigated. This paper reports the results of the experiment study of the inert gas influence. In order to exclude the uncertainties caused by the factors other than inert gas, we measured the 253.7nm UV radiation efficiency instead of luminous efficacy. The results may be useful for the development and improvement of the compact fluorescent lamps.

2 EXPERIMENT

Experiment lamps with inside diameter of 10mm were a straight tube with length of 200mm and DC operated. The central part of the tube was sealed by a fused silica tube with same diameter by graded glass for UV measurement. The cathode of the lamp was a double coiled tungsten filament coated with electron emitter and the anode was a nickel cylinder. In order to avoid anode oscillation, which might interfere with measurements, auxiliary discharge was applied between anode and auxiliary cathode, which was installed inside anode cylinder. Two water jackets, separately mounted at both ends of the lamp to leave its central part for UV measurement, were used for controlling the mercury pressure. The lamps were evacuated by conventional lamp manufacturing procedures and then filled sufficient amount of mercury and 3 Torr (at 20°C) inert gas mixtures with various percentages.

Optical measurement system consisted of a focus lens, two diaphragms, a monochrometer and a photomultiplier. The measurement system was calibrated by a standard tungsten ribbon lamp (in visible) and a deuterium lamp (in UV). Using this optical measurement system we obtained radiance B of the lamp in normal directions to the lamp surface. The measurement was carried out at wavelength 253.7nm, and visible 404.7nm, 435.8nm and 546.1nm. From measured radiance B, we may obtain radiant emittance

$$R = \beta \pi B \qquad (1)$$

Where β is a correction factor, for Lambert radiator, $\beta = 1$. Since the lamp is a non-Lambert radiator, the β was not available for CFL. Approximately we adopt the values by Koedam et al for T12 lamps[4]. $\beta = 0.615$ for 253.7nm, $\beta = 0.315$ for visible triplet. The radiation power per unit arc length is

$$W = \beta\pi^2 DB / \rho \qquad (2)$$

Where D is the inner diameter of the lamp, D=10mm; ρ is the transmittance of the fused silica wall, ρ =0.80 for 253.7nm, ρ =0.80 for visible.

The axial electric field E was measured by two floating probes inserted in the discharge column. The input power per unit arc length may obtain by multiply E and discharge current I.

3 RESULT AND DISCUSSIONS

The results shown here is an optimum wall temperature, in which the 253.7nm have highest efficiency. The optimum wall temperature was about 47°C despite of the difference of the gas mixtures. We believe it was because the optimum wall temperature was determined mainly by mercury vapor and tube diameter.

Table I

	253.7nm efficiency	Visible efficiency	Axial electric field V/cm
ϕ 10mm 100% Ar	65	6.0	1.87
ϕ 10mm 90% Ar 10%Ne	65	6.5	1.89
ϕ 10mm 80% Ar 20%Ne	66	7.0	1.93
ϕ 10mm70% Ar 30%Ne	65	7.5	1.97
ϕ 10mm 75% Ar 25%Ne	63	6.0	1.75
ϕ 10mm 50% Ar 50%Ne	63	5.5	1.64
ϕ 10mm 25% Ar 75%Ne	63	5.5	1.49
ϕ 36mm 100% Ar	65	4.0	0.81

In table I showed the 253.7nm and visible efficiency and axial electric field in various gas mixtures at optimum wall temperature (47°C) and discharge current 200mA. It is showed the addition of neon caused the rise of the axial electric field and krypton the decrease. This is because the electron mobility is lower in neon and higher in krypton than in argon. Gas mixtures only have a little influence on the 253.7nm efficiency. Therefore the content of the gas mixtures could bd used to modify the lamp electric parameters.

4 REFERENCE

[1]L.E. Vrenken, J.I.E.S. Vol. 7, P.154 (1978).

[2]A.Bouwknget, J.I.E.S. Vol. 11, P.204 (1982).

[3]J.Kreher, D.Foerster, W.Meier, 5-International Symposium on the Science and technology of Light Sources, P.55, York-England.(1989)

[4]M.Koedam, A.A.Kirthof, and J.Rienmens, Physica, 29,P.565, 1961.

A 09

EVAPORATION CHARACTERISTICS OF Zn-Hg FLUORESCENT LAMP DOSES BY THERMOGRAVIMETRIC ANALYSIS

T.R. Brumleve, S.C. Hansen, P.W. Lehigh, D.A. Stafford and K.S. Wilcox
APL Engineered Materials, Inc., 2401 N. Willow Rd., Urbana, IL 61802 USA

1 INTRODUCTION

Zinc-mercury (Zn-Hg) amalgam spheres provide a means of accurately dosing small amounts of mercury in fluorescent lamps. Zn-Hg provides Hg vapor pressure comparable to that of pure Hg [1], and is beginning to see significant commercial application. The available Hg content and the rate of Hg loss are important parameters for lamp and process design. Thermogravimetric analysis (TGA) has been employed in the study of the vaporization behavior of single Zn-Hg amalgam spheres. Results are reported for both atmospheric pressure and vacuum (1.5-1.7 torr) conditions, and parameters for the accurate determination of total Hg content of single Zn-Hg amalgam spheres are determined. In addition, weight gain by re-absorption of Hg vapor by depleted Zn-Hg spheres is also reported under simulated lamp conditions.

2 EXPERIMENTAL

2.1 Continuously recorded weight loss

A Perkin Elmer TGA-7 instrument was operated in two modes: (1) atmospheric pressure under 30 cm^3/min of Ar purge gas through the top balance inlet, and (2) under a vacuum of about 1.5 to 1.7 torr of Ar. A small purge rate of 10-20 cm^3/min was maintained during vacuum operation to protect the sensitive components of the balance chamber from exposure to Hg vapor. This gas flow resulted in a constant and stable offset of about 7% in the positive direction. The accuracy of the initial and final weights was not affected by this offset since both weights were recorded to ±0.001 mg under the standard 1 atm conditions. The TGA-7 was used in both linear temperature programmed and isothermal modes. Balance accuracy was confirmed with an independent Mettler microbalance. Temperature calibration was performed using Curie point standards.

2.2 Multiple sphere weight loss experiments

Rapid, total weight loss of individual Zn-Hg spheres was performed in groups of 100-200 spheres. The spheres were placed in labeled, shallow holes in graphite blocks. The blocks were then placed in a large glass tube and heated under a flow of Ar gas of about 100 cm^3/min at a pressure of 1.5 torr. The optimum temperature program was determined to be a ramp from 30 to 300 C at 10 C/min followed by a hold at 300 C for 90 minutes. Individual sphere weights were recorded to a precision of ±0.001 mg before and after the heat/vacuum treatment.

2.3 Re-absorption of Hg

A tubular balance vessel, 15 mm o.d. by 40 cm long, was constructed of borosilicate glass, loaded with 5 grams of 8 mg 50-50 wt% Zn-Hg spheres on one side, and sealed off under a pressure of 1.5 torr Ar. A large fraction of the Hg (typically 700 mg) was driven to the opposite side by applying a large temperature gradient. The re-absorption of the Hg vapor by the depleted Zn-Hg spheres was then recorded by measuring the balance deflection with the left and right sides at equal temperatures.

3 RESULTS AND CONCLUSIONS

3.1 Continuously recorded weight loss

Single sphere samples of uncoated 8 mg Zn-Hg of 50-50 wt% nominal composition were employed for all runs. Both fresh and aged (6 months) samples of the same lot were studied. At atmospheric pressure with a linear temperature program from 30 to 400 C at 20 C/min no weight loss was observed. Increasing the maximum temperature to 460 C caused the spheres to boil violently and explode. We conclude that the dense Hg vapor does not effectively mix with the Ar purge gas and prevents Hg evaporation. Quantitative measurement of Hg loss and Hg content required vacuum conditions. Under vacuum conditions (1.5-1.7 torr), temperatures were maintained below 400 C to prevent vaporization of Zn metal. The temperature program shown below in Figure 1 (30 C to 350 C at 10 C/min, hold at 350 C for 30 min) was determined to be optimum for the TGA-7 instrument. As illustrated in Figure 1, fresh Zn-Hg samples showed a rapid loss in weight of about 3-4% at about 90-100 C followed by a second larger weight loss, occurring at about 13 to 23 minutes into the run. Less than 0.02 wt% loss was observed over the last 15 minutes of the run. Results of 16 runs of both fresh and aged spheres yielded a mean weight loss of 50.3% with a relative standard deviation of 0.70% (chemical analysis 49.9% Hg). We conclude that the vacuum TGA method above is suitable for accurate determination of Hg content. Isothermal runs from 60 C to 140 C were also performed.

3.2 Multiple sphere weight loss

Weight loss results under the experimental conditions noted in Sec. 2.2 above for 500 spheres are shown in the run chart in Figure 2. The mean weight loss was 50.2% with a relative standard deviation of 1.1%. The few results outside the $\pm 3\sigma$ range are attributed to measurement error, rather than to variation in the Hg content of the spheres. This method provides rapid, accurate determination of Hg content, and indicates a consistent Hg content from sphere to sphere.

3.3 Re-absorption of Hg

Initial results indicate that re-absorption of Hg vapor by depleted Zn-Hg spheres is possible when free Hg and spent spheres are held at identical temperatures of 70 C. Of the initial 700 mg of Hg driven off of a 5 g sample, 147 ± 7 mg was re-absorbed in a 216 hour time period. This is consistent with the slightly lower thermodynamic Hg activity predicted for the Zn-Hg liquid phase [1]. Re-absorption of Hg is not expected to affect the Hg vapor pressure or Hg-releasing capacity of the Zn-Hg material.

4 REFERENCES

[1] T.R. Brumleve, S.C. Hansen and L.G. Kaczorowski, *Proc. Electrochem. Soc.*, High Temp. Lamp Chemistry III, ed. by J.M. Ranish and C.W. Struck, **93(16)** 235-256 (1993).

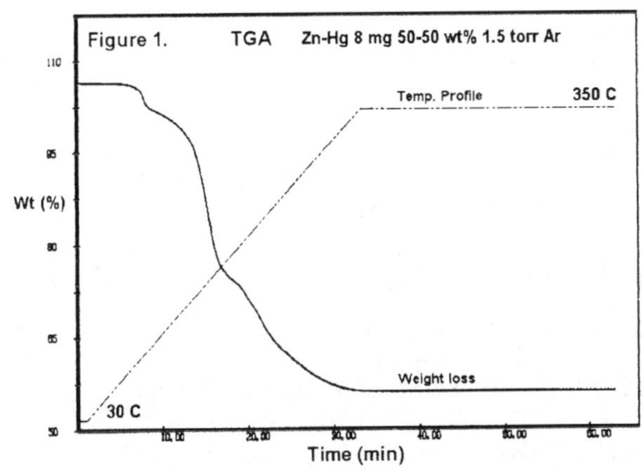

Figure 1. Typical vacuum TGA response and temperature profile for a single Zn-Hg sphere.

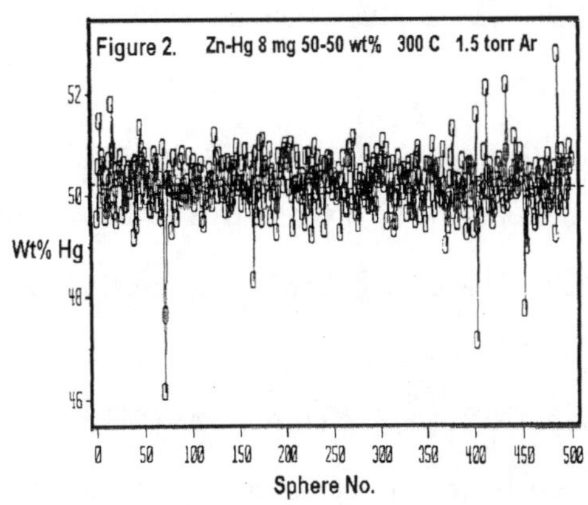

Figure 2. Total Hg weight loss under vacuum for 500 individual Zn-Hg spheres.

A 10

THE INFLUENCE OF CURRENT WAVE CREST RATIO ON THE FLUORESCENT LIGHT LIFE

Yu Xiaoping

Physics Department of Jinan Educational College

37 Gannan East Street Jinan P.R. CHINA Post Code : 250001

1 INTRODUCTION

The paper analysed the reasons of the influence of the current wave crest ratio on the fluorescent light life and verified the results quantitatively from the experimental angle.

2 TECHNICAL DETAILS

The current wave crest ratio refers to the crest value divides by virtual .In regard to the requirement for this target, every country have set up relevant standards.The requirement for the ratio in Chinese standard GB/T15144 is the same as in the standard IEC929,which is not more than 1.7 The influence of the current wave crest ratio on the light life forms major basis for drawing up the standard.The experts of the same occupation have been paying close attention to the standard and putting forward relevant theory basis.

According to the heat electron emission theory, the relation between the evaporation speed of the electron emission material in the light negative pole and the heating point temperature in the same place meets Arrhenius equation:

$$V = C \cdot \exp(-Q/RT)$$

C-constant Q-activation energy T-negative heating point temperature R-gas universal constant.

The equation illustrates that the evaporation speed of the electron emission material in the negative pole is highly sensitive to the heating point temperature of the negative pole.The higher the T, the quicker the V. However, the heating point temperature of the negative pole is determined by the filament current of the negative pole.The bigger the light current wave crest ratio, the bigger the current crest value throw the filament and also the higher the heating point temperature in the negative pole the too higher temperature of the negative pole necessarily leads to the much more evaporation of the electron emission material, the reduction of the emission capability in the negative pole and the shorter light life finality.

Knowing from the above- mentioned analyses, the current wave crest ratio is very important to the fluorescent light life. The exist of the high frequency pulse current makes the heating point temperature reaching very high instantly. It's very unfavorable to the light life.

In order to verified the influence of the current wave crest ratio on the light life quantitatively, we use the method of light accelerated-ageing in the negative pole, which is to engage in the contrast experiment of the switch life-span to the different kinds of wave crest ratio fluorescent light. The experimental way is: we use 30 energy-saving lights and every10 is about the same ratio.The light power is 9w. In order to ensure the identity of the verified condition the lights we used are in the same factory and producted in the same time. In order to ensure the light input-voltage is in the range of 220V, we use 220V alternating voltage stabilizer.furthermore,we control the switch on or off by using the cycle timer and set up the switch on for one minute, the switch off for tow minutes,till the light can't switch on. The verified results presented in the attached list.

Tube No	M=1.93-1.96	Tube No	M=2.2-2.4	Tube No	M=1.55-1.6
1	3360	11	2680	21	7200
2	3120	12	3030	22	7680
3	2640	13	2760	23	6240
4	2880	14	2640	24	8160
5	3210	15	2970	25	5760
6	2070	16	2490	26	7440
7	3600	17	1980	27	7920
8	3060	18	2910	28	5520
9	2970	19	2550	29	8250
10	2010	20	2420	30	8010
Average	2892	2643		7218	

* M-light current wave ratio.

3 CONCLUSIONS

From the above-mentioned analyses and experimental data, it indicated that the bigger the light current wave crest ratio, the shorter the light life. So, from the angle of protecting the light negative pole and prolonging the light life, we should decrease the light current wave crest ratio as far as possible.

A 11

VARIABLE COLOR TEMPERATURE FLUORESCENT LAMP

J. Ravi and J. Maya
Matsushita Electric Works R&D Laboratory
216 W. Cummings Park, Woburn, MA 01801, USA

1 INTRODUCTION

The variable color temperature fluorescent lamp is an attractive light source prospect for ambiance control, mood setting and task lighting applications. Recent studies have reported on the research efforts to realize a single lamp construction with a variable color temperature feature [1, 2]. This paper presents a new scheme for color temperature change. Like the previous references, it also consists of a pulse excited mercury-rare gas discharge utilizing phosphor combinations. The key difference is that, instead of also generating UV radiation from the rare gas, only the mercury radiation is altered to selectively excite particular phosphors and thereby effect color change.

2 PRINCIPLE OF OPERATION

The predominant radiation in a low pressure mercury discharge is from the 254 nm resonance line which is the mainstay of all fluorescent lamps. In addition, the mercury discharge produces the 185 nm and 365 nm radiation in the UV band and also visible lines, the most prominent of which are the 407, 435 and 536 nm spectral lines. Lamp phosphors are optimized for maximum excitation at 254 nm and the excitation curves also encompass the 185 nm wavelength although at a much lower level.

Higher energy states of mercury can be excited when the discharge is operated on a pulse drive and this will alter the intensity ratios of the other lines (185, 365, 407 nm, etc.) with respect to the 254 nm line. Thus, a method of color control becomes possible with a combination of a suitable pulse drive and phosphors that are selectively excited by the different UV lines.

There are major difficulties that need to be overcome in order to make this approach practical. These are

i. availability (or synthesis) of phosphors that will have narrow excitation spectrum for the different mercury UV lines.
ii. the pulse drive scheme required to alter the mercury line intensity ratios to a sufficient degree.

In addition, there are also issues such as the best way to coat the phosphor (blend or layers), lamp life, luminous efficacy, etc.

3 LAMP CONSTRUCTION AND RESULTS

After a search of available phosphors for general lighting and specialty applications, one phosphor was found ($(Y, Ba)_2SiO_5$: Ce) which appeared suitable. Its excitation spectrum peaks near 365 nm and is negligible at 254 nm. The emission peak is centered at 420 nm (blue). This phosphor was blended with a standard green/red lamp phosphor combination ($LaPO_4$: Ce, Tb + Y_2O_3: Eu) in several ratios (20, 50, 80 percent, etc.). Many lamps were made using these phosphor blends.

The lamps were driven with various waveforms created using a function generator and an amplifier. Driving with a sinusoidal waveform of 50 kHz is equivalent to operating the lamps on a commercial high frequency ballast. Pulse burst waveforms of various shapes (sinusoidal, triangular and square) and duty cycles (40 to 100 percent) were also applied. The objective here was to alter the intensity ratio of 365 nm to 254 nm radiation to change the amount of light emission form the blue phosphor. Greater color temperature change was obtained with triangular pulse bursts having fast rise times. During pulse burst operation, the lamp power was kept the same or slightly higher than during CW operation. A color temperature change of nearly 1700 K (4000 K to 5680 K) was obtained with a phosphor blend of 80 percent $(Y, Ba)_2SiO_5$: Ce.

4 IMPROVED SCHEME FOR COLOR TEMPERATURE CHANGE

The color temperature change described above is a demonstration of the principle of using selective phosphors in conjunction with altering the intensity ratios of the mercury lines. Better results can be expected if the ratio of 185 nm to 254 nm line intensities is utilized as a means of control.

This scheme will require a new phosphor to be synthesized that has a narrow excitation spectrum centered at 185 nm. Further, the application of the phosphors as layers would be preferable to mixing them in a blend. It can be expected that the custom phosphor that absorbs 185 nm radiation will transmit the 254 nm radiation. Hence, the new phosphor should be coated on top of the standard lamp phosphor that absorbs both 254 nm and 185 nm radiation. As an example, the custom phosphor would emit blue or blue/green light while the standard phosphor will emit red/green light. The color temperature may be raised continuously from a warm temperature by suitably pulsing the mercury discharge. We await development of such a custom phosphor in order to realize the variable color temperature fluorescent lamp.

5 CONCLUSION

A new method of changing the color temperature of a fluorescent lamp has been demonstrated. A better method of color control is proposed but this requires the development of a new phosphor.

6 ACKNOWLEDGMENT

The authors acknowledge the significant contribution of Dr. M. Shea to this study and the many discussions with Dr. John Waymouth.

7 REFERENCES

[1] M. Aono, M. Jinno, M. Kubo and R. Itatani, *Proc. Seventh Int. Symp. on the Sci. & Tech. of Light Sources.,* edited by R. Itatani and S. Kamiya (Illum. Engg. Inst. of Japan, 1995).

[2] S. Tanimizu and A. Kougami, *Proc. Seventh Int. Symp. on the Sci. & Tech. of Light Sources.,* edited by R. Itatani and S. Kamiya (Illum. Engg. Inst. of Japan, 1995).

A 12

PLASMA CONDUCTIVITY ESTIMATION IN INDUCTIVELY COUPLED ELECTRODELESS DISCHARGE BY EQUIVALENT CIRCUIT

Yoshio Watanabe
Kanagawa University, Rokkakubashi, Kanagawa-ku, Yokohama, 221, Japan

1. INTRODUCTION

One of the method for investigating the inductively coupled electrodeless discharge system is to represent the electrodeless discharge characteristics by an equivalent circuit. The equivalent circuit using a distributed constant circuit has been presented in the previous paper[1], where the subjected discharge tube has a column structure with an excitation coil wounded around the tube column. In the present paper, an another type of the inductively coupled electrodeless discharge tube is subjected.

2. EQUIVALENT CIRCUIT

The system diagram for the inductively coupled electrodeless discharge is shown in Fig.1. The employed discharge tube consists of an inner and an outer tube in coaxial configuration. Ar of 67Pa and Hg are filled. The excitation coil with 19 turns is inserted inside the inner tube, and is connected to a high frequency (3-28MHz) power source through a capacitor, C_s, and a matching circuit. The circumferential discharge is formed in the space between the inner and the outer tube.

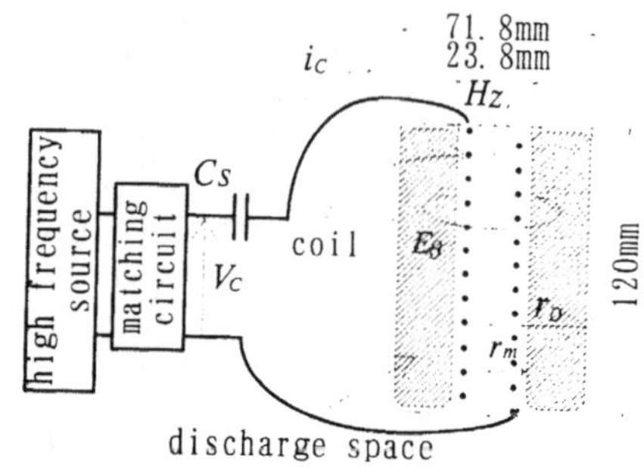

Fig.1 System diagram for the electrodeless discharge

The equivalent circuit for the region of the discharge tube is derived using Maxwell equations. The tube is assumed to be infinite and cylindrical coordinates (r, θ, z) are employed. The following equations are derived for the axial component of magnetic field, H_z, and the azimuthal component of voltage, $V_\theta(=2\pi rE_\theta)$, under the assumption of $\partial/\partial z=\partial/\partial\theta=0$ and $\partial/\partial t=j\omega$.

$$-\frac{\partial}{\partial r}H_z = \frac{\sigma}{2\pi r}V_\theta + j\omega\frac{\varepsilon}{2\pi r}V_\theta \equiv g_\theta V_\theta + j\omega C_\theta V_\theta \quad (1)$$

$$-\frac{\partial}{\partial r}V_\theta = j\omega\mu 2\pi r H_z \equiv j\omega L_\theta H_z$$

where ω: angular frequency of the power source σ: conductivity, ε: permittivity in vacuum and μ: permeability in vacuum. From eq.(1), the equivalent distributed constant circuit diagram for H_z- V_θ is obtained with the boundary condition, $V_\theta=0$ on the axis and $H_z=0$ at the outside of the outer tube, and shown in Fig.2.

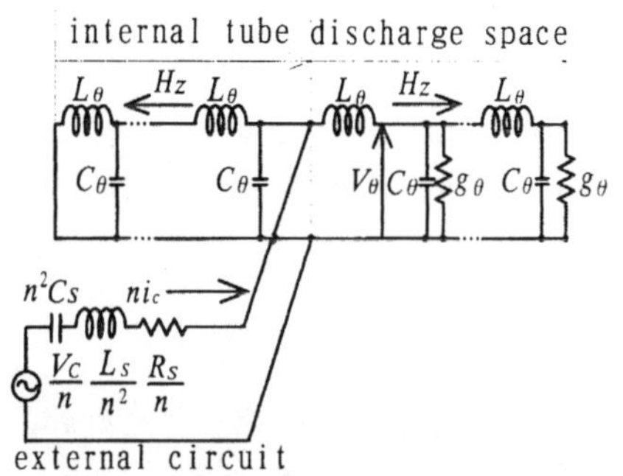

Fig.2 Equivalent distributed constant circuit diagram

The equivalent circuit for the excitation coil is derived and also shown in Fig.2, where L_s: the inductance of the lead wire, R_s: the resistance of the lead wire plus the excitation coil per unit turn including the skin effect, i_c: input current, V_C: applied voltage per unit meter of the tube length and n: the number density of windings.

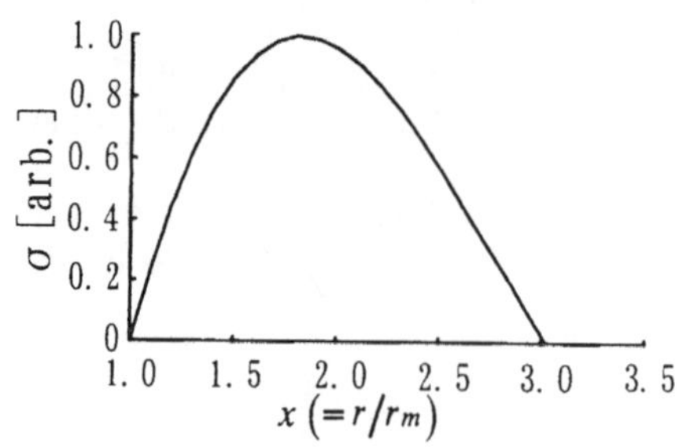

Fig.3 Calculated plasma conductivity profile

Plasma conductivity distribution profile in the discharge tube is determined by the following equation.

$$\frac{\partial^2}{\partial x^2}\sigma + \frac{1}{x}\frac{\partial}{\partial x}\sigma + \frac{B}{x}\sigma = 0, \quad (2)$$

where $x=r/r_m$ (r_m: inner tube radius) and B is the constant. The calculated profile is shown in Fig.3. The validity of this profile is examined by comparing with the profile of excited mercury atom density [2].

3. MEASUREMENT

V_C was measured as a function of i_c at each operating frequency, as shown in Fig.4. No discharge takes place in the small i_c region where V_C increases proportionally with i_c. When i_c is increased, the circumferential discharge is produced and V_C jumps. Thereafter that, V_C increases with i_c. V_C at i_c =0.5A when no discharge is formed are replotted as a function of operating frequency, f, as shown in Fig.5. The V_C-f curve shows LC resonance characteristics. The equivalent circuit shown in Fig.2 is simplified to (L_s+L_0)-C_s-R_s series circuit in the case of σ =0 where L_0 is self inductance of the excitation coil. L_s and R_s are determined to fit the calculated curve with the measured data. The calculated curve is also shown in Fig.5.

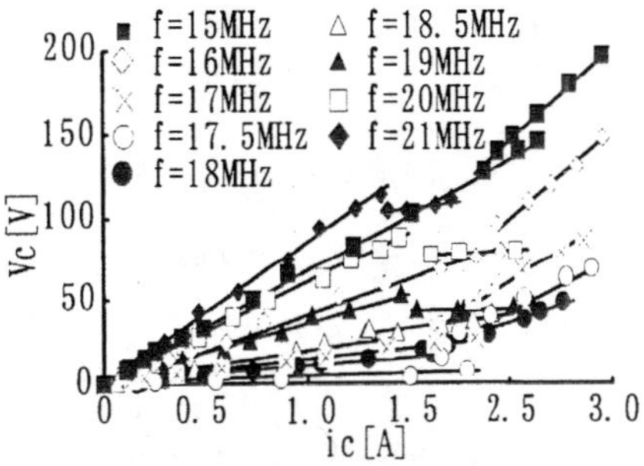

Fig.4 Measured V_C as a function of i_c at each operating frequency

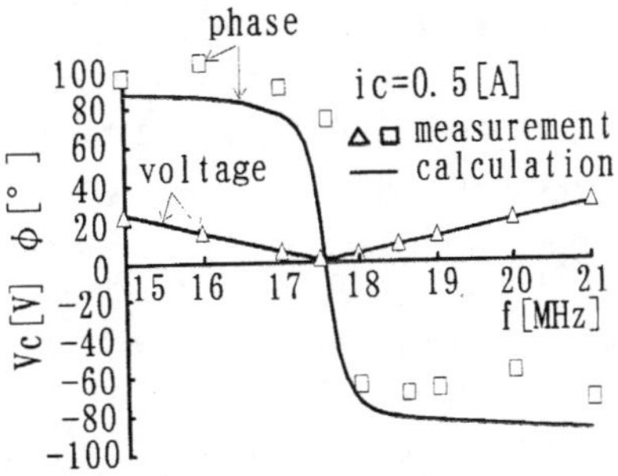

Fig.5 V_C at i_c=0.5A as a function of operating frequency, f, without discharge

4. ESTIMATION OF PLASMA CONDUCTIVITY

Plasma conductivity is estimated using the model with the measured data. The maximum plasma conductivity, σ_{max}, by which the calculated V_C by the model is equal to the measured one existed in the discharge region is determined.

Typical distribution profile of V_θ, σ and current density, $i_\theta(=V_\theta/g_\theta)$ with obtained σ_{max} are shown in Fig.6. The induced current in the plasma, i_D, per unit meter of tube length is also calculated with obtained σ_{max}. The σ_{max} as a function of i_D is shown in Fig.7. The obtained σ_{max} lie on the curve expressed by

$$\sigma_{max}=0.007814 i_D^2+0.4836 i_D \quad (3)$$

The σ_{max} is independent of the operating frequency.

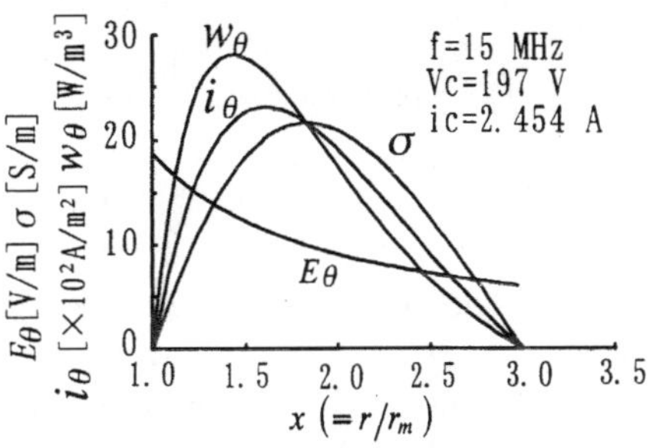

Fig.6 Distribution profile of each plasma quantity

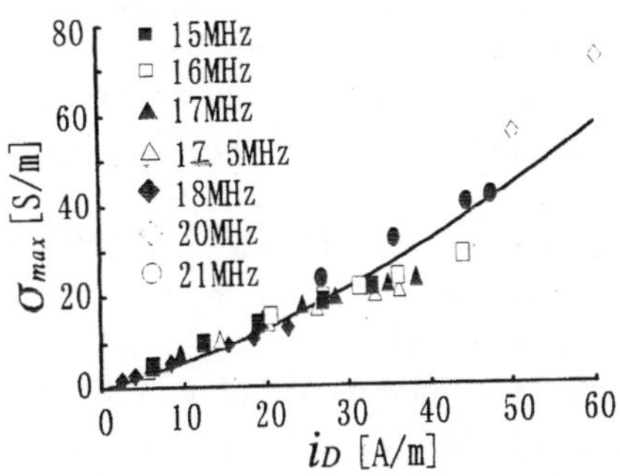

Fig.7 Estimated σ_{max} as a function of induced discharge current, i_D

5. CONCLUSION

An equivalent circuit model for inductively coupled electrodeless discharge with double tubes is derived. Using the model, the dependence of plasma conductivity on the induced discharge current is estimated. The estimated plasma conductivity is independent of the operating frequency.

6. REFERENCE

[1] Y.Watanabe: Jpn.J.Appl.Phys. Part 1, 35 (1996) 4498

[2] Y.Watanabe and M.Yasuda: Proc. of 23rd ICPIG, Vol.1, I-170 (1997)

A 13

THE PREDICTION OF MERCURY PARTIAL PRESSURES ABOVE AMALGAMS USING MTDATA

G.M.Forsdyke[1], S.A.Mucklejohn[1] and A.T.Dinsdale[2]

[1]GE Lighting Europe, Melton Road, Leicester LE4 7PD, UK

[2]National Physical Laboratory, Teddington, Middlesex TW11 0LW, UK

1 INTRODUCTION

The mercury vapour pressure is one of the key parameters which determine the luminous flux of fluorescent lamps. To maintain optimum output from compact fluorescent lamps (CFL) it is often necessary to incorporate an amalgam into the design to control the mercury partial pressure. The temperatures in these lamps are too high to allow the use of pure mercury which needs to be at 313 K (40°C) to give the required vapour pressure, p(Hg) = 0.8 Pa. At higher temperatures, and hence higher mercury vapour pressures, the luminous output falls due to self absorption by mercury atoms. The amalgams are designed to lower the mercury partial pressure and to keep it relatively stable over a broad temperature range. Selection of an inappropriate amalgam, however, can give the lamp a prolonged warm-up time which, from the customer's perspective, is undesirable. The early work on selecting amalgams for high temperature environments is reported in elegant studies by Bloem *et al.* [1,2] and Franck [3]. To predict the variation of mercury partial pressures above amalgams the phase equilibria must be taken into account because control of the mercury pressure can be exercised only where 3 phases (solid, liquid and vapour) co-exist.

2 MTDATA

MTDATA (4) has been used to provide a more detailed understanding of the factors governing the temperature dependence of the mercury partial pressure above amalgams. MTDATA, the NPL databank for metallurgical thermochemistry, has extensive databases for a variety of alloy systems, chemical substances, oxide systems and aqueous solutions. Data for multicomponent amalgam systems were predicted from data for binary subsystems using well tested and reliable models and the phase equilibria calculated for a range of temperatures and compositions. The quantity of data stored for a given system is often very small - usually data for each relevant phase for each element as a function of temperature and a small number of coefficients representing the change in thermodynamic properties in binary systems on mixing. Established models are then used to extrapolate the data from the component binary systems into ternary and higher order systems incorporating any ternary interaction parameters where required. The way in which the binary thermodynamic data are represented and extrapolated depends on the nature of the phase. For amalgam systems it is sufficient to use the Redlich-Kister formalism [5] for the binary systems with the Muggianu method [6] for extrapolation into the multicomponent systems. MTDATA has implemented software for handling a wide range of models such as the ideal gas phase model, a multiple sublattice model for compound phases with ranges of homogeneity and the quasichemical model for complex oxide liquids. It is important to note that combination of data for different binary systems requires use of compatible data for the elements. Data for the elements recommended by SGTE, Scientific Group Thermodata Europe [7,8] were used in this study.

3 CRITICAL ASSESSMENT OF DATA

Key components of a thermodynamic database are the binary mixing data representing the changes in thermodynamic properties as a function of composition between two elements. These data are derived by a process of critical assessment in which all the relevant experimental data for a given system *e.g.* phase diagram, enthalpies of mixing, enthalpies of fusion, heat capacities, vapour pressures and emf measurements, are combined in an optimisation process to find a small number of coefficients which represent all the available experimental information. This process is not simply a curve fitting exercise - the process must be carried out with the benefit of experience and expertise otherwise the data obtained may lose their significance and their applicability in extrapolation to multicomponent systems. Most critical assessments are now carried out using computer optimisation software such as PARROT (9) and that developed by Lukas (10,11).

4 CALCULATION OF PHASE EQUILIBRIA FOR AMALGAMS

A commonly used amalgam for compact fluorescent lamps is the {Bi+In+Hg} system [1,2]. An amalgam with the composition x(Bi) 0.53; x(In) 0.47 plus 6 weight per cent Hg has mercury partial pressures within 90 per cent of the optimum over the temperature range 325 to 393 K [2]. The {Bi+In+Hg} system has been successfully modelled using data sets for the {Bi+Hg}, {Bi+In} and {In+Hg} binary systems and taking into account experimental information on the ternary system. Data for the three binary systems were taken from references [12], [13] and [14] respectively. The model allows the mercury partial pressure to be calculated as a function of temperature over the entire composition range. Having validated the modelling approach using a well characterised system, the models were extended to other systems of potential interest.

Using the modelling techniques outlined here, many systems have been assessed for their suitability as amalgams in compact fluorescent lamps and a substantial database of binary, ternary and quaternary alloys has been established. After validation by lamp experiments, a number of new amalgam systems for fluorescent lamps have been identified.

It will soon be possible to link MTDATA with computational fluid dynamic (CFD) packages to model the interactions of geometry, temperature and chemical composition in lamps.

5 REFERENCES

[1] J.Bloem, A.Bouwknegt and G.A.Wesselink, Philips Tech. Rev. **38**, 83 (1978/9).
[2] J.Bloem, A.Bouwknegt and G.A.Wesselink, J. Illumin. Eng. Soc. 141 (1977).
[3] G.Franck, Z. Naturforsch. **26A**, 150 (1971).
[4] R.H.Davies, A.T.Dinsdale, T.G.Chart, T.I.Barry and M.H.Rand, High Temp. Science **26,** 251 (1990).
[5] O.Redlich and A.T.Kister, Ind. Eng. Chem. **40,** 345 (1948).
[6] Y.M.Muggianu, M.Gambino and J.P.Bros, J. Chim. Phys. **72,** 83 (1975).
[7] A.T.Dinsdale, CALPHAD **15,** 317 (1991).
[8] A.T.Dinsdale, J. Phys. Chim. **90,** 190 (1993).
[9] B.Jansson, *PhD Thesis*, Royal Institute of Technology, Stockholm, Sweden (1984).
[10] H.L.Lukas, E.Th.Henig and B.Zimmermann, CALPHAD **1**, 225 (1977).
[11] H.L.Lukas, J.Weiss and E.Th.Henig, CALPHAD **6**, 229 (1982).
[12] P.Y.Chevalier, Unpublished work (1986).
[13] P-Y. Chevalier, CALPHAD **12**, 383 (1988).
[14] S.A. Mucklejohn and A.T. Dinsdale, Unpublished work (1994).

A 14

SHEATH LOSSES OF CAPACITIVE FLUORESCENT LAMPS OPERATED AT LOWER RADIO FREQUENCIES (RF)

Rainer Hilbig

Philips Research Laboratories, Weisshausstrasse 2, D-52066 Aachen, Germany

1. INTRODUCTION

Lamp life and (partly) lumen maintenance of fluorescent lamps are determined by the lamp electrodes. That is why electrodeless fluorescent lamp concepts have been introduced into the market (Philips: QL-Lamp, GE: GENURA-Lamp, OSRAM: ENDURA-Lamp). All these examples use inductive energy coupling at operation frequencies of several hundreds of kilohertz up to a few megahertz. The coupling losses that can be below 10% are comparable with the electrode losses in lamps with inner electrodes.

It has been shown in the past [1] that capacitively coupled fluorescent lamps also can be very efficient when compared to lamps with inner lamp electrodes. But this high efficacy must be gained by operating the lamps at frequencies $\nu \geq 13$ MHz which can result in lower electronic gear efficiency and increased electromagnetic interference (EMI). Efficient lamp operation at lower frequencies seemed not to be possible due to the dependence of the capacitive coupling losses (sheath losses) P_{sh} on $\nu^{-5/2}$ [2,3].

In the present report we discuss the sheath losses of a capacitive tubular fluorescent lamp in the lower radio frequency region $\nu = 1\text{-}30$ MHz. The model descriptions that are supported by experiments demonstrate that with decreasing frequency the sheath losses do not increase as strongly as expected in the literature [2,3].

2. MODEL DESCRIPTION OF SHEATH LOSSES

2.1. General expression for the sheath serial resistance in cylindrical lamps

In the past C. Beneking [2] derived an expression for the sheath serial resistance R_s that was restricted to RF-frequencies $\nu \gg 10$ MHz. We give here a more general expression for R_s.

$$R_s = \alpha \cdot X_p^2 \cdot \frac{\sqrt{I}}{I + \alpha^2 \cdot X_p^2} \qquad \alpha = \left(\frac{3 \cdot \varepsilon_0 \cdot A}{5}\right)^{1.5} \cdot \left(\frac{3 \cdot \sqrt{p_g}}{2 \cdot k}\right) \cdot \sqrt[4]{\frac{1}{2}} \cdot \omega^{2.5} \qquad (1)$$

$X_p = (\omega \cdot C_p)^{-1}$: reactance of the parallel circuit diagram;
I: current coupled into the lamp;
k: ion mobility constant [2]
{Ar^+ in Ar: $k_{Ar+}=103.75\ m^{3/2}Pa^{1/2}/(V^{1/2}s)$}.

ε_0: vacuum permittivity;
A: area of the coupling surface;
p_g: buffer gas pressure;
$\omega = 2\cdot\pi\cdot\nu$, ν: operating frequency;

The capacitance C_p in X_p is the plasma capacitance (sheath capacitance), which is in general not known. For tubular lamps (inner diameter d_i) with conductive surfaces at both outer ends of the tube we fit the absolute value of X_p and its functional dependence on the lamp parameters and frequency with the help of experimental results:

$$|X_p| = 2583 \cdot \frac{d_i^{0.4}}{\varepsilon_0 \cdot A \cdot p_g^{0.6} \cdot \omega^{1.66}}$$

2.2. Frequency and lamp diameter dependence of sheath voltage

The power loss P_{sh} in one sheath is directly connected to R_s from eq. 1 by $P_{sh} = R_s \cdot I^2$. Here the sheath voltage $U_{sh} = 2 \cdot R_s \cdot I$ is used as a measure for the sheath losses of a lamp. - In standard fluorescent lamps the electrode losses are commonly given by the electrode fall voltage. -

Figure 1 shows the calculated frequency dependence of the sheath voltage for different lamp currents on a log-log scale ($d_i = 8$ mm, $p_g = 1000$ Pa Argon, $A = 7.5$ cm^2). At high frequencies (>20 MHz) U_{sh} is proportional to $\nu^{-3/2}$ while for $\nu < 5$ MHz the slope is proportional to $\nu^{-1/2}$. With increasing lamp current I, U_{sh} increases. There also exists a slight influence of I on the ν-dependence of U_{sh}.

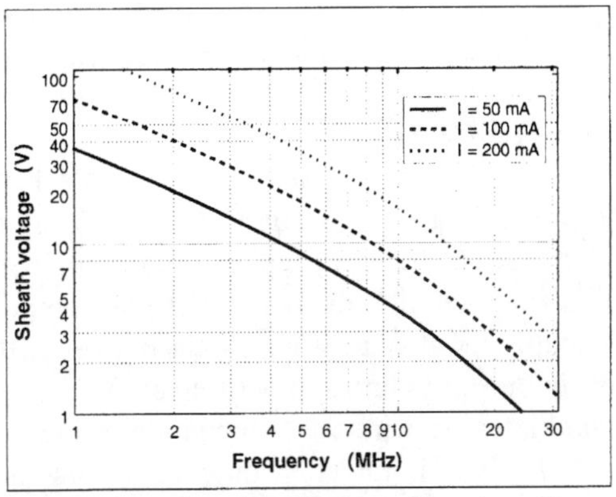

Figure 1: Sheath voltage as function of frequency for three lamp currents

Figure 2 plots U_{sh} as function of the inner diameter of the fluorescent tube ($\nu = 3$ MHz, $A = \pi \cdot d_i$ 3 cm^2). The Argon buffer gas pressure is varied with the diameter ($p_g \cdot d_i$ = const.) to sustain sufficient efficacy of the fluorescent lamp column. Again U_{sh} increases with current I. For all currents the voltage is lowest for the narrowest tubes; for example, $U_{sh}(d_i=3$ mm, $I \leq 0.1$ A$) \leq 15$V. The sheath voltage has a maximum at $d_i \approx 25$ mm:

For higher frequencies the maxima of U_{sh} are more pronounced, their position is shifted to smaller diameters. The absolute value of U_{sh} decreases with increasing ν at all diameters.

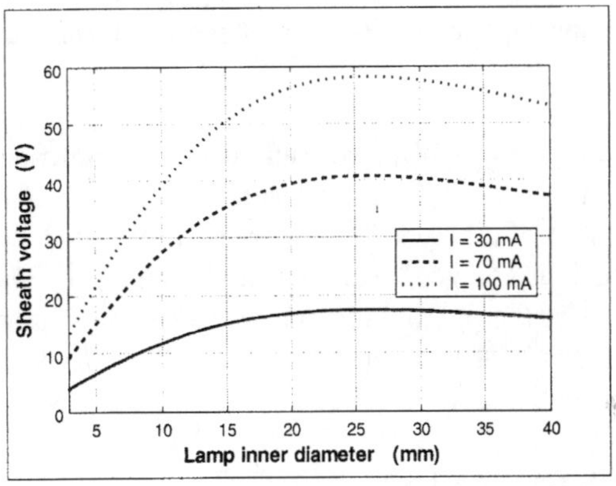

Figure 2: Sheath voltage as function of d_i for three currents ($\nu = 3$ MHz, $p_g = 10$ Pa·m/d_i.).

3. CONCLUSIONS

A quantitative description of the sheath losses (in terms of the voltage U_{sh}) of capacitive fluorescent lamps operated at lower RF-frequencies has been derived from simple modeling that was supported by experiments. U_{sh} can be minimized by choosing a suitable fluorescent lamp diameter for a given lamp current. Even at low RF-frequencies ($\nu \approx 3$ MHz) the losses can correspond to less than $U_{sh} = 15$ V.

4. REFERENCES

[1] V.A. Godyak, R.B. Piejak, B.M. Alexandrovich, *Proc. 6th Int. Symp. Light Sources,* ed.: L. Bartha and F.J. Kedves, Budapest 1992

[2] C. Beneking, J. Appl. Phys. **68**, 4461 (1990)

[3] V.A. Godyak, R.B. Piejak, B.M. Alexandrovich, IEEE trans. Plasma Science **19**, 660 (1991) and references therein

A 15

Time resolved near-IR emissions from fluorescent lamps

Rolf S. Bergman
General Electric, 1975 Noble Rd, Cleveland, OH, 44112, USA

1. INTRODUCTION

Many devices, such as remote controls for TV and various tracking systems, have been introduced that communicate information using LEDs as the source and a silicon photodiode as the receiver. The LEDs are designed to send signals in the near infrared region of the electromagnetic spectrum, around 900 nm, and the silicon photodiode receivers are usually made with a receiving window sensitive to wavelengths between 700 and 1100 nm. It has been found that the presence of fluorescent lamp systems, either linear or compact types, operated on high frequency ballasts, could interfere with the IR communication in a number of ways. The results presented are from a study aimed at understanding the characteristics of the interfering emission.

2. EXPERIMENTAL SET-UP

Two types of measurements were obtained. First, the spectral irradiance between 400 and 1600 nm were obtained from both the positive column and cathode region of a number of fluorescent lamps, operated with a number of different ballast schemes, using an Optronics OL750 spetroradiometer.

Secondly, time resolved measurements of fluorescent lamp emissions in the near IR were obtained by recording the amplified signal from a visible-light-blocking silicon photodiode. For this test a number of different F32T8 lamps were tested. Excitation of the lamp was through an ENI power amplifier coupled through a hand-wound impedance matching transformer. The frequency dependence of the photodiode-detected IR lamp emission from both the positive column and the cathode region, was investigated between 20 kHz and 1.5 MHz. The dependence of the signal on distance from the lamp and the lamp warm-up emission characteristics were also obtained.

3. RESULTS OF MEASUREMENTS

Figure 1 shows the spectral irradiance in the near-IR region, from both the positive column and from the cathode region of a standard 32Watt T8 (26 mm OD) lamp, where the detector was placed 5 cm from the axis of the lamp. Note that the emission from the center region (positive column) contains mostly mercury (Hg) lines, as noted, while the emission from the cathode region is largely due to argon with reduced emission from mercury. This increase in argon spectra is attributed to the high electric field in the cathode fall. The spectral sensitivity of the photodiode normally used in communication systems is superimposed on the spectra. Thus the detector has good sensitivity to many of the argon lines and the 1014 nm Hg line.

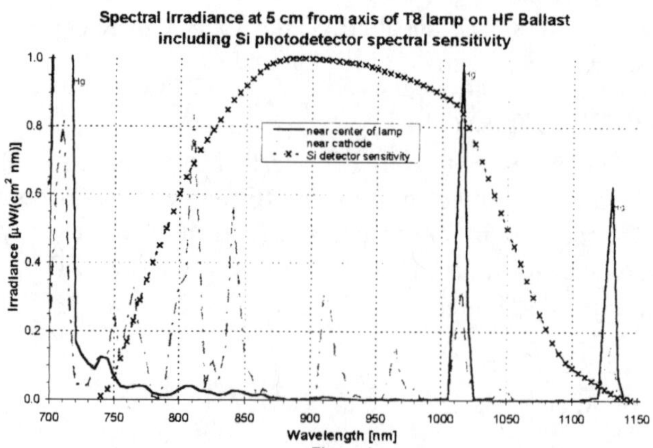

Figure 1

Further, the spectra, based on our measurements, are independent of the source excitation type and frequency, lamp manufacturer and even lamp type. Spectra of compact T4 diameter fluorescent lamps are very similar to those shown, although it is difficult to separate the cathode region from the positive column region due to the tubing being folded back on itself.

The results of time resolved measurements of the near-IR emission from fluorescent lamps are shown in **Figure 2**. Here the receiver weighted irradiance, indicated by the photodiode sensitivity curve shown in **Figure 1**, from either the positive column or cathode region of the lamp was recorded on a oscilloscope. From the scope trace the percent modulation, i.e., ([maximum - minimum]/ maximum) was calculated and plotted as a function of frequency.

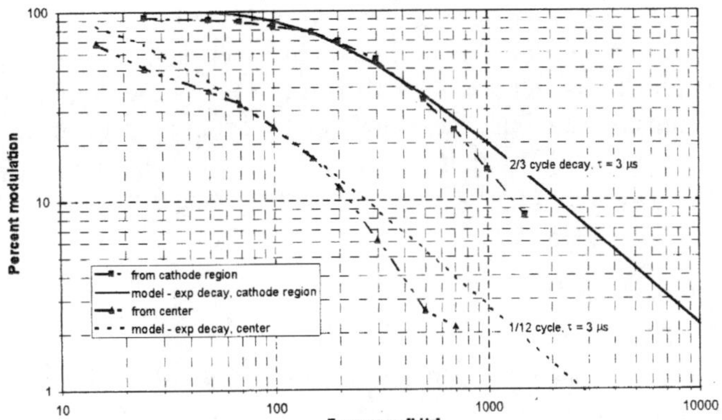

Figure 2

Note that the decrease in modulation depth with frequency for the emission from the cathode region is much slower with frequency than that from the positive column. This is due mainly to the fact that the IR emission generation processes in the electrode region occur only when the electrode is a cathode, while those in the positive column occur with the electric field in either direction. Assuming an exponential decay process for the IR emission due to collisional decay of the electron energy, a decay time constant of around 3 µseconds gives a reasonable fit.

4. CONCLUSIONS

Infrared emission from fluorescent lamps, in the wavelength region used by IR communication devices, are due to high level atomic energy level transitions of either or both argon and mercury. For operating lamps argon emission is due mostly to the cathode fall electric field and is thus found mostly near the cathode. Mercury emission, and only the 1014 nm line is of significance, is distributed throughout the lamp discharge.

Interference with communication devices in the infrared are normally due to the modulation of the IR emission, which becomes a problem when the lamp operating frequency or its first few harmonics are near the operating frequency of the communication device. It is shown that the modulation of the fluorescent lamp emission decreases with frequency and between 1 and 10 MHz largely disappear. This is attributed to the inability of the electron energy to change significantly at or above this frequency. Modulated emission from the cathode region of the lamp is the likely cause of interference in the 50 kHz to 1 MHz region.

5. ACKNOWLEDGEMENTS

Thanks to Ed Hammer, GEL for pointing out the need for the work, to Al Bush, GEL, for building the necessary equipment and to Robert DeCaro, and his people at Novalog, Inc. for sending the parts to build the photodiode detector device.

A 16

VARIABLE COLOUR TEMPERATURE FLUORESCENT LAMPS

L.P. Bakker, G.M.W. Kroesen and F.J. de Hoog
Eindhoven University of Technology, P.O. Box 513, 5600 MB, Eindhoven, The Netherlands

1 INTRODUCTION

It is very interesting from a technical point of view to be able to change the colour temperature of a lamp, while maintaining the white light black body radiation spectral distribution. The colour temperature is decisive for the perceptional impact of light on people. In general, fluorescent lamps are used in times when high intensity light is needed and incandescent lamps are used when dimming is required. Recently a new lamp concept has been introduced [1]: a tubular fluorescent lamp of which the colour can change from cool-white to warm-white. This lamp concept is investigated.

2 LAMP DESIGN

We start with a fluorescent lamp that is filled with 3000 Pa neon and 0.8 Pa mercury vapour. In combination with the fluorescent powder, the mercury vapour produces cool-white light. When the neon atoms are excited, they produce red light that can be mixed with the cool-white light, giving a warm-white colour. By changing the amount of neon light, the colour temperature of the lamp can be put anywhere in the range from 2500 K to 4000 K. The main problem of this concept is the excitation of the neon atoms. The excitation potential of neon is much larger than the excitation potential of mercury. In a normal fluorescent lamp, the neon atoms won't be excited at all. A way to excite the neon atoms is to produce high-energy electrons in the discharge. In a capacitively coupled radio frequency discharge these electrons can be generated. This rf discharge can be struck between two aluminium electrodes which have been glued to the outside of the tube, as depicted in figure 1. In the sheath region of this discharge, high-energy electrons are produced that are capable of exciting the neon atoms. Note that there are two discharges present in the tube, the normal discharge (axial positive column), and the rf discharge.

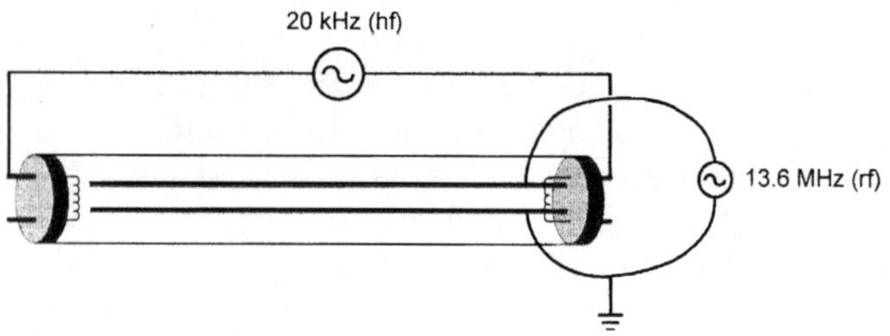

Fig. 1: Schematic drawing of the fluorescent lamp, with the additional electrodes

3 RESULTS

In order to make a first characterisation of the combined discharge, optical and electrical measurements are carried out. The rf discharge impedance, the voltage current characteristic of the normal discharge, and the colour of the lamp are measured. The rf discharge is operated at 13.6 MHz. The power input to the rf discharge varied from 0 to 17 Watt. The power input to the normal discharge varied from 0 to 6 Watt. Figure 2 shows the correlated colour temperature of the lamp as a function of the two input powers, (1) P_{hf} for the normal discharge, and (2) P_{rf} for the rf discharge. From this figure it is clear that the colour of the lamp changes indeed. However, the colour change is too small at high values of the normal discharge power P_{hf}. The amount of red light produced by the rf discharge is small and independent of the power input to the normal discharge. Only at low power input to the normal discharge, this small amount of red light will result in a significant change of the correlated colour temperature of the lamp.

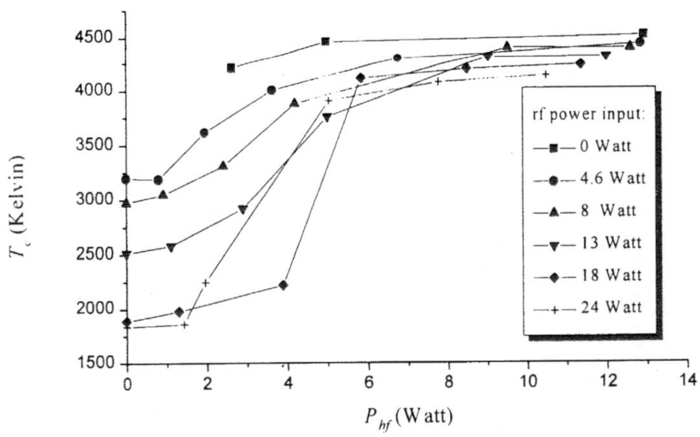

Fig. 2: The correlated colour temperature of the lamp.

The most important mechanism in this discharge lamp is the competition between neon and mercury atoms to be excited or ionised. We have a lot of neon atoms with a high excitation potential and few mercury atoms with a low excitation potential. The electron energy in the sheath of the rf discharge is not high enough to produce a sufficient amount of neon light at a reasonable rf power input level. Therefore, other solutions have to be found to produce the necessary amount of red light.

4 CONCLUSIONS

The lamp concept introduced in this work does have a perspective to lead to a new industrial product, but many technological improvements still have to be implemented. The biggest problem of the lamp is the low amount of red light, which is produced by the rf discharge. Instead of the sinusoidally shaped rf voltage, other pulse shapes could lead to a more efficient production of red light. Another solution could be to use a semi-dc glow discharge with electrodes inside, instead of outside the tube. In the cathode fall of this dc discharge, the neon ions will carry the current. In the negative glow adjacent to this region, there could be a more efficient production of red light.

5 ACKNOWLEDGEMENTS

This research is supported by the Technology Foundation STW and by Philips Lighting Eindhoven.

6 REFERENCES

[1] R. Itatani, M. Kimoto, M. Ryoko, "Multicolour gas discharge tube lighting apparatus - includes gas discharge tube containing mercury vapour and neon and having a photoluminescent material coated on its inner surface", US patent application 5410216A 950425.

A 17

TQS - TOTAL QUALITY SHIELD A LOW MERCURY DOSING AND GETTER PRODUCT FOR FLUORESCENT LAMPS INDUSTRY

S.P. Giorgi - M. Righetti
SAES GETTERS S.p.A., Viale Italia 77, I-20020 Lainate, ITALY

INTRODUCTION

Fluorescent Lamps technology has been recently characterised by a growing interest in environmental protection considerations. This technological trend derives from the innovative R&D work carried out during recent years, together with the different environmental regulations continuously issued world-wide.

These regulations have forced the fluorescent lamps manufacturers to look for more environmentally friendly products and to seriously consider ambient pollution both for the production sites and for the lamp disposal. This leads to the attempt to reduce the needed amount of Hg for operation of fluorescent lamps and to the adoption of precise reliable and safe means of Hg dosing.

In this frame, one of the main issues is mercury reduction in Fluorescent lamps.

Modern day Fluorescent Lamps can be manufactured with low Hg doses: in Halophospors Lamps 8 mg seem to be sufficient, while for the new Three Band Phosphors Lamps generation 4 mg appear to be more then satisfactory. These results have been achieved by introducing processes, which reduce Hg consumption within the operating lamps [1], [2], [3].

All the big Fluorescent Lamps manufacturers are now using different Hg dosing systems capable to at least reduce the quantity of Hg in the lamps to less than 10 mg.

Improved mechanical dosing systems for droplet and capsules, amalgams and High Yield GEMEDIS have been fully characterised and used in different production during these last years. The target of such small quantities of Hg imposes not only reliability in dosing but also high controlled production steps.

TQS: A NEW PRODUCT FOR FLUORESCENT LAMPS INDUSTRY*

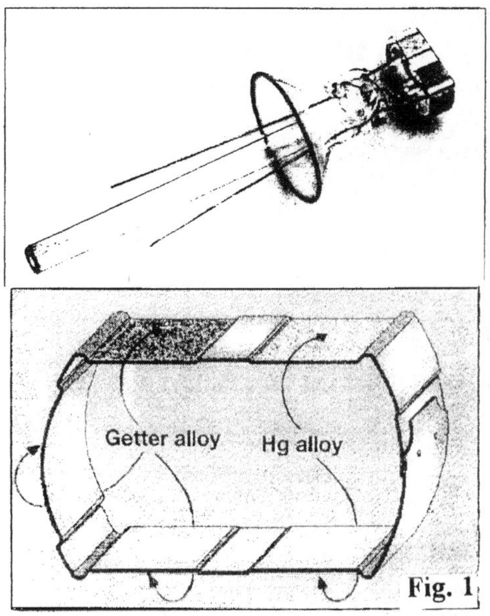

Fig. 1

High Yield GEMEDIS products (HYG and HYG-HT) [4] have been introduced in 1995 as a novel Hg dispenser capable to release the total Hg content during activation. This product has been developed starting from the characteristics of the well-known GEMEDIS [5]: a combination of the Ti_3Hg intermetallic compound with the Zr-Al getter alloy. HYG is a proven technical solution to the new market requirements for a reliable and more environmentally friendly product. A special configuration of the HYG dispensing material has been recently developed for easy use in all those lamps mounting the shield around the cathode.

This is the so-called Total Quality Shield (TQS) as shown in Figure.1. The TQS can be used to replace one of the two metallic shields currently mounted in FL. TQS is formed from a wide nickel plated iron strip which has been coated with different getter and High Yield Hg releasing compounds.

A typical TQS for T8 lamps, releasing 4.5 mg of Hg, is characterised by having 2 tracks of High Yield Hg compounds and 3 tracks of getter material. All these materials are coated on pre-formed tracks on one side of the base metal. In this way it is possible to ensure good uniform and reliable

coating of the strip. The strip is then pre-cut at 5-mm bands and then formed and welded into a closed electrical loop to get the final TQS configuration.
The Total Quality Shield integrates innovative production technologies that allow the achievement of a broad range of Hg yields from 8 mg to below 1 mg per lamp.
Typical TQS Hg yields are:

Product	Hg content	Tolerance
TQS - 8	8.0 mg	+/- 0.8 mg
TQS - 4.5	4.5 mg	+/- 0.5 mg
TQS - 1	0.7 mg	+/- 0.1 mg

All the manufacturing steps of TQS are continuously monitored by on line checking systems to guarantee a total quality product in terms of Hg release, gettering characteristics and dimensional parameters.

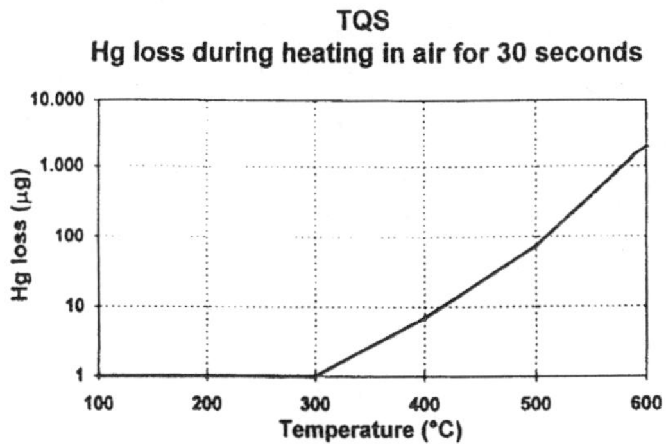

Figure 2

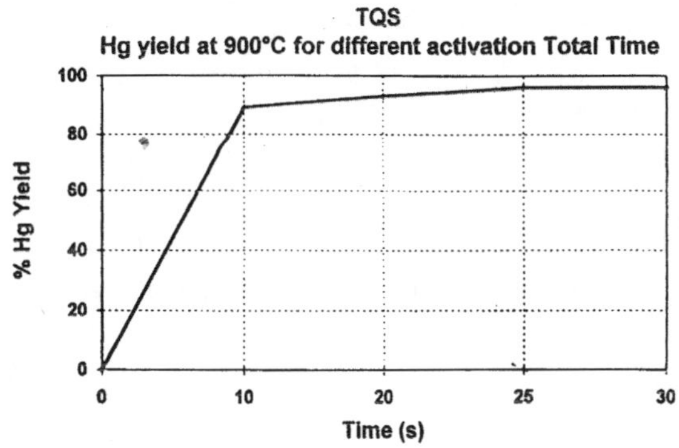

Figure 3

TQS can be automatically mounted around the cathodes during lamp production. The lamps can be safely processed (evacuated, baked and gas filled) because of the stability of the TQS in air (Fig. 2). Only after sealed off the TQS should be heated by induced eddy currents for a very short time, for example between 10 and 15 sec, to the required temperature to release Hg content and to activate the getter. Fig. 3 shows a typical yield curve of a 4.5 mg TQS revealing the quick Hg release.

CONCLUSION

TQS is an innovative Hg dosing and gettering device for fluorescent lamps industry capable to accurately dose Hg from 8 mg down to less than 1 mg in all those lamps mounting the shields around the cathodes. TQS yield characteristics have been proved to be very reliable and controlled.

REFERENCES

[1] H. Tomioka, T. Higashi and K. Iwama, Seventh International Symposium on the Science & Technology of Light Sources, p 323 (1995)
[2] Published PCT Application WO 96/06451
[3] Published PCT Application WO 96/06452
[4] S.P. Giorgi, A. Schiabel, C. Boffito, Seventh International Symposium on the Science & Technology of Light Sources, p 321 (1995)
[5] P. della Porta and E. Rabusin, Proc. 6th Internl. Vacuum Congr. p 45 (1974)

* TQS has been optimised and finalised in close co-operation with OSRAM GmbH Munich and related production departments.

A 18

THE DYNAMICAL BEHAVIOUR OF PULSED LOW-PRESSURE DISCHARGES

R.Devonshire, G.R.Harris, T.J. Healey, D.A.Stone[1], and R.C.Tozer[1]

High Temperature Science Laboratories, Chemistry Department, University of Sheffield, Sheffield, S3 7HF, UK.

&

[1]Electronic & Electrical Engineering Department, University of Sheffield, Sheffield S1 3JD, UK

Inspired by the work of Itatani and co-workers [1] we have developed a pulsed light source with near- UV output characteristics suitable for use as a backlight in a novel liquid crystal display system [2]. The ratio of 365 nm to 254 nm radiation from a low pressure Hg-rare gas discharge was increased by factors of at least 100 when the discharge was driven by repetitive, short duration (μs), high instantaneous power, electrical pulses as compared to its being driven using 50 Hz ,or higher frequency, sinusoidal ballasts. These studies of pulsed discharge behaviour have been extended to include non-Hg systems and computer models of the discharge and the power supply-lamp interface are being developed to support the experimental work.

The Figure below shows the voltage, *V*, and current, *I*, traces measured at the terminals of a Hg-rare gas lamp subjected to repetitive 2 μs voltage pulses. *V* rises rapidly before falling back to a plateau value as *I* builds up; it collapses abruptly when the pulse is terminated. After an initial short pulse, believed to be related to the charging of connecting leads, *I* rises to a plateau value before falling abruptly as the pulse is terminated. Longer pulses are characterised by a similar initial behaviour but a longer plateau period. Nanosecond time resolved and steady-state emission spectroscopy have revealed important features of the Hg excited states dynamics.

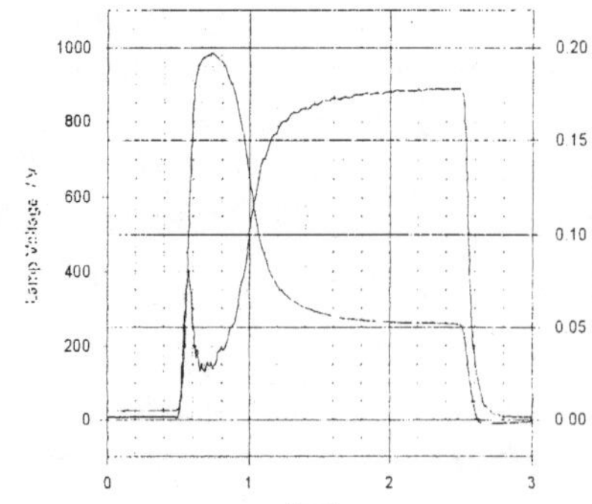

The behaviour of the Hg-rare gas lamp-pulsed power supply system has been explored for a range of lamp designs, a wide range of lamp fill conditions and compositions, and different electrode designs. The drive electronics have evolved through several design phases.

Currently, the experimental work is being extended to other discharge systems, in particular, the low pressure deuterium (D_2) gas discharge. Our special interest in the D_2 system is because we have available in the group a well developed computer-based model of the low pressure D_2 discharge which has intrinsic time-dependent capabilities [3]. The model simulates the "mobility-limited" regime of dc and low frequency discharges (*few MHz at a few torr*), where neglect of inertial terms in the momentum equations allows the particle fluxes to be used directly in the continuity equations [4]. The model is designed for easy addition of new chemical species and reactions. The number of time-dependent particle diffusion equations for ions, ground and excited states of atoms and molecules is in effect arbitrary; in practice it is limited only by computer resources and, more importantly, the availability of suitable reaction cross-sections. The model includes energy equations for the electron and gas temperatures, and a diffusion equation for the current density.

The Figures below demonstrate the time dependent capability of the code. The results refer to a 7 mbar, 2 mm radius, D_2 gas discharge using a model incorporating 7 species and 12 reactions. They show the response of the electric field, E, the electron energy, T_e, and the atom density, F_D, to a 2 μs rise time current pulse.

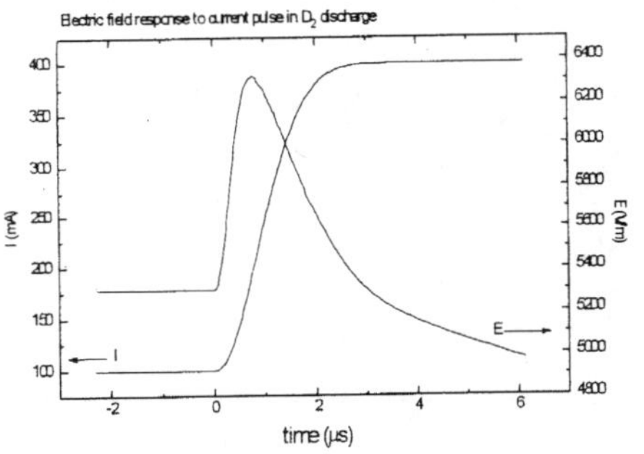

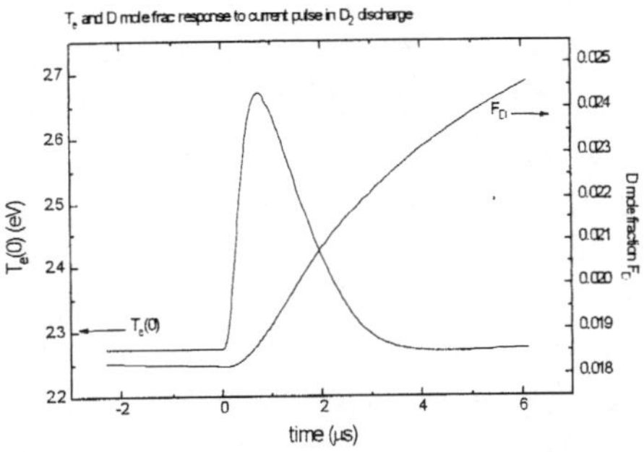

Predictions of the model for both dc and pulse driven D_2 discharges are being compared with the results of spatially resolved D_2 vibrational CARS measurements of the temperature distribution in operating D_2 lamps and with results from other spectroscopic and electrical diagnostics.

The structure of the model makes it readily adaptable to other discharges and it is planned to do this, firstly, for the low pressure Hg-rare gas discharge. This system is chosen not only because of our newly available experimental observations but also because of the availability of extensive and reliable modelling data and its general interest and technological importance. We are also developing an electrical model of the pulse driven plasma that will be compatible with SPICE and an in-house circuit modelling package, SIMNET [5].

Our overall objective is to establish a modelling environment for pulsed low-pressure discharges. When combined with powerful experimental diagnostics it will help identify the plasma operating conditions, pulse profiles, etc., necessary to achieve target plasma performance characteristics relating to efficiency, radiant output, etc. The programme will provide a firm basis for a fundamental understanding of the processes involved when a steady state discharge is subjected to a short duration voltage pulse.

References

[1] M.Aono, M.Jinno, M.Kubo and R.Itatani, *J.Light & Visual Environment*, 3(1989)1.

[2] B.Needham, *Electronic Engineering*, April 1998, p43.

[3] G.R.Harris, A.Hartley and R.Devonshire, *7th Int.Symp.Science & Tech. Light Sources*, Kyoto, Japan (1995), p.219.

[4] G.G.Lister, *J.Phys.D: Appl. Phys.*, 25(1992)1649.

[5] SIMNET, by M.Jenkins and K.Mitchell (Electronic and Electrical Engineering Dept., University of Sheffield) under a BRITE-EURAM contract with Philips, Eindhoven.

A 19

CHARACTERISTICS OF ELECTRODELESS FERRITE-FREE FLUORESCENT LAMPS OPERATED AT FREQUENCIES 1-15 MHz

Oleg Popov and Jakob Maya
Matsushita Electric Works R&D Laboratory,
216 West Cumming Park, Woburn, MA 01801, USA

1. INTRODUCTION.

The paper deals with electrical and photometrical characterization of electrodeless inductively-coupled fluorescent lamps with a reentrant cavity operated at frequencies, f = 1 - 15 MHz, and RF power, Prf = 30-80 W. All lamps were ignited and maintained using an induction coil without a ferrite core. Lately there have been many publications describing design, performance, and electrical and photometrical characteristics of electrodeless lamps operated at frequencies of 2.65 MHz (with a ferrite) and 13.56 MHz (without ferrite) [1-3]. However, some lamp characteristics such as coil starting and maintaining voltages, Vst and Vm, coil starting and maintaining current, Ist and Im, and lamp power efficiency, η = Ppl/Plamp, (Ppl is the RF power absorbed by the plasma, Plamp is the RF power delivered to the lamp) have not been reported within the large range of frequency and RF power.

2. EXPERIMENTAL SET UP.

We made, processed and studied ferrite-free fluorescent lamps having a reentrant cavity and filled with rare gas such as Ar and Kr at various pressures; mercury vapor pressure was maintained at $\simeq$4-8 mTorr by controlling the amalgam temperature. (The design and performance of these lamps were described in our patents [4])

The induction coil inserted in the reentrant cavity was made from silver coated copper wire. We used several coils with different number of turns that allowed to vary the coil inductance from 1.2 μH to 9 μH. An aluminum cylinder was positioned between the coil and the cavity walls. The cylinder was welded to the lamp base that worked as a lamp heat sink. The plasma generated heat was removed by the cylinder and redirected to the lamp sink. The cylinder also worked as a Faraday shield; to reduce eddy currents the cylinder had a few axial slits.

The RF line consisted of the signal generator, the RF amplifier, the RF forward and reflected power meter, and the matching network. The coil current was measured with the current transformer and the voltage across the coil was measured with the oscilloscope. The coil voltage and current and phase shift between them were also measured with a vector voltmeter. The coil resistance, Rcoil, coil reactance, X_L, coil inductance, Lcoil, and coil quality, $Q = X_L/Rcoil$ were measured using an HP network analyzer.

All measurements were done at room temperature, Tamb=20-25°C. The photometrical data presented were measured at conditions of maximum lamp light output where the amalgam temperature was between 80°C and 110°C, using an integrating sphere and computerized CCD detection system.

3. RESULTS AND DISCUSSION.

The lamp started by the ignition of a capacitive discharge near the turns with a high RF potential. Typically, the coil voltage, Vc, required for the ignition of the capacitive discharge was $\simeq$ 200 V. The coil transition voltage and current, Vtr and Itr, were determined as the highest value of Vc and Ic at the transition from the capacitive to the inductive discharge. It was found that Itr increased and Vtr decreased as the driving frequency decreased. The maintaining (at chosen RF power) coil voltage, Vm, and coil current, Im, had similar dependencies on the driving frequency as their transition values have. It was found that at the high coil quality, $Q = \omega L/R_{coil} \gtrsim 200$, the coil power losses were very small, $P < 3$ W. Indeed, the decrease of f requires the increase of Im to maintain RF field in the plasma needed to maintain inductively-coupled discharge. However, the coil resistance, R_{coil}, decreases as frequency decreases. As a result, the coil power losses, $P = (Im)^2 R_{coil}$, increase slowly with f, thereby the lamp power efficiency, η , remains high, $\eta > 94\%$, even at the lower end of the frequency range. The relative light output (RLO) had the same dependence on the coil Q's as the lamp power efficiency. The transformer model of the inductive discharge [5] was found to be in good agreement with our experimental data.

4. REFERENCES

[1] U.S.Patent 4,568,859 by H.Houkes, J.W.Denneman, and P.Postma, Feb.4, 1986.

[2] M.Shinomiya, K.Kobayashi, M.Higashikawa, S.Ukegawa, J.Masuura, and K.Tanigawa, J. Ill. Eng. Soc., 44, winter (1991)

[3} D.O. harmby, IEEE Proc. A, 140 (6) 465 (1993)

[4] U.S.Patent 5,621,266 by O.Popov, J.Maya, and E.K.Shapiro, Apr.15, 1997; U.S.Patent 5,698,951 by J.Maya, O.Popov, and E.K.Shapiro, Dec.16, 1997; U.S.Patent 5,723,947 by O.Popov, J.Maya, K.Kobayashi, and E.K.Shapiro, March 3, 1998.

[5] R.B.Piejak, V.A.Godyak, and B.M.Alexandrovich, Plasma Sources Sci. and Technol. 1, 79 (1992)

A 20

ON THE VALIDITY OF SELF-CONSISTENT MODELLING OF THE POSITIVE COLUMN OF FLUORESCENT LAMPS

Georges Zissis[1], Hartmut Lange[2], D. Porras[1], Peter Hardt[2]

[1] CPAT, 118 rte de Narbonne 31062 Toulouse, France
[2] INP, Robert Blum Str 8-10, 17489 Greifswald, Germany

1. Introduction

Lots of experimental and theoretical studies concerning low pressure mercury-argon discharge exist. The last few years, important progresses, concerning not only the increase of light efficiency but more generally better knowledge about gas discharge behaviour, have been achieved. These researches mainly turn to more and more complete models which allow the prediction of optimal operating conditions of lamps. Furhermore, the experimental data available in the literature concern mainly the classic T12 and T8 lamps. Thus, these experimental values of electric field and emitted power at 254 nm are used widely for model validation. Then the theoretical models are used in order to predict the optimal operating conditions of new devices (e.g. compact and thin tubes). The crucial question arising from this procedure is what are the limitations of our theoretical calculations. The objective of this paper is to illustrate the domain of validity of a rather self-consistent collisional-radiative model by using absolute measurements of emitted power at 254 and 185 nm from thin tubes for several values of the arc current (200 mA -1.5 A).

2. The model

The Self-Consistent Collisional - Radiative model allows the determination of various characteristic quantities of the discharge solving the population and electronic energy conservation equations coupled to quasi-neutrality condition, perfect gas low and Ohm's low. In order to treat effectively this equation set we need the following simplifying assumptions: The positive column is assumed to be axisymmetric and therefore all calculations have been performed using cylindrical co-ordinates. In this model, we consider for Hg atom a six level scheme (6^1S_0, $6^3P_{0,1,2}$, 6^1P_1, 7^3S_1) and the mercury ion. In this work we include also a very simplified 3 atomic level for argon (Ar, Ar^*, Ar^+). As the average energy of Ar^* atomic levels is about 11.65 eV direct ionisation from the ground state is assumed negligible under the conditions considered in this study (the mean electron energy is about 1-2 eV). Interactions such as quenching, associative and Penning ionisation have been taken into account.

3. Experimental set-up

The discharge vessels were cylindrical glass tubes with a piece of quartz in the middle part of the positive column. Both electrodes were heated and the discharge was supplied with a commutated dc current of 10 Hz. The two resonance lines of mercury were measured simultanously with the aid of narrow-band interference filters and two phototubes. For the determination of the radiated power per column length of the two resonance lines the radiance normal to the emitting suface of a small area of the positive column was measured and than intgrated over the cylindrical surface.The calculation takes into consideration the deviation from a diffuse radiator [1]. The calibration was done relative to measurements by Koedam [2] with a 36 mm discharge.The whole arrangement is enclosed in a box which was continously flushed with nitrogen in order to prevent the 185-nm radiation from being absorbed by atmospheric oxygen. The temperature of a small bulge is controlled by water flow from a thermostat. This serves as an appendix creating the coldest spot which is used to control the mercury density.

4. Results - Discussions

We use our calculation code in the case of a 7 mm diameter tube filled with 4 Torr pure argon in order to calculate the emitted power at 254nm and 185 nm Hg resonance lines. This calculation has been carried out for four arc currents (I1=1.5 A, I2=600 mA, I3=400 mA and I4=200 mA) and several values of the cold spot temperature varying in the range of 10 to 80 °C. Figures 1 and 2 summarise the calculation results and the experimental measurements. In these plots we observe that concerning the 254 nm line the code's predictions are valid for currents up to 600 mA and this for the compete cold spot temperature range studied. The code fails for high current values. This is probably due to the fact that at these high current cases the 3-body and radiative recombination is no more negligible. However, if the results for the 254nm could be considered as rather satisfactory (comparing to other positive column models), the obtained values for the 185nm line are surprising. In fact the code predicts that the emitted power should pass through a maximum however the experiment show a "saturation" effect. At this moment seems rather difficult to explain why this difference appears.

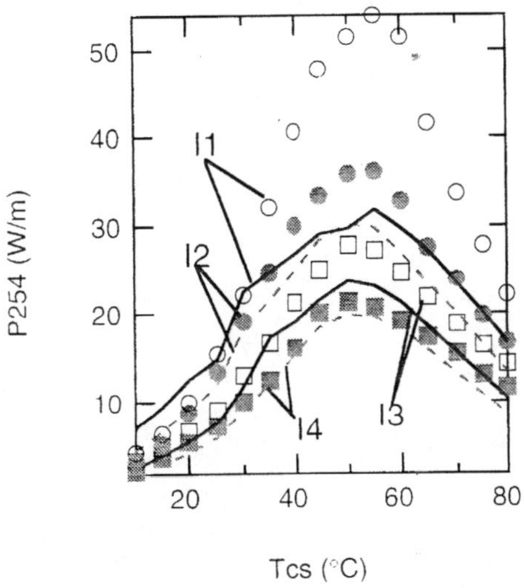

Figure 1

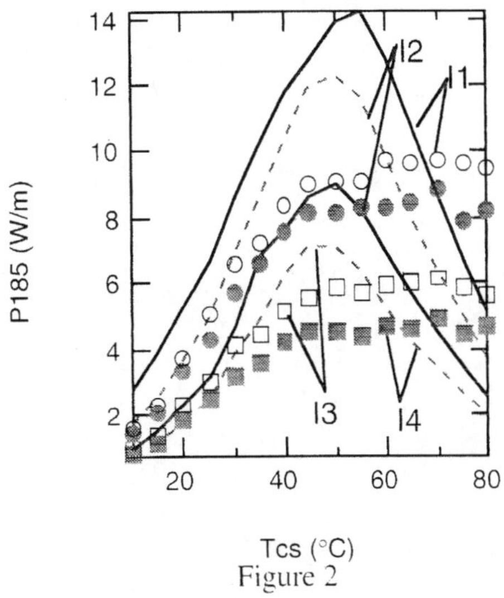

Figure 2

5. Acknowledgements

This work has been done in the frame of Procope Franco-German co-operation project n°97091

6. References

1. P. Anderer, J. Appl. Phys. **69**, 1224 (1991)
2. M. Koedam et. al., Physica **29**, 565 (1963)

Incandescent Lamps

B 01

LINE VOLTAGE TUNGSTEN HALOGEN GENERAL PURPOSE LAMP HAVING A 51 MM INTEGRAL REFLECTOR

Lieven De Weerdt , Robert De Ceuster
Sylvania N.V. Industiepark 13, B-3300 Tienen, Belgium

1 INTRODUCTION

Halogen lamps with integral reflector of diameter 51 mm are common use for general lighting applications over the last decade. Reason for this success is the small fixture size for those lamps. All those lamps operate at low voltage and require a transformer. Unfortunately the freedom of designing fixtures is limited by the presence of this transformer. The challenge for the authors was to develop a halogen reflector lamp with similar dimensions as the existing low voltage reflector lamp but directly operating at line voltage without the use of any transformer. The light output and life expectancy should be double of a comparable incandescent reflector lamp.

2 TECHNICAL DETAILS

2.1 Burner development

All existing line voltage lamps with an integral reflector were using a 50 mm long burner with an axial filament (see figure 1). In order to fit into a 51 mm small reflector, the burner construction had to be completely changed. The desired compact size of the burner was obtained by "folding" the filament in a W - shape. The result was a very compact burner, which dimensions were comparable with low voltage burners (see figure 1). In combination with a hard glass envelope a new light source was born to withstand the high temperature load during operation in a 51 mm reflector. Life tests showed that it is feasible to make a reliable 2500 h lamp with this new technology.

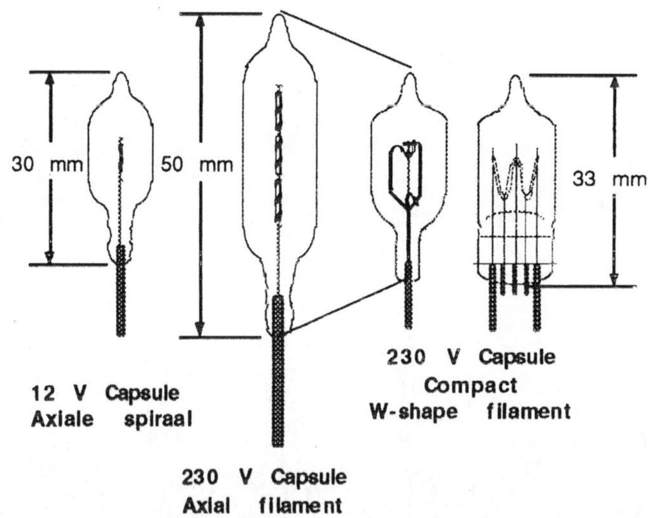

figure 1 : comparison between low voltage, axial line voltage and compact line voltage burners

2.2 Reflector development

In order to achieve an optimal optical performance, a new reflector was designed for this compact burner. Therefor our existing patent [1] for faceted reflectors was used. Two beam angles 25 and 50 ° were developed in order to cover a large area of applications. Light intensities with a 50 W compact burner are summarised in the table 1 below.

	Light Intensity	Beam angle	mean life
	Candela	degrees	hours
50 W flood reflector	1150	25	2500
50 W wide flood reflector	500	50	2500

table 1 : optical performance of reflector with 50 W compact burner

	Light Intensity	Beam angle	mean life
	Candela	degrees	hours
R 50 40 W	530	25	1000

table 2 : optical performance of incandescent 50 mm reflector lamp

Comparing those values with an incandescent reflector lamp having the same beam spread of 25 degrees (see table 2), one can conclude that the targets for doubling life and light output were reached.

2.3 Lamp cap development

Examining all existing lamp caps, like screw base E 14, bayonet base B 15, learned us that none of them allowed a compact lamp. On the other hand, using the two pin lamp cap from the low voltage lamp reflector lamp would result in a compact but unsafe lamp. Conclusion was that a new base had to be developed.
From the early start of development it was also planned to have as well as an aluminised as a dichroic coated version in the range. Thermal analysis on structural components of fixtures and environment learned us that an aluminised lamp is not interchangeable with a dichroic coated version. This led us to develop two new lamp caps, designated as GU 10 for aluminised and GZ 10 for dichroic coated reflector lamps. Those caps are a kind of very compact bayonet bases which enable save and easy use for the consumer (see figure 2). The GU 10 system with a chamfered base was designed to provide for non-interchangeability with GZ 10 based lamps.

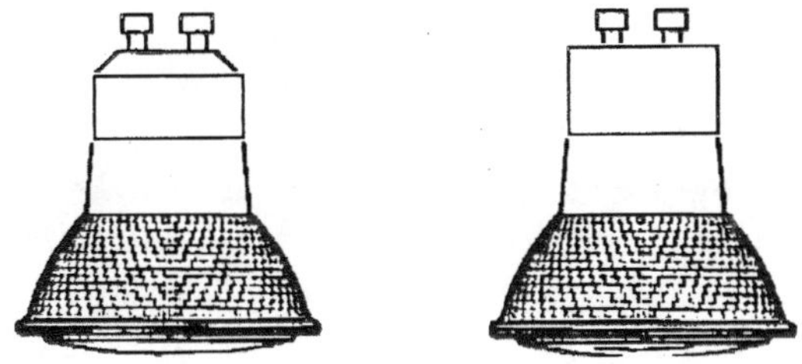

figure 2 : Halogen line voltage 51 mm reflector lamp with GU 10 (left) and GZ 10 (right) lamp cap.

3 CONCLUSIONS

This line voltage lamp with integrated 51 mm reflector is a new compact light source which enables fixture designers to develop new small luminaires without transformer. Comparison with existing incandescent lamps showed that this halogen source is twice as efficient and life expectancy is more than double.

4 ACKNOWLEDGEMENTS

Thanks to Anton Cornelje (Lumiance) and Alan Hall (Concord) for their detailed reports on temperature measurements.

5 REFERENCES

[1] United States Patent, Patent Number : 5,272,408 Date of Patent : December 21, 1993

B 02

APPLICATION OF SMALL ANGLE NEUTRON SCATTERING METHOD FOR CHARACTERIZATION OF POTASSIUM BUBBLES IN K-Si-Al DOPED TUNGSTEN

Péter Harmat[1], Attila Nagy[2], Ottó Horacsek[1]

[1]Research Institute for Technical Physics and Materials Science, POB. 49 Budapest H-1525, Hungary

[2]General Electric Lighting Tungsram Rt., Vaci ut 77 Budapest H-1340, Hungary

1 INTRODUCTION

The crucial factor in the high temperature strength of doped tungsten wires is the potassium bubble structure that develops during the thermomechanical processing of the sintered ingot. While in the recrystallized wire the bubbles are spherical (equiaxed), their precursors in the intermediate products (swaged rods and as-drawn wires) are elliptically elongated (uniaxed). This elongated shape above a critical length-to-width (aspect) ratio is required for bubble multiplication providing the dispersion strengthening of the wire [1]. As the degree of dispersion depends on the processing history, design of the lamp grade material needs a suitable shape control of the deformed potassium bubbles during processing.

2 EXPERIMENTAL

2.1 Anisotropic small angle neutron scattering

To obtain a comprehensive picture of the bubble morphology in different stages of the thermomechanical processing, small angle neutron scattering (SANS) experiments were used for the characterization of the bubble structure in doped tungsten. The SANS method is able to reveal inhomogeneities in the material, e.g. second phase particles, voids, up to size region of 100 nm. Using a typical arrangement of SANS equipment [2], the neutrons having wave nature at 0.4-1.5 nm wave length create a scattering intensity distribution on a two dimensional position sensitive detector (PSD) array except a small central region which is shielded from the direct beam. From these data a volume concentration or a size of the particles / bubbles can be determined as an average value for a macroscopical sample volume.

In the majority of materials investigated by SANS the scattering pattern is statistically isotropic, e.g. in case of second phase particles [3]. Such assumption simplifies the evaluation [4]. In our case, however, elongated bubbles having the same orientation are produced during wire drawing. If the orientation of the elongated bubbles is perpendicular to the incident neutron beam an anisotropic scattering can occur and the equal intensity contours of the map representation of the scattering pattern on the two dimensional PSD array are no longer spherical. It can be shown [5] that oriented particles having exactly ellipsoid shape give ellipsis contours in this map representation because the ellipsoid form of the particle / bubble makes possible a simple mathematical evaluation. In this way the map representation of the scattering pattern allows direct visualization of the shape of such particles / bubbles.

2.2 Results

Samples of K-Al-Si doped tungsten wire and bar were tested by SANS equipment having two-dimensional PSD of 64x64 pixels, 1 cm^2 each at 1.2 nm wave length and 4.82 m PSD-sample distance. The samples were selected from different stages of thermomechanical processing, from swaging to wire drawing and were cut into 25 mm long straight pieces bundled parallel in 1-5 layers (depending of the bulk diameter) with the same orientation. The wire axes were positioned always perpendicularly to the incident neutron beam.

The SANS measurement on these samples after several steps of processing revealed remarkable anisotropy, as expected. The anisotropy increases in the progress of processing. A decreased but also well expressed anisotropy has been found on the partly annealed samples, while the anisotropy vanished after perfect annealing of the sample at 2700 K 300 sec. Typical scattering maps can be shown on the Fig. 1. The appearance of this anisotropy is in accordance with the previous knowledge and models, e.g. by Moon and Koo [1].

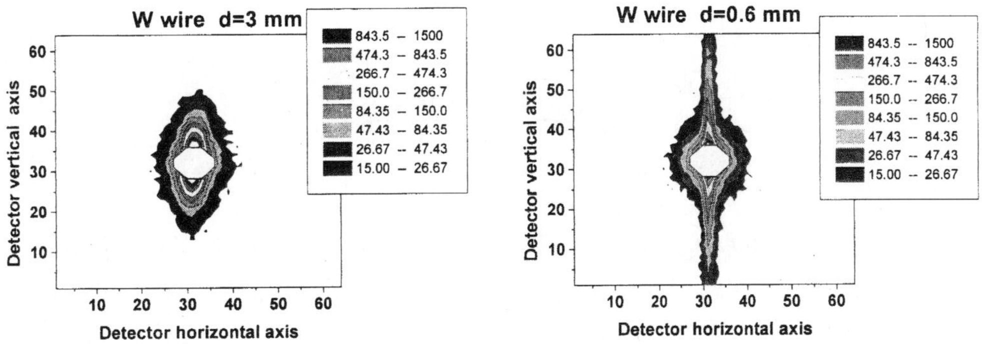

Fig. 1: SANS map of doped tungsten wire of 3 mm (left) and 0.6 mm (right) diameter.

2.2 Discussion

As a first approximation the isointensity contours show elliptical shape reflecting an average bubble shape in the sample. The isointensity contours on the scattering maps are, however, not simple ellipses and thus can not be explained by one set of similar ellipsoids. The peculiar shape of the isointensity lines indicates the presence of other component among the strongly deformed elongated ellipsoids.

Such scattering intensity distribution can be given by modeling the scattering intensity directly on the 64x64 pixel of SANS array. Two sets of particles are considered: 1) uniform sized elongated ellipsoids with half axes vr_a, r_a, r_a (v is the aspect ratio) with the percentage of ellipsoids p, 2) spheres with percentage $1-p$.

Our simple model provides a fairly good agreement between the measured and the calculated scattering patterns. It gives direct evidence that - as a result of size dependent deformability of potassium pores - among the highly elongated ellipsoids there is a remarkable amount of spherical bubbles without any deformation even after wire drawing [6].

3 CONCLUSIONS

SANS seems to be a suitable technique to characterize the morphology of the bubbles in doped tungsten. The method is sensitive enough to reveal the presence of a small amount of spheres beside the large concentration of elongated bubbles.

4 REFERENCES

[1] D. M. Moon and R. C. Koo, Metall. Trans. **2,** 2115-2122 (1971)

[2] G. S. Bauer and G. F. Crawford, *The spallation neutron source SINQ. A new large facility for research at PSI* (Paul Scherrer Institut CH-5232 Villigen, Swiss, Oct 1994)

[3] R. W. Hendricks, J. Schelten, W. Schmatz, Philos. Mag. **30,** 819 (1974)

[4] Lord Rayleigh, Proc. Roy. Soc. London **A 84,** 25 (1911)

[5] G. Porod, *in: Small Angle X-ray Scattering*, edited by O. Glatter, O. Kratky (Academic Press, London, 1982)

[6] O. Horacsek, M. Menyhárd and J. Lábár, High Temp. Mat. and Proc. **14,** 207-213 (1995)

B 03

CREEP AND CAVITATION IN VIBRATING COILS OF INCANDESCENT LAMPS

István Gaal[1], Péter Harmat[2] and Csaba L. Tóth[2]
[1]Research Institute for Technical Physics and Materials Science, P.O.Box 49, H-1525 Budapest
[2]GE Lighting TUNGSRAM, Váci ut 77, H-1340 Budapest

1 INTRODUCTION

The present work analyses the creep performance of vibrating coils from a macro-mechanical point of view. Our starting point is the simple fact that a coil elongates in both half periods of the vibration, while the sign of its bending is opposite in the subsequent half periods. This work proves that the stretching force connected with the elongation of the coil governs the stress in the wire. Thus, in a vibrating coil the magnitude the actual stress changes periodically in time without the reversion of its sign and consequently the incandescent coil deforms by cyclic creep [1] during vibration. The most important consequence of this mode of deformation is that even a vibration with a *very high frequency* will be able to cause not only creep damage but also sag, as *the time average of stress is different from zero.*

2. FORCES ARISING IN COILS DURING TRANSVERSAL VIBRATION

Let us consider a horizontal coil with clamped ends that is bent and stretched upon vibration in the vertical X Y plane. This motion is governed by the body forces arising from gravitation and vertical lamp acceleration. In order to determine the stretching force, F, and the normal force, G, acting in a coil, the following equations have to be solved (e. g. [2]):

$$S = \sqrt{X_\xi^2(\xi) + Y_\xi^2(\xi)} = 1 + \frac{F(\xi)}{D_C}, \qquad \left[\frac{X_\xi Y_{\xi\xi} - Y_\xi X_{\xi\xi}}{S^3}\right]_\xi = -\frac{G(\xi)}{A_B}$$

$$F\frac{X_\xi}{S} - G\frac{Y_\xi}{S} = T, \qquad F\frac{Y_\xi}{S} + G\frac{X_\xi}{S} = wa\xi \tag{1}$$

Here the X axis is horizontal, the index ξ denotes derivation, X(ξ), Y(ξ) are the coordinates of the coil middle line at a prescribed resultant vertical acceleration a, and ξ is the coordinate along the straight coil (at a=0) after thermal elongation. Thus, ξ changes between 0 and L(1+e), where L is the distance between the clamps, and e is the thermal dilatation of tungsten between the ambient and the operating temperature. T is the constant horizontal component of the stretching force, which has to be determined from the boundary condition according to which the ends of the coils are clamped at fixed points. D_C is the stretching stiffness factor of the coil [3], while A_B is its bending stiffness factor. A_B was determined from that critical length at which a vertical rod looses its stability under its own weight, if it is clamped at the lower and free at the upper end by making use of the relation [2]:

$$A_B = \Lambda^3 wg / 8, \qquad (2)$$

where Λ is the critical length, w is the mass of the coil for unit length, and g is the gravitational acceleration. *The most important results of the numerical solution of equations (1) are* :

(i) With a high accuracy the stretching force, F, changes along the coil as: $F = T[1+0.5(aw\xi/T)^2]$.

(ii) With a very good accuracy $T = (awL)(6e)^{-0.5}$, i. e. T decreases with increasing thermal dilatation (see also [3]).

(iii) The maximal value of G is equal to $T(6e)^{0.5}=awL$, i.e. T is four times larger than G, if e amounts to 0.01 (as usual [3]). Thus, the coil creep is practically governed by the stretching force.

(iv) The bending has its own contribution to the total *elastic* deflection. When h denotes the elastic and thermal sag due to the stretching at $L>>\Lambda$, then at a=4g the additional elastic sag due to bending amounts to 0.125h and 0.0625h at $L=\Lambda$ and at $L=4\Lambda$, resp., while at a=g the additional sag amounts to 0.76h and 0.48h at $L=\Lambda$ and at $L=4\Lambda$, resp. (This means, the *visual observation* of the coil shape will *reveal marked bending* even in cases, where the bending force is relatively small with respect to the stretching force.)

3. THE CYCLIC CREEP PERFORMANCE OF NON-SAG TUNGSTEN

The average stress governs the creep [1], if the vibration frequency of the stress is very high with respect of the characteristic times of the creep processes. If the vibration frequency is very very low, the average strain rate of the cyclic creep will be given by the relation[1]:

$$e_t = \frac{1}{T}\int_0^T [k\sigma(t)]^n dt, \qquad (3)$$

(In (3) n is the creep exponent of the stationary creep under constant stress.) The measured average creep rate after sufficiently long period of cyclic creep can be either lower or higher than e_t [1]. An experimental study of the cyclic creep at 2600 K on straight non-sag tungsten wires [4] proved the following facts: when a constant stress σ_h (64MPa < σ < 78 MPa) was applied for short periods (e.g. 2 sec) and subsequently σ_l=3MPa was applied for longer periods (e.g. 100 sec) periodically, then the actual creep rate was with a factor of three lower than e_t (determined by static creep), while the cyclic creep rate approached the value of e_t, when the cyclic creep consisted of 10 sec periods under constant stresses of 78 and 3 MPa, resp.

REFERENCES

[1] W. Blum, *Materials Science and Technology,* edited by R:W. Cahn, P. Haasen and E. J. Kramer (VCH Weinheim 1992) Vol 6 p. 384, 391 and 397

[2] L. D. Landau and E. M. Lifshitz, *Theory of Elasticity*, (Pergamon Press, Oxford, 1970)

[3] E.Pink and I. Gaal: *Metallurgy of Doped/Non-sag Tungsten*, edited by E. Pink and L. Bartha (Elsevier Applied Science, London, 1989) p 209-231

[4] P. Harmat and I. Gaal : *Proc. 12th Int. Plansee Seminar, edited by* H Bildstein and H. M. Ortner (Metallwerk Plansee, Reutte, 1989) Vol.1. 861-868

B 04

TUNGSTEN IMPROVEMENT FOR SHOCK-RESISTANT LAMPS

F.J. M. Mertens
Philips Lighting B.V. Maarheeze, The Netherlands.

INTRODUCTION

Many players in the field of producing lamps use a doped tungsten wire for the whole range of incandescent lamps. Critical halogen lamps have the requirement of a very coarse recrystallized wire structure with interlocking crystals (swallow-tail type). Low-temperature lamps are not very sag sensitive, so the wire structure does not require a coarse structure. A special case in the family of low-temperature lamps is a traffic signal lamp. This specific lamp suffers from a serious problem: premature filament failure which leads to a lot of replacement costs. The development of a special type of tungsten "fit for use"in this application was the objective of the present investigation.

PROBLEM ANALYSIS

Broken filaments of traffic signal lamps (40 Watt/ 220 V) which failed in field tests, were examined in a SEM. Several important features were established:
(1) The fracture was always in the vicinity of a support or clamp.
(2) The fracture always occured in a metallographic transition zone, more specifically: The transistion between a partly and a completely recrystallized structure.
(3) The fracture zone always showed thermal attack (indicating an increased level of contaminations).

Since the supports act as heat sinks, there is a the temperature gradient between the center of the filament and the filament part clamped in the support , which can amount as much as 500° C. During the recrystallization process of the filament, this can result in a transition area separating a non- or partly- recrystallized area near the support and a complete recrystallized area some distance remote from the support. The location ,where the transition is occurring , depends on the recrystallization temperature of the filament in relation to the heat sink temperature gradient. Transition areas are known for their brittle nature. Moreover, if the transition zone is also a critical zone with respect to vibration of the filament (e.g. caused by heavy traffic), the chance of failure increases.

PROBLEM SOLUTION

From the SEM investigation the following hypothesis was posed:
"The chemical attack is a consequence of an increased level of accumulated impurities, as a result of the sweeping effect of impurities during the recrystallization process. This phenomenon decreases locally the wire diameter and in combination with vibrations it results in filament fracture".
Following this hypothesis two solutions are possible; either prevention of a transition zone and/or shifting the vibration sensitive area closer to the support. Thus to prevent filament fracture, a material recrystallizing over a longer distance and closer in the vicinity of the support would be recommendable. Such a property of the filament is determined by the dope distribution in the drawn tungsten wire.

The actual doping is part of the tungsten powder processing phase. Doping of T.B.O. (tungsten blue oxide) with K, Al and Si results in the incorporation of dopant phases into the tungsten grain during hydrogen reduction. Only this incorporation guarantees that sufficient K remains in the sintered ingot. The incorporated K in the sintered bar is present in very small voids or bubbles. The amount and distribution of these bubbles is the beginning of a continuous redistribution (breaking-up ,etc.) during subsequent rolling, swaging and drawing operations, resulting in rows of tiny potassium containing bubbles.

RESULT

Starting from a standard doped tungsten powder lot, we were successful in obtaining a structure, that does not show a transition zone at all, by changing process parameters influencing the bubble distribution. This material has been tested in the application of traffic signal lamps and in other low-temperature lamps (so called 'Philinea' lamps), which have severe requirements on vibration sensitivity. Regarding the traffic signal lamps, the results of a field test were actually very good: the standard material had a 2% early failure percentage, the newly designed tungsten material had a 0.1% early failure percentage. Other field tests confirmed these results. The results concerning the new material applied in 'Philinea' applications are shown in Figs. 1 and 2.

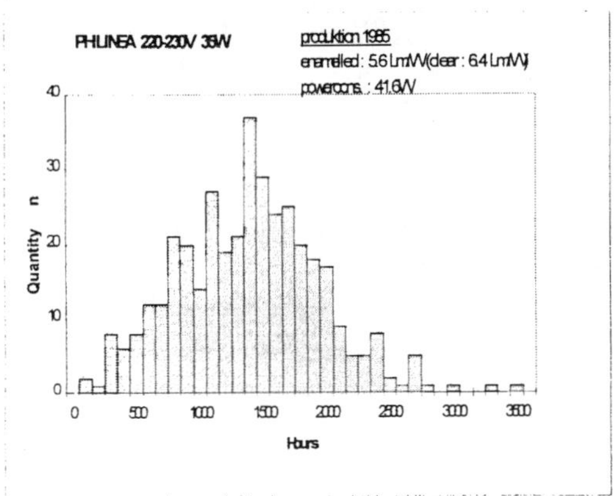

Fig. 1: Standard doped tungsten

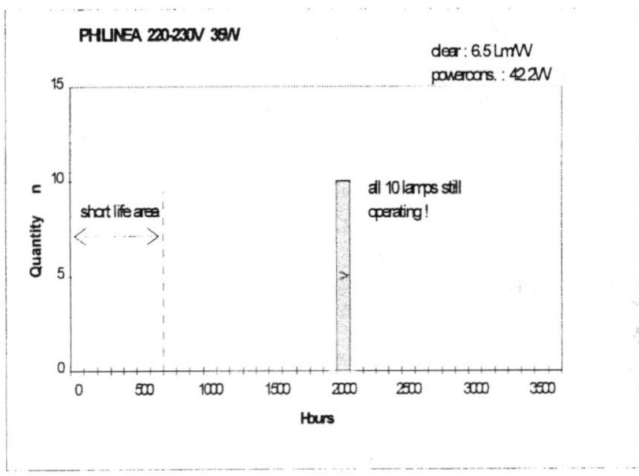

Fig. 2: New developed doped tungsten

B 05

THEORETICAL ANALYSIS OF SPECTRAL POWER DISTRIBUTIONS FROM MICROCAVITY RADIATORS

Seishi Sekine, Kohji Kashiwagi, Masatoshi Ueno, Masashi Ohkawa
Department of Electrical and Electronic Engineering, Faculty of Engineering, Niigata University, Ikarashi 2-8050, Niigata 950-2181, JAPAN

1 INTRODUCTION

At the 5th International Light Source Symposium in 1989, Jon F. Waymouth proposed "Microcavity Lamp" based on "Cavity Quantum Electrodynamics Principle" as one of the next generation lamps and estimated that a tungsten radiator whose surface is perforated by square waveguides 0.35 μm across and 7 μm deep, with walls 0.15 μm thick, would permit luminous efficacy of 60−80 lm/W at operating temperature of 2000−2100 K [1]. This dramatic potential, however, has not been experimentally demonstrated yet. Therefore we analyzed theoretically spectral power distributions from microcavity radiators and estimated their luminous efficacy to verify the Waymouth Hypothesis.

2 ANAYTSIS

2.1 The modes of cavity with walls of imperfect conductor

We analyzed the modes of cavity resonator surrounded by the walls of imperfect conductor and obtained a result as follows. The resonance wavelength λ in the cavity with the walls of imperfect conductor of skin depth δ is longer than the resonance wavelength λ_0 in that with those of perfect conductor by $(\delta/2L)\lambda_0$, i.e., the wavelength of electromagnetic waves resonating between the imperfect conductive walls with the skin depth δ and a separation L equals that of those resonating between the perfect conductive walls with the separation $[1+(\delta/2L)]\lambda_0$.

2.2 The skin depth of tungsten

From analysis of Maxwell's equations and the motion equation of free electron in the tungsten, we calculated the skin depth of tungsten, δ, using a value of emissivity measured by De Vos [2]. A graph of spectral distribution of δ is shown in Fig. 1. The shorter the wavelength, the deeper the skin depth in the range of wavelength from visible to near ultraviolet light and the skin depth at 2000 K is about 83 nm at a cut-off wavelength of 700 nm.

2.3 Radiant emittance and the luminous efficacy

With the Fresnel-Kirchfoff formula, spectral radiant emittance from microcavity, $M_e(\lambda,T)$, given by

$$M_{MC}(\lambda,T) = M_e(\lambda,T)\sqrt{1-(\lambda/\lambda_c)^2} \quad (1)$$

where $M_e(\lambda,T)$ is the spectral radiant emittance by Planck's radiation law and λ_c is the cut-off wavelength of microcavities. Fig. 2 shows an example of tungsten microcavity radiator perforated by cylindrical waveguides that have a diameter of 327 nm for the cut-off wavelength of 700 nm and a width of wall of 166 nm, which is twice as thick as δ, at the thinnest parts. Fig. 3 shows the

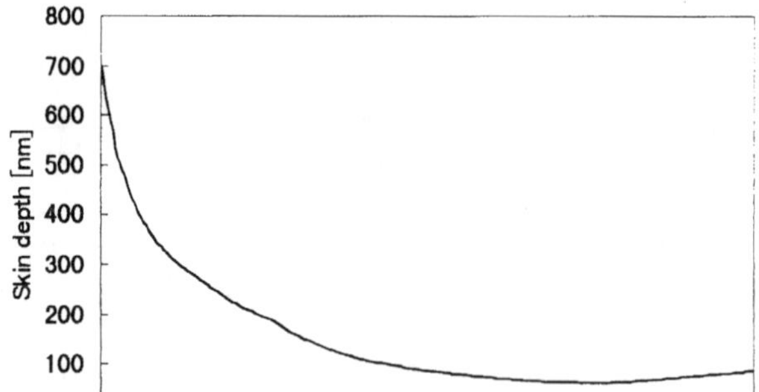

Fig.1 Spectral distribution of the skin depth of tungsten

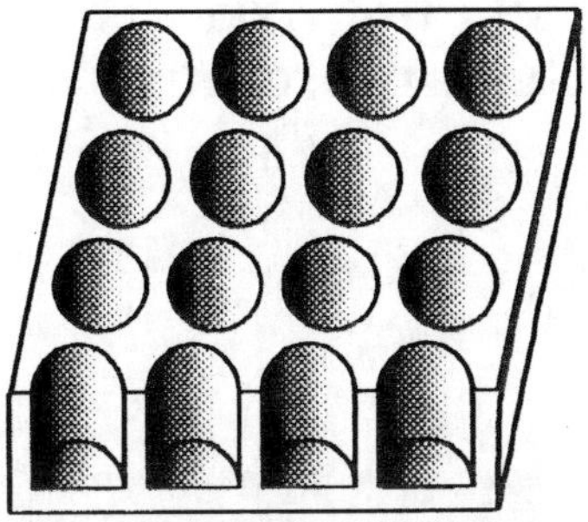

Fig. 2 Tungsten microcavity Radiator perforated by cylindrical waveguides

the thinnest parts. Fig. 3 shows the spectral radiant emittance from microcavity provided that the cavity is vacuous and 2000K and has λ_c of 700nm. The luminous efficacy is about 123 lm/W and the color temperature about 2500 K, which is higher than the operating temperature of 2000 K. Fig. 4 shows the spectral radiant emittance from tungsten microcavity radiator shown in Fig. 2. The luminous efficacy is about 68 lm/W, which accords with the value estimated by Wamouth. Fig. 5 shows the relation between the cut-off avelength and the luminous efficacy on the microcavity. The maximum luminous efficacy of tungsten microcavity is about 180 lm/W at the wavelength of 620 nm and that of perfect conductor microcavity is about 260 lm/W at the same wavelength. The shorter the cut- off wavelength, the higher the color temperature of radiation, naturally.

3 CONCLUSION

Tungsten microcavity radiator would permit the luminous efficacy estimated by Waymouth, with a higher color temperature than the operating temperature. The more conductive the radiator, i.e., the shallower the skin depth of radiator, the higher the luminous efficacy

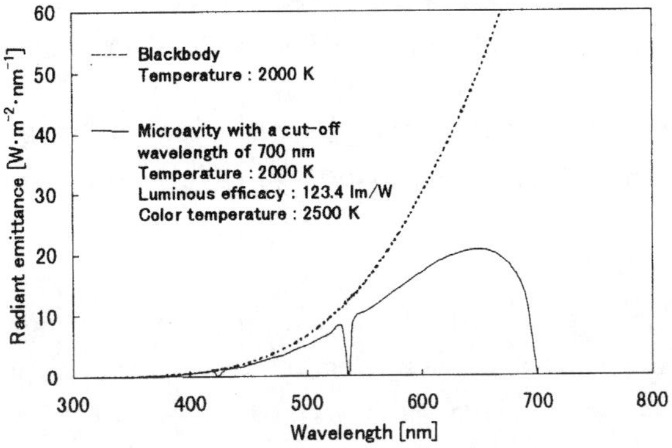

Fig. 3 Spectral radiant emitance from the microcavity with a cut-off wavelength of 700 nm

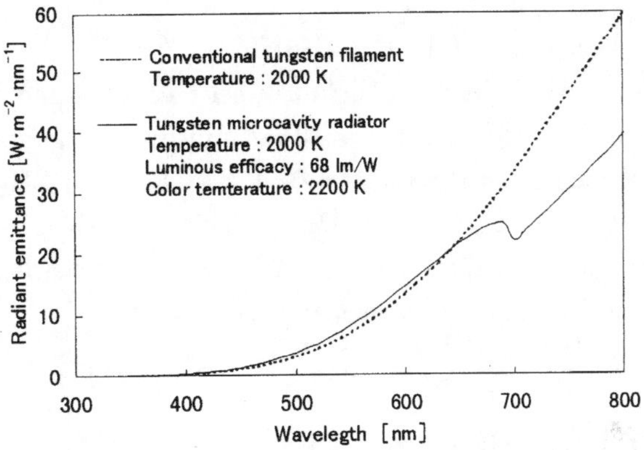

Fig. 4 Spectral radiant emittance from the tungsten microcavity radiator

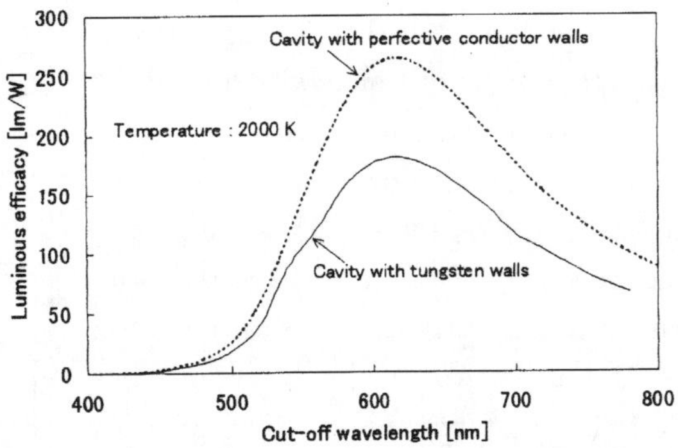

Fig 5. Relation between the cut-off wavelength and the luminous efficacy on the microcavity

4 REFERENCES

[1] J. F. Waymouth, J. Light & Vis. Eng., Vol. 13, No. 2, pp. 51-68(1989).

[2] J. C. De Vos. Physica [holland], Vol. 20, pp. 690-714 (1956).

B 06

THE INFLUENCE OF SOL—GEL DERIVED INTERFERENCE MULTI—LAYER COATINGS AT DIFFERENT TEMPERATURE ON HALOGEN SUNLIGHT SPECTRUM

Zhi—guo Lu, Xiao—yun Hu, Su—juan Zhang, Tao Wang, Ai—hua Gao
Department of physics, Northwest University, Xi'an 710069 P. R. China

1. INTRODUCTION

The halogen sunlight deposited with sol—gel interference multi—layer coatings has been used widely as energysaving lamp, fogproof lamp, signal lamp, and so on. Relatively speaking, this sol—gel derived coatings is very convenient and economical for illuminant industry, therefore, in the sol—gel optical field, many efforts have been made to study optical properties of this kind of coatings. In this paper, we take the columned halogen sunlight deposited with fogproof coatings for example. When lighting, the high temperature of bulb will affect optical properties[1] of sol—gel derived coatings and then change the spectrum of halogen sunlight.

2. EXPERIMENTAL CONDITIONS

Suppose the H4 halogen sunlight, we selected TiO_2 and SiO_2 as film material and deposited them alternatively up to 11 layers on the lamp by sol—gel process. In this experiment, we derived TiO_2 coatings from the solution of $Ti(OC_4H_5)$ and CH_3CH_2OH, derived SiO_2 from the solution of $(C_2H_5O)_4Si$ and CH_3CH_2OH by dipping—pyrolysis method in which the dipping speed was controled from 0.785mm/s to 3.975mm/s. The coatings had been designed as long wave transmitted, the alterant wavelength was located around 510nm.

3. EXPERIMENTAL RESULTS

3.1 The influence of heat temperature to thickness and refractive index of single — layer coatings.

As the glass plate deposited with single—layer TiO_2 coatings was heated with molybdenum foil, we observed the surface temperature with the infrared thermometer, and at the same time measured the refractive index by the ellipsometer. Thus we got:

1>Thickness (D) of coatings decreased logarithmly with increasing heat temperature (T). See also fig. 1. And under the above condition they abide by the following function $D = 139 - 35 * \log(T)$

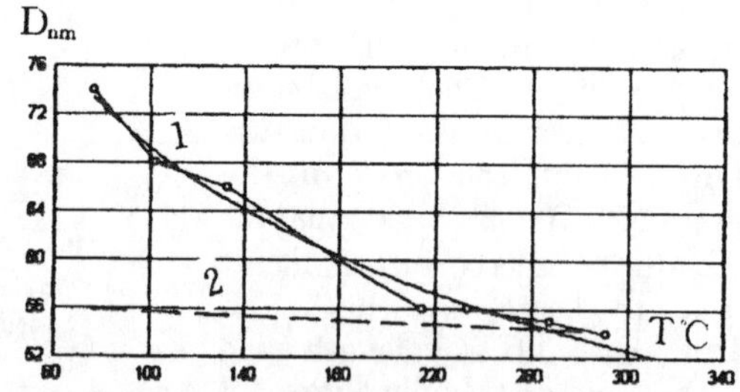

Fig. 1: Effect of heat temperature (T) on thickness (D) for TiO_2 sol—gel derived coatings (1)non—annealed (2)annealed at 300℃

2>Refractive index value (n) increased linearly with increasing heat temperature (T). Shown in fig. 2

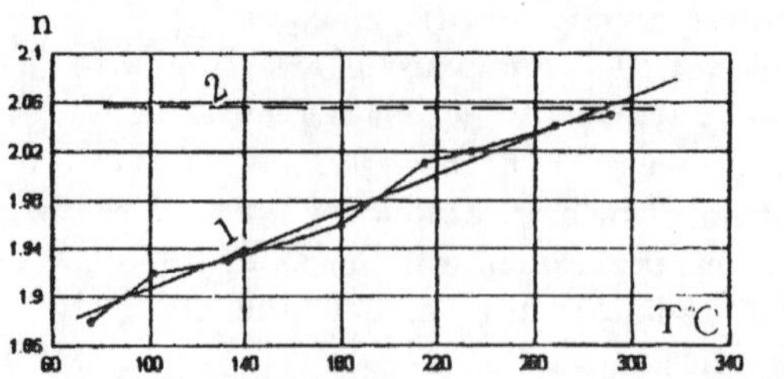

Fig. 2: Effect of heat temperature(T) on referactive index (n) for TiO_2 sol—gel derived coatings (1)non—annealed (2)annealed at 300℃

When it comes to SiO_2 sol—gel derived coatings, We observed that in the same way just as well as TiO_2 and found that to this kind of coatings, the influence of temperature is very faint.

3.2 **The spectrum movement of halogen sunlight with multi-layer coatings**

Measured the spectrum of halogen sunlight both non-coated and coated. we got the relative movement of spectrum ($\Delta\lambda$) at different filament temperature which caused the change of coatings' optical properties (fig. 3) and the movement of transmissivity spectrum ($\Delta\lambda$) at different coatings' temperature (fig. 4). Both of two movement reduced linearly with increasing temperature. The above phenomenon indicate that optical properties of coatings changed with temperature. [2]

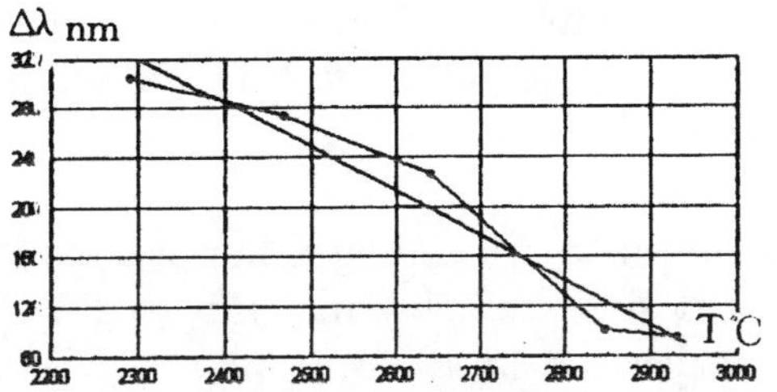

Fig. 3: Relative movement of spectrum($\Delta\lambda$) caused by the faliment temperture (T) which result in the change of coatings' optical properties

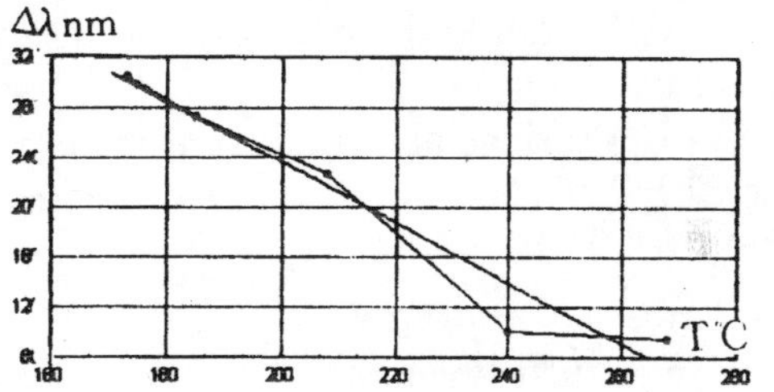

Fig. 4: The movement of transmissivity spectrum ($\Delta\lambda$) caused by the change of coatings' temperature(T).

4. DISCUSSION

We have talked over that thickness and refractive index of TiO_2 sol-gel derived coatings varied along with increasing heat temperature, but after annealing treatment at certain temperature, both of thickness and refractive index of coatings almost remain at annealing state if temperature is lower than the annealings (See fig. 1 and fig. 2). Otherwise, they will changed again. In order to keep the stability of spectrum, the coatings of lamp should be annealed. [3]

5. ACKNOWLEDGMENT

We thank to the educational committee of Shaanxi province for their subsidization.

6. REFERENCES

[1] Hiroshi Hirashima, Takao Kusaka, Structure of TiO_2 sol-gel coatings, Proc. SPIE, **1758**: 67-76 (1992).

[2] L. L. Hench, G. P. LaTorre, S. Donovan, J. Marotta, E. Valliere Properties of gel-sillica optical matrices with 4.5nm and 9.0nm pores, Proc. SPIE **1758** 94-104 (1992).

[3] Rung-Ywan Tsai, Fang-Chong Ho, Mu-Yi Hua, Annealing effectson the properties of indiumtin oxide films coated on soda lime glasses with a barrier layer of TiO_2-SiO_2 composite films, Opt. Eng, **36**(8): 2335-2340 (1997).

B 07

Deposition of Multi-layer for Halogen Lamps by LPCVD

Shinji Suzuki, Toshiyasu Kojima, Seishin Shimaoka
Technology Center, Lighting Group, Matsushita Electronics Corp.
1-1 Saiwaicho, Takatsuki, Osaka 569-1143, Japan

1. Introduction.

Luminous efficacy of halogen lamps can be improved by coating multi-layer on the outside of glass bulb, because infrared radiation reflected by multi-layer heat the filament located on the focus of glass bulb again. Furthermore, it can control the chromaticity of halogen lamps, so that color temperature of lamps can be varied. The multi-layer consists of alternative high and low refractive index. The low pressure chemical vapor deposition(LPCVD) process is adapted in order to coat the multi-layer, because it can apply also on base glass with complex shape. The material with high and low refractive index have to coat at same reactive temperature.

This paper describes about the materials and the source material to form multi-layer by LPCVD.

2. Experiment.

2-1 Coating multi-layer.

Titan dioxide (TiO_2), tantalum pentaoxide (Ta_2O_5), and silicone dioxide (SiO_2) is studied as materials to form multi-layer, because of having suitable refractive index and heat resistance more than temperature of halogen lamp walls. The source materials are indispensable in order to coat these materials by LPCVD. The condition required to these source materials is as following. 1)To be vapor easily, it is desiable to have the vapor pressure more than 10mmHg at 150℃. 2)To have decomposition temperature less than 600℃ which is heat resistance of pinch seals of halogen lamps. 3)To be Safety. 4)To be stable at vaporize temperature.

So we studied the following method, 1)SiO_2 reacted from tetra ethoxy silane(TEOS) and oxide, 2)SiO_2 reacted from TEOS and ozone, 3)SiO_2 by thermal decomposition of diacetoxy ditertiary butoxy silane (DADBS), 4)TiO_2 by thermal decomposition of tetra isopropyl titanate(TiPT)、 and 5)Ta_2O_5 by thermal decomposition of penta ethoxy

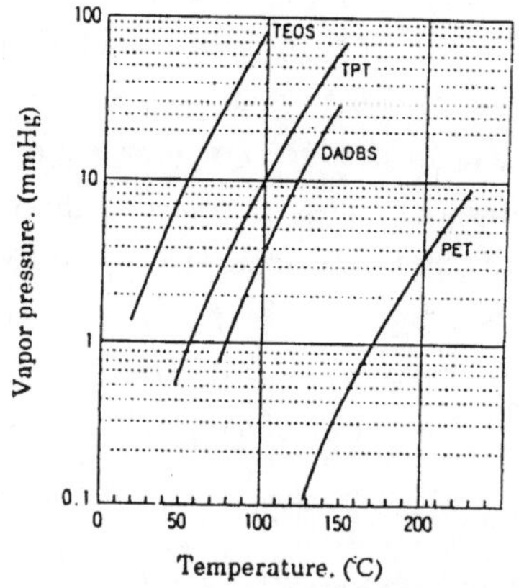

Fig.1 Vapor pressure of source materials.

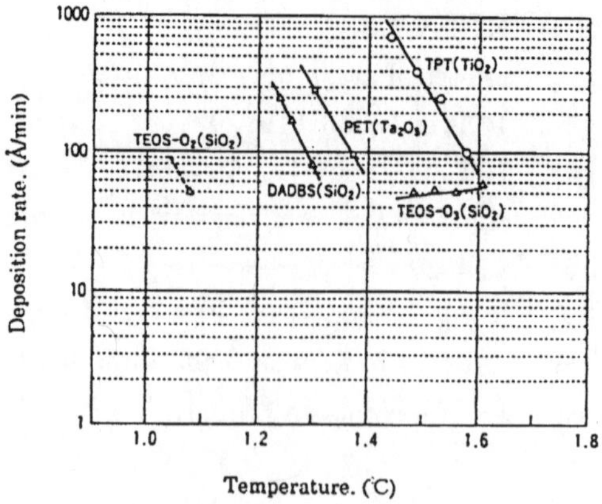

Fig.2 Deposition rate of thin films.

tantalate (PET).
In fig.1 vapor pressure of these source materials is shown as function of temperature centigrade. The vapor pressure of PET is very low so that carrier gas has to be flown.

2-2 Results and discuss.
In fig.2 deposition rate is shown as function of reactive temperature with various method.

2-2-1 SiO_2 layers from TEOS and oxide.
Reactive temperature of this method is about 700℃, so that halogen lamps may be burst at this temperature range. This method is not accepted.

2-2-2 SiO_2 layers from TEOS and ozone.
The slant of this system's line in fig.2 is negative, this is own to that ozone trend to resolve with rising temperature. The fault of this system is that deposition rate is slow.

2-2-3 SiO_2 layers from DADBS
Although a high purity DADBS is difficult to obtain, this method has superior deposition rate and the reactive temperature than the other method to form SiO_2 layers.

2-2-4 TiO_2 layers from TiPT.
The layers have anatase crystal structure and faintly whiten. The refractive index of this layer is 2.35 at 1000nm of wave length.

2-2-5 Ta_2O_5 layers from PET.
The layers have initially amorphous crystal structure, but after annealing more than 600℃,these layers transit β-Ta_2O_5 crystal. The refractive index of the layers are 2.2 at 1000nm of wave length.

3. Conclusions

The most suitable combination of the layers under same reaction temperature and which give high and low refractive index is as follows.
1)TiO_2 layers from TiPT and SiO_2 layers from TEOS and ozone.
2)Ta_2O_5 layers from PET and SiO_2 layers from DADBS.
Though deposition rate of the system for no.1 is small, it can be applied to filter with thin layer ,e.g. yellow filter using motor car.
Infra red reflective layers with 19 layers are coated by the system for no.2, so that luminous efficacy of single ended halogen lamp has been increased up to 60%.

4. References

[1] M Okada et al : Characteristics of tantalum oxide thin film by CVD and by sputtering. Electric Chemistry Vol.53,No.2,p109-113(1985)
[2] K Fujii et al : Silicon dioxide deposition by atmospheric pressure and low temperature CVD using TEOS and Ozone. L. Electrochem. Soc.37, No.9, p2883(1990).
[3] Chemical Industry editing. : CVD handbook. P243(1991).

B 08

INVESTIGATION OF HfN COATINGS ON TUNGSTEN FOR INCANDESCENT LAMP EFFICACY IMPROVEMENT

Laurence Bigio
General Electric Corporate Research and Development Center
1 Research Circle, Niskayuna, N.Y. 12309 U.S.A.

1 INTRODUCTION

The goal of this work was to investigate the use of thin hafnium nitride (HfN) coatings on the tungsten filaments of standard incandescent (GLS or A-line) lamps with the goal of improving efficacy by 10-15% with no loss in lamp life and very little added cost. Translated to ~10% energy savings for equal lumen output, this would have a large impact on total energy usage in the US or Europe simply due to the large number of such lamps in service. Coatings were made using chemical vapor deposition techniques [1].

2 TECHNICAL DETAILS

2.1 Spectral and Total Emissivity of HfN

The spectral and total emissivity of HfN was measured in the temperature range 2600-2800 K. Spectral emissivity curves were determined by comparing the spectral emission of HfN coated on one side of a flat tungsten ribbon with that of tungsten operating at equal thermodynamic temperature from the other side of the flat ribbon. Figure 1 shows the spectral emissivity results relative to tungsten. The desire for an incandescent lamp material would be to have a higher emissivity in the visible and a lower emissivity in the infrared region compared to tungsten. As can be seen in the figure, the selectivity (ratio of visible to infrared emissivity) is reduced as the temperature increases.

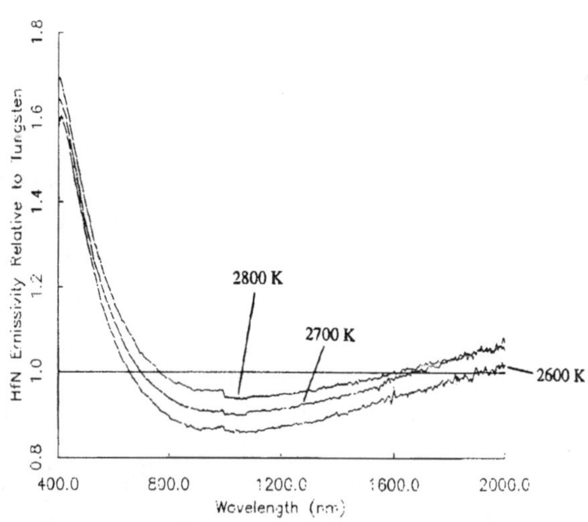

Fig. 1: HfN emissivity relative to tungsten

2.2 Total Emissivity of HfN

Total emissivities ranging from 0.290-0.314 for the temperature range 2600-2900 were determined by comparing total powers required to reach known equal temperatures for plain tungsten and for HfN-coated tungsten ribbons (2-sided). The temperatures were measured using an optical pyrometer at 615 nm, knowing the spectral emissivities for both tungsten and HfN (see above) at this wavelength. The results are given in Table 1 which also lists for comparison the values derived from a visual extrapolation of data by Marple [2].

Table 1: Total emissivity of HfN and tungsten

Temperature (K)	Measured* ε_{tot} for HfN	ε_{tot} derived from Marple spectral data	ε_{tot} for tungsten
2600	.290	.275 (5.2% off)	.311
2700	.305	.285 (6.6% off)	.318
2800	.306	.294 (3.9% off)	.323
2900	.314	.304 (3.2% off)	.329

* Estimated uncertainties ~3.5%

2.3 Vaporization Rates for HfN

Vaporization rates for HfN were measured separately in the presence of 600 torr (at room temperature) of the following gases: pure nitrogen, a 50/50 mix of nitrogen and krypton, a 25/75 mix of nitrogen and argon, and a 5/95 mix of nitrogen and argon. This was done by operating HfN-coated ribbons as well as HfN-coated coiled filaments in the temperature range 2600 - 2800 K in sealed bulbs, and using a combination of methods to determine the vaporization rates including weight-loss, observation of resistance-rise, measurements of Hf on the bulb walls, and observations of the coated filaments themselves. Values ranged from 5.4×10^{-10} g/cm^{2}sec for HfN in pure N_2 at 2600 K to 2.8×10^{-8} g/cm^{2}sec for the 5/95 N_2/Ar mix at 2800 K. This translates to coating thickness loss rates ranging from 1.4-73.1 microns/1000 hours, respectively.

2.4 Efficacy and Color Temperature for HfN-Coated Lamps

Single coil 34V filament lamps were made with and without HfN coatings on the tungsten filament. The lamp efficacies and correlated color temperatures (CCT) were measured and compared in the range from 2600-2800 K. Table 2 shows the percent increase in efficacy and CCT for the HfN-coated lamps compared to the uncoated lamps.

Table 2: Efficacy and CCT results for 34V filament lamps

Temperature (K)	% Increase in Efficacy	Increase in CCT (K)
2600	15.4	169
2650	14.3	169
2700	12.5	164
2750	10.8	159
2800	10.0	154

These results showing a decreased efficacy improvement for higher temperatures are consistent with the reduced emissive selectivity for HfN discussed above.

3 CONCLUSIONS

It is believed that the demonstrations described here represent the first time a CVD-coated tungsten filament showing an efficacy increase has been made into a sealed lamp and burned successfully for several hundred hours. Unfortunately, from a mechanical viewpoint, given the wire resistance and diameter needed for 120V operation, it was determined that a maximum coating thickness that could be tolerated is ~5 microns, with 3 microns being a more reasonable upper bound. Thicker coatings tended to leave the filament too brittle for practical use. Therefore, given the vaporization rate mentioned above, a 3-5 micron coating would only last ~300-500 hours, too short for the desired life goal of ~1000 hours.

4 ACKNOWLEDGEMENTS

This work was supported by funding from the New York State Energy Research and Development Authority, the Empire State Electric Energy Research Corporation, Long Island Lighting Company, New York State Electric and Gas, and the Electric Power Research Institute.

5 REFERENCES

[1] L. Bigio, C. Briant, R. Wilson, J. Ackerman, P. Meschter, P. Kosky, M. Gigliotti, H. Witting, and J. Myers, Final Report to the New York State Energy Research and Development Authority for contract No. 3034-IABR-BR-94.

[2] Marple, D.T.F., GE Memo Report P-123, 1957.

HID Lamps

C 01

Electrical Field Strength and Electrode Voltage Drops of an ac Mercury High-Pressure Discharge

Heinz Schöpp, Gottfried Hartel, Helmut Hess, and Lothar Hitzschke*
Institute of Low-Temperature Plasma Physics, Robert-Blum-Str. 8-10,
17489 Greifswald, Germany
* OSRAM GmbH, Hellabrunner Str. 1, 81536 München, Germany

1 INTRODUCTION

Mercury is the basic system of a group of light sources. This type of discharge is of high practical importance and often used for assessing the progress obtained by changing discharge parameters. In the process of developing a method for the extrapolation of the temperature [1] using the field strengths, there were found some inconsistencies in assumptions about the electrode sheaths drop voltage and the column voltage. The aim of this work was to get reliable electrical field strengths.

2 METHOD AND EXPERIMENTAL SET-UP

The determination of field strengths E is possible for discharges where only the length of the positive column l is varied (e.g. [2]). With the assumption that the electrode drops are additive ($U_{(C+A)}$), the time-dependent total voltage U of the discharges have been measured for i different lengths of the positive column

$$U_i\,(t) = U_{(C+A)\,i}\,(t) + E\,(t)\,l_i \tag{1}$$

By linear regression $E\,(t)$ can be calculated.

Table 1 Datas for used lamps and ballasts

lamp No.	electrode separation [mm]	volume [ml]	Hg-amount [mg]	induct. ballast [mH]	input-power [W]
1	27.8±0.1	5.5±0.1	26.0	171	242
2	27.5±0.1	5.3±0.1	25.5	171	240
3	40.9±0.1	7,8±0.2	38.0	256	360
4	41.1±0.1	7.5±0.2	36.5	256	363
5	55.6±0.1	10.3±0.2	49.8	319	473
6	55.7±0.1	10.5±0.2	51.1	319	468
7	83.0±0.1	16.1±0.2	77.8	467	694
8	82.7±0.1	15.7±0.1	76.3	467	677

The essential point for the determination of the field strengths is the similarity of lamps with only different interelectrode distances [3]. This requires lamps with high demands on the design, geometry and electrical parameters. The lamps must have the same properties as, e.g. radial temperature distribution, amount of mercury per volume, volume per electrode region, electrode falls and arc contraction regions. To reduce uncertainties between the different interelectrode distances two lamps were used for each l. The same discharge currents are obtained by using different inductive ballasts in series. Table 1 gives the parameters and the ballasts for the lamps with an outer diameter of 18.6 mm.

The lamps are vertically operated on an ac power supply. The spectral properties of the discharges are checked concerning corresponding radial temperature profiles and the line profile of the Hg I 546 nm. The radial temperature profiles are determined from measurements of the absolute side-on intensity of optically thin Hg I lines (577/579 nm). The emission coefficient as a function of radius has been obtained using an Abel inversion (cf. [4]). The pressure was determined with the hook method [5]. The axis temperatures of all lamps are identical within an error of less than ±5% (6070 K - 6140 K, 6 bar). Also, the measured radiance of the Hg I 546 nm lines and their profiles are identical within the limits of the CCD array. The arc contraction regions are controlled by a video-camera in current maximum. As a result, only inessential differences were observed. Figure 1 gives the electrode regions for different *l* (28, 56, 83 mm) in current maximum for comparison of arc diameters and electrode regions. The time-resolved discharge currents and voltage characteristics are monitored with an oscillograph. Figure 2 shows time resolved currents and total voltages for the different lengths *l*.

3 RESULTS

Executing a linear regression according to (1) for the total voltages, a linear behaviour is obtained which is shown for the current maximum time in Figure 3. The derivative is delivering *E*. The regression makes it possible to determine electrode drops $U_{(C+A)}$ too. In Figure 4 a reignition peak can be observed. After this peak a remarkable minimum occurs. The time-averaged voltage of 16.5 V well agrees with 15 V given by [3]. Nevertheless, the approach to treat $U_{(C+A)}$ time-independent can only be considered as a crude approximation.

Figure 1: Arc diameter and electrode region in current maximum

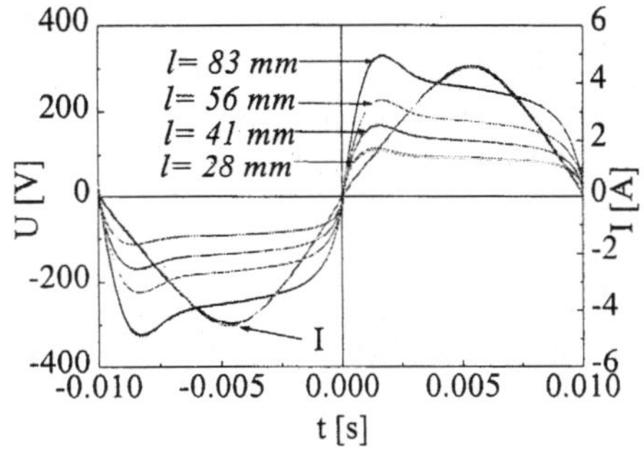

Figure 2: Total voltage and current for different electrode distances

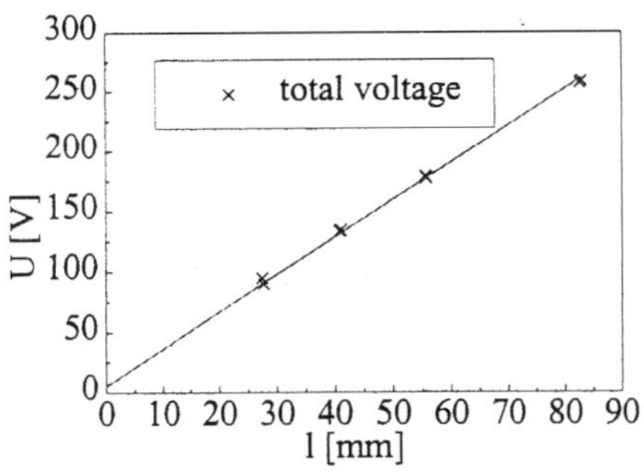

Figure 3: Total voltage at current maximum

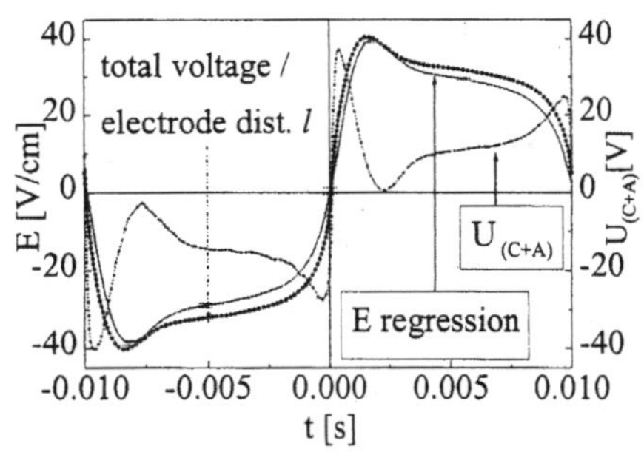

Figure 4: Field strength dependence on time

4 ACKNOWLEDGEMENTS

This study was supported by the BMBF.

5 REFERENCES

[1] G. Hartel, H. Schöpp, J. Phys. D: Appl. Phys. 29 (1996) 2881-2884

[2] P. L. Denbigh, D. O. Wharmby, Lighting Res. & Techn., Vol. 8, 3, 1976, 141-145

[3] W.Elenbaas, The High Pressure Mercury Vapour Discharge, N.-Holland Pub. Comp., Amsterdam, 1951

[4] R J Zollweg, J J Lowke, R W Liebermann J.Appl. Phys. Vol.46, No.2, 1975 S.650-659

[5] E. Kindel, C. Schimke, M. Kettlitz, H. Schöpp, J. Phys. D: Appl. Phys., 1998 in print

C 02

PERFECT COLOUR RENDITION FROM A PULSED HIGH PRESSURE CESIUM LAMP

Jianping Liu*), Klaus Günther*), Heinrich Kaase), Felix Serick**)**

*) OSRAM GmbH, Nonnendammallee 44-61, D-10625 Berlin
**) Institut für Elektronik und Lichttechnik der TU Berlin, Einsteinufer 19, D-13587 Berlin

1. Introduction

A perfect colour rendition CRI of a light source at a given colour temperature CCT requires a spectral distribution which is proportional to the shape of the black body radiation of this temperature in the visible spectral range. This may be realized by the radiation of solid surfaces in incandescent lamps for a CCT up to 3000 K. The main disadvantage of this is the poor efficacy. At higher temperatures, the use of cluster radiation with a very smooth spectral distribution [1] was proposed. However, the development of such systems is complicated due to the necessity that the discharge lamp is operated electrodeless using a microwave excitation. The radiation distribution of metal halide lamps is typically characterized by an optimized pseudocontinuum of spectral lines showing gaps and pronounced maxima which lead to colour distortions of the illuminated objects.

2. Cesium discharges

The use of cesium high pressure discharges as light sources was first proposed by FINKELNBURG and PETERS [2] because of its favourable spectral distribution essentially consisting of a smooth recombination continuum in the visible. Thus, the very first proposed HPS lamp used a high pressure cesium discharge in fact [3]. Unfortunately, the claimed efficacy of more than 40 LPW could by far not be realized because of the energy consuming intense resonance lines in the IR at 852 nm and 894 nm (Fig. 1).

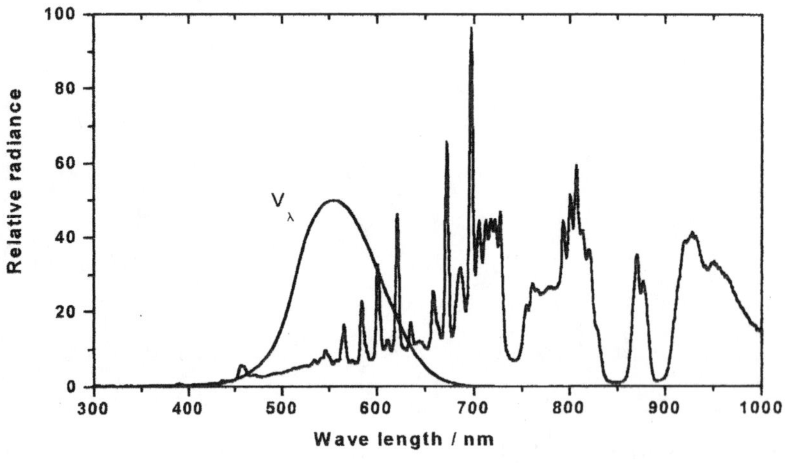

Fig. 1. Spectral distribution of a stationary high pressure cesium discharge at 80 W (CCT = 2 700 K, CRI = 98, efficacy = 8.0 LPW). V_λ indicates the eye efficiency curve.

3. Pulsed operation

To increase the efficacy it is necessary to raise the continuous radiation compared to the lines. As an estimate, we can use the following expression for the relation of continuous and line radiation $\varepsilon_{cont} / \varepsilon_{line}$ in a thermal plasma:

$$\frac{\varepsilon_{cont}}{\varepsilon_{line}} \propto T \cdot \exp\left(-\frac{\chi - E_a}{kT}\right)$$

where T is the plasma temperature, χ the ionisation energy, E_a a characteristic excitation energy of the lines and k Boltzmann's constant. This relation strongly increases with T which can be raised by means of a boosted energy input. To avoid an overload of the arc tube this must be accomplished by a pulse mode operation as used in pulsed HPS lamps (see e.g. [4]). It should be noted that the effect of pulsing a cesium lamp is completely different from the one in sodium: While a HPS discharge shows an increasing CCT at nearly constant efficacy, the efficacy of a high pressure cesium lamp increases considerably at nearly stable CCT.

A second positive effect of pulsing the cesium discharge is a further rise of the optically thin continuum in the visible relative to the optically thick IR resonance lines which only show an increase of the self absorption.

4. Results

Fig. 2 shows the spectrum of the same lamp in the pulsed mode operation at the same average electric power input: In the visible region the emission follows the Planck distribution at T = 3850 K almost perfectly resulting in a CRI = 98.5 [5] . Note the strongly suppressed resonance lines compared to the stationary spectrum in Fig. 1.

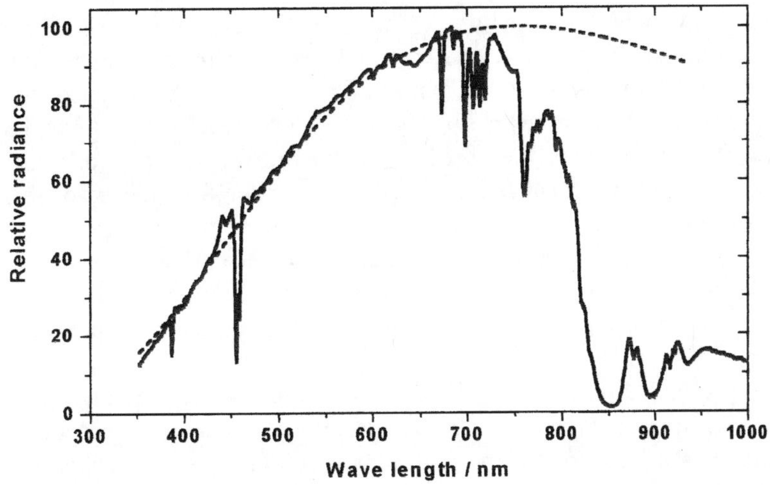

Fig. 2. Spectral distribution of the pulsed cesium lamp at an average power of 80 W resulting in an efficacy of 42 LPW. Dashed line: Planck distribution at T = 3850 K.

5. References

[1] R. SCHOLL and G. NATOUR, Invited papers of XXII ICPIG, AIP Press Woodbury New York 373-82, 1995

[2] FINKELNBURG and PETERS, Handb. Physik XXVIII, Springer 1957

[3] K. SCHMIDT, US Patent 2971 110, 1960

[4] K. Günther, Proc. LS:7, Kyoto 1995, p.93

[5] J.Liu, Thesis, Technical University Berlin 1998.

C 03

Radiation Output of an AC High-Pressure Mercury Vapor Discharge Considering Radiation Transport

Gottfried Hartel, Heinz Schöpp, Helmut Hess, and Lothar Hitzschke*
Institute of Low-Temperature Plasma Physics, Robert-Blum-Str. 8-10,
17489 Greifswald, Germany
* OSRAM GmbH, Hellabrunner Str. 1, 81536 München, Germany

1 INTRODUCTION

The radiation output of high-pressure discharges is strongly coupled with the radiation transport of optically thick lines. Comparison of modelling and experiments in mercury results in better understanding of the basic system of many applications in general lighting.

2 MODEL

The power-balance equation and the radiation-transport problem for a cylindrically symmetric high-pressure mercury arc are simultaneously solved without any additional approximation for the radiation. Axial variations as well as convection have not been taken into account. The side-on intensities of the different mercury spectral lines in the visible and the near uv have been calculated and compared with experimental values which are absolutely measured on 240 W high-pressure mercury lamps. The starting point was the current through the lamp which can be measured with high accuracy.

To ensure the reliability of the model, also the calculated and measured temperature profiles were compared. The pressure and the temperature distribution have been simultaneously determined from side-on radiance measurements of Hg I 577/579 nm lines [1], using assumptions about the heat flux to the wall, measured values of the wall temperature, the current through the discharge and the electrical field strength. Further the dependence of the electrical conductivity on temperature and pressure is based on cross section data from [2]. The electrical conductivity strongly depends upon the temperature; since small changes of the assumed pressure in the arc always results in changes of the corresponding temperature, the measured current and field strength could be used to determine quite accurately the pressure via Ohm's law.

For a complete understanding of radiation transfer reliable sets of transition probabilities and line broadening constants are necessary. Values given by different authors have been compared [3]. The influence of the radiation on the power balance could be determined by calculation of the net-emission coefficient.

3 RESULTS AND DISCUSSION

The verification of the measured side-on intensities by model calculation is shown in Figures 1 and 2. The total spectral line intensities and distributions agree with calculations within the limits given by the error of the temperature measurement. Deviations are mostly smaller than 10%.

In Figure 3, the radial distribution of the temperature is shown. Experimental curves are determined in an ac discharge at the time of maximum current and current zero from the centre (r = 0) up to r = 0.4 cm by a spectroscopic method. The results of the calculations for current maximum and current zero well agree with the temperature distributions derived from the experiments. A dc arc with the maximum electrical current shows a temperature which is lower in the axis and higher in outer parts of the cross section (I (dc)). In Figure 4 the line contributions to the side-on intensities during current maximum are shown. The whole intensity is emitted in nearly equal parts by

- the resonance lines 185/254 nm,
- the other optically thick lines and
- the remaining optically thin lines.

For understanding the power balance mechanism, the knowledge of resonance line parameters is very important. Here essential success could be achieved.

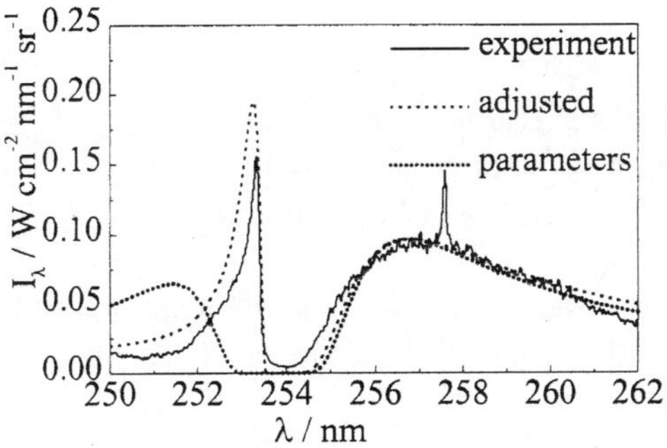

Figure: 1 Comparison of measured and calculated resonance line, parameters taken from [4]

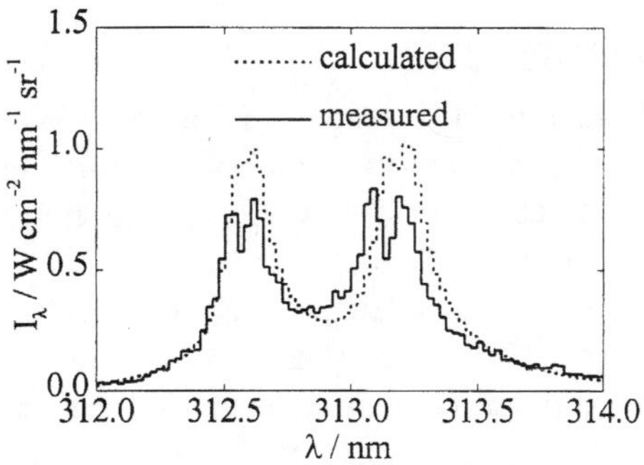

Figure: 2 Comparison of measured and calculated optically thick lines, broadening constants c_R , c_W from [5], c_S adjusted

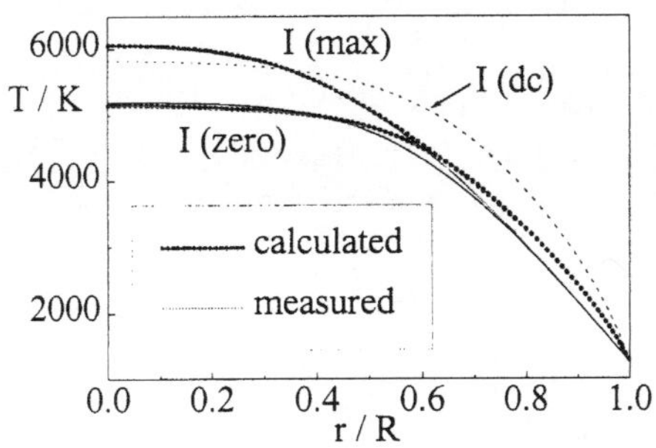

Figure: 3 Radial temperature distribution measured and calculated for current maximum and zero

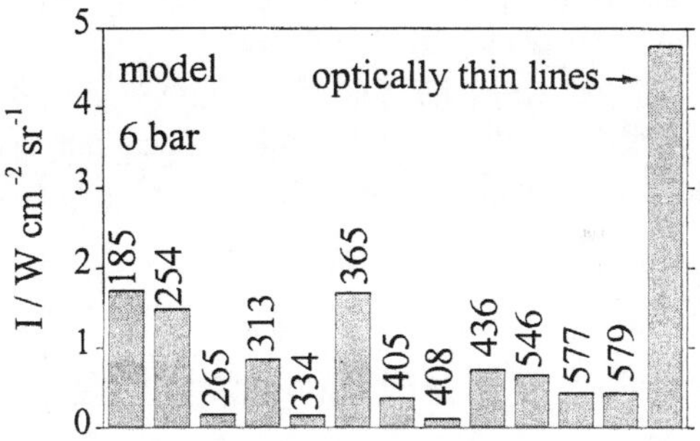

Figure: 4 Comparison of radiation output of different lines

4 ACKNOWLEDGEMENTS

This study was supported by the BMBF.

5 REFERENCES

[1] G. Hartel, H. Schöpp, J. Phys. D: Appl. Phys. **29** (1996) 2881-2884
[2] J.P. England, M.T. Elford, Aust. J. Phys. **44** (1991) 647
[3] H.-P. Stormberg, R. Schäfer, J. Appl. Phys. **54** (1983) 4338-4347
[4] R. Schäfer, H.-P. Stormberg, J. Appl. Phys. **60** (1986) 1263-1268
[5] R.W. Liebermann, pers. communication 1995

C 04

IGNITION PROCESS OF HPS LAMPS WITH HIGH XENON PRESSURE

A. Lembcke, K. Günther
OSRAM GmbH, Nonnendammallee 44-61, D 13625 Berlin

1 INTRODUCTION

The performance of HPS Lamps can be improved using a higher Xenon pressure in the arc tube. Because this requires a higher ignition voltage which cannot be provided without difficulties it is necessary to find effective ignition methods. Therefore a better understanding of the basic processes is needed.

In contrast to metal halide lamps the arc in HPS lamps is wall stabilized. Especially the ignition process is dominated by the influence of the tube walls on the distribution of the electric field strength and on the production or annihilation of charge carriers.

2 MEASUREMENTS

2.1 Investigation in the ignition process of HPS lamps

The typical HPS lamp consists of an arc tube made of PCA (see Fig. 1), two electrodes reaching into the arc tube, and an external metallic ignition aid. The ignition aid is usually contacted to one of the electrodes, but can be also „floating„, i.e. not contacted. This assembly is contained in an outer bulb made of glass.

The lamp is ignited with the help of a high voltage pulse which is applied to one of the electrodes. It is an inductively coupled decaying oscillating voltage of some 500 kHz superimposed by a signal of about 5 Mhz originating from stray capacities and inductances.

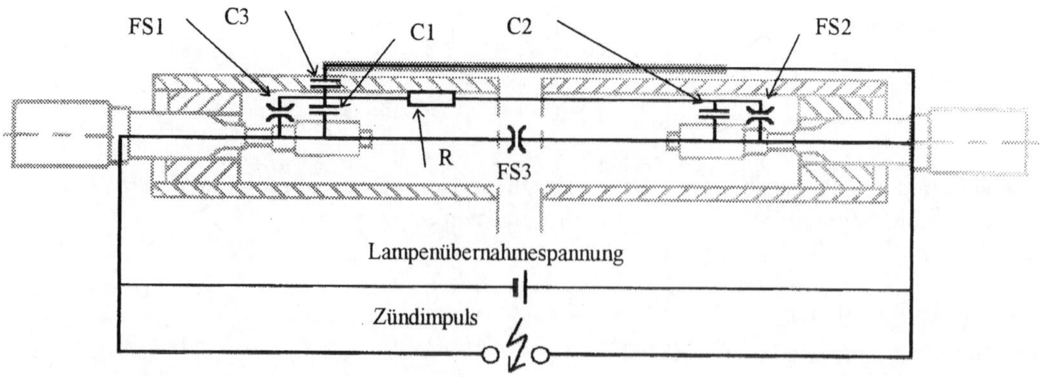

Fig. 1: Typical HPS arc tube including an equivalent electric circuit for the interpretation of the measurements.

To examine the events during breakdown high resolution digital oscilloscopes were used to record the voltage waveforms and the current response of the lamp. The interpretation of these data was confirmed by High speed frame and Streak records of the breakdown and streamer processes with a time resolution of 10 ns in the arc tube made from sapphire wall material. A photomultiplier signal was used to investigate the sequence of the consecutive phases of the ignition process.

2.2 Steps of ignition [1]

The following steps should be distinguished during the process from the insulating discharge tube to the establishment of a conducting plasma between the electrodes:

1. The first step in the breakdown process is a dielectric barrier discharge between one of the electrodes and the ignition aid initiated by the electrical field between the electrode and the ignition aid resulting in a surface charge on the inner surface of the tube wall below the ignition aid.
2. Next a streamer [2,3] is developed along the arc tube wall towards the second electrode supported by the electric field of the ignition aid.
3. A discharge takes place from the inner tube wall surface to the second electrode; i.e. an electrical connection between the two electrodes is completed.
4. Triggered by the free electrons of the discharge along the wall a discharge with a much smaller impedance takes place between the electrodes and short circuits the discharge along the wall.

3 CONCLUSIONS

- Different steps of the breakdown process in HPS lamps were found and the model could be sustained by some experimental proof.
- The discharge was modelled in an electrical circuit and some of the values were estimated in experiments:
  - the capacity of the streamer to the ignition aid is 0.3 pF per millimeter of streamer length and the width of the surface charge distribution is 1.6 mm at 0.6 mm ceramic wall thickness and 2.7 kV ignition voltage.
  - the capacity increases proportionally to the ignition voltage and can be estimated as

$$C \approx 0.1 \frac{pF}{mm \cdot kV} \cdot U \cdot l$$

- Some actual HPS lamp designs and experimental results were examined using this model and the results explained.
- In case of a noncontacted ignition aid the high voltage is applied between the electrodes of the lamp, and the potential of the ignition aid is determined by a capacitive voltage divider consisting of C_1 and C_2 (see Fig. 1). As a result, the breakdown voltage increases by more than 500 V compared to the contacted ignition aid. In this case the existing limits for the ignition voltage can be only maintained by a lowering of the Xenon fill pressure, of course, with unfortunate consequences for efficacy and maintenance.

4 REFERENCES

[1] A. Lembcke, *Zündung von Natrium-Hochdrucklampen*, Diplomarbeit, Technische Universität Berlin, 1997

[2] Gallimberti et al, *Proc. 7th Int. Symp. On High Voltage Engineering.*, Dresden, Paper 41.10., 1991

[3] N. Allen, Private information, April 1997

C 05

A FURTHER STEP TO A COMPLETE DESCRIPTION OF THE HIGH-PRESSURE DISCHARGE IN MERCURY

Helmut Hess[1], Gottfried Hartel[1], Heinz Schöpp[1], and Lothar Hitzschke[2]
[1]Institut für Niedertemperatur-Plasmaphysik, Robert-Blum-Str. 8 - 10, D-17489Greifswald
[2]OSRAM GmbH, Hellabrunner Str. 1, D-81536 München

1 INTRODUCTION

The high-pressure discharge in mercury is the basic system for a group of light sources of high practical importance. Besides the discharge in pure mercury, there is a variety of additives which are applied e.g. in metal-halide lamps for improving the properties of the emitted light. Nevertheless, the mercury discharge is still used as a benchmark system for assessing the progress by changing discharge parameters.

The last attempts to describe the high-pressure mercury discharge were made about 15 years ago [1,2]. At that time, some approximations were used which restrict the general applicability of the obtained results. For the electrical circuit, Kirchhoff's rule was used with an artificial (constant) term taking into account the voltage drop in the electrode regions. For the radial temperature distribution, parabolic profiles were considered only, and a comparison between experiment and calculation was made for the temperature profiles in the dc case only whereas in the ac case only the axis temperatures were compared.

The improvements in the treatment of the high-pressure mercury discharge consists in the following: Instead of Kirchhoff's law the measured arc current is used to start the calculations. The radial temperature distribution and its development during an ac cycle is calculated and compared with corresponding measurements. The electrical field strength is calculated, and by comparison with the measured one conclusions about the reliability of the electrical conductivity used are possible.

2 EXPERIMENT

The measurements have been carried out on 240 W quartz lamps with an inner diameter of 1.6 cm. For the radiation measurements, a combination of a spectrograph, a multi-channel plate (MCP), and a CCD array has been used. The side-on radiance of the optically thin lines Hg I 577 and 579 nm has absolutely been determined, and the radial temperature distribution has been derived after an Abel inversion. The pressure in the discharge has simultaneously been determined with the temperature by an iteration procedure via Ohm's law using the electrical conductivity.

The electrical field strength as well as the electrode sheaths voltage drop were determined using lamps which differ in the electrode distance only. The voltage was measured at identical currents, and from a plot voltage versus electrode distance the field strength can be determined from the slope and the electrode sheaths voltage drop from the extrapolation to zero length.

3 MODEL

For describing the high-pressure mercury discharge, it is assumed that the input power by Joule's heat is balanced by losses caused by heat conduction and radiation. The model is one-dimensional, i.e. only radial variations are taken into account. The discharge plasma is midway between the optically thin and thick limits for most of the spectral lines. Therefore the radiation transport has to be carefully considered. The model calculations start with the measured arc current and consider the time dependence of the arc development. The measured electrode sheaths voltage drop has to be taken into account to reproduce the measured voltage. Consistent sets of broadening constants and transition probabilities are proposed.

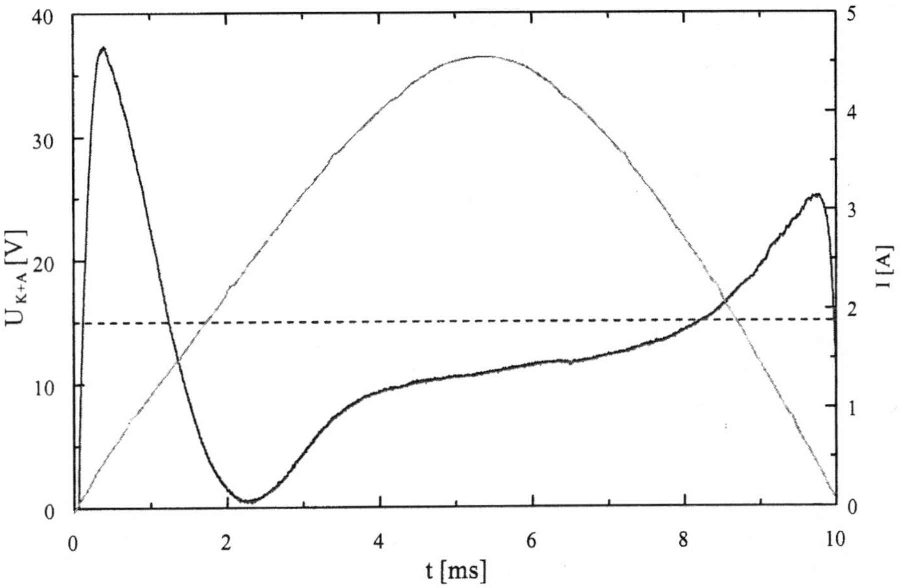

Fig. 1: Electrode sheaths voltage drop U_{K+A} (heavy) during one cycle of a 50 Hz ac discharge; the dashed line corresponds to the often used constant value from Elenbaas. Further the current I (light) is shown.

4 RESULTS

The side-on radiances of the most prominent spectral lines were absolutely measured. Their total intensities as well as the spectral intensity distribution agree within ± 8 per cent with the calculations and are so within the limits given by the errors in the temperature measurement.

An agreement between measured and calculated temperatures could only be reached if the ac character of the discharge is considered. A dc calculation for the same current shows a lower value in the center and higher values (a „shoulder") in the outer region. These differences can obviously be explained by a relaxation time for smoothing out the temperature distribution of the order of one ac period.

The measured electrical field strength could only be reproduced by the calculations if more recent values for the electrical conductivity are used [3]. However, this was not sufficient to reproduce the measured voltage and here especially the so-called reignition peak which occurs after current zero. The electrode sheaths voltage drop proved to be the „missing link" between the measured voltage and the calculated field strength. As can be seen in Fig. 1, where the measured voltage drop is shown, it has an impressive time dependence.

Finally, the pressure in the discharge could be determined by a measurement of the total current and the electrical field strength via Ohm´s law using the dependence of the chosen electrical conductivity from temperature and pressure.

5 REFERENCES

[1] J.J. Lowke, R.J. Zollweg, and R.W. Liebermann, J. Appl. Phys. **46**, 650 (1975)
[2] H.P. Stormberg and R. Schäfer, J. Appl. Phys. **54**, 4338 (1983)
[3] J.P. England and M.T. Elford, Austr. J. Phys. **44**, 647 (1991)

C 06

INVESTIGATION OF THE CATHODE PLASMA SHEATH IN HIGH-PRESSURE SODIUM DISCHARGES

Manfred Kettlitz and Ralf Wendt
Institut für Niedertemperatur-Plasmaphysik e.V. an der Ernst-Moritz-Arndt-Universität Greifswald, Robert-Blum-Straße 8-10, D-17489 Greifswald, Germany

1 INTRODUCTION

High pressure sodium discharges are very interesting for lighting because of their high efficacy. A main problem of the efficacy and life time of such lamps are energy losses at the electrodes. The processes in the plasma sheath in front of the electrodes are responsible for heating and erosion of the electrodes. To explain the energy transport to the cathode knowledge of the dimension of that area, temperature profiles and particle number densities is necessary. The aim of the investigation is a better understanding of the processes in the transition zone between the arc and the cathode in high-pressure sodium discharges.

2 EXPERIMENT

The pure sodium arcs are ignited in sapphire discharge tubes with an inner diameter of 4.6 mm and a length of about 70 mm. The tungsten electrodes are mounted in niobium caps which are sealed to the ends of the sapphire tube [1]. To observe the cathode plasma spectroscopically the cathode spot was fixed to the tip of the electrode by shifting the electrode windings backwards (fig. 1). The discharge tube is filled with 4 mg sodium and 40 mbar argon as ignition gas. The arc discharges are established in vertically mounted lamps within an evacuated envelope with sapphire and ZnSe windows. The discharge was driven with a dc power between 50 and 150 W and the vapour pressure is controlled by the spot with the lowest temperature in the lamp. This spot is in general located in one of the niobium caps.
The temperatures of the sapphire wall, the electrodes, the niobium end caps and the electrical connections are measured with different pyrometers and thermocouples. The plasma temperature is determined spectroscopically.

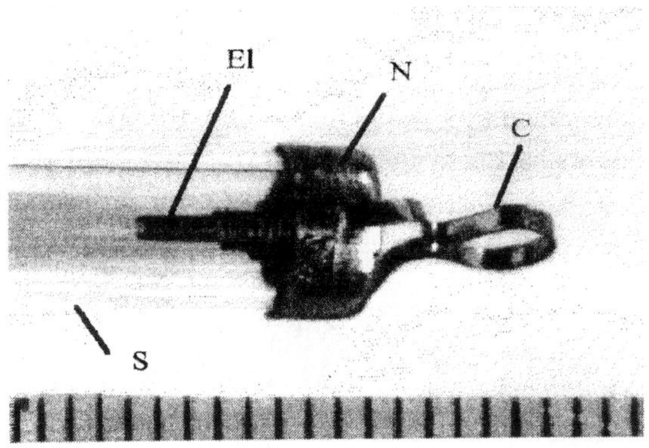

Fig. 1: Sodium arc tube with indicated mm-scale: S - sapphire tube, El - tungsten electrode, N - niobium cap, C - electrical connection.

The spectroscopic measurements are performed in the visible spectral range. The side-on intensity of the discharge is imaged to the entrance slit of a 1-m-spectrograph with a grating of 600 grooves/mm. The data are recorded by means of an intensified CCD camera system. For an absolute radiation calibration a tungsten ribbon lamp is used.
From the absolute side-on radiance the emission coefficient is obtained carrying out an Abel inversion [2]. Knowing the discharge pressure and the atomic data in further processing the radial temperature profile is calculated using the optically thin atomic sodium lines at 498 nm [3]. These results are checked by radiation transport calculations of the Na D lines. Calculated spectral distributions are fitted to measured side-on intensities varying the temperature profile.

3 RESULTS

At an input power of 70 W the pressure in the lamp is 60 kPa, the temperature in the arc centre is 3150 K and that of the wall 930 K. The temperature curves of the plasma for different distances z from the cathode are shown in figure 2. In the arc centre they are around 3150 K from 0.5 to 9 mm to the cathode. A radial contraction of the plasma takes place for distances lower than 2 mm towards the cathode. For larger distances shape and absolute values of the temperature profiles do not vary.

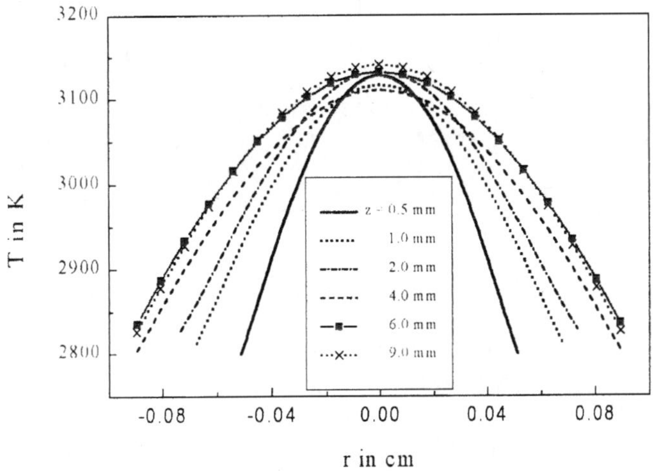

Fig. 2: Temperature profiles at different distances z of the cathode for a sodium discharge of 70 W.

4 ACKNOWLEDGEMENT

This work is supported by the BMBF (FKZ: 13N7109).

5 REFERENCES

[1] H. Schöpp, Vakuum in Forschung und Praxis **4**, 233 (1996).
[2] G. Pretzler, Z. Naturforsch. **46a**, 639 (1991).
[3] J.J. de Groot and J.A.J.M. van Vliet, *The High-Pressure Sodium Lamp* (Kluwer, Deventer, 1986).

C 07

LUMINOUS FLUX MAXIMISATION OF MERCURY-FREE HIGH PRESSURE SODIUM LAMPS

Carlo Vlekken and Rudy Geens
Sylvania Lighting International, Industriepark Soldatenplein 13, B-3300 Tienen, Belgium

1 INTRODUCTION

The development of high pressure xenon technology made it possible to make mercury-free high pressure sodium (HPS) lamps with a luminous flux in between the standard and the high pressure xenon lamp. This is achieved by replacing the mercury pressure present in a conventional sodium/mercury/xenon lamp by a supplemental xenon pressure and by making longer and narrower discharge tubes for operation on standard ballast systems [1]. Since luminous efficacy depends on discharge tube diameter [2], the need for longer and narrower discharge tubes decreases the luminous efficacy. The optimisation of the discharge tube dimensions and xenon pressure for obtaining maximal luminous flux on standard ballast systems was experimentally worked out and will be further described in detail.

2 TECHNICAL DETAILS

2.1 Experimental set-up

Mercury-free HPS lamps are made with discharge tube inner diameters ranging between 3.35 and 4.91 mm, arc lengths ranging between 30 and 100 mm and xenon fill pressures ranging between 200 and 700 mbar.

2.2 Measurement method

For all lamps, the temperature of the discharge tube wall was measured as a function of the lamp input power using the D-line pyrometry method [3]. From these measurements, the lamp input power is calculated to obtain a predetermined wall temperature goal. At this lamp-specific determined lamp input power, i.e. at this predetermined wall temperature goal, the lamp electrical and photometrical characteristics are measured as a function of the sodium D-line reversal width. This is achieved by artificial heating the cold spot of the discharge tube. During this artificial heating, the lamp input power is kept constant by remotely adjusting the line input voltage of the lamp circuit.

2.3 Calculation method

Plasma electric field and luminous plasma are calculated from the experimental data. The measured lamp voltage is the sum of the electrode fall and the discharge voltage. The electrode fall can be calculated from the measured lamp current and electrode rod diameter [4]. The discharge voltage is the product of the electric field strength and the distance between the electrodes. The luminous plasma efficacy is obtained by dividing the measured luminous flux by the power dissipated in the discharge column. The measured lamp input power is the sum of the power dissipated in the discharge column and the electrode power, which is calculated as the electrode fall times the lamp current.

The calculated plasma electric field and luminous plasma efficacy associated with the predetermined wall temperature goal are plotted and fitted as a function of the discharge tube inner diameter, the sodium D-line reversal width and xenon fill pressure. The plasma electric field decreases with increasing discharge tube inner diameter and increasing sodium D-line reversal width. The influence of the xenon fill pressure on plasma electric field is different for small and high values of the sodium D-line reversal width. For small values of the D-line, an increase in xenon fill pressure causes an increase in plasma electric field whereas for high values of the D-line a decrease of the plasma electric field is observed. The luminous plasma efficacy increases with increasing discharge tube inner diameter and increasing xenon fill pressure at fixed sodium D-line reversal width. The influence of the sodium D-line reversal width on the luminous plasma efficacy is rather small with a broad maximum around a reversal width of 11nm.

The lamp-specific wall loading associated with the predetermined wall temperature goal (the wall loading is obtained by dividing the lamp input power by the inner area of the discharge tube wall between the electrodes) is plotted and fitted as a function of the discharge tube inner diameter and xenon fill pressure. The wall loading necessary to obtain the predetermined wall temperature goal increases with increasing discharge tube inner diameter (as a result of a higher discharge tube wall thickness) and increasing xenon fill pressure.

For all lamp wattage types, maximisation calculations of luminous flux are then performed by optimising the discharge tube dimensions and xenon fill pressure. These calculations make use of the obtained dependence of the plasma electric field, luminous plasma efficacy and wall loading on discharge tube inner diameter and xenon fill pressure. For each lamp wattage type, the discharge tube dimensions and xenon fill pressure are calculated so that the luminous flux is maximised taking into account some restrictions on dimensions and xenon fill pressure. The xenon fill pressure has to be chosen so that the lamp ignition characteristics comply with the minimal lamp ignition conditions. The discharge tube dimensions has to be chosen so that the nominal lamp electrical characteristics are obtained at standard ballast systems, so that the discharge tube fits into the lamp outer envelope and so that the predetermined wall temperature goal and a maximal sodium D-line reversal width are not exceeded in order to obtain a good life performance of the lamps.

3 CONCLUSIONS

The optimisation of the discharge tube dimensions and xenon fill pressure has been worked out to obtain mercury-free high pressure sodium (HPS) lamps with a luminous flux in between the standard and the high pressure xenon lamp.

4 REFERENCES

[1] R. Geens and E. Wyner, IEE Proceedings-A **140**, No. 6, 450 (1993)
[2] J. Waymouth and E.. Wyner, J. illum. Eng. Soc. **10**, 237 (1981)
[3] J.J. de Groot, IEE Conf. Publ. **90**, 124 (1972)
[4] J.J. de Groot and J.A.J.M. van Vliet, *The High-Pressure Sodium Lamp*, 274 (1986)

C 08

INTENSITY DISTRIBUTIONS OF LINE SPECTRA EMITTED FROM INDUCTIVELY-COUPLED METAL-HALIDE-LAMP

Makoto Hamamoto[1], Tomoaki Hashimoto[1], Shingo Higashisaka[2], and Shigeaki Wada[2]

[1]Faculty of Engineering, Oita University, Oita 870-1192 Japan

[2]Lighting R&D Center, Matsushita Electric Works, Ltd., 1048 Kadoma Osaka 571-8686 Japan

1 INTRODUCTION

The intensity distributions of line spectra were measured to investigate the effect of jacket glass tube for inductively-coupled metal-halide-lamp. Since the results, obtained for the lamp having jacket glass tube, have already been published [1, 2], the results for the same kind lamp without jacket glass tube will be shown and compared with the former.

2 EXPERIMENTAL AND RESULTS

The intensity distributions of line spectra (437.447nm:Sc ion, 479.152nm:Sc atom, 615.423nm:Na atom), emitted from inductively-coupled metal-halide-lamp (tube diameter 27mm, tube height 10 mm; NaI:16.3mg, ScI_3:4.1mg, Xe:26.7kPa) were measured for both axial and radial directions with moving the lamp by a x-y-z translational stage and with varying RF power (150-300W) as a parameter.

i) Asymmetric distributions are observed at low RF power less than 200 W for Sc ion and Sc atom as shown in Figs. 1, 3 and 4, otherwise the distributions are rather symmetric as shown in Figs. 2, 3, 4 and 5.

ii) Intensity of Sc ion increases with increase of RF power up to 200 W, then decreases up to 300 W (Fig.3). This differs from the monotonic decrease of that for the lamp having jacket.

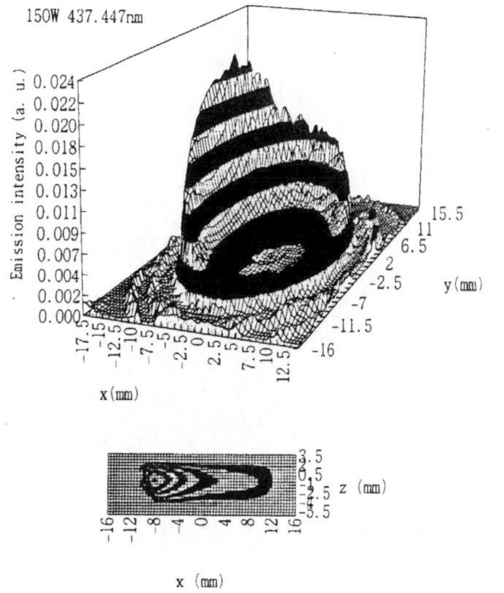

Fig. 1. Emission intensity distributions of 437.447nm line of Sc ion observed in axial (top) and radial (bottom) directions [RF power 150W].

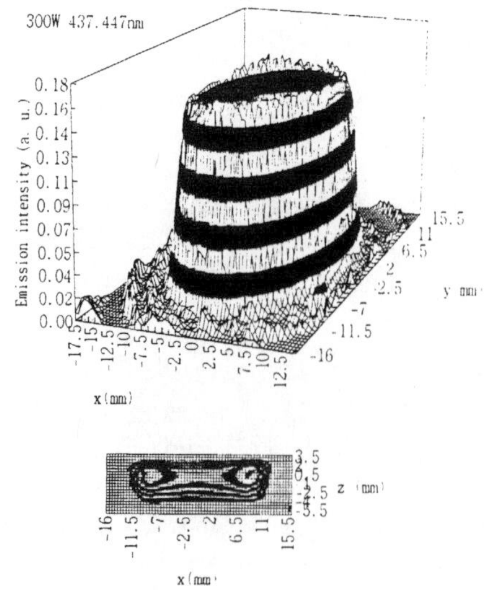

Fig. 2. Emission intensity distributions of 437.447nm line of Sc ion observed in axial (top) and radial (bottom) directions [RF power 300W].

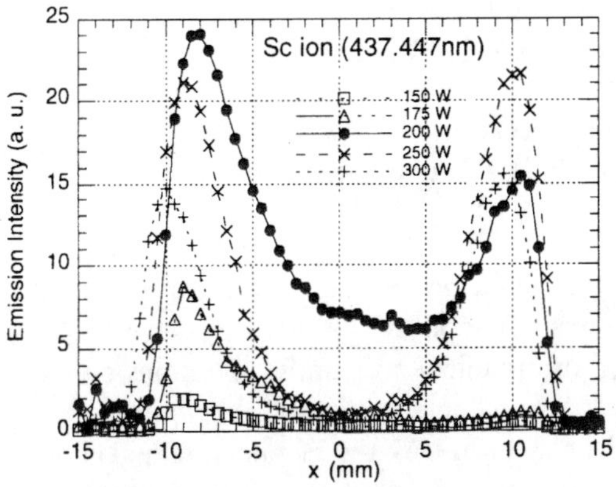

Fig. 3. Emission intensity distributions of 437.447nm line of Sc ion [axial].

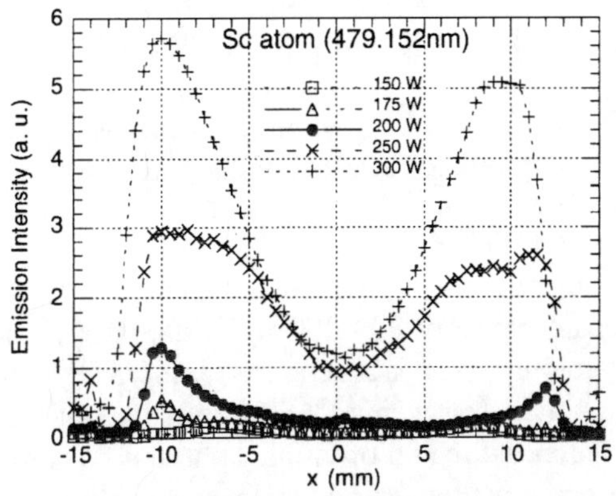

Fig. 4. Emission intensity distributions of 479.152nm line of Sc atom [axial].

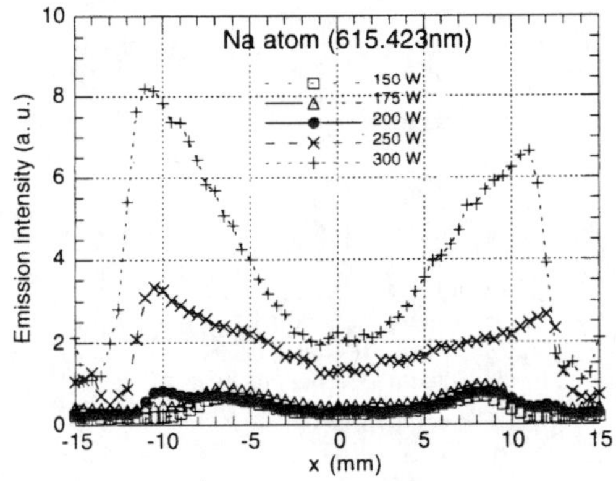

Fig. 5. Emission intensity distributions of 615.423nm line of Na atom [axial].

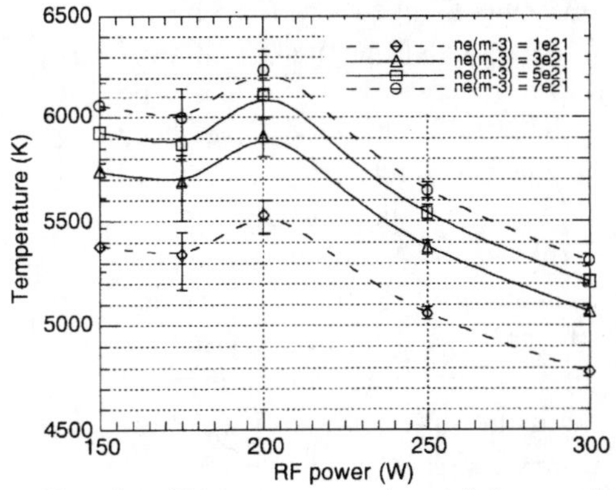

Fig. 6. Temperature estimated from ratios of spectral line intensities at their peaks.

iii) Intensities of Sc atom and Na atom do increase largely with increase of RF power only beyond 200 W. This differs from the monotonic increase of those over the whole range of RF power for the lamp having jacket.

iv) Temperature, estimated from ratios of Sc atom and Sc ion lines at their peaks, are higher by from 300 to 1,000 K than that for the lamp having jacket assumed the same electron density (Fig.6).

4 ACKNOWLEDGMENTS

The authors would like to thank Messrs. Taka. Yamaguchi, Tomo. Yamaguchi, T. Higashijima and Y. Tooda for their technical help.

This work was supported in part by a Grant-in-Aid for Scientific Research from Ministry of Education, Science and Culture.

5 REFERENCES

[1] M. Hamamoto, S. Higashisaka and S. Wada, *Proc. LUX PACIFICA '97, Nagoya, 1997*, pp. B83-B88 (1997).

[2] M. Hamamoto, S. Higashisaka and S. Wada, Trans. Inst. Electr. Eng. Jpn. [in Japanese], **118-A**, 49 (1998).

C 09

PRESSURE DEPENDENT UV-C EMISSION FROM HIGH-PRESSURE MERCURY LAMPS

Michael Lambrecht, Wolfgang Heering
Lighting Technology Institute (LTI), University of Karlsruhe, Germany

1 INTRODUCTION

Long cylindrical high pressure mercury discharge lamps with rated power from 1 kW to 40 kW and a specific power of 80 - 300 W/cm are used in many industrial applications as for instance UV curing of colours, inks, lacquers and plastics, UV oxidation of organic compounds, pollutions etc. and UV synthesis of chemicals. The influence of mercury pressure on the spectral distribution is well studied in the region of visible light, but concerning the region of UV it is still uncertain. This work presents the pressure and lamp current dependencies of spectral distribution and efficiency of UV radiation in high pressure mercury discharge lamps.

2 SPECTRAL UV MEASUREMENTS

Lamp tubes were filled with different amounts of mercury and operated at different specific powers between 100 W/cm and 300 W/cm. Spectral intensities have been measured in the UV range from 175 nm to 400 nm as well on horizontally as on vertically burning arcs. A strong dependence of UV-C radiance on mercury pressure and direction of emission has been observed, obviously due to the large contribution of resonance radiation to the UV-C. For comparison, the UV spectra of two mercury lamps that have equal geometry but different working pressure are shown in Fig. 1, both operated at equal specific power of 200 W/cm and measured with equal spectral responsivity at spectral bandwidth of 0.6 nm. The direction of spectral measurement was vertical onto the bottom of the horizontally burning lamps. The enhanced reabsorption by the thicker cold zone between arc and wall is evident and has a strong influence on UV-C efficieny. Spectral measurements under inert conditions have revealed the strong self-reversal of the 185 nm resonance line, from 180 nm to 188 nm at a mercury pressure of 2 bar and from 180 nm to 192 nm at 6.3 bar. The spectral range of complete reabsorption depends also on the upward displacement of the arc and the viewing direction. The lamps were operated by a new electronic ballast at a rated power of 4 kW. The electronic gear that converts line frequency into high frequencies of more than 75 kHz was developed to control lamp current, lamp power and radiation output. By varying the frequency it is possible to dim the lamp, even down to 10 %

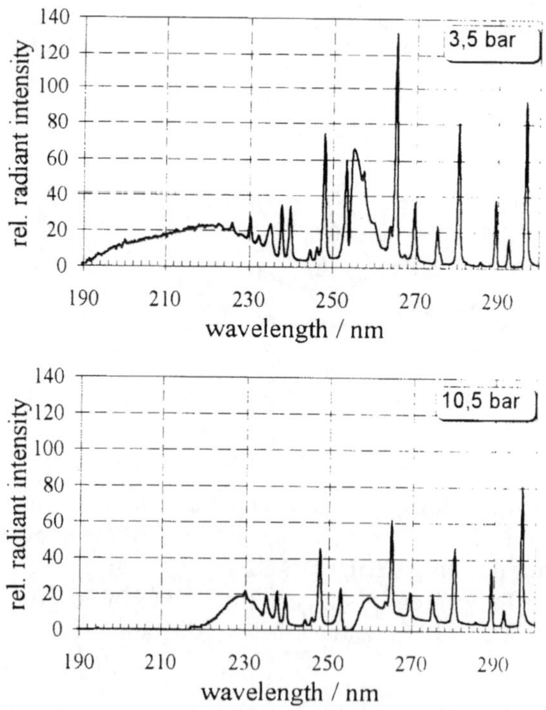

Fig. 1: Spectra of two lamps with equal diameter and arc length operated with equal power but filled with different amounts of mercury.

of its rated power, and so to adapt the UV-C output to the changing requirements of industrial processes.

3 LAMP MODELLING

A numerical model has been developed to describe the plasma of a high-pressure mercury discharge operated at an electronic ballast in different dimming states. The high frequency operation of the discharge allows to neglect the time dependence of the electrical plasma conductivity. By using a simplified radiation model, the temperature profile, lamp voltage and mercury pressure can rapidly be computed for a given lamp current, mercury filling and bulb diameter. Based on the calculated temperature profil, radiation transport is implemented into the model to determine the total spectral radiant power leaving the plasma as a function of wavelength between 185 nm and 4000 nm. For this the transition probabilities and the broadening constants of [1,2] were used. So changes in the spectral composition of the emitted radiation caused by dimming or different mercury pressure are computable. Some results of such computations are shown in Fig. 2. The much higher mercury density in the dark annular space of the discharge lamp can explain the substantial higher reabsorption of resonance radiation and the emission of optically thick UV lines some of which are selfreversed. The simulation of lamp dimming gives decreased widths of temperature profile so that the relative contribution of UV-C to the total radiant flux decreases quite strongly with reduction of lamp current and lamp power.

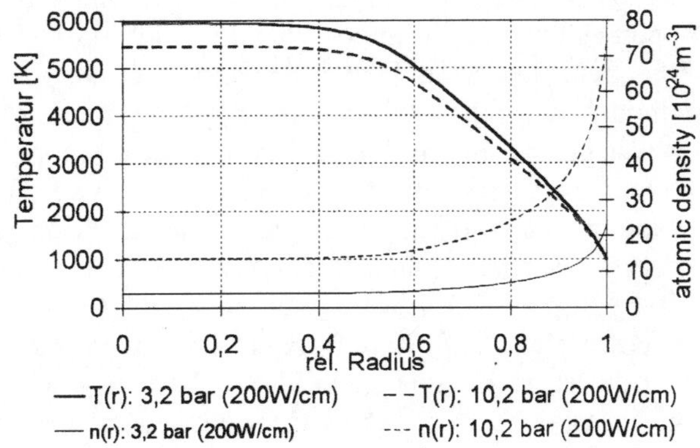

Fig. 2: Calculated temperature profiles and atomic densities for two lamps with equal geometry but different mercury filling.

4 CONCLUSIONS

Measured and calculated electrical and spectral data agree very well. Because of some uncertainties in the radiation processes that are dominant in the spectral range between 200 nm and 245 nm there are some differences between measured and calculated side-on spectral radiances in this region. The axial symmetry of horizontally burning arcs assumed in the calculation is nearly only given for lamps operated with rather high currents and pressures sufficiently below 10 bar. Otherwise a three dimensional model that also takes into account convection is necessary to describe a wall-stabilized horizontally burning arc. Generating UV-C radiation depends on mercury pressure and on current. The efficiency of line radiation in the whole UV range as well as the spectral efficiency of radiation below 220 nm of the investigated medium pressure mercury lamps were observed to increase with decreasing pressure and thus with increasing current at the same specific power.

5 REFERENCES

[1] H.-P. Stormberg, R. Schäfer, Time-dependent behavior of high-pressure mercury discharges, J. Appl. Phys. **54** (8), (1983)

[2] E.R. Jr.Mosburg, M.D. Wilke, Some new transition probabilities for mercury I, J. Quant. Spectrosc. Radiat. Transfer **19**, 69-81 (1977)

C 10

DETERMINATION OF THE POWERFLUX INTO LAMP CATHODES

Rudolf Bötticher[1] Woldemar Bötticher[2] and Dirk Windelberg[1]
1 AG Qualität Institut für Mathematatik Universität Hannover
Welfengarten 1 D 30167 Hannover Germany
2 Abteilung Plasmaphysik Institut für Atom- und Molekülphysik Universität Hannover
Callinstr. 38 D 30167 Hannover Germany

1 INTRODUCTION

It would be very helpful for the development of HID lamps to have a model which can predict all lamp properties from an ab initio calculation. At present the limited quantitative understanding of the transition sheath between cathode and arc plasma is the main impediment for an ab initio model. Recently some insight into the cathode sheath properties of high current welding arcs was gained by the use of different sheath models [1- 3]. Whether these models are valid at the low currents typical for lamps is an open question.

The direct experimental verification of the predicted sheath structures is extremely difficult due to the necessary spatial resolution. One has therefore to measure the integral properties:

- total voltage drop over the sheath (including effects of the arc constriction)
- total net power flux Q_o flowing from the plasma into the cathode.

This contribution deals with the possibilities to determine accurately the total effective power flux Q_o from measurements of the surface temperature of the cathode or by measuring the power flux Q_u leaving the discharge-vessel via the current feed-through.

2 ELECTRODE AND ARC GEOMETRY

The present investigation deals with a cylindrical rod cathode with a plane front-face. This configuration is chosen in order to get an arc configuration which can easily investigated experimentally. A similar configuration has been used by Tielemans and Oostvogels [4]. A recent experiment is presented in this conference [5]. In this experiment the temperature of the cathode at the feed-through is hold at a temperature T_u by a cooling liquid which is controlled by a thermostat. Thus Q_u at this point can be measured. The gas flow in the arc is negligible. The gas pressure p (Argon) and the arc current I are in the range: p = 1-5 bar and I = 1-5A. A typical cathode diameter is 2R= 1.5mm. The cathode length L is defined as the distance between the front-face and the position of the isotherm T_u . L is between 10mm and 30mm.

3 MODEL CALCULATIONS

3.1. Elements of the power-balance

The power-balance is made up by the contributions:

Inside the cathode : Heat flux density $q = -\lambda(T)\,\mathrm{grad}\,T$ and Joule heating $w = \sigma_{el}(T)E^2$.

Fluxes through the surface of the cathode: Thermal radiation $q_{rad} = \varepsilon_{SB}(T)\,\sigma_{SB}\,T^4$; thermal flux Q_u leaving the cathode through the isotherm at T_u and the effective power flux density q_o from the arc into the cathode.

It is assumed that q_o enters the cathode in the same area A through which the current flows. q_o shall include the power due to ions hitting the cathode, radiation from the arc column and heat flow due to the temperature gradient in the cathode sheath. This heating fluxes are to be reduced by the cooling due to emission of electrons. In a complete model q_o and A is to be calculated from the model of the sheath. Here for A and q_o assumed values are used.

3.2. Results of two-dimensional model calculations

The purpose of this calculations is to show the effect of variations of the area A at otherwise constant parameters . For these calculations the finite element program ANSYS has been used. The graphs in Fig. 1 were calculated for a diffuse arc with A = full front face (2) and a constricted arc having a spot diameter of 200μm (1). The constant parameters are:

2R= 1.5 mm, L = 24mm, $Q_0 = q_o A$ = 37 Watt and T_u= 293K. Shown are the radial temperature distribution T(r,x=L) on the front face and the temperature distribution T(r=R,x).

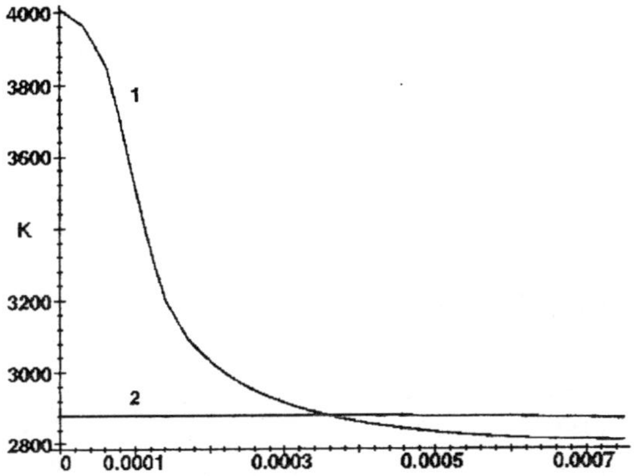

FIG. 1a T(r, x=L)

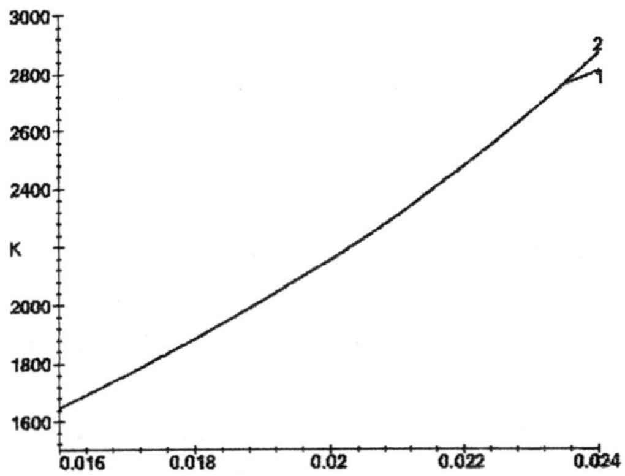

FIG. 1b T(r=R, x)

From Fig.1b one learns that <u>T(r=R, x> 1mm) does not depend on the chosen A</u>. This opens the possibility to determine Q_o without exact knowledge of the current density distribution on the front face from a measurement of T(r=R,x). For x>1mm the radial temperature differences turn out to be smaller than some K. Therefore T(r=R,x) can be calculated with an one dimensional model with appropriate methods and transient calculations become very fast. The results of calculations with and without Joule heating in the cathode body differ by a few K only. This is due to the low current and different to welding arcs. Lamp cathodes are therefore a better tool to study the quality of sheath models.

3.2. Accuracy of the determination of Q_0

Starting from measured values of Q_u and T_u the model predicts Q_o and T_o. The table below gives some selected results for two electrodes with L=24mm, 2R=1.5mm and L=7mm, 2R=0.8mm with T_u = 293 K. T_o is between 2700K and 3000K. If one varies $\lambda(T)$ uniformly by ±1% only the predicted values of Q_o vary by about ±3,5% for L=24mm and about ±0.5 % for L=7mm.The potential error of 3-5% of the thermal conductivity reported in [6] yields due to the strong non-linearity of the model that proper determination of Q_o from Q_u requires that L<10R. Similar calculations show that for this conclusion variations of the emission coefficient are less critical. But the error of $\varepsilon_{SB}(T)$ may be much larger.

Q_u/W	T/K	Q_o/W		Q_u/W	T_o/K	Q_o/W	
25.5	2734	36.2		19	2673	21.3	
27.5	2831	39.7	L=24mm,R=0.75mm	20	2851	23.1	L=7mm,R=0.4mm
29.5	2939	43.4		21	3031	25.1	

An application of the described methods to the results given in [5] will be shown in the poster.

4 ACKNOWLEDEMENT

This work is part of a researcn program sponsored by the Bundesminister für Bildung und Forschung (Vorhaben FN 7106/0). Full report: www.tkh.de/zb1.ps

5 REFERENCES

[1] J. J. Lowke, R. Morrow and J. Haidar J.Phys.D Appl. Phys. **30**, 2033 (1997)
[2] J. Wendelstorf, I. Decker, H. Wohlfahrt and G. Simon In: H.Cerjak (ed) Mathematical modeling of weld phenomena 3, The Institute of Materials, London 1996
[3] M. S. Benilov and A. Marotta J. Phys. D Appl.Phys. **28,** 1869-1882 (1995)
[4] P. Tielemans and G. Oostvogels Phillips J. Res. **38,** 214-221, (1983)
[5] J. Mentel et. al. Ruhr-University-Bochum, Contribution to this conference
[6] J. G.Hust and A.B.Lankford Standard Reference Materials :
National Bureau of Standards Special Publication 260-90 (1984)

C 11

MODELLING THE SPECTRUM OF A S_2 HIGH PRESSURE DISCHARGE

Achim Körber

Philips Research Laboratories, Weisshausstr. 2, D 52066 Aachen, Germany

1 INTRODUCTION

High power microwave discharges in S_2 vapour at pressures of several bars have been discovered as a highly efficient light source several years ago [1]. The discharge spectrum is originating mainly from bound-bound transitions in the S_2 B $^3\Sigma_u^- - X\ ^3\Sigma_g^-$ band system. At such high S_2 pressures re-absorption is an essential mechanism shifting the spectral maximum from the UV into the visible region [2]. With the help of the available spectroscopic information and some reasonable assumptions it is possible to *quantitatively* simulate the visible spectrum of this "Sulfur Lamp" by solving the *radiation transport equation* for all observed vibronic bands.

2 ASSUMPTIONS

Since we are dealing with a plasma of several bars pressure, *Local Thermal Equilibrium* (LTE) is an adequate approximation. The spherical quartz discharge vessel is usually rotated to obtain a homogeneous discharge [2]. Therefore we assume a *spherically symmetric plasma* suggesting a *cubic temperature profile* T(r)

$$T(r) = T_{max} - (T_{max} - T_w) \cdot \left(\frac{r}{r_w}\right)^3$$

where: T_w = wall temperature (1200 K)
r_w = inner vessel radius
T_{max} = maximum plasma temperature
(→ free model parameter !).

We consider the *330 vibronic bands* connecting the niveaus v' = 0...9 and v" = 0...32 of the S_2 B-X band system for which *Franck-Condon-Factors* ($fcf_{v'v''}$) are known from the literature [3]. Each vibronic band is treated as one single "transition" with a *rectangular line profile* of *fixed frequency width* $\delta\nu_{vib}$ (→ free model parameter !). $\delta\nu_{vib}$ is the sum of the widths of all rotational lines within a vibronic band.

3 CALCULATION PROCEDURE

The *population densities* $n_{v'}$ (T) and $n_{v''}$ (T) of the vibrational levels v' and v" of the S_2 B and X states, respectively, are obtained from the dosed amount of sulfur, a thermodynamic equilibrium calculation and a Boltzmann population distribution over the vibrational levels. The *transition probabilities* $A_{v'v''}$ are evaluated from the lifetime $\tau_{vf'}$ of an arbitrary, fixed level vf' of the B-state (here: $\tau_{3,4} \approx 19$ nsec [4]) and the $fcf_{v'v''}$. The *spectral emission coefficient* $j_{v'v''}$ (T) and the *spectral absorption coefficient* $k_{v'v''}$ (T) is calculated according

$$A_{v'v''} = \frac{\nu^3_{v'v''} \cdot fcf_{v'v''}}{\tau_{vf'} \cdot \sum_{vf''} \nu^3_{vf'vf''} \cdot fcf_{vf'vf''}} \qquad j_{v'v''}(T) = \frac{n_{v'}(T) \cdot A_{v'v''} \cdot h \cdot \nu_{v'v''}}{4\pi \cdot \delta\nu_{vib}} \qquad k_{v'v''}(T) = \frac{c^2 \cdot n_{v''}(T) \cdot A_{v'v''}}{8\pi \cdot \delta\nu_{vib} \cdot \nu^2_{v'v''}}$$

where: $\nu_{v'v''}$ = transition frequency, c = velocity of light and h = Planck's constant.

The one-dimensional *radiation transport equation* for the *spectral radiance* $L_{v'v''}$ (x)

$$\frac{dL_{v'v''}}{dx}(x) = j_{v'v''}(T) - k_{v'v''}(T) \cdot L_{v'v''}(x)$$ where: x = co-ordinate along line of sight

is integrated yielding the radiance at the discharge surface $L_{v'v''}(x_{wall})$ which is transformed into the *total emitted spectral power* $P_{v'v''}$ (Fig. 1) by integration over the solid angle and multiplication with the discharge surface. Finally the frequency spectrum of $P_{v'v''}$ is convoluted with a "typical" rotational intensity spectrum (i.e. an normalised exponential function with damping constant $\delta v_{rot} = (B_B - B_X) \cdot k_B T_{rot} / (h \cdot \overline{B}) = 760\ cm^{-1} \cdot c$ for $T_{rot} = 4000$ K) to yield the ro-vibronic spectrum P(λ) (Fig. 2).

4 RESULTS

The spectrum of a discharge with r_w = 16.5 mm containing 1.63 mg / cm^3 sulfur and 200 mbar (cold) Ar at 760 W input power can be well described by adjusting the free model parameters to T_{max} = 5000 K and $\delta v_{vib} = 14\ cm^{-1} \cdot c$ (Fig. 1).

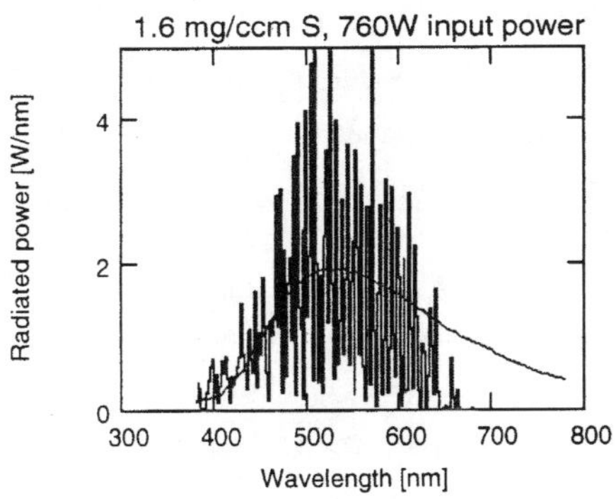

Fig. 1: Simulated vibronic spectrum $P_{v'v''}$ and experiment ("smooth" curve)

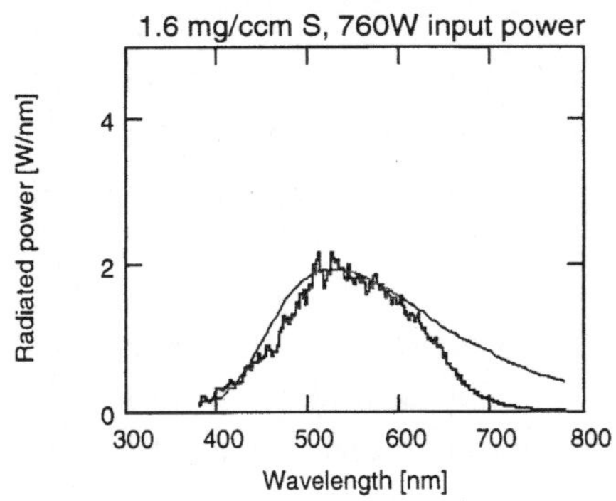

Fig. 2: Simulated ro-vibronic spectrum P(λ) and experiment ("smooth" curve)

5 CONCLUSIONS

It is possible to simulate the spectrum of the "Sulfur Lamp" *quantitatively* with only two free parameters: T_{max} and δv_{vib} yielding physically meaningful values of these quantities. In the red an IR the experimental spectrum is higher indicating contributions of continuum radiation (e^- -S^+ and S-S recombination) which are not included in the simulation. The response of the S_2 spectrum to parameter variations (S_2 pressure, input power, plasma radius) may also be studied with the described model.

6 REFERENCES

[1] J.T. Dolan, M.G. Ury and C.H.Wood, 6th Int. Symp. on the Science and Technology of Light Sources, Budapest (1992).
[2] B.P. Turner, M.G. Ury, Y. Leng and W.G. Love, J. Illum. Eng. Soc. **26** (1), 10 (1997).
[3] W.R. Anderson, D.R. Crosley and J.E. Allen Jr., J. Chem. Phys. **71** (2), 821 (1979)
[4] K.A. Meyer and D.R. Crosley, J. Chem. Phys. **59** (4), 1933 (1973)

C 12

Longitudinal Acoustic Mode Structure in a Ceramic High Intensity Discharge Lamp

Jerry Kramer
Philips Research
Briarcliff Manor, NY 10598 USA

Acoustic resonances and the resulting arc instabilities are a major obstacle to the use of high frequency ballasts for HID lamps. Although the frequencies of acoustic resonances can be determined experimentally, assigning these modes can be difficult because of the non-uniform shape of the discharge tube and the non-uniform temperature distribution in the arc tube. HID lamps have been developed recently with ceramic envelopes which are cylindrical in shape. The wave equation can be solved for the acoustic modes in a cylinder with a uniform temperature distribution. The longitudinal mode frequencies are at $f_l = nC_l/2l$, where n is the number of nodes, C_l is the speed of sound in the longitudinal direction and l is the inner length of the arc tube. The azimuthal and radial modes are at $f_r = \alpha C_r/\pi d$ where α corresponds to the zeros of the first derivative of the Bessel function J_1 of order 1, C_r is the speed of sound in the radial plane and d is the inner diameter of the arc tube. The three lowest order modes are the 1st azimuthal (α = 1.84), the 2nd azimuthal mode (α = 3.05) and the 1st radial mode (α = 3.83). Radial - longitudinal combination modes are also allowed at $f = (f_r^2 + f_l^2)^{1/2}$.

A 39 W ceramic HID lamp with 5 mm ID and 6 mm IL was operated in vertical orientation. The acoustic resonances below 110 kHz power frequency were determined by operating the lamp with a low frequency (500 Hz) square wave driver with a high frequency sine wave of variable frequency and voltage superimposed. Arc motion was observed visually at 51, 67, 81, 94 and 103 kHz. These resonances were assigned to the 1st azimuthal mode (51 kHz), the 2nd longitudinal mode (67 kHz), the 2nd azimuthal mode (81 kHz), and the 1st radial mode (103 kHz). The experimental ratio of frequencies of the first three azimuthal and radial modes are 1 : 1.59 : 2.02 in good agreement with the theoretical ratios of 1 : 1.66 : 2.08. The assignment of the resonance at 94 kHz is not obvious. Calculated frequencies of combination modes are at 83.4 kHz (1st azimuthal - 2nd longitudinal), 87.5 kHz (2nd azimuthal - 1st longitudinal) and 104.5 kHz (2nd azimuthal - 2nd longitudinal) and the 3rd longitudinal mode is predicted at 100.5 kHz.

The longitudinal mode structure of the resonances at 67 and 94 kHz was investigated by optical emission spectroscopy. Spectra from 360 to 620 nm were taken at different axial positions with a 1/4 m spectrograph and gated CCD. The mode structure was determined under high frequency ballast conditions by driving the lamp with a high frequency sine wave from a function generator and a power amplifier. (With a pure sine wave the power frequency is equal to twice the current frequency.) The power frequency was swept slowly (0.5 s) from low to high frequency and back again (sawtooth) either through the 67 kHz resonance (60 to 70 kHz) or toward the 94 kHz resonance (82 to 92 kHz). (The 94 kHz resonance had a sharp onset in frequency and extinguished the lamp.) Spectra were taken either at 67 or 92 kHz and at a lower frequency (no resonance) in the sweep as a control.

The major changes in the emission spectrum induced by the resonances were seen in the self-reversed sodium (590 nm) and thallium (535 nm) peaks and the optically thick mercury line at 546 nm. At the acoustic resonance frequency the sodium and thallium emission lines become more (or less) self-reversed. The changes in the emission intensities for the resonance at 67 kHz, shown in Figure 1, have two nodes confirming the assignment of the 2nd longitudinal mode. The axial dependence of the changes in emission intensities for the resonance at 94 kHz followed an approximately sinusoidal pattern with a single node. A single node is consistent with a mode that contains the 1st longitudinal mode and the resonance at 94 kHz was assigned to the 2nd azimuthal - 1st longitudinal mode.

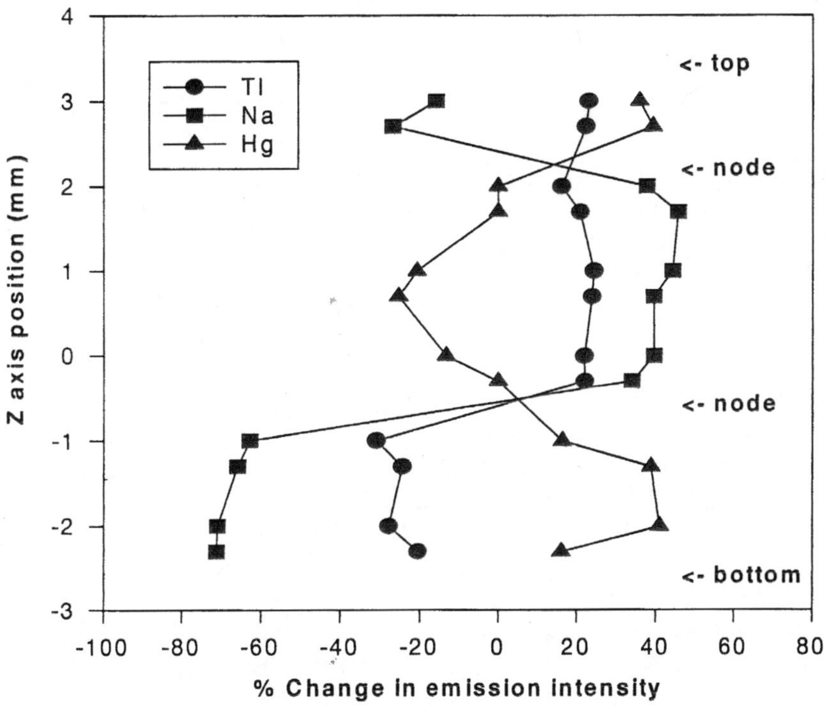

Figure 1

C 13

Light Sources for Video Projection

Holger Mönch and Günther Derra

Philips GmbH Forschungslaboratorien, Weißhausstr.2, D-52066 Aachen

1 Video Projection

The key components of a light valve projector [1] are the displays (transmissive LCD or reflective mirrors DMD) and the light source (HID lamp). The optical system separates and recombines the 3 primary colours and focuses the light onto the screen.

The efficiency of the optical system reaches for the best projectors $\approx 25\%$ (60% typical display efficiency; 80% polarisation efficiency incl. polarisation recovery; 60% color separation and correction and 90% lens transmission). This efficiency has to be multiplied with the amount of light from the source which can be collected on the display. The possibilities to collect maximum light from an arc lamp are discussed in this paper.

Pushed by size and cost reduction projection displays become much smaller (1993: 3" diagonal, accepted light cone half angle 7-10°; 1998: 1-1.3" diagonal, 5-7° light cone half angle). Smaller displays require smaller lamp arcs to keep the collection efficiency high. The development of displays with 1" diagonal relies on the development of lamps with $< 1.5mm$ arc length.

2 Etendue

High optical demand (small displays) corresponds to small Etendue number. The Etendue $E = A \cdot \Omega$, also called the optical extent [2] characterizes an optical system throughput and is the product of the aperture area A and the acceptance solid angle Ω. A typical display with 3:4 aspect ratio, 1" diagonal and 7° acceptance cone has e.g. an Etendue of

$$E = 310mm^2 \cdot 2 \cdot \pi \cdot \int_0^{7^\circ} \sin\vartheta \cdot d\vartheta = 14.5mm^2 \cdot sr .$$

The light flux Φ through an optical system is given by $\Phi = E \cdot L$ with the luminance L. Along the optical path of a system the luminance is at best conserved or decreased. If light flux shall be conserved the Etendue may only increase through an optical system.

The element of minimal Etendue (usually the display) defines a limiting box around the source. If the Etendue of the display in the above mentioned example is followed back to the source it can be shown that only light which is generated inside a projected arc area of $A \approx 1.2mm^2$ can be used in the system. The more light is generated inside this limiting box the brighter the display will be. Therefore short arc and high luminance are the main issues for projection lamps.

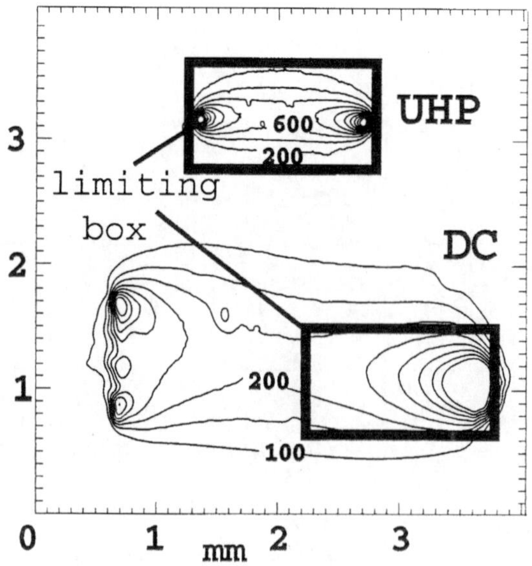

Figure 1: Luminance Distribution of UHP100W 1.3mm lamp (top) and DC250W 2.5mm lamp (bottom). The unit of measure is Mcd/m^2. Note the 4 x higher luminance contour scale of the UHP lamp.

3 Luminance

Figure 1 shows the luminance of a Philips UHP 100W lamp [3] with 1.3mm arc size compared to a 250W-DC metal halide (mainly Indium) lamp [4] with 2.5mm arc size. Only ≈ 80% of the total lamp lumens originate from the arc as shown in Figure 1 and the rest comes from the electrodes, from relatively dark but large plasma regions or is reflected at the wall. Due to its short arc and its high luminance the UHP lamp with only 100W will give maximum light inside the limiting box (and on the screen) of the above mentioned system.

Based on the luminance measurement (in generally for several wavelengths) in Figure 1 the performance of the lamp in a PTV system can be simulated [5]. It is an advantage of the UHP lamp with only mercury as radiating species that the luminance distribution is nearly independent from the wavelength.

4 Collectable Light and System Performance

A reflector covers as much as possible the solid angle around the lamp arc to catch maximum light. Therefore deep parabola (parallel beam) or elliptical (second focus) reflectors are used. Practical measurements of collectable light vs. Etendue are relatively independent from the reflector type because the Etendue can be defined for a parallel beam (cross section and cone angle) as well as for the second focus of an elliptical reflector (diameter and angle of incidence).

Figure 2 shows measurements of collectable lumens for the two lamps mentioned above. For displays ≤ 2.0" the UHP lamp will give more light on the screen than the 250W lamp. The screen lumens of a projection system can be calculated from the collected lumens in Figure 2 by multiplication with the optical efficiency of 25%.

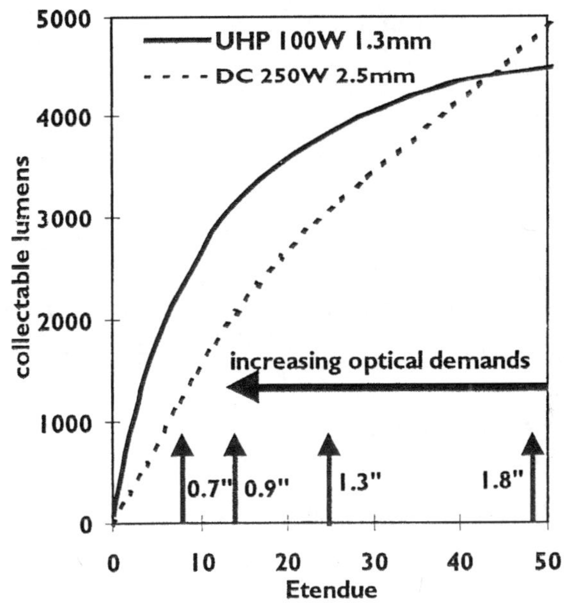

Figure 2: System brightness for projection lamp types.

5 Conclusions

Light valve projectors use small displays with high optical demands on the light source. Arc length and luminance are much more important than lamp power. The Philips UHP 100W lamp gives optimum light collection for actual and future projection systems.

6 References

[1] D.Armitage, Design issues in liquid-crystal projection displays, in Projection DisplaysII, Proc.SPIE 2650

[2] M.Brennesholtz, Light collection efficiency for light valve projection systems, in Projection Displays II, Proc.SPIE 2650

[3] E.Schnedler and H.v.Wijngaarde, Ultrahigh-Intensity Short-Arc Long-Life Lamp System, in SID95

[4] T.Higashi, Long-Life DC Metal-Halide Lamps for LCD Projectors, in SID95

[5] J.Schweyen, K.Garcia and Ph.Gleckman, Geometrical Optical Modelling Considerations for LCD Projector Display Systems, in Projection Displays III, Proc.SPIE 3013

C 14

MERCURY-FREE HPS LAMP WITH HIGH CRI AND ITS ONE APPLICATION ON PLANT GROWTH

Naoki Saito[1], Katsusuke Murakami[2], Kimitoshi Horaguchi[3]
Atsunori Okada[1], Koji Nishioka[1]

1 Matsushita Electric Works, LTD., 1048, Kadoma, Osaka 571-8686, Japan
2 Osaka Prefecture University, Gakuencho 1-2, Sakai, Osaka 599-8570, Japan
3 Matsushita Electric Industrial Co., Ltd., 3-4, Hikaridai, Kyoto 619-0237, Japan

1. INTRODUCTION

In recent years, technologies for pollution-free light-sources are developed widely. In continuation of this trend, we reported mercury-free HPS lamp with high CRI[1,2] last year.

In this paper, we show the results of experiments and the characteristics of the mercury-free HPS 150 W lamp with high CRI operated with choke type ballast and present its one application on plant growth for supplemental lighting.

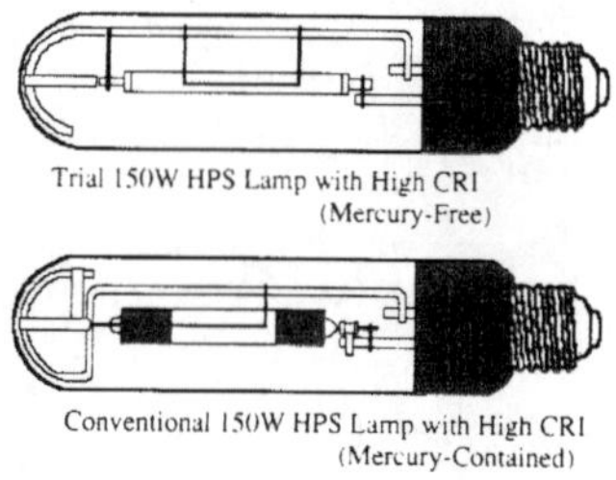

Figure1. Lamp construction

2. EXPERIMENTS AND RESULTS

The 150 W lamp construction for experiments is shown in Figure 1, compared with that of conventional lamp contained mercury.

From the data of various inner diameter, arc length and xenon pressure, CRI can be indicated on a curve, as is shown in Figure 2. Where E is average potential gradient (= lamp voltage / arc length) and d is inner diameter of the arc tube. Maximum value of CRI was expected over 80.

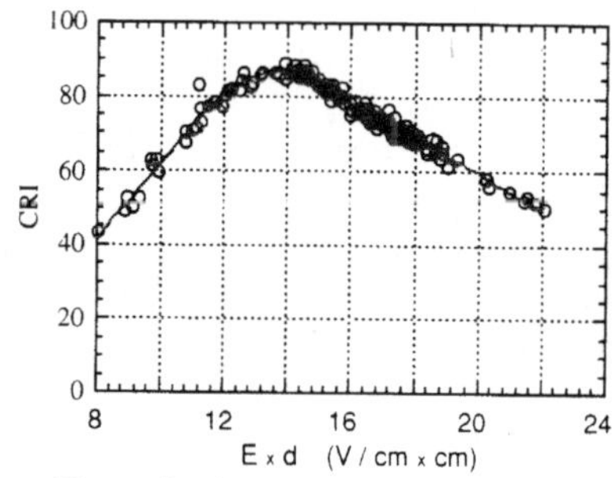

Figure 2. Average potential gradient(E) x d vs. CRI

Figure 3 shows the relationship between inner diameter of the arc tube and luminous flux. Luminous flux goes down with the increase of lamp voltage and decreases with the increase of inner diameter of the arc tube.

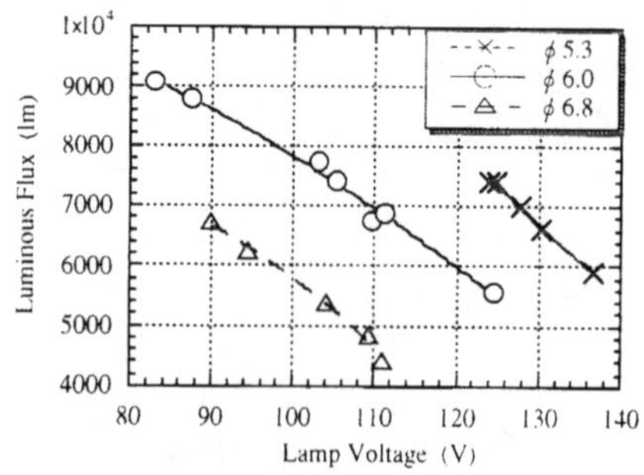

Figure 3. Luminous flux with the variation of inner diameter of the arc tube

The final size of the trial 150W arc tube is as follows; inner diameter is 6.0 mm and the arc length is 40 mm. Only sodium metal 3 mg and xenon 4×10^4 Pa (25°C) are sealed in the arc tube.

Lamp characteristics of 150 W lamp operated with choke type ballast are shown in Table 1. Nearly the same electrical and optical initial characteristics as those of conventional mercury-contained HPS lamp are realized. Due to the slight deviation from the Planckian locus (Duv), the light seems to be a little greenish compared with that of conventional lamp.

Table 1. Characteristics of 150W lamp

	Trial Lamp	Conventional Lamp
Mercury	Free	Contained
Lamp Voltage (V)	100	100
Lamp Wattage (W)	150	150
Luminous Flux (lm)	8,000	7,000
CRI	85	85
Color Temp. (K)	2,500	2,500
Duv	+0.006	0
Life Time (h)	12,000	9,000
Color Shift	Yellowish	Pinkish

Spectral power distribution of the trial 150 W lamp at the color temperature of 2,500 K is shown in Figure 4, compared with that of mercury-contained conventional lamp with inner diameter of 8.0 mm.

The emission of the trial lamp is specially stronger at 550-570 nm and weaker near 450 nm than that of conventional lamp. This may be the main reason why the deviation from Planckian locus is +0.006 and the color hue seems to be a little greenish. As for the region of 600-780 nm, the spectral power distribution of the trial lamp is stronger compared with that of conventional lamp, except that of near 680 nm.

3. ITS ONE APPLICATION

This mercury-free HPS lamp with high CRI operated with choke type ballast is, of course, suitable for general lighting, such as indoor lighting, flood lighting, street lighting and so on. In addition to it, this lamp seemed to be useful for seedling growth, because it has strong emission in the range of 600-800 nm. It is estimated that R/FR is an important factor for plant growth. We made 400 W reflector type lamp. These lamps were settled at the greenhouse for transplant production of JA Shinano Asahi in Nagano, and used for supplemental lighting with every 2 hours before sunrise or after sunset.

PPFD of the lamp was 80μ mol $m^{-2}s^{-1}$ at the plant bed. R/FR was 0.98 on the average of 20 lamps. R/FR of this lamp can be changed by the control of lamp voltage, as is shown in Figure 5.

The results of lettuce and broccoli growth were shown in Table 2[3]. It increased both in weight and height by supplemental lighting, compared with those of only daylight. It became clear that it was much effective with that of before sunrise, compared with that of after sunset.

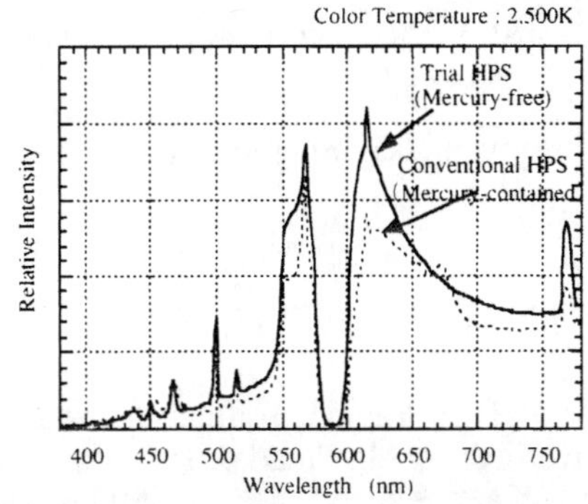

Figure 4. Spectral power distribution at color temperature 2,500K

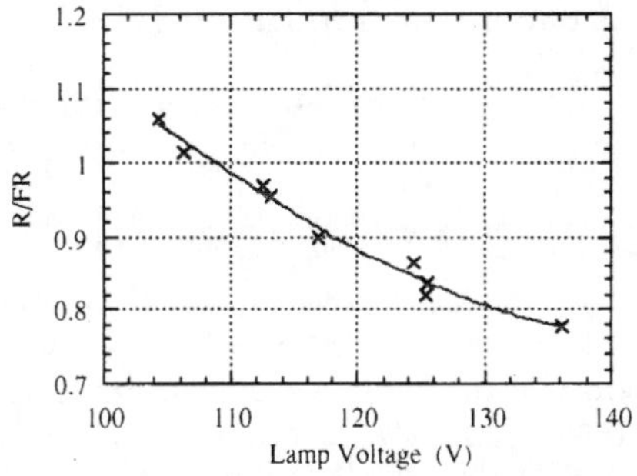

Figure 5. Lamp voltage vs. R/FR

Table 2 Growth of lettuce and broccoli by supplemental lighting

		Lettuce			Broccoli		
	Days of treatment	Before sunrise	After sunset	Comparison	Before sunrise	After sunset	Comparison
Weight (g)	9	0.11	0.11	0.10	0.17	0.17	0.16
	16	0.27	0.26	0.24	0.38	0.35	0.27
	22	0.31	0.26	0.26	0.53	0.43	0.33
Height (cm)	9	3.31	3.15	2.88	3.27	3.21	3.16
	16	5.02	4.95	4.64	7.14	6.61	5.57

4. CONCLUSIONS

The mercury-free trial HPS lamp with high CRI operated with choke type ballast showed nearly the same electrical and optical characteristics as those of conventional mercury-contained HPS lamps with high CRI. This mercury-free HPS lamp with high CRI operated with choke type ballast is expected for street lighting, flood lighting and so on. If the color point of mercury-free HPS can be closer to Planckian Locus or on it, it will be used widely. This lamp is also suitable on seedling growth for supplemental lighting. In this experimental period, fine days continued. The effectiveness wasn't so much clear, especially with lettuce. We'll try again this year.

5. ACKNOWLEDGEMENTS

Authors acknowledge to Dr. J. Maya and Dr. J. Ravi of Matsushita Electric Works R&D Laboratory for their useful bits of advice. We also acknowledge to Mr.Shiotani and his members of Matsushita Electronics Corporation for making plant growth lamps, and thank to Mr. T. Morita and Mr. S. Kosaka of Matsushita Electric Works for promoting the applicant on plant growth.

6. REFERENCES

[1]. N.Saito, et al., Preprint of 1997 Ann. Conf. of I.E.I. of Jpn. No.36
[2]. N.Saito, et al., Proc. of 3rd Lux Pacifica. B-71 (1997)
[3]. K.Murakami, et al., Proc. 1997 Ann. Meeting Kinki Branch Soc. Agric. Meteorol. Jpn. No.1

C 15

ECOARC : A NEW CONCEPT FOR ENERGY REDUCTION

Rudy Geens, Johan Cox and Carlo Vlekken
Sylvania Lighting International, Industriepark Soldatenplein 13, B-3300 Tienen, Belgium

1 INTRODUCTION

High Pressure Sodium lamps as retrofit for High Pressure Mercury Vapour lamps have been on the market for quite some time. These lamps offer a higher luminous output at a reduced energy consumption. The lumen gain is realised thanks to the much higher luminous efficacy of the HPS lamps while the power reduction is due to the lower power factor of the HPS lamp. These lamps usually apply a Ne-Ar Penning gas mixture in order to start the HPS lamps without the use of an ignitor on the HPMV socket.

This paper describes a further step in the enrage reduction without sacrificing luminous output and life by applying a new concept.

2 TECHNICAL DETAILS

The lamp power of an HID lamp is determined by the ballast and by the lamp voltage through the so-called ballast curve [1]. This curve gives the relationship between the lamp power and the lamp voltage. Figure 1 shows the ballastcurve of a HPMV lamp (HSL) and of an HPS lamps for HPMV retrofit (SHX). The operation point of the HPMV lamp is centred around 130 V. As can be seen the power of the SHX lamp is reduced due a vertical shift of the ballast curve.

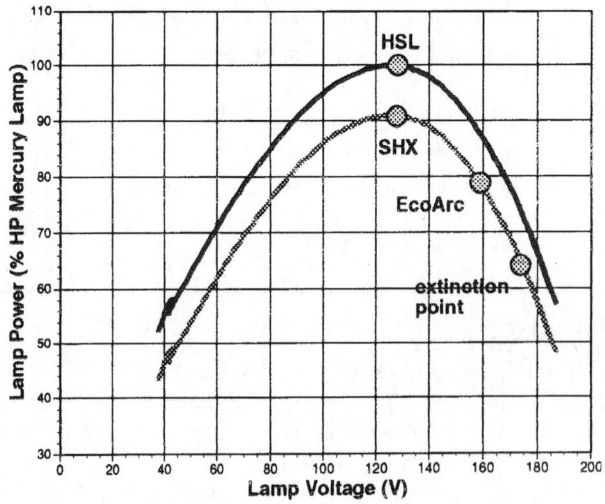

A further power reduction can be obtained by positioning the lamp voltage at a lower or a higher lamp voltage. However, if the lamp operates at a voltage below approximately 100 V, the current drawn becomes excessively high with a risk if overheating the ballast. On the other hand, at a high lamp voltage the lamp life will be limited because of the fact that the lamp voltage rises over life so that the extinction point will be reached sooner.

A solution to this problem is realised by making the HPS lamp unsaturated and positioning the lamp voltage at an increased value. A power reduction of up to 25 % of the HPMV value has thus be obtained. The unsaturated design [2] ensures that the lamp voltage can not rise over life. In order to maximise the luminous flux, Xe is used as the starting gas. An internal starter is mounted inside the lamp to ensure reliable starting of the lamp on HPMV positions. The Xe fill gas also ensures that the life of these lamps exceeds the life of the older Ne-Ar based technology.

This lamp concept only works if the reactions of sodium inside the arc tube are reduced to a minimum. This is realised by applying extremely good processing techniques and by using the right materials and lamp design. A further demand to the lamp is that the lamp-to-lamp voltage variation is very narrow. This is required in order to limit the power variation and to make sure that every individual lamp remains well below the extinction point.

3 CONCLUSIONS

A lamp family called Ecoarc offering reduced energy consumption at better luminous efficacy and life is developed. The family consists of a 98 W for replacement of an 125 W HPMV lamp, a 190 W for replacement of 250 W HPMV and a 295 W for replacement of a 400 W HPMV.

4 REFERENCES

[1] J. De Groot and J. Van Vliet, The High Pressure Sodium Lamp, Philips technical Library (1986)
[2] R. Geens and E. Wyner, IEE proceedings-A **140**, N0. 6, 450 (1993)

C 16

Life Performance of Inductively Coupled Electrode-less Metal Halide Lamp

Akira Itoh, Kozo Uemura, Toshihiko Ishigami, Akihiro Inoyue
Toshiba Lighting & Technology Corp.
1-201-1, Funakoshi-cho, Yokosuka-shi, Kanagawa 237-8510, Japan

1. Introduction

The inductively coupled electrode-less metal halide lamps (ICMH) are expected to show high efficacy and good life performance because they don't have electrodes.1) RE(rare earth metal)I3-NaI doses and rare gases are in the ICMH arc tubes without mercury. However, after several hundreds hours operation, some free iodine gas generated in their arc tubes. As iodine gas is apt to capture free electrons, it caused their ignition failure. It was tried to add limited quantity of some metal halide additives in their arc tubes to decrease free iodine gas.

2. Arc tube design of the ICMH

A rated wattage of the arc tubes is 300W. The arc tubes are made of fused quartz and their outside diameter is 29mm. A ignition probe is put on the arc tube surface. A main dose is a mixture of ScI3-NaI (weight ratio = 1:5) or NdI3-NaI (1:0.7). Combinations of CsI, TlI and SnI2 are put as additives. The xenon gas pressure in the arc tube is 16,600Pa.

3. Measurement of amount of free iodine gas

Amount of free iodine gas is estimated from absorption depth by iodine gas at 510nm wavelength.2) The arc tubes are heated up to 250°C in a furnace to fully vaporize solid iodine. Then a light output from a tungsten halogen lamp is irradiated to the heated arc tubes to measure the absorption depth.

4. Life Performance of the ICMH

Limited quantity of CsI, TlI and SnI2 additives are added to the arc tubes. Amount of free iodine gas is measured for the several arc tubes which have the different doses and additives after their 1000 hour operation. A result is shown in Fig.1. Their doses and the additives are listed on Table 1. CsI additive in the arc tube B & C and TlI additive in the arc tube C have some effect to reduce amount of iodine gas. On the other hand, SnI2 additive completely suppresses generation of iodine gas and has nothing to do with kinds of the rare earth metals and their quantity. Their lumen maintenance are shown in Fig.2. The arc tube D,E and F are operated on 400W loading and their life times in Fig.2 are multiplied by an acceleration rate. The acceleration rate for 400W operation is deduced to be 2.71 from comparison of amounts of iodine gas between on 300W operation and on 400W one. Though all arc tubes show no wall-blackening, no devitrification and excellent lumen maintenance, the arc tubes without SnI2 additive (A,B,C) were not able to survive over two thousands hours. On the other hand, those with it (D,E,F) are still operating.

5. Role of the additives

Both of NaI and ScI3 can generate iodine gas. Sodium gas of NaI is easily ionized and can migrate through a wall made of fused quartz. Scandium gas of ScI3 can react with it to compose scandium-silicates. Each contribution to the iodine gas generation is estimated through a furnace heat test and it suggests that mainly NaI contributes to it and ScI3 does as much as a fifth of NaI.

CsI additive is supposed to decrease the arc temperature and the wall temperature as a result. As for TlI and SnI2, their iodine partial pressure is high and it brings about decrease of the sodium and scandium partial pressures at the tube wall. Especially, SnI2 additive has quite high iodine partial pressure and is apt to composes metal halide complexes with NaI and ScI3. A simple calculation indicates that sodium partial pressure in the arc tube D is approximately a hundredth of the one in the arc tube A at 1073K and scandium partial pressure is as well a thousandth. Further more, SnI4 is more stable than SnI2 at a room temperature. Even if iodine gas is liberated, SnI2 absorbs it to compose SnI4; SnI2+I2→SnI4. Accordingly, more quantity of the SnI2 additive is required to make the ICMH life time longer. On the other hand, increase of SnI2 weight ratio to the main doses brings about decrease of their efficacy, which is shown in Fig.3 in the case of the ScI3-NaI mixture. The smaller weight ratio is preferable to make their efficacy higher.

6. Conclusion

In the REI3-NaI dosed ICMH, generation of iodine gas is completely suppressed by adding limited quantity of SnI2. It reduces the RE gas and sodium gas partial pressures at the tube wall and their loss rates as a result. Even if iodine gas is liberated, SnI2 absorbs iodine gas to compose SnI4. Accordingly, those arc tubes do not fall into an ignition failure and are expected to have an average life time of several ten thousands hours.

Reference

[1] Stewart C. N. et al : "*Inductively Coupled HID Lighting System*", LS6 (1992) 78:L

[2] Suenaga M.. et al : "*Deterioration of Metal Halide Arc Tubes due to Sodium Iodide at High Temperature*", J. Light & Vis. Env., Vol.2, (1978)

Table 1 Doses and Additives(weight ratio)

No.	ScI3	NdI3	NaI	CsI	TlI	SnI2
A:	1.0	0.0	5.0	0.0	0.0	0.0
B:	1.0	0.0	5.0	0.3	0.0	0.0
C:	1.0	0.0	5.0	0.3	0.3	0.0
D:	1.0	0.0	5.0	0.0	0.0	0.06
E:	1.0	0.0	5.0	0.0	0.0	0.9
F:	0.0	1.0	0.7	0.0	0.0	0.1

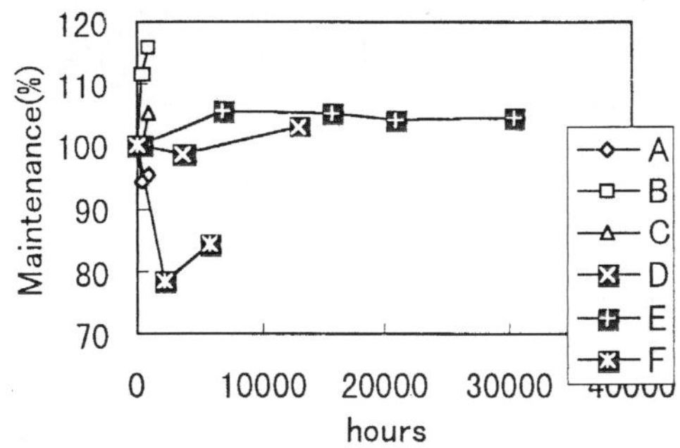

Fig.2 Lumen Maintenance of the ICMH

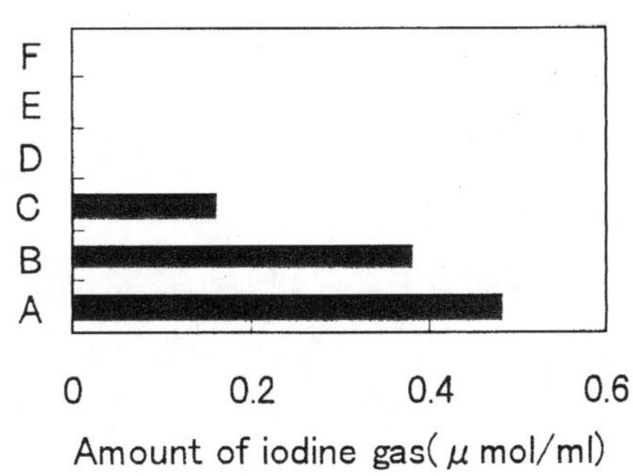

Fig.1 Amount of Iodine Gas at 1000 hours

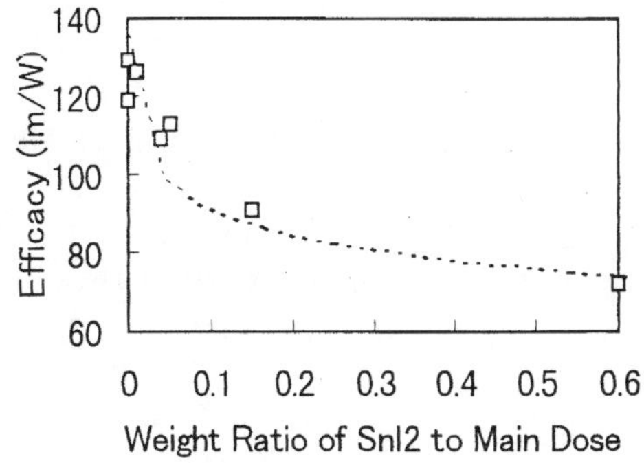

Fig. 3 Decrease of Efficacy with SnI2

C 17

HPS LAMPS WITH INTEGRATED ANTENNA

Yolanda J.J. Dams[1], Huub A.M. Coenen[1], Rob A.J. Keijser[2]
[1]Philips Lighting, Turnhout, Steenweg op Gierle 417, B2300 Turnhout, Belgium
[2]Philips Lighting B.V., P.O. Box 80020, 5600 JM, Eindhoven, The Netherlands

1. Introduction

HPS lamps with a high xenon pressure become more and more popular as a result of their increased luminous efficacy. In order to ignite such lamps at starting voltages not exceeding those specified by IEC or ANSI an ignition aid is necessary [1]. Mostly such an ignition aid is a simple wire, providing a direct contact between one of the lead wires and the outer surface of the ceramic HPS arc tube. After completion of the ignition process a bimetal switch will open, so as to avoid large voltage gradients across the arc tube wall in the steady state. The construction of the ignition wire or antenna is mostly complex, adds to the lamps cost price and, incidentally, bad contacts can lead to ignition problems during lamp life.
The present paper gives a new approach of implementing an ignition aid in HPS lamps that is both simple and reliable.

2. Arc tube concept with integrated antenna

In the new concept the antenna is incorporated in the arc tube by means of a tungsten trail on its outer surface. Electrically the antenna is floating, i.e. there is no longer a direct contact with one of the lead wires (figure 1).

Fig. 1: HPS concept with integrated antenna.

Various tests have been done with different geometry's. Both the length and the width of the W-trail have been varied. It has been found that the width of the W-trail can be kept small so that light losses are minimal. The antenna appears to be optimal when it extends over the full length of the arc gap. Best ignition results have been obtained with additional rings around the arc tube surface at both ends (see figure 1).

3. Practical results

At equal xenon pressure, the breakdown voltage with the new antenna concept is somewhat higher because of the absence of an electrical connection with one of the lead wires. On the other hand, the spread in ignition voltage is far less as compared with the standard antenna construction (figure 2). This is due to the limited reproducibility of the antenna contact with the arc tube in the latter case.

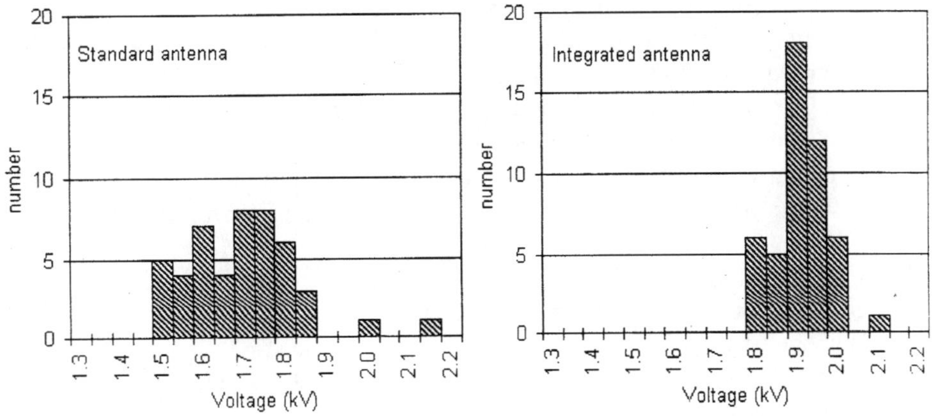

Fig. 2 Ignition voltage distributions for HPS lamps with standard antenna and with integrated antenna.

Our tests also indicate that the reliability of ignition over lamp life is much better. The old configuration is much more sensitive to degradation of materials such as the bimetal switch, leading to lamp ignition failures over life.
Also the absence of extra weld joints leads to an increased reliability.

An additional benefit of the new antenna design for lamp performance is that the (warm) re-ignition time is considerably reduced. Some typical numbers for various HPS types have been collected in Table 1.

Table 1: Re-ignition times (in seconds) for various HPS lamp types.

	100 W	150 W	250 W	400 W
integrated antenna	17	3	29	42
standard antenna	27	12	53	68

4. Conclusion

The concept of an integrated antenna has been proven to be a reliable and simple aid to ignite HPS lamps with high xenon pressure.

Reference:
1. De Groot, J and Van Vliet, J. ; "The High Pressure Sodium Lamp", Philips Technical Library, (1986).

C 18

DIMMING OF CERAMIC METAL HALIDE LAMPS

Rob A.J. Keijser
Philips Lighting B.V., P.O Box 80020, 5600 JM Eindhoven, The Netherlands

1. Introduction

Over the last years there has been a growing interest in dimming possibilities for various types of gas discharge lamps. Driving force is that considerable energy savings could result. For HID lamps in general and metal halide lamps in particular there are two major issues connected with lamp operation at reduced power.

The first one is the dependence of luminous flux and spectral distribution of the emitted light on the lamp power level. A second important issue concerns the possible consequences of MH lamp operation at low power levels for lamp performance over life. Earlier studies [1,2] have reported on results obtained for specific types of quartz metal halide lamps.

Recently, HID explorations have led to the introduction of a new generation of metal halide lamps in ceramic envelope with light technical properties superior to those of the traditional quartz metal halide lamps [3].

The present paper aims to extend the exploration of metal halide dimming to also include these new ceramic lamp types.

2. Lamp explorations

We have carried out an investigation of both the the light technical and electrical behaviour of various types of quartz and ceramic metal halide lamps at reduced lamp power levels. Different power reduction methods have been applied (variable transformer, phase-cutting systems).

It has been found that in most cases the variation of light technical properties only depends on the lamp power level and that the dimming method applied has no significant influence on this relation. A characteristic result obtained, comparing the variation of luminous flux, colour temperature and colour rendering index with relative power level for comparable type quartz and ceramic metal halide lamps is shown in fig. 1.

It is found that both the decrease in luminous flux and the change in colour temperature and colour rendering are considerably less for the ceramic type lamps.

The explanation for this reduced sensitivity for the reduction of the lamp power can be found in the fact that the metal halide vapour pressures (a sodium - thallium - rare earth iodide mix in both lamp types) are considerably higher in the ceramic lamps.

Regarding the other aspects of dimmed operation, our results indicate that concerns and limitations for ceramic and quartz metal halide lamps are similar. Lamp operation at low power can lead to extra blackening as a result of excessive evaporation or sputtering of tungsten from the electrodes. It also appears that, for dimming systems operating at 50/60 Hz , beyond a certain point improper electrode functioning can provoke visible flicker at the mains frequency.

References

1. Smith, D. and Zhu, H. Properties of High Intensity Discharge Lamps Operating on Reduced Power Lighting Systems., J.Illum.Eng.Soc. Vol. 22 (nr. 2) pp. 27-39 (1993).
2. Gibson, R.G. Dimming of Metal Halide lamps., J.Illum.Eng.Soc. Vol. 23 (nr. 2) pp.19-25 (1994).
3. Seinen, P.A., High Intensity Discharge Lamps with Ceramic Envelopes., 7th Int. Symp. on the Science and Technology of Light sources. Kyoto (1995).

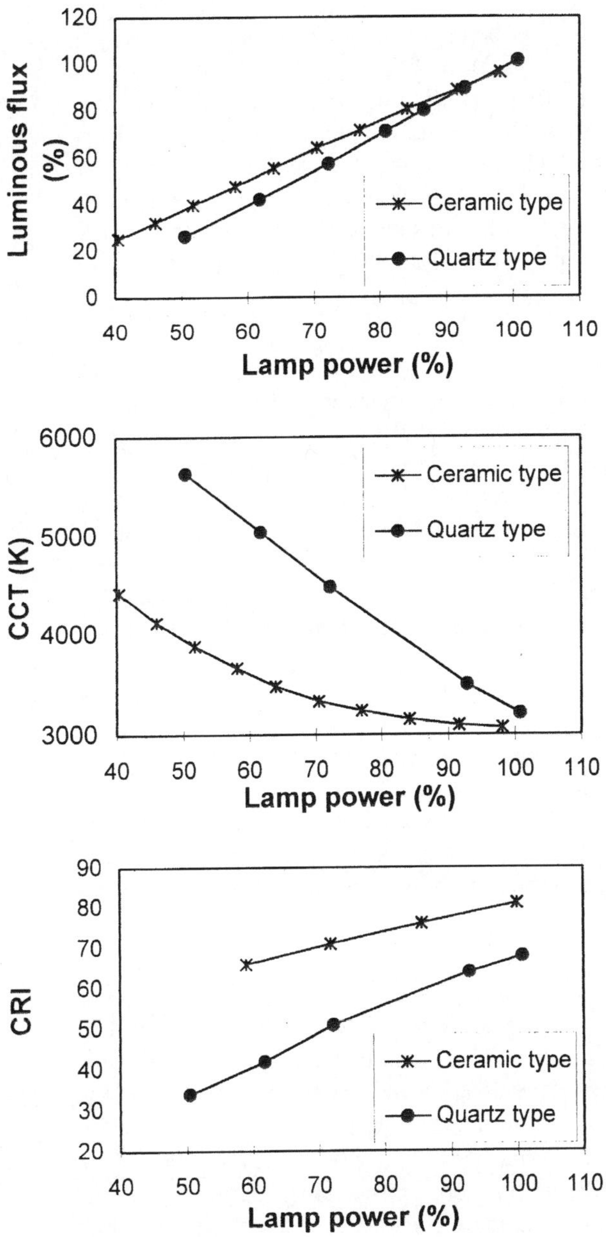

Fig.1 Comparison of the variation of luminous flux, colour temperature and colour rendering with lamp power for 70 Watt 3000K metal halide lamps.

C 19

Arc-Straightened Operation of Short Arc Metal Halide Lamps Containing HoI3-InI for LCD Projectors

M. Horiuchi, K. Miyazaki, S. Kominami, K. Takahashi and M. Takeda
Lighting Res. Lab., Matsushita Electric Industrial CO., LTD.
3-4 Hikaridai, Seika, Soraku, Kyoto, 619-0237 Japan

1. Introduction

An arc of a horizontally operated HID lamp is bended upward by convection, which causes overheating of a fused quartz envelope of the lamp. Consequently, the envelope is distorted and devitrification of the envelope occurs. In a metal halide lamp for LCD projectors, especially, a rapid growth of devitrification and distortion of the envelope remarkably shorten lamp life. Therefore the bend of the arc should be reduced as possible. It has been known that the bend of the arc can be reduced by operating the lamp at certain frequencies (Arc-Straightened Operation)[1]. In this paper, we describe characteristics of lamps horizontally operated by Arc-Straightened Operation for LCD projectors. Our experiments show that straight arc lamps have an advantage over arc-bending lamps in terms of lamp life.

2. Experiments

In this study a new 200W short arc metal halide lamp were prepared . It has a almost spherical envelope made of fused quartz with the highly load of 60W/cm2, the inner diameter of 10.4mm and the arc length of 3mm. HoI3, InI, Hg and Ar are contained in the envelope.

We have already reported that an arc is straightened by standing waves formed in a envelope when a lamp is operated with the ac current at the frequency *fc* given by the following formula [2,3] :

$$fc = k \times V/D \ (KHz)$$

where V(m/sec) is a sound velocity in the envelope, D(mm) is an inner diameter of the envelope and $k \times V$ equals about 210.

First of all, we determined the frequency *fc* of the HoI3-InI lamp at which the arc was straightened, and measured characteristics such as envelope temperatures and a life of the lamp whose arc was straightened (called the straight arc lamp). The characteristics were compared with those of the conventional HoI3-InI lamp whose arc was bending. In a life test we measured a maintenance factor of a spatially average illuminance on the screen to which light emitted from the lamp was projected. In the projection system the lamp was arranged within a reflecting mirror. Also, X-ray photographs of the envelope through the mirror were taken with respect to burn time.

3. Results

Figure 1 shows a comparison of temperature distribution on the envelope between the straight arc lamp and the conventional one. The straight arc lamp is operated by a 16KHz sine wave ac current and the conventional lamp by a 250Hz rectangular wave one. It is shown that by straightening the arc, the envelope tends to be more isothermal. In the straight arc lamp the temperature at the top portion of the envelope decreases from 890℃ to 810℃, while there is an increase of approximately 20 degrees at the middle and the bottom portions, respectively. The difference in temperature between the hottest and the coldest portions of the straight arc lamp decreases to about 80 degrees as compared with about 150 degrees of the conventional one.

Figure 2 shows the relative screen illuminance maintenance of straight arc lamps and conventional ones. When a lifetime is defined as the time at which the spatially average illuminance on the screen decreases to 50% of the initial one straight arc lamps have a lifetime more than 6000 hours. This value is twice or more than a lifetime (2000~ 4000hours) of conventional lamps. We consider that the long lifetime resultes from suppression of devitrification of the envelope. This is because devitrification of the envelope strongly affects a convergence of the reflecting mirror. As a result the illuminance on the screen is decreased remarkably.

Typical X-ray photographs of envelopes of these lamps at 2000 hours are shown in Figure 3. By straightening the arc the envelope distortion can be suppressed.

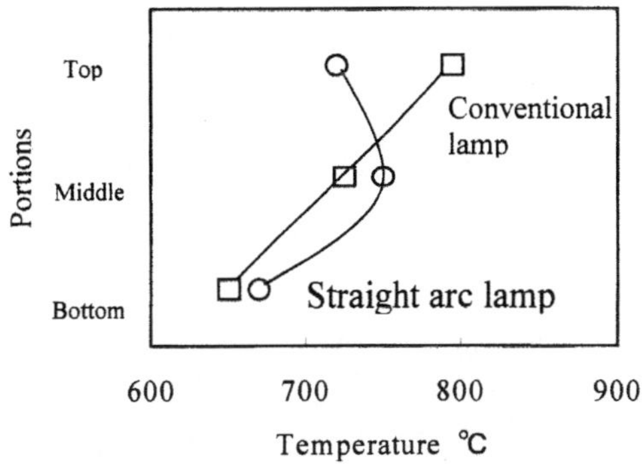

Fig.1 Envelope temperature distributions of the straight arc HoI3-InI lamp and the conventional one.

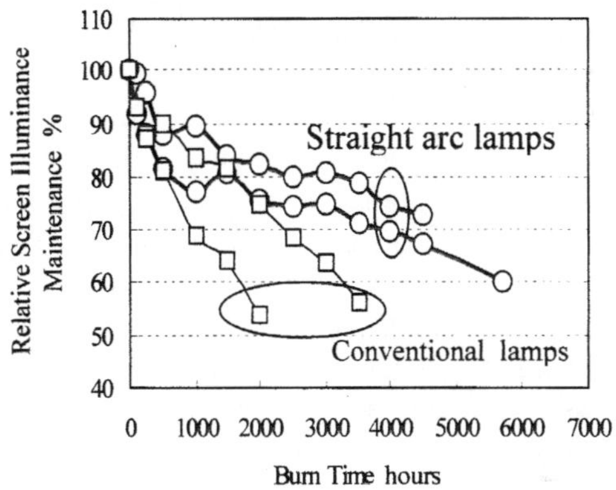

Fig.2 Screen illuminance maintenance of straight arc HoI3-InI lamps and conventional ones.

4. Conclusion

Characteristics of the HoI3-InI lamp which arc is operated to be straightened are described. It is seen that there are great advantages of suppression of the envelope devitrification and distortion in straightening the arc. We will apply this Arc-Straightened Operation to shorter arc lamps for LCD projectors and smaller miniature lamps such as discharge automotive headlamps.

5. REFERENCES

[1] D.C. Fromm and J. Seehawer, Lighting Res. Technol. 27(1), pp17-26(1995)

[2] M. Horiuchi, S. Kominami, K. Takahashi,K. Miyazaki and M. Takeda, Light Sources and Related Systems, 96-16 (1996)

[3] K. Miyazaki, M. Horiuchi, S. Kominami, K. Takahashi, S. Horii and M. Takeda, National technical Report, Vol.43 No.2, pp154-161 (Apr. 1997)

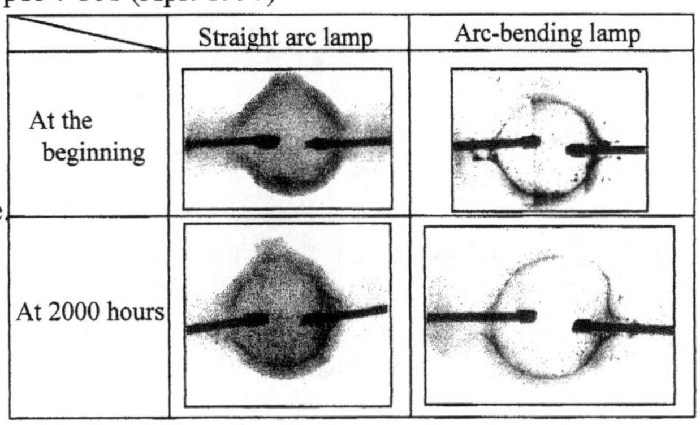

Fig.3 Typical X-ray photographs of envelopes of the straight arc HoI3 -InI lamp and the conventional one

C 20

THE EVOLUTION OF LOW WATTAGE METAL HALIDE LAMPS WITH CERAMIC ARCTUBES

S.A.Mucklejohn

GE Lighting Europe, Melton Road, Leicester LE4 7PD, UK

1 INTRODUCTION

The potential advantages of ceramic arctubes for metal halide discharge lamps have been recognised for many years. It is only since the mid-1990's, however, that such lamps have been commercially available [1]. The development of metal halide lamps with ceramic arctubes has a history dating back over 20 years. The first detailed report of a lamp that had commercial potential was published in 1982 [2]. This lamp contained a sodium-tin mixed halide dose {Na+Sn+Cl+I} in a polycrystalline alumina (PCA) arctube with cermet closures and leadthroughs. Ceramic arctubes offer three key advantages over fused silica arctubes: (i) higher operating temperatures giving increased luminous efficacy and colour rendering properties; (ii) greatly reduced sodium loss from the plasma thus giving stable colour through the life of the lamp; (iii) improved control over arctube dimensions which gives rise to low colour spread and which also helps with the control of lamp voltages during production.

The kinetics of chemical interactions between polycrystalline alumina and some metal halide systems are such that arctubes can operate at temperatures up to ~1150°C compared to ~950°C for fused silica. The high operating temperatures of lamps with ceramic arctubes, however, do give rise to many potential materials problems. Thus, the early development of metal halide lamps with PCA arctubes was beset by difficulties with corrosion which offset the performance advantage over fused silica arctubes. The current designs of low wattage metal halide lamps with ceramic arctubes have addressed the materials problems that in the past have limited life and lumen maintenance to levels below that acceptable to the market.

2 PERFORMANCE

The progress in lamp performance from the early 1980's to the mid 1990's can be appreciated from the data listed in Table 1 which compares the properties of two 150 W lamps. Although reference [2] does not contain a comprehensive list of performance parameters, it does provide sufficient information to show that although the lamps of the 1980's and the 1990's have similar luminous efficacies at 100h, the more recent lamps show immense advances in colour uniformity, colour stability and colour rendition properties. Similarly these lamps with ceramic arctubes show substantial improvements in performance over the most recent lamps with fused silica bodies. The latter typically not only have a larger colour spread at 100h but also show a colour shift of >500K at 2000h. The former have efficacies approximately 20% higher than the fused silica counterparts.

3 SUMMARY

Birkhofer (3) in his paper presented in 1986 outlined possible future developments for low wattage metal halide lighting systems. These developments were expressed as economy and design requirements. The majority of these features, quantified in Table 1 for a 150W lamp, are provided by the new generation of metal halide lamps with ceramic arctubes. Future developments will include colour temperatures other than 3000K and extending lamp life. Hot re-ignition will remain a severe technical challenge because of the thermal characteristics of metal halide discharges in ceramic arctubes.

Economy
* high luminous efficacy
* good colour stability
* long life with good lumen maintenance
* low production costs (lamp)
* low system cost (ballast, ignitor, luminaire)

Design
* low wattages 20 to 150W
* compact form and high luminance
* CCT 3000K and >4000K
* CRI>75
* universal burning position
* small, lightweight ballasts & ignitors
* quick warm-up & hot re-ignition
* safety

The introduction of metal halide lamps with ceramic arctubes into the market has marked a significant change in the performance of low wattage metal halide lamps. It also marks the beginning of a new era of customer expectations in terms of colour consistency both in terms of uniformity and stability. These improvements will change the currently accepted perception of metal halide lamp performance and will inevitably lead to major challenges for light source manufacturers.

Table 1: Comparison of lamp performance

Performance	1982 *(a)*	1994 *(b)*
power	150W	150W
lumen output at 100h	13500 lumens	14000 lumens - vertical 13500 lumens - horizontal
efficacy	90 lumens per watt	93 lumens per watt
lumen maintenance	not given	85% at 2000h 70% at 5000h
CCT	3800K	3000K
chromaticity coordinates	x = 0.390; y = 0.385	x = 0.435; y = 0.400
colour uniformity	~ +/- 350K at 100h	< +/- 150K at 100h
colour spread	not given	< +/- 200K at 5000h
R_a	>70	>80
life *(c)*	not given	6000h
operating position	vertical only	universal - depending on jacket type

Notes
(a) Data from reference [2]
(b) Data from Philips booklet 'Mastercolour CDM lamps and gear' (1994)
(c) Life defined as hours to 50% survival

4 REFERENCES

[1] S.Carleton, P.A.Sienen and J.Stoffels, J. Illumin. Eng. Soc. **26**, 139 (1997).
[2] K.E.Brown, A.G.Chalmers & D.O.Wharmby, J. Ilumin. Eng. Soc. **11**, 106 (1982).
[3] H-P.Birkhofer, Lighting in Australia 14 (1987).

C 21

THE LIFE OF A TWINARC HIGH PRESSURE SODIUM LAMP

Rudy Geens
Sylvania Lighting International, Industriepark Soldatenplein 13, B-3300 Tienen, Belgium

1 INTRODUCTION

High pressure sodium lamps with two arc tubes mounted in parallel in a vacuum outer jacket have been on the market for quite some time as safety lamps in case of a power interruption. Recently, Sylvania introduced the SHP Twinarc family. These lamps combine two high pressure xenon (Super) arc tubes in parallel

In view of the extended life of a super lamp compared to the standard HPS lamp, this combination projects an extremely long life. It is usually assumed that the life of a dual arc tube lamp is about doubled with respect to the single arc tube equivalent .

2 TECHNICAL DETAILS

When a Twinarc lamp is started, the high voltage of the ignitor is applied over both arc tubes. In the case of an unused lamp, the pulse breaks down both arc tubes and brings them in a discharge characterised by a glow mode during the current reversal part of the AC period. The main AC current is drawn by one of the arc tubes while the other arc tube draws a keep-alive current of variable amplitude. Due to the high reignition voltage peaks associated with a glow mode and occurring at every current reversal, the identity of the arc tube drawing the main current can switch at every half cycle. At the instant that -in one of the arc tubes- an arc discharge can be sustained through the current reversals, the other arc tube can then no longer be reignited so that it cools down and extinguishes.

At every start-up the ignition process is repeated and the identity of the arc tube that eventually reaches full operation is largely a statistical matter. The time dependence of the probability of an arc tube to become the active arc tube has been studied. The study shows that the duty is statistically divided over both arc tubes. This essentially means that a Twinarc lamp ages two times slower than a one arc tube lamp and also that the lumen maintenance is twice as good. However, due to the fact that both arc tubes initially go into a glow mode, also the non-active arc tube plays a role at every start. Therefore, life and maintenance are not exactly doubled.

The failure rate distribution of the dual arc tube lamp is mathematically modelled as the integrated product of two single arc tube distributions. The failure rate at time t is the sum of all combinations of two distributions which give a combined arc tube life equal to t. If the single arc tube distribution is labelled $f_S(t)$, the combined failure rate can be written :

$$f(t) = \int_0^t f_s(\tau)\, f_s(t-\tau)\,d\tau$$

The single arc tube failure rate $f_S(t)$ is evaluated as follows. From the above it is clear that life and maintenance of the Twinarc lamp can not be modelled as two lamps starting in turn. Rather, we will use a model for the failure rate based upon two fatigue components : fatigue through repetitive starting (failure rate distribution f_{rs}) and fatigue by continuous burning (failure rate distribution f_{cb}). The f_{rs} rate is applicable for lamps which are continuously switched on and off but have a negligible number of burning hours. The f_{cb} rate is applicable for lamps burning continuously without being switched off. It is assumed that the failure rate components can be combined into the single arc tube distribution f_S as follows :

$$f_s(t) = R_{cb}(t)f_{rs}(t) + R_{rs}(t)f_{cb}(t)$$

where

$$R_{cb}(t) = 1 - \int_0^t f_{cb}(\tau)d\tau \text{ and } R_{rs}(t) = 1 - \int_0^t f_{rs}(\tau)d\tau$$

are the fractions of survivals at time t for the respective components.

The repetitive starting component of one arc tube in a dual arc tube lamp is evaluated by means of a rapid cycling test of Twinarc lamps. In this test the actual burning time must be negligible (i.e. 30 min. on, 30 min. off). The failure rate distribution f_{rs} is obtained by making a Weibull analysis of the test data.

The continuous burning component was not initially evaluated since it involves long and costly life testing. Therefore it was calculated from the available information of the failure rate distribution of SHP Super lamps at the standard 11 hr on, 1 hr off cycle $f_0(t)$ and from the Weibull function $f_1(t)$ obtained from a separate rapid cycle test of SHP Super lamps. By means of the combined failure procedure explained above, the continuous burning fatigue component of one arc tube is fitted as a Weibull distribution

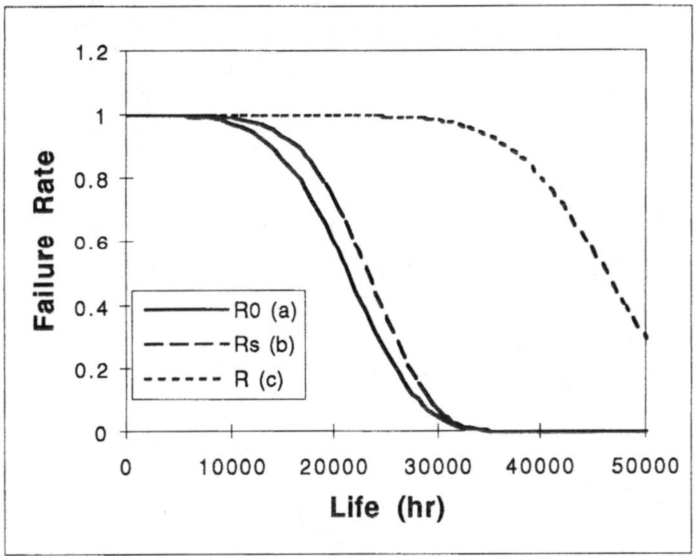

Fig. 1: Median survival curves of the SHP-TS 70W lamp (a) compared to the single arc tube median survival curve (b) of the SHP-TS 70W Twinarc . Also shown is the survival curve of the SHP-TS 70W Twinarc (c).

The analysis was carried through for the case of SHP-TS 70W Twinarc lamps. The result is summarised in the comparison of the survival curve of the SHP-TS 70W lamp and the single arc tube SHP-TS 70W Twinarc (Figure 1). This figure also shows the survival curve of the complete SHP-TS 70W Twinarc.

3 CONCLUSIONS

The data shows surprisingly that the fatigue of repetitive starting component of Twinarc lamps is negligible. Hence the results from the analysis indicate that starting of a Twinarc arc tube is facilitated by the presence of the second arc tube. This means that starting of a Twinarc lamp hardly influences the failure rate. Due to the fact that for a Super lamp, the repetitive starting component is much more important, the average life of a Twinarc lamp is actually slightly more than twice the life of the corresponding Super lamp. So it must be concluded that that the life of a Twinarc lamp is longer than the sum of the individual lives of the two arc tubes.

Excimer Devices

D 01

HIGH POWER EXCILAMPS

Victor F. Tarasenko, Mikhail I. Lomaev, Aleksei N. Panchenko,
Victor S. Skakun, Eduard A. Sosnin

High Current Electronics Institute, SB of Russian Academy of Sciences,
4, Akademichesky Ave., Tomsk, 634055, Russia

1 INTRODUCTION

In the recent years, interest in design and development of powerful spontaneous radiation sources (excilamps) operating in the UV and VUV regions has grown [1-4]. These sources are considerably simpler in design and operation as compared to lasers and emit in wider range of wavelengths. The most efficient excilamps on transitions of rare gas dimmers R^*_2 and rare gas monohalides RX^* can find wide industrial application. Therefore improving of their output parameters is of great importance. In this paper, output parameters of excilamps with different geometry pumped by continuous glow discharge, high-pressure volume discharge with UV-preionization and pulsed barrier discharge are presented. The current article reviews the experimental results, pumping generators and excilamps peculiarities, and their highest output parameters, as well.

2 EXPERIMENTAL SETUP

Design of the coaxial, planar and cylindrical excilamps under investigation is described in [2-4]. The coaxial and cylindrical excilamps were made from quartz tubes of high quality. Diameter of the tubes was 10-60 mm and its length was up to 90 cm. The gap between the tubes in coaxial excilamp was changed from 2.5 to 8 mm. Water cooling of the inner tube or electrodes was used. Glow discharge in planar lamp was initiated in a gap between solid cathode and mesh anode. Diameter of the electrodes was 7 cm, the gap was varied from 1 to 3 cm. UV radiation passed through high quality quartz plate. Profiled electrode 12 cm in diameter and CaF_2 output window were used in excilamps pumped by a high-pressure volume discharge with UV-preionization. In the case of a barrier discharge, ceramic plate was placed on the potential electrode and the gap was reduced to 0.8-8 mm.

The electrodes of glow discharge excilamps were coupled to DC power supply providing discharge current up to 1 A and voltage across the discharge gap up to 10 kV. AC voltage was used in some experiments. Discharge current I and voltage across the excilamps U were monitored using DC or pulsed millimeters and kilovoltmeter. Input power was calculated as $P=I\times U$. Two types of pulsed generators were used for high-pressure volume discharge with UV-preionization and barrier discharge formation. The first one produces voltage pulses with duration varied in the range from 50 to 200 ns and amplitude up to 40 kV. The second one produces sinusoidal voltage pulses with amplitude up to 10 kV and pulse repetition rate up to 20 kHz.

Rare gas (Xe, Kr, Ar) or rare gas (Xe, Kr, Ar, Ne and He) - halogen (F_2, CH_3Br, Cl_2, HCl, I_2, NF_3) gas mixtures were prepared in a separate gas-container.

3 RESULTS

Design and operating parameters of powerful excilamps with different discharge geometry pumped by glow discharges, high-pressure volume discharge with UV-preionization and barrier discharge was studied. Intense radiation of Ar_2*, Kr_2*, Xe_2*, ArF*, KrBr*, Cl_2, KrCl*, KrF*, XeI*, XeBr*, XeCl*, XeF*, I_2 and IBr molecules was obtained in rare gas or in rare gas - F_2

(CH_3Br, Cl_2, HCl, I_2, NF_3) mixtures. Excilamps with high spatial uniformity of the output, narrow emission line and high gas lifetime were developed.

It was shown that the efficiency of barrier discharge excilamps pumped by sinusoidal pulses with duration of several tens microseconds can be sufficiently improved. Radiation power density at λ~222 and 308 nm, total UV output power and efficiency of 3.6 W and 9.7 %, respectively, were obtained under pumping by AC current with frequency of 20 kHz.

Under pumping by UV-preionized high-pressure volume discharge maximal pulsed radiation densities of 5 kW/cm^2 were obtained in mixtures Ne-Xe-HCl (λ~308 nm) and He-Kr-F_2 (λ~250 nm).

Coaxial excilamp pumped by a glow discharge with the gap of 8 mm was found to emit the highest average power. Output power at λ ~222 nm up to 200 W (specific output power P_{sp}=1 W/cm^3) was obtained. Efficiency with respect to input power as high as 14% was demonstrated. Similar results were obtained in Xe-Cl_2 mixture at λ ~308 nm. Output power of about 500 W was demonstrated when three excilamps operated in parallel. It was shown that the efficiency of luminescence of exciplex molecules KrCl* up to 30% can be obtained in high-voltage glow discharge and positive column of glow discharge. High efficiency of KrCl* and XeCl* excilamps is caused by high rate of exciplex molecules formation in harpoon reaction and low rate of nonradiative decay of these molecules in glow discharge plasma. Peak radiation power (up to 3 W) was obtained in Xe: CH_3Br = 3:1 gas mixture at total pressure of 4 Torr. Therewith the radiation efficiency with respect to the input power was as high as 2,5%. Radiation power in Kr:CH_3Br = 3:1 gas mixture at total pressure of 6 Torr does not exceed 0,85 W. In this case spectral half-width of KrBr* molecules B-X band (λ ≈ 206 nm) was observed to be no more than 2,5 nm.

Additionally Xe-NF_3 and He-Xe-NF_3 mixtures were studied. Working pressure was lower than 12 Torr, otherwise discharge constriction took place. Discharge current over 30 mA and voltage across the lamp 5-6 kV were obtained for tubes 40 cm long and 9 mm in diameter. Discharge voltage increased up to 7 kV when He was added into the mixture. In that case peak average output power over 20 W and efficiency up to 6% were obtained. Notice that the rate of degradation of working mixture containing NF_3 was higher as compared to that of KrCl and XeCl-excilamp mixtures.

4 ACKNOWLEDGEMENTS

This work has been made under partial financial support from Russian Foundation for Fundamental Research, Project No 96-02-16668-a.

5 REFERENCES

[1] "Science &Technology of Light Sources", *Proceedings of the 7th International Symposium, Kyoto, 27th-31th August,* 1995.

[2] M.I. Lomaev, A.N. Panchenko, V.S. Skakun, E.A. Sosnin, and V.F. Tarasenko, "High-average-power exciplex flashlamps," *Proc. of the Intern.conf. Atomic and Molecular Pulsed Lasers*, Tomsk, Russia. 1995. V.2619. P.294-302.

[3] V.F. Tarasenko, M.I. Lomaev, A.N. Panchenko, E.A. Sosnin, "Coaxial, cylindrical and planar UV excilamps, pumped by glow or barrier discharge," *Proc. of the International conference Photonics West'97*, USA, 1997, Vol. 2992, pp. 24-34.

[4] E.A. Sosnin, M.I. Lomaev, A.N. Panchenko, V.S. Skakun and V.F. Tarasenko, "Glow and Barrier Discharge Efficient Excilamps," *Proc. of the Intern.conf. Atomic and Molecular Pulsed Lasers*, Tomsk, 1997, Vol. 3403.

D 02

CONTINUOUS WAVE VACUUM ULTRAVIOLET LIGHT SOURCES BASED ON EXCIMER HALIDES

Andreas Ulrich[1], Jochen Wieser[1], Hossein Dahi[2], and Daniel E. Murnick[2]
[1]Fakultät für Physik E12, Technische Universität München,
James Franck Str. 1, D-85747 Garching, Germany
[2]Department of Physics, Rutgers University, 101 Warren Street, Newark, NJ07102, USA

1 INTRODUCTION

Efficient light sources in the vacuum ultraviolet (VUV) spectral region are of growing interest for applications such as photochemistry and photolithography. Developments in the semiconductor industry using deep ultraviolet photolithography are summarized in ref. [1]. Excimer emission from pure rare gases and rare gas halides is considered for light sources in this field due to high efficiencies and short wavelengths. Extremely precise and elaborate imaging optics has to be designed for each photolithographic system. Light sources where quartz optics can still be used ($\lambda \geq 160$nm) are therefore likely to be used in the next step towards shorter wavelengths, shorter than the so called i-line at 365nm.

A light source which can produce excimer light contiuously and under well controlled conditions has recently been developed [2]. Here we descibe its application for rare gas halide excimer systems, in particular ArCl* emitting at a wavelength of 175nm.

2 EXPERIMENT

The excimer light source is schematically shown in Fig.1. A continuous 10-20keV low energy electron beam is formed in an evacuated section of the setup and focused through an extremely thin nonstoichiometric silicon nitride (SiN_x) ceramic entrance window into a gas filled section where light emssion is produced by collisional excitation.

The SiN_x windows are typically 300nm thick and have an aperture size of $1 \times 1 mm^2$. They are leak tight and can withstand pressure differentials up to 2000hPa. The foils are produced using standard solid state technology starting with a silicon wafer which finally provides stable frames for easy handling of the foils. Metal coatings are used to protect the foils from reactive etching when rare gas halide mixtures are used as the light emitting gas.

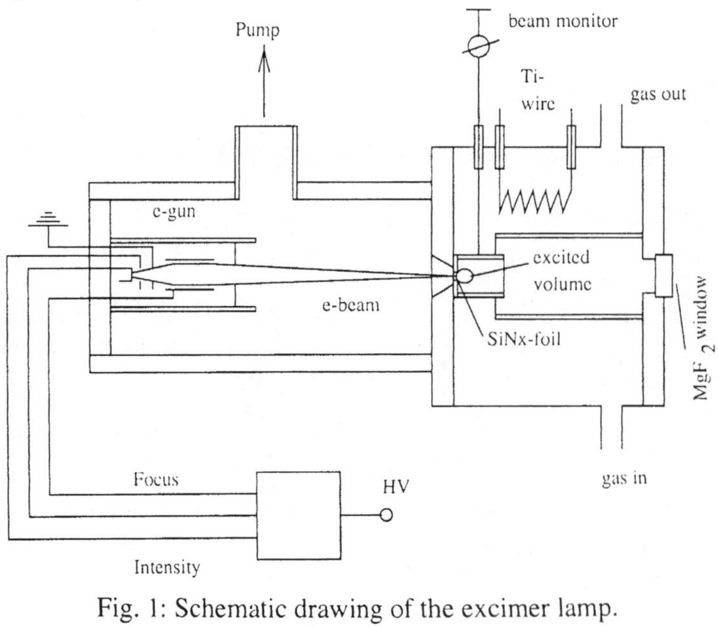

Fig. 1: Schematic drawing of the excimer lamp.

The electron beam loses less than 10% of its energy in the thin entrance foil and is stopped in the gas exciting a small region near the entrance foil. The energy deposition depth e.g. of 15keV electrons in 1000hPa Ar is 1mm [3, 4]. Electron beam currents up to 50μA can be sent through a 1mm^2 aperture. For 20keV beams this corresponds to 1W total beam power and a power deposition per unit area of the entrance foil of 100W/cm^2.

3 ARGON - CHLORIDE EXCIMER EMISSION

Energy transfer from rare gas excimer molecules leading to excitation or ionization of other species and subsequent light emission is a well known process. It is used for a whole class of laser systems using hard ionizing radiation for pumping of dense gas mixtures in which gas kinetic processes lead to selective excitation of specific levels and population inversion.

Despite high efficiencies which can be achieved for light emission on the excimer continua they have not been widely used as light sources because continuous and homogeneous excitation of dense gases is technically not easy to achieve. The system described above has now been used to study excimer halide formation by mixing low concentrations (<1%) of a halogen gas in argon.

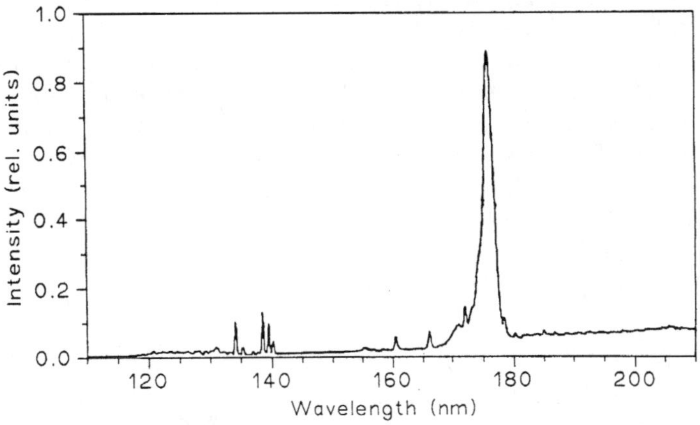

Fig 2: E-beam excited Ar (1bar), Cl_2 (2hPa) emission.

Detailed experiments on Cl_2 in Argon, for example, allow study of $ArCl^*$ production yielding a bright lamp at 175nm with an estimated efficieny of electron beam energy into light exceeding 5%. The pressure and concentration dependence of the spectral output which includes ClI lines near 138nm and Ar_2^* radiation at 129nm provides information on the excitation kinetics. At pressures near 1000hPa and Cl_2 concentrations near 0.2% the $ArCl^*$ band at 175nm is the only bright spectral feature in the 100 to 250nm region (Fig. 2). The output intensity scales with electron beam current as expected from previous work. Experiments on other excimer halides and rate equation modeling of the collision kinetics are in progress.

4 ACKNOWLEDGEMENTS

Work has been funded by NSF Grant No. CTS 94-19440, NATO, INTAS, A. v. Humboldt foundation. SiN_x foils were manufactored at Lucent Bell Laboratories and Fraunhofer Institut für Festkörpertechnolgie München.

5 REFERENCES

[1] Ruth DeJule, Semiconductor International, **Feb. 1998**, 54 (1998)

[2] J. Wieser, D.E. Murnick, A. Ulrich, H.A. Huggins, A. Liddle, and W.L. Brown, Rev. Sci. Instr. **68**, 1360 (1997)

[3] M.J. Berger and S.M. Seltzer, Tables of energy losses and ranges of electrons and positrons, (NASA Report No. SP-3012, Washington, 1964)

[4] A. Cohn and G. Caledonia, J. Appl. Phys. **41**, 3767 (1970)

D 03

COAXIAL AND PLANAR EXCILAMPS PUMPED BY BARRIER DISCHARGE

Eduard. A.Sosnin, Victor. S.Skakun, Victor .F.Tarasenko
High Current Electronics Institute, 4, Academichesky Ave., Tomsk, Russia, 634055.

1 INTRODUCTION

There is a number of applications for incoherent eximer UV-radiation such as photochemistry, curing and surface modification. Large body of papers on characteristics of dielectric barrier discharge exilamps (DBDL) have been published [1-4]. Althought the DDBL has been studied for the long time, relationship between the efficiency and pumping regime has not been discussed widely. In this study we investigate which regime is most usefull for efficient operation of DBDL.

2 EXPERIMENTAL SETUP

Coaxial and planar lamps were used. The coaxial lamp consist of two concentric transparent quartz tubes with foil inserted into the inner tube and external mesh electrodes. Diameter of the outer tube is varied from 5 to 6 cm and its length is 10 cm. The gap d between the tubes is changed from 7 to 9 mm. In the case of planar lamp the output window diameter and d are 8 cm and 3 mm, respectively. Moreover planar DBDL has only one dielecric electrode with $\varepsilon \sim 1000$, covered metal electrode. The barrier discharge lamps was driven by three types of power supplies (PS). Characteristic parameters of the DBDL PS are shown in Fig.1.

	Voltage pulse FWHM duration (s)	Voltage pulse amplitude (kV)	Voltage pulse polarity, PRR (Hz)
1	up to $2 \cdot 10^{-7}$	up to 30	one sign, 100 - 400
2	$\sim 5 \cdot 10^{-8}$	5 - 16	opposit sign, 0.5 - 12
3	$\sim 1{,}2 \cdot 10^{-5}$	2 - 4	opposit sign, 22

Fig.1. Characteristic parameters of power suppliers for DBDL.

Pure rare gases, binar gas mixtures Xe-(Cl_2, HCl) and triple mixtures Ne(He)-Kr(Xe)-(Cl_2, HCl) were studied in our experiments. Gas pressure was varied in the range from 0,4 to 1,2 atm. Procedure of average radiation power mesurement was similary to [3].

3 EXPERIMENTAL RESULTS AND DISCUSSION

Characteristic waveformes of discharge current, voltage across the gap and UV spontaneous emission obtained with different PS are shown in Figs.2,3. Under pumping by ns pulse PS (lines 1 and 2 in Fig.1) the current sign is generally changed more than once while the voltagr pulse still in progress; that is electrical energy transfering to load is unoptimal. Under pumping by μs pulse PS (lin. 3 in Fig.1) allows to extend the time of electrical energy transfer to the load. The duration of the enegy transfer at fixed pd is determed by the minimal applied voltage necessary to initiate the discharge and the point, where extremum of the voltage is reached.

Lut us compare UV radiation power density and efficiency in Ne-Xe-HCl ($\lambda \sim 308$ nm) mixture under pumping by of 2nd and 3rd PS. At 12 kHz in first case ~ 0.4 mW/cm^2 and ~ 0.4% and at 22 kHz in second case ~ 14 mW/cm^2 and 9% were obtained, respectively. This comparsion is possible, because output flux is approximately proportional to the kHz range of PRR. Similar values were obtained in another working mixtures.

Low efficiency of ns-voltage pumping regime utilizing is also assosiated with high output power density per one particle of the gas. It is conceivable that in that regime turnes on the quenching of the exciplex molecule channel.

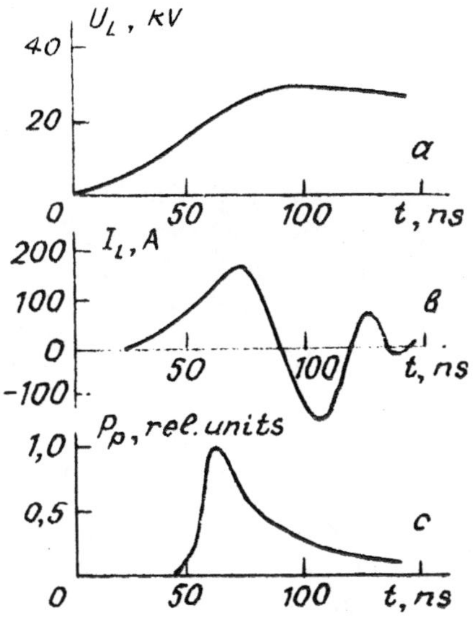

Fig.2. Oscilloscope traces of voltage (a), discharge current (b), UV radiation (c) obtained under pumping by PS 1.

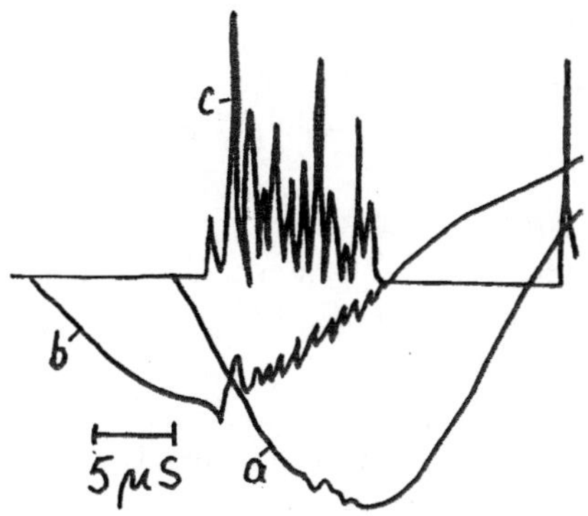

Fig.3. Oscilloscope traces of voltage (a), discharge current (b), UV radiation (c) obtained under pumping by PS 3.

Anoter reason is the gas overheating in narrow region of the filament caused so-called "discharge channel area radiation" to be dominant. It is known that the maximum contribution to the radiation in DBD filament made so called "surface discharge" region of filament [4]. Going to pumping by sinusoidal μs voltage pulses we can increase both surface discharge contribution and UV-output. For example, filaments in the optimal in term of peak UV-output mixtures Ne-Xe-Cl_2 (*pd* ~ 300 cm·Tor) includes little or no the narrow channel areas and the filaments takes a form of two cone, based on the dielectric surfaces.

Under pumping of pure rare gases (Ar,Xe) and its mixtures with He,Ne by PS 3 in planar lamp significant gain in UV-output was also obtained as compared to pumping from 1 and 2 generators.

4 CONCLUSIONS

The advantages of 10-μs sinusoidal voltage for pumping of the DBDL are presented. The efficiency up to 10% on both XeCl* and KrCl* molecules is achieved.

5 ACKNOWLEGMENTS

This work has been made under partial financial support from Russian Foundation for Fundamental Research, Project No 96-02-16668-a.

6 REFERENCES

[1] G.A.Volkova, N.N. Kirillova, E.N. Pavlovskaya, and A.V. Yakovleva, *Zh. Priklad. Specroskopii.* **41**, 681-695 (1984).

[2] B.Eliasson and U.Kogelschatz, *Appl. Phys. B.* **46**, 299-303 (1988).

[3] V.F.Tarasenko, M.I.Lomaev, A.N.Panchenko, E.A.Sosnin, *Proc. of the Intern. conf. Photonics West'97*, USA, Vol.2992, 464-466 (1997).

[4] S.Kaiser, H.-P.Popp, *Proc. of the 7-th Int. Symp. on Science & Technology of Light Sourses*, Japan, **46:P**, 163-164 (1995).

D 04

VUV-RADIANCE OF HIGH PRESSURE HOLLOW CATHODE DISCHARGES IN XENON

Hartmut Lange[1], Ahmed El Habachi[2] and Karl H. Schoenbach[2]
[1] Institut für Niedertemperatur-Plasmaphysik, Robert-Blum-str. 8-10, 17489 Greifswald, Germany
[2] Physical Electronics Research Institute, Old Dominion University, Norfolk, VA 23529, USA

1 INTRODUCTION

Medium pressure hollow cathode discharges are micro discharges between metal foils less than one milimeter apart, and with submillimeter diameter openings in cathode and /or anode.These discharges in xenon were found to be strong sources of a continous radiation even at vacuum-UV wavelengths. Measurements of the spectral radiance of this rare gas excimer radiation are communicated and comparision is made with the radiance of the xenon resonance line in low pressure gas discharges.

2 EXPERIMENT

The electrode geometry of the high pressure hollow cathode discharge consists of a cathode which contains some kind of a hollow and an arbitrarly shaped anode [1-2]. In this experiment the diameter of the cathode opening in the hollow cathode discharge geometry is on the order of 100μm. The discharge operates in a direct current mode up to atmospheric pressure.The discharge chamber with a LiF-window is mounted at one branch of a VUV radiation detection system [3] and the radiation of the plasma around the hollow cathode is detected by an evacuable scanning monochrometer. The setup was sensitive to wavelenghts in the range of 120 - 200 nm. The radiance L_n was measured directly at the discharge axis. An aperture of 0.6 mm diameter is placed very closed to Lif-window. A mirror forms an image of this cylindrical area on the entrance slit of the VUV-monochromator. For the calibration we compare the radiance of the discharge with the radiance of a radiometric source standard, measured by means of the same optical components which is mounted on the other branch of the VUV radiation detection system. Our radiation standard was a deuterium lamp (Cathodeon V04) calibrated by the Physikalisch Technische Bundesanstalt (PTB).

3 RESULTS

Fig. 1 shows the absolute spectral radiances of the micro hollow cathode discharge in xenon for various values of the gas pressures in the discharge chamber.These radiances were measured at a discharge current of 2 mA. The strongest emission, found at 172 nm, is in the shape of a continuum that can be attributed to transitions between the weakly bond excited molecular states $^1\Sigma^+_u$ and $^3\Sigma^+_u$ of Xe_2 and the ground state.The formation of these excimers is caused by in a three-body collision process of the lowest excited atomic states with two xenon atoms. Fig 2. shows the VUV-radiances which were integrated over all wavelengths as a function of pressure at a current of 2 mA. At higher discharge currents than shown, the discharge area outside the hollow cathode becomes higher. As a consequence of this behavior the radiance can decreased (curve b in Fig 2).
As an excample of another VUV-radiation source the radiance of the xenon 147-nm resonance line in the positive column of a low pressure gas discharge is presented in Fig. 3. A comparison of these radiances shows that radiance of the resonance line is a magnitude lower than that in Fig. 2, but the latter has only a very small radiating area. The possibility to form arrays of these discharges allows the generation of larger areas.

4 REFERENCES

[1] K. H. Schoenbach, M. Ciocca, A. El-Habachi, W. Shi, F. E. Peterkin and Th. Tessnow, Proc.

12-th Int. Conf. on Gas Discharges & Application, **Vol. I**, 280 (1997)

[2] K. H. Schoenbach, R. Verhappen, T. Tessnow, and F. E. Peterkin, Appl. Phys. Lett. **68**, 13 (1996)

[3] H. Lange, Proc. 12-th Int. Conf. on Gas Discharges & Application, **Vol.II**, 468 (1997)

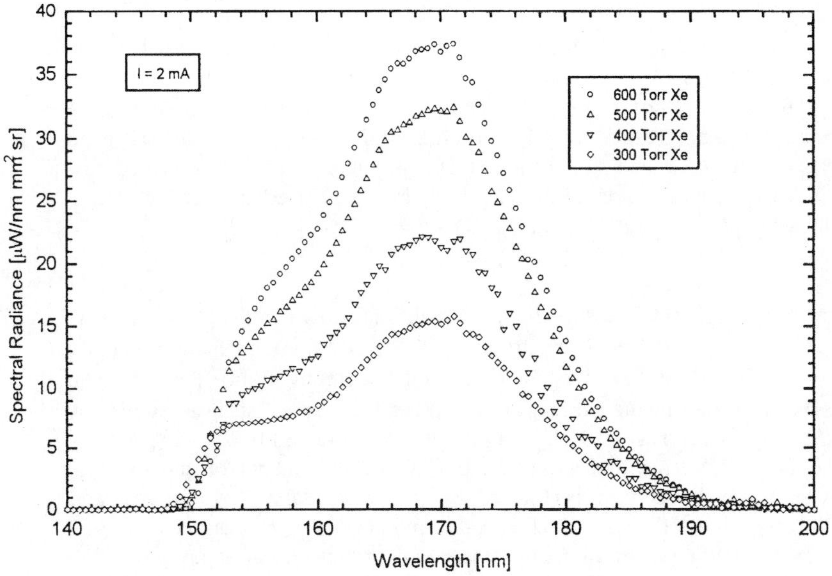

Fig. 1: Spectral radiance of xenon excimer continuum for various values of filling pressures of micro hollow cathode discharges

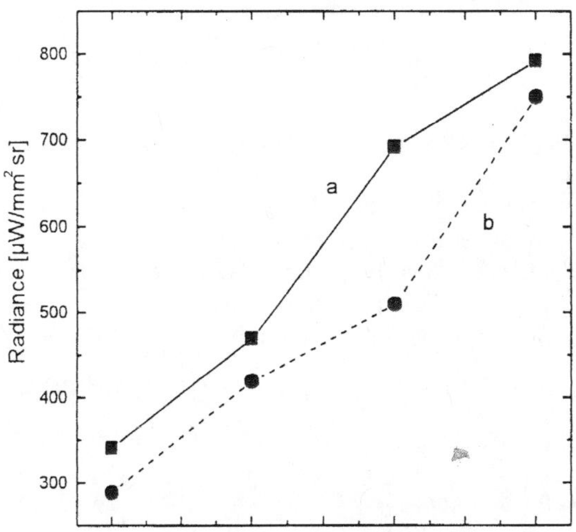

Fig. 2: VUV-radiance vs xenon pressure for two various discharge currents of the hollow cathode discharge (curve a: 2 mA, curve b: 4 mA).

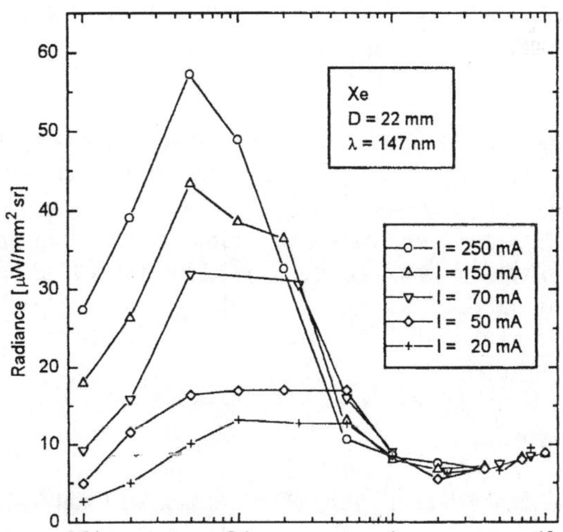

Fig. 3: Radiance of the xenon resonance line vs pressure, measured in the positive column of a low pressure glow discharge in a cylindrical glass tube [3].

D 05

Microhollow Cathode Discharge Excimer Lamps

Robert H. Stark, Ahmed El-Habachi, Wenhui Shi, and Karl H. Schoenbach
Physical Electronics Research Institute, Old Dominion University, Norfolk, VA 23529, USA

1. Introduction

Microhollow cathode discharges [1,2] are high pressure, direct current gas discharges between hollow cathodes with circular openings of approximately 100 µm diameter and a hollow or solid anode. These discharges, operated in either xenon or argon were found to be strong sources of excimer radiation at 172 nm and 130 nm, respectively [3]. In order to utilize the micro discharges in excimer lamps, they need to be arranged in arrays. Efforts to form discharge arrays are described, and the potential for high power excimer lamps is discussed. Besides being used as excimer lamp, applications in flat panel displays, plasma reactors and plasma cathodes are considered.

2. Experiment and Experimental Results

The electrode geometry of a microhollow cathode-anode system is shown in Fig. 1. Generally, molybdenum foils of 100 µm thickness are used as electrodes, with holes drilled either mechanically or by means of a pulsed laser. Cathode hole sizes range from several 100 µm to 30 µm. The dielectric is mica. Arrays are presently fabricated in the same way, but other techniques, such as etching is considered to form microelectrode structures. The VUV radiation is observed end-on, and spectrally resolved by means of a 0.5 m VUV spectrograph [4]. Absolute measurements of the excimer emission from discharges in xenon have been performed by means of calibrated sources, and calibrated detectors [5]. The results showed a continuous increase in intensity with pressure up to 400 Torr (Fig. 2). The efficiency (radiative power of the excimer emission/electrical input power) at optimum pressure was found to be between 6% and 9 % (Fig. 3). The radiative power of an individual discharge at 400 Torr in xenon at a voltage of 210 V and a current of 3 mA is consequently 50 mW. The intensity increases linearly with current in the range of observation (1mA-6mA). Experiments with an array of microhollow cathodes (Fig. 4) showed that parallel operation is possible if certain conditions (which at time due to patent considerations cannot yet be revealed) are satisfied. The discharges are stable and do not interfere with each other for distances on the order of the hole diameter.

3. Discussion

Combining the results of the single hole and array measurements allows us to predict the achievable radiative power density of a microhollow cathode discharge array. In an array consisting of 100 µm hollow cathodes, spaced 100 µm apart the power density of excimer emission at 172 nm is 125 W/cm^2. Assuming that 50% of this radiation can be utilized, the power density is still more than 60 W/cm^2. At this power level the thermal loading would be for dc operation very high and either efficient cooling would be required or pulsed operation needs to be considered. Experiments with the goal to increase the efficiency, and consequently reduce the thermal loading at the same radiative power are underway.

4. References

[1] K. H. Schoenbach et al, Appl. Phys. Lett. 68, 13
[2] K. H. Schoenbach et al, Plasma Sources Sci. Technol. **6**, 468 (1997)
[3] K. H. Schoenbach et al, Proc. Gas Discharges and Their Applications, Greifswald, Germany, 1997, p. 280.
[4] A. El-Habachi and K.H. Schoenbach, Appl. Phys. Lett. **72** , 1 (1998).
[5] A. El-Habachi and K.H. Schoenbach, Appl. Phys. Lett., submitted

Fig. 1

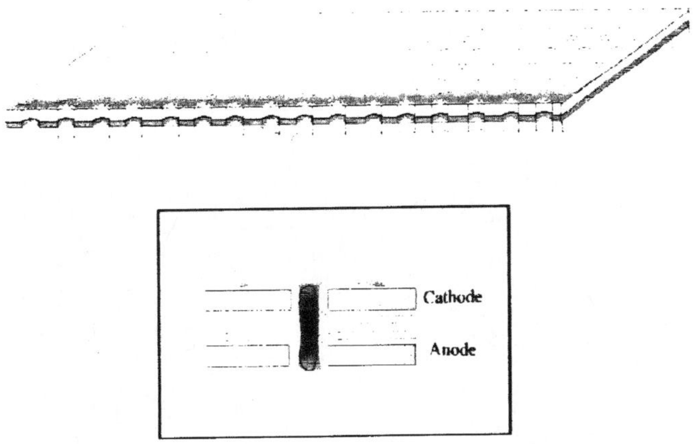

Fig. 2

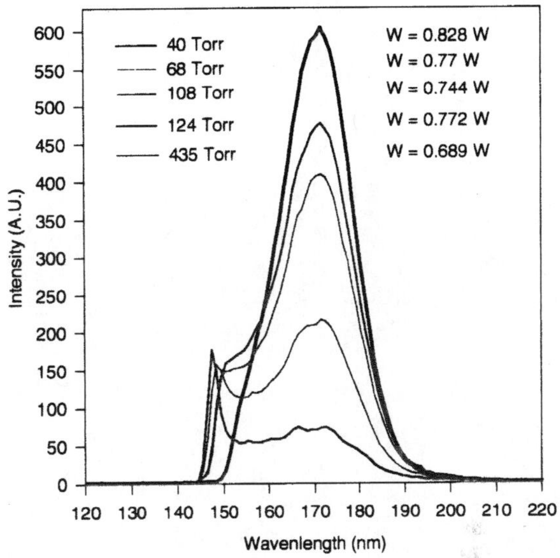

Fig. 3

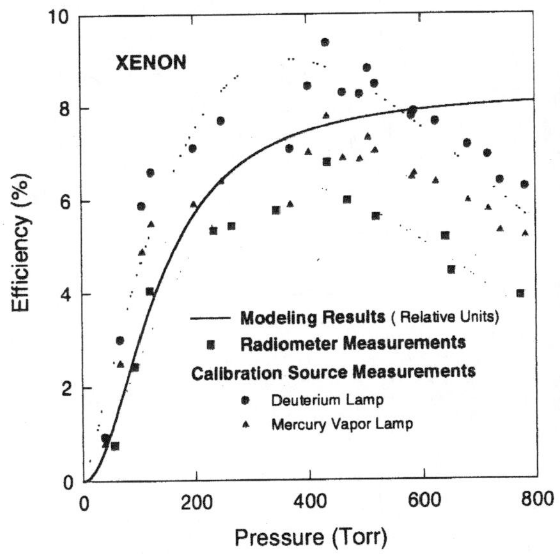

Fig. 4

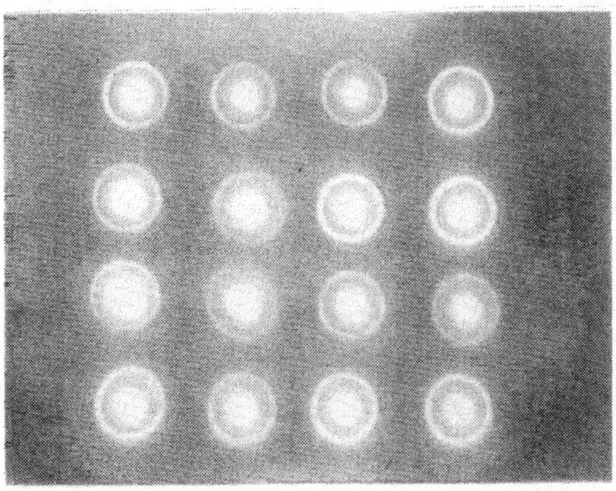

Fig. 1 Schematics of microhollow cathode-anode array
Fig. 2 Excimer intensity at various gas pressure in xenon
Fig. 3 Efficiency of the excimer radiation sources versus pressure.
Fig. 4 Parallel operation of microhollow cathode discharges

E

Dielectric Barrier Sources

E 01

VUV SPECTRUM OF BARRIER DISCHARGE IN Xe-Kr MIXTURE.

G.N.Gerasimov, G.A.Volkova,G.N.Zvereva
S.I.Vavilov State Optical Institute,Birzhevaya line 12, St-Petersburg,199034 Russia

It is well known that in krypton plasma with small xenon impurity effective energy transfer takes place from lighter to havier gas.Addition of Xe to Kr bring about changes in Kr spectrum: Xe resonance line,moleclar continuum of Xe_2^* and $KrXe^*$ molecular emissions appear [1,2]. The aims of our work were to continue rare gas mixture spectra investigations by means of spectrometer with higher resolution (0.02 nm) and to study the possibility of inverse population in rare gas mixtures.

During experiments three barrier discharge sections with 3 mm discharge gaps and sinusoidal 2,5 kv tension were used.

In Fig.1 spectrum of krypton discharge with 1% Xe impurity is shown at the 120-200 nm wavelength range. It is evident that as compared to pure Kr spectrum, at this spectrum krypton emission intensity decreases several times.There are two maxima in emission spectrum: at 156 nm and at 172 nm. The first one is assigned to $KrXe^*$ molecular emission, the second - to Xe_2^* molecule. With Xe concentration increasing , Xe_2^* molecular emission becomes dominating in discharge spectrum.

In Fig.2 the results of plasma component intensity calculations in Kr-Xe barrier discharge are presented.Plasma component densities were found by means of solving kinetic system equations . The choice of the electric field shape and main discharge parameters such as electron density (10^{14} cm^{-3}) and discharge duration (10 ns) was based on the work [3].The gas temperature was taken to be equal to 300°K, pressure - 300 torr. In Fig.2 it is possible to notice faster krypton component intensities decreasing after discharge end and slower changes of xenon and $KrXe^*$ component values.It is of interest to note rather high intensity values of Xe_2^* and $KrXe^*$.

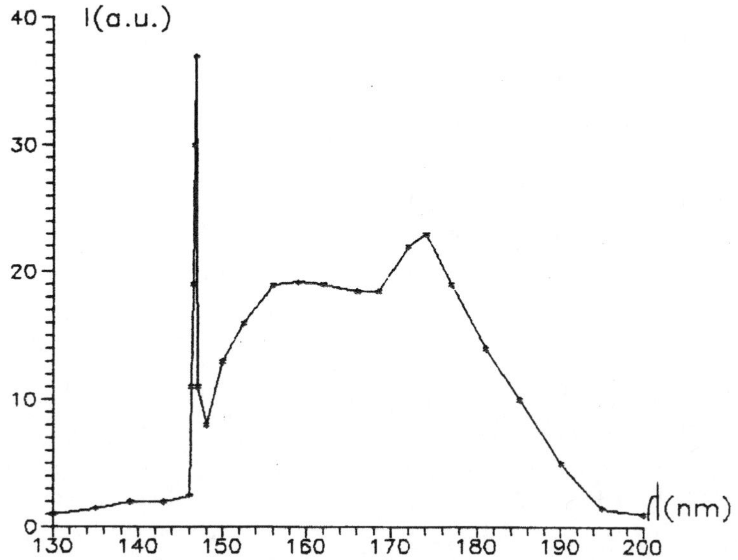

Fig.1 Barrier discharge spectrum of Kr with 1% Xe impurity, P=300 torr.

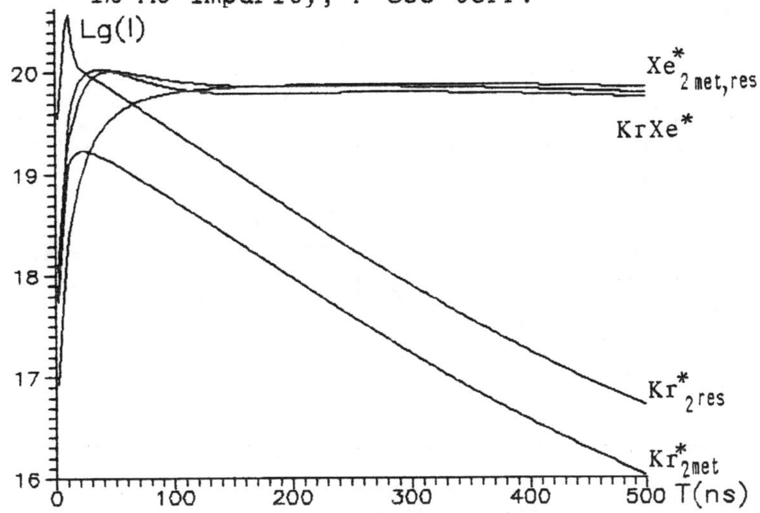

Fig.2 Calculations of Kr with 1% Xe plasma component emission intensities.

Investigations of emission spectra at 147 nm range at higher resolution and the comparisions of them with absorbtion spectra of $KrXe^*$ molecule make it possible to assign this emission to $KrXe^*$ molecule. Consequently, emission amplification at this region may belong to $KrXe^*$ molecule.

REFERENCES

[1] A.Gedanken, J.Jortner, B.Raz, A.Szoke, J. Chem. Physics, v.57, N8, p.3456, (1972)

[2] J.D.Cook, P.K.Leichner, Phys. Rev. , v.43, N3, p.1614 , (1991)

[3] B.Eliasson, U.Kogelschatz, Appl.Phys.B, v.46, pp.299-303, (1988)

E 02

EXPERIMENTAL INVESTIGATIONS OF PHOSPHOR COATED XENON BARRIER DISCHARGES

Jürgen Dichtl, Rainer Kling, and Manfred Neiger
LTI, University of Karlsruhe, Box 6980, D-76128 Karlsruhe (FRG),
E-mail: Manfred.Neiger@etec.uni-karlsruhe.de

1 INTRODUCTION

Dielectric barrier discharges (DBD) in rare gases like Xe and Kr are efficient uv radiation sources, emitting excimer bands around 172 nm (Xe_2*) and 146 nm (Kr_2*). Combined with appropriate phosphors (especially 2-quantum phosphors, if available) they offer new possibilities for Hg-free fluorescent lighting. The rare gas filling is not only a desirable substitute for Hg fillings, but it also adds the advantage of instant light emission. Other advantages are flexibility in geometric shape of discharge bulbs and considerably higher fill pressure (0.1 - 1 bar). According to previous work [1-4], Xenon DBD's are more efficient excimer radiators than lighter rare gases, with the added advantage of the longest uv wavelength (172 nm) of all rare gases (except Radon).

2 EXPERIMENTAL SETUP

Discharge tubes of rectangular cross-section (3-8 mm gap spacing, 50 mm width) were used for our investigations. Electrodes are metal grids on the outside of the two flat walls facing each other. Both discharges without and discharges with a thin phosphor layer deposited on the inside walls were investigated. The phosphors used were VUV phosphors mixed to a composition to emit white light when excited by Xe_2* radiation. To be able to easily change the Xe pressure, some of the discharge bulbs were not sealed, but were connected instead to a larger ballast tank, also filled with Xenon. Both sinusoidal and quasi-rectangular, bipolar voltage waveforms were used to drive the discharges (up to 14 kV peak-peak voltage and 1 to 50 kHz driving frequency). The lumen light flux of phosphor coated lamps was measured using a calibrated integrating sphere. A short time exposure framing camera (PCO-DICAM, 5 ns to 1 ms exposure time) was used to monitor different spatial configurations of individual microdischarges for different operating conditions, such as Xe pressure, gap spacing, voltage waveform, -amplitude, -frequency, and with or without phosphor coating.

3 EXPERIMENTAL RESULTS AND CONCLUSIONS

Figure 1 shows measured luminous efficacies of Xe-DBD's with phosphor coating versus specific electrical input power for both sinusoidal and rectangular voltage waveforms (5 mm gap, 300 mbar Xe pressure, 8 kHz operating frequency). Maximum luminous efficacies are considerably lower for sinusoidal excitation than for rectangular one, 20 lm/W instead of 45 lm/W in case of Fig. 1. The reason for this considerable improvement can be seen from the photographs in Fig. 2(a-c). For sinusoidal voltage (Fig. 2a) the discharge is strongly filamentory, i.e. individual microdischarges exhibit a constricted channel diameter in the gap and rather far reaching surface discharges with diameters considerably larger than the channel diameter (factor 5 and more). In case of rectangular voltages (Fig. 2b and Fig. 2c) the discharge looks different. In this case the individual microdischarges are much more diffuse, with a ratio of surface discharge diameter and channel diameter only slightly higher than one (see Fig. 2b), or the whole discharge volume is filled almost homogeneously by a single microdischarge (see Fig. 2c). As the charge transferred is almost identical in all 3 cases, we conclude that electron densities in Fig. 2a, i.e. for sinusoidal voltage, are considerably higher than those in Fig. 2b and Fig. 2c (rectangular voltages). We further conclude, that the average E/N-values are highest for the case of a single microdischarge filling the discharge volume completely (Fig. 2c), due to low electron densities in this case. On the other hand, average E/N-values are lowest for the case of strongly filamentory discharges like in Fig. 2a, because of much higher electron densities inside the constricted discharge channels. Additional benefical effects of lower electron densities are fewer loss processes, e.g. electron collisional destruction of excited Xe atoms and less local gas heating.

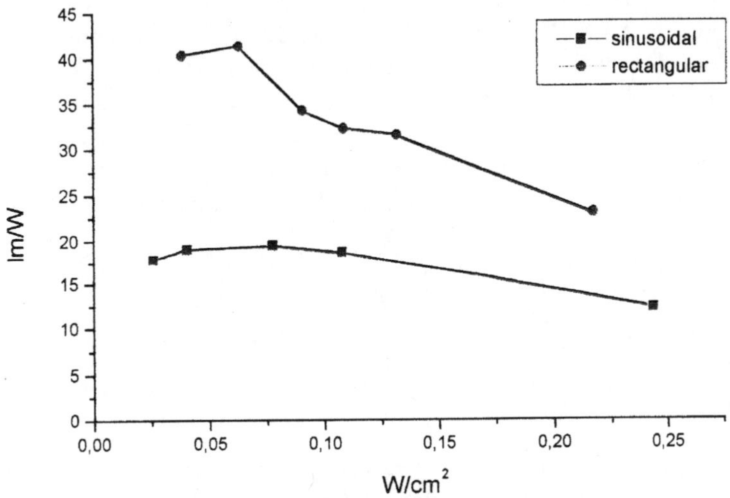

Fig. 1: Luminous efficiacy versus specific electric input power for sinusoidal (lower curve) and rectangular (upper curve) driving voltages (d = 5 mm, p_{Xe} = 300 mbar, f = 8 kHz).

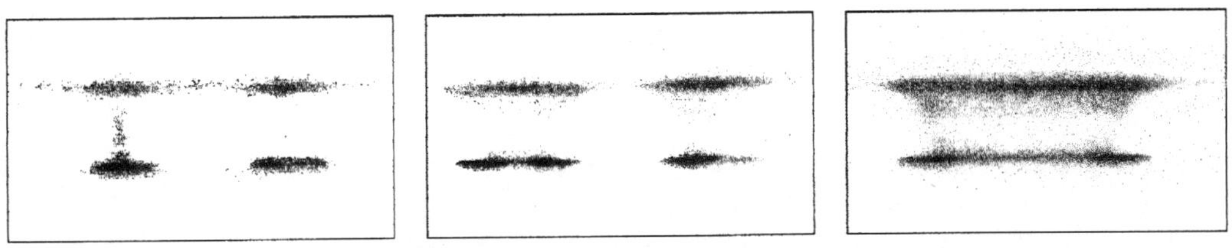

Fig. 2: Short time exposure photographs (anode on top) of Xe-microdischarges (d = 3 mm, p = 100 mbar, f = 20 kHz, exposure time 20 μs)

(a) sinusoidal (b) rectangular (c) rectangular at 20 % higher voltage than (b)

5 REFERENCES

[1] K. Stockwald, PhD Dissertation, LTI / University of Karlsruhe (1991)
[2] K. Stockwald and M. Neiger, Contributions to Plasma Physics, 35 (1), p. 15-22 (1995)
[3] F. Vollkommer and L. Hitschke, Phys. Blaetter, 53 (9), p. 887-889 (1997)
[4] R. Kling, PhD Dissertation, LTI / University of Karlsruhe (1997)

E 03

Investigations on the secondary electron emission coefficient in dielectric barrier discharges

S.Götze, S. Müller
Institute of Low-Temperature Plasma Physics
Robert-Blum-Straße 8-10, 17489 Greifswald, Germany

Introduction

Dielectric Barrier Discharges (DBD`s) are generated in gaps by means of AC-voltages where at least one electrode is covered by a dielectric layer.
Usual fields of technical application for DBD`s are radiation sources (powerful coherent infrared radiation in CO_2 lasers, incoherent VUV or UV excimer radiation in excimer lamps).
The properties of the plasma source are determined by choosing parameters, especially the external ones. There are a lot of important parameters like pressure, frequency, sustaining voltage, gap distance, gas mixture or the nature of the used dielectric materials.
The secondary electron emission coefficient (γ-coefficient) at the cathode is determined by the dielectric materials.
The γ-coefficient consists of γ_i , γ_m and γ_p , where γ_i means the electron emission due to ions, γ_m due to metastables and γ_p due to photons.
The present paper deals with investigations on the effective γ-coefficient of different dielectric barrier materials which are partly evaporated.

Experimental Details

In the experiments different dielectric barrier materials were used: glass, glass ceramic and evaporated layers of MgO, Al_2O_3 and CeO_2. The gas mixture consisted of $He/N_2/CO_2$. The total pressure range was 75-760 Torr.
The discharge was driven by a sinusoidal power amplifier in a range between 10 and 350 kHz.
The experimental setup of the discharge cell can be seen in Fig.1.
The effective γ-coefficient was estimated from the breakdown criterion given by the Townsend theory. Applying this theory (only for the ignition) it was possible to calculate the coefficient relatively.

Experimental results

The time behaviour of the discharge is pulse-shaped as can be seen from the discharge current (Fig. 2). Nevertheless, a nearly homogeneous optical discharge structure could be observed in a pressure range between 100 and 200 Torr (Fig. 3). Just beyond 200 Torr the appearance of the discharge changed into the well known filament like type (Fig. 4).
Fig. 5 shows the calculated relative γ coefficients versus the reduced electric field for various dielectric layers and the gas mixture ratio $He:N_2:CO_2$ / 8:3:1. As seen a DBD forming between MgO evaporated layers has the largest coefficient whereas the use of pure quartz layers lowers γ. Additionally, the γ-coefficient shows a strong E/p dependence in the lower range of the reduced electric field as one can see from Fig. 5.
Furthermore, the gas mixture mentioned above was replaced by the new composition of $He:N_2:CO_2$ / 4:3:1. The result is shown in Fig. 6.
The increasing concentration of the rare gas He results in a higher γ. This is an indication of the considerable influence of the He ions on the secondary electron emission yield.

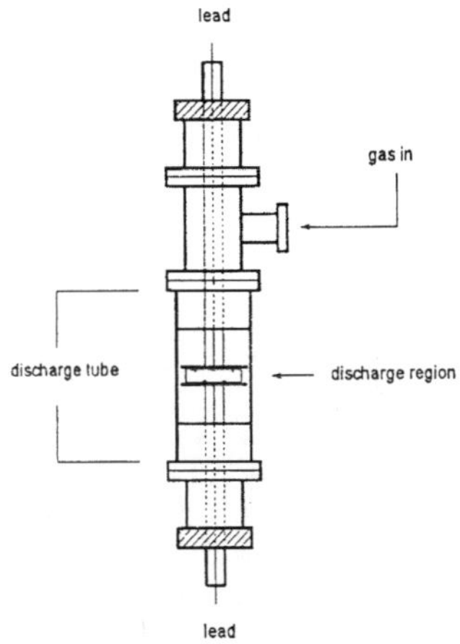

Fig.1 discharge tube

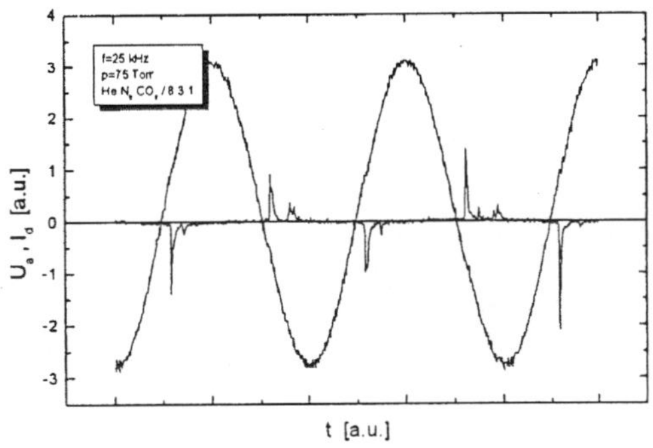

Fig.2 temporal behaviour of the discharge current

Fig.3 homogeneous discharge structure

Fig.4 filament like discharge structure

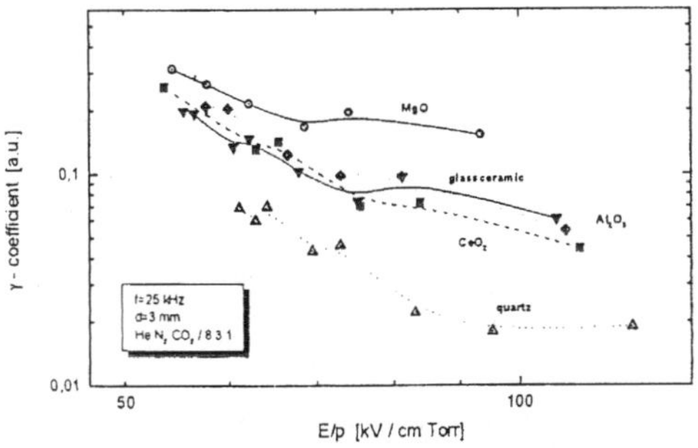

Fig.5 γ versus E/p for several dielectric layers

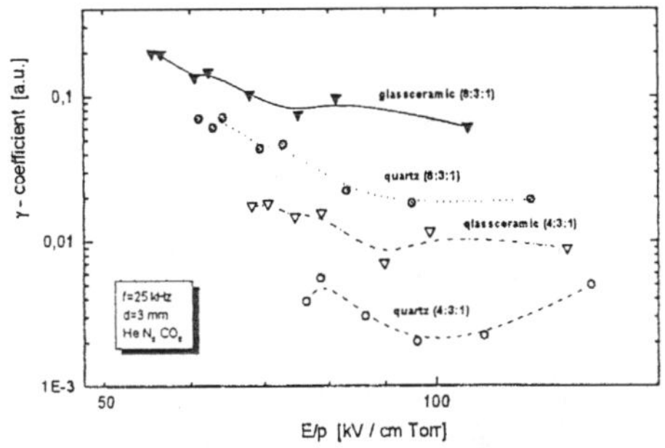

Fig.6 γ versus E/p for several dielectric layers

F

Displays

F 01

High-Luminance, High-Efficiency, and Simple-Structured Flat Discharge Lamp for LCD Backlightings

T. Shiga, K. Hirayama, K. Hashimoto, S. Mikoshiba, S. Shinada*

The University of Electro-Communications, 1-5-1, Chofugaoka, Chofu, Tokyo, Japan
*Heating & Lighting Division, Hitachi, Ltd., 888, Fujihashi, Ome, Tokyo, Japan

1 Introduction

Liquid crystal displays with diagonal sizes ranging from 4 to 7 inches are widely used for car navigation systems. Because of severe environmental conditions for these displays, many requirements are assessed to their backlights. Among them are high luminance (over 10,000 cd/m^2), good luminance uniformity (better than 80 %), dimming capability, and low power consumption.

A flat discharge lamp with insulated electrodes have been developed[1]. Although a luminance of 10,000 cd/m^2 was achieved, the luminous efficiency was only 12 lm/W. In this paper significant improvements of luminance and luminous efficiency will be reported.

2 Flat Discharge Lamp

2.1 Lamp structure and specifications

Figures 1 (a) and (b) show the plan and cross sectional views of the flat discharge lamp. It consists of a front glass plate and a rear glass vessel. An active area is 108 × 75 mm (5.2-in.-diagonal). On the inner side of the front glass plate is a couple of thick-film Ag electrodes which are covered with a 0.06 mm-thick dielectric layer. Tri-color-phosphor is deposited on both front and rear glasses. The lamp contains a minuscule amount of mercury with a buffer gas of argon+10% krypton at a total pressure of 13.3 kPa. The gap of the discharge space is 2.4 mm. Since the electrodes are insulated from the discharge by the dielectric layer, sputtering of the electrodes by ion bombardments and chemical reactions of Hg with the electrodes do not occur. Therefore the lamp is expected to have a long lifetime.

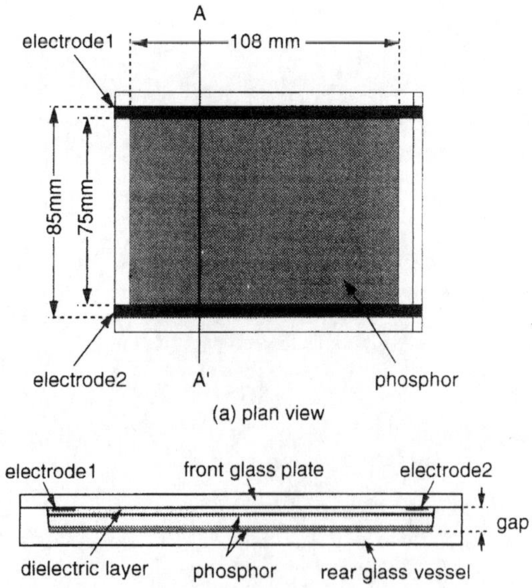

(b) cross-sectional view at A-A' plane of plan view

Fig. 1 Structure of flat discharge lamp

Each of the electrodes is driven by trains of square pulses. The pulses have an identical amplitude V, pulse width τ, and pulse interval T, but there is a 180 °phase difference between them. If the drive conditions are set adequately, a uniform and stable ac-discharge expands over the entire volume. Ultraviolet radiation from the discharge excite the phosphor.

2.2 Optical characteristics

It was found that the luminance and luminous efficiency could be improved by optimizing the drive pulse conditions. The luminous efficiency achieved a maximum value at a duty $\tau/T = 50$ %. Figure 2 shows the luminance and luminous efficiency at various pulse intervals as functions of drive voltage when $\tau/T = 50$ %. They increase with the drive voltage. For the pulse intervals ranging between 60 and 100 μs, luminous efficiency do not differ very much. Namely, as the pulse interval becomes shorter while luminance is kept constant, luminous efficiency decreases. This is due to a saturation of ultraviolet radiation intensity with respect to discharge current which takes place because of electronic de-excitation of the resonance level which is trapped in the discharge volume. The luminance of 30,000 cd/m^2 and luminous efficiency of 50 lm/W were achieved at the pulse voltage 680 V and interval 60 μs.

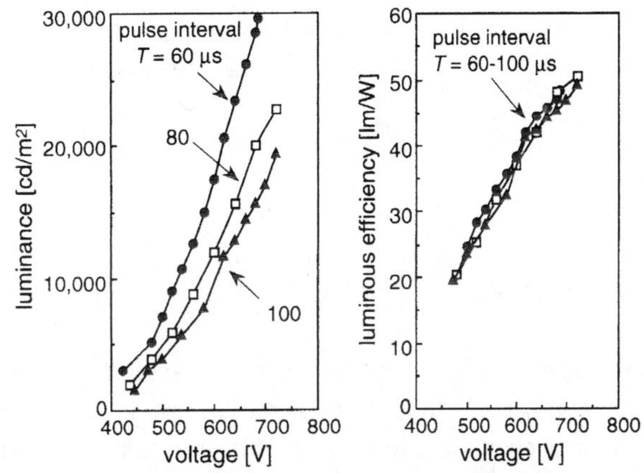

Fig. 2 Luminance and luminous efficiency *vs.* pulse voltage

Dimming of the lamp luminance can be performed by reducing the voltage, by increasing the pulse interval, or by the combination of these. One can see from Fig. 2 that luminance can be decreased from 30,000 to 1,200 cd/m^2 (4 % of the peak value) when the pulse voltage is 450 V and interval is 100 μs.

Figure 3 is a luminance distribution at nine regions on the lamp surface. A diffuser sheet was not employed. Drive pulse voltage, interval, and duty were 500 V, 60 μs, and 50 %, respectively. Upper and lower numbers indicate luminance in cd/m^2 and their percentages with respect to the average. Luminance uniformity, defined by the ratio of the minimum to maximum luminance is 92%.

lamp surface

29,070 (96.6)	29,130 (96.8)	29,770 (99.0)
31,130 (103.5)	31,230 (103.8)	31,600 (105.0)
29,960 (99.6)	29,120 (96.8)	29,750 (98.9)

Fig. 3 Luminance distribution in the lamp

3 Conclusions

A simple-structured, 5.2-in.-diagonal low pressure Hg flat discharge lamp was developed for car-navigation LCD backlightings. Excellent optical characteristics were obtained by optimizing the drive pulse waveform; luminance of 30,000 cd/m^2 and luminous efficiency of 50 lm/W were achieved. The uniformity is 92 % without the use of a diffuser sheet. The luminance can be dimmed down to 4 % of the peak value by changing both drive pulse amplitude and pulse interval.

Acknowledgments

The authors would like to express their thanks to Dr. K. Kinameri of Hitachi, Ltd. for his support. They are also grateful to "Research for the Future" program (JSPS-RFTF96R12501) conducted by The Japan Society for the Promotion of Science, for financial assistance.

Reference

[1] T. Shiga, K. Igarashi, S. Mikoshiba, S. Shinada, *Proc 16th Internat. Display Res. Conf.*, 233, 1996.

F 02

Characteristics of the VUV Emission from an AC PDP Cell

Ki-Woong Whang, Heui Seob Jeong, Jeong Hyun Seo, Joong Kyun Kim, and Cha Keun Yoon

School of Electrical Engineering, Seoul National University
Seoul 151-742, Korea

1 INTRODUCTION

We have been trying to understand the fundamental discharge physics of the PDP(Plasma Dislay Panel) plasma. Since the plasma volume is very small, it is very difficult to measure the spatio-temporal characteristics of the plasma parameters such as the particle densities and the various emissions.

In this work, we measured the VUV(Vacuum UltraViolet) and IR(InfraRed) emission from a single discharge cell to elucidate the discharge physics of the surface discharge type AC PDP because the spatio-temporal variation of the 147nm VUV emitted from $Xe^*(^3P_1)$, 828nm IR emitted from Xe^{**} and 173nm VUV emitted from Xe_2^* contains the information about the rare-gas discharge reactions[1,2] and compared with the results of time-dependent, 2-dmensional fluid simulation.

2 EXPERIMENTAL SETUP AND RESULTS

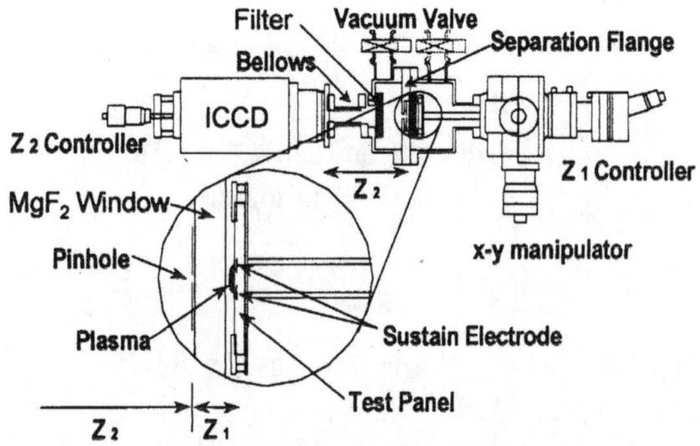

Fig. 1 Schematic diagram of VUV imaging system.

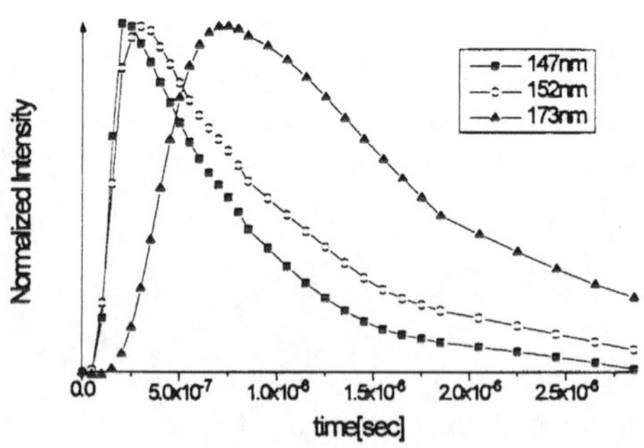

Fig.2 Temporal variation of normalized VUV emission

A direct observation of the VUV from a single surface discharge type AC PDP cell by the gated, image intensified CCD camera is done to obtain the time-dependent behavior of VUV emission.

Fig.1 shows the schematic diagram of VUV imaging system. This system consists of ICCD and test chamber which houses a test panel holder. The ICCD can detect the time resolved monochromatic emission using narrow band pass filters.

The surface discharge type AC PDP cell is driven by 83.3KHz pulses(25%duty) and contains He-Xe(1%) mixture at 300Torr. Fig.2 shows the time variation of 147nm, 152nm and 173nm spectral line intensities during one pulse on-time. As soon as the 180V pulse is applied, the intensity of 147nm reaches to its maximum within about 300nsec and decays with two decay time constants(0.54usec, 16.9usec) while those of 150nm and 173nm rise more slowly and decay with longer time constants.

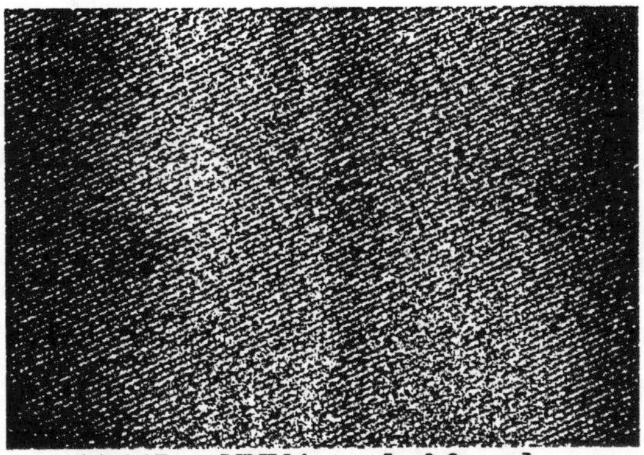

(a) 147nm VUV image[t=0.3usec]

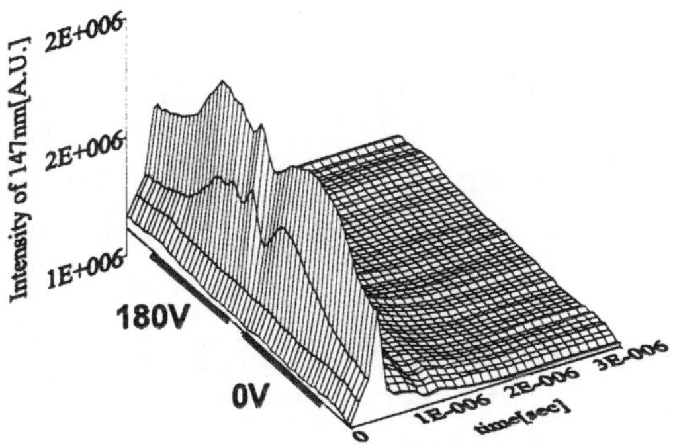

(b) Spatio-temporal variation of 147nm VUV emission

Fig.3 147nm VUV images at 0.3usec (a) and Spatio-temporal variation of 147nm VUV emission(b) [gate time=0.1usec, sustain voltage=180V]

Fig. 3(a) shows the 2 dimensional images of 147nmVUV emission from a single surface type AC PDP cell, and Fig 3(b) shows the spatio-temporal variation of the 147nm VUV emission. The parallel sustain electrodes are running from top to bottom in Fig. 3(a). The VUV radiation comes out from both of the anode and cathode area. A numerical simulation using a time dependent, 2-dimensional fluid equations has been carried out and their results are used to explain the experimental results.

3 CONCLUSIONS

The spatially, time dependent behavior of VUV emission from a single PDP cell has been obtained using an ICCD. The resonance line(147nm) and excimer continuum show different rising and decay behavior. Qualitative explanation of the experimental results could be obtained from the numerical simulation.

4 REFERENCES

[1] A. L. Schmeltekopf and F. C. Fehsenfeld, J. Chem. Phys., vol.53, Num.8, pp3173-3177, Oct. 15,1970

[2] W. P. Sholette etal, J. Chem. Phys. ,vol.36, pp. 3368-3373, Num.12, June 15, 1962

[3] K.W.Whang etal, IDW '97, pp531-534, Nov.19-23, 1997

F 03

Ar PRESSURE DEPENDENCE OF PLASMA PARAMETERS IN Ar-Hg DISCHARGE USED FOR LIQUID CRYSTAL DISPLAY BACKLIGHTING

Miki Goto, Kiyoshi Ohtani and Toshihiko Arai
Kanagawa Institute of Technology, 1030 Shimo-ogino, Atsugi 243-0292, Japan

1. INTRODUCTION

A lamp of extremely narrow inner diameter (usually 2-6mm) has been recently developed for liquid crystal display (LCD) backlighting and its importance is undoubtedly increasing.
We made measurements about Hg vapor pressure dependences of plasma parameters and particle densities in a fluorescent lamp (Ar filling pressure 3 Torr) of 4mm inner diameter used for LCD backlighting, before[1,2].

In this work, the UV radiation intensity, electric field strength, electron temperature and the mercury ion density on the tube axis of Ar-Hg discharge lamp were measured with variations of Ar pressure.

2. EXPERIMENT

The discharge tube was made of quartz glass of 4mm inner diameter. The tube was placed in a water bath to control Hg vapor pressure by controlling the water temperature.
The electric field strength and the electron temperature were measured using the discharge tube with two cylindrical probes. The Hg ion density was measured in the tube with two cathodes to changed the plasma length for using a absorption method[3]. The UV radiation intensity of the Hg(I)253.7nm line was measured by its tube.

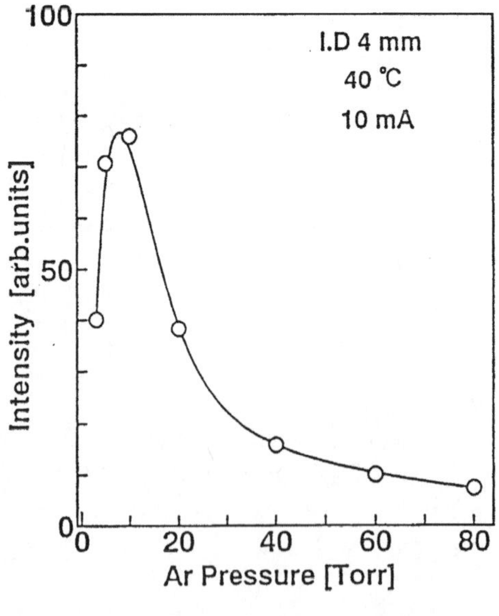

Fig. 1: The Hg(I) 253.7nm line intensity as a function of the Ar pressure.

3. RESULTS AND DISCUSSION

Figure 1 shows the Ar pressure dependence of Hg(I) 253.7nm line intensity from 4mm inner diameter at the discharge current of 10mA and the wall temperature of 40 ℃. Hg(I) 253.7nm line intensity increases with the Ar pressure, has a peak at around 10 Torr, and then decreases slowly. The intensities of the Hg(I) 253.7nm line have maxima at about 10 Torr for LCD backlighting and 3 Torr for devided by Ar pressure P, and the electron classical lamp.

Figure 2 shows the Ar pressure dependences of the electric field strength devided by Ar pressure P, and the electron temperature measured using a double probe method. Both E/P and T_e decrease with the Ar pressure. These values of E/P and T_e in LCD backlighting are higher about 2 times than that in classical lamp.

Figure 3 shows the Ar pressure dependences of Hg ion density, Ar ion density, electron density and Ar_2 molecular ion density. Hg ion density was measured with absorption method, has a peak at around 10 Torr. The variation of measuring Ar line intensity can be assumed to be roughly proportional to that of the Ar ion density. The Ar ion density has the minimum at around 10 Torr, and about constant above 20Torr. The electron density was calculated by value that was measured with single probe method at Ar 3 Torr. The electron density is given by

$$n_e = n_e(3)\,\frac{\mu_e(3)E(3)}{\mu_e E},$$

where n_e is the electron density, μ_e is the electron mobility, and E is the electric field strength. They are obtained at a specified pressure. $n_e(3)$, $\mu_e(3)$ and E(3) are obtained at a pressure of 3 Torr. The Ar_2 molecular ion density is given by

$$n(Ar_2^+) = n_e - n(Hg^+) - n(Ar^+),$$

where $n(Hg^+)$ is the Hg ion density and $n(Ar^+)$ is the Ar ion density. Here Hg_2^+ molecular ion is negligibly small. The electron density and the Ar_2 molecular ion density increase with Ar pressure. Therefore, ions in more than Ar pressure 20 Torr are dominated by the Ar_2 molecular.

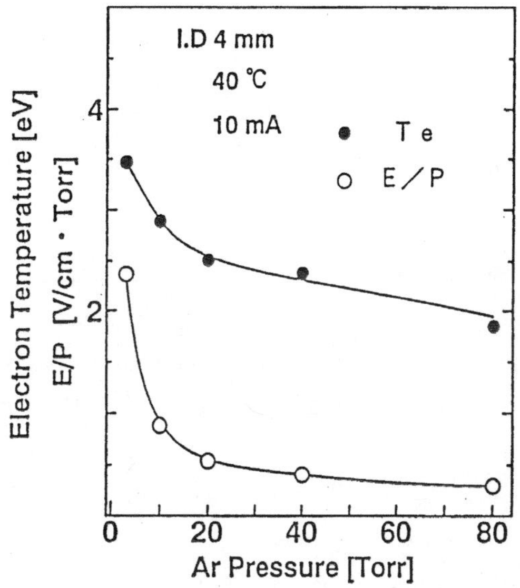

Fig. 2: The electron temperature and E/P (electric field strength / pressure) as a function of the Ar pressure.

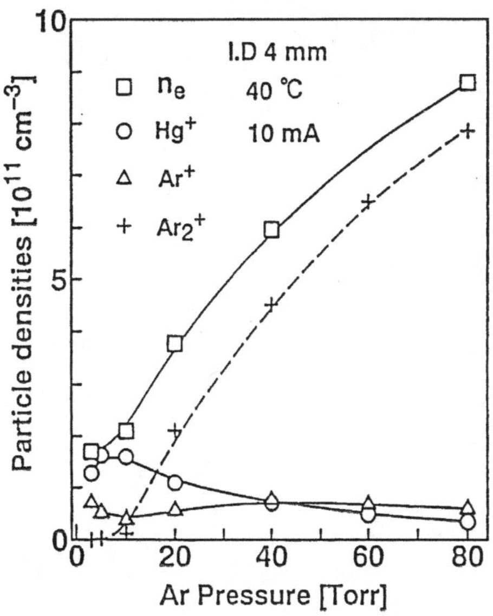

Fig. 3: The particle densities as a function of the Ar pressure.

4. REFERENCES

[1] M.Goto, T.Sakai and T.Arai, Jpn. J. Appl. Phys. **33**, L896 (1994).

[2] M.Goto and T.Arai, Jpn. J. Appl. Phys. **34**, L1074 (1995).

[3] T.Goto, M.Mori and S.Hattori, Appl. Phys. Lett. **29**, 358 (1976).

Scanned Archives - Best Available Copy

F 04

The Townsend's coefficient in Helium/Xenon- and Xenon/Neon-mixtures

M. Otte, S. Pfau and J. Rohmann[1]

Institute of Physics, E.-M.-Arndt University, Domstr. 10a, 17487 Greifswald, Germany

[1]Institute of Low Temperature Plasma Physics, R.-Blum-Str. 8–10, 17489 Greifswald, Germany

Recently the first 40" full color plasma displays have been on the market. Without a doubt this is only the beginning of a new modern technology in this field. The filling gas in plasma displays are mainly mixtures of Xenon and Helium or Xenon and Neon. These mixtures have in comparison to pure gases the advantage to have a much smaller breakdown voltage [1]. First of all this is caused by the Townsend's ionization coefficient, which can be serveral orders of magnitude higher in the mixture than in pure gases. On the example of He/Xe a few typical results of the calculated Townsend's ionization coefficient as a function of the mixture ratio $x = n_{Xe}/(n_{Xe} + n_{He})$ are shown. An overview of the course of the Townsend's ionization coefficient of the mixtures shows Fig. 1.

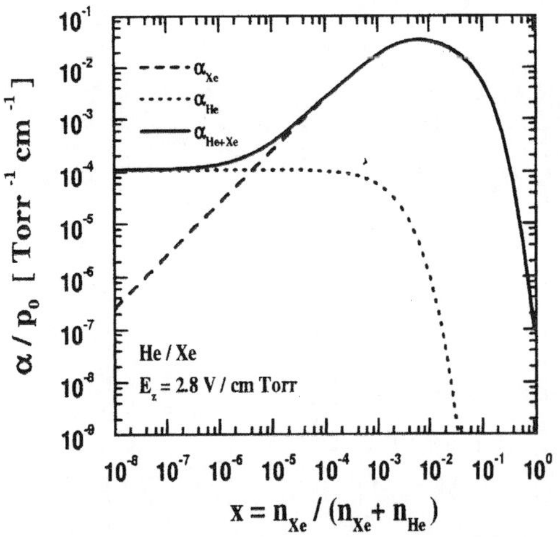

Figure 1: Townsend's ionization coefficient in a He / Xe mixture

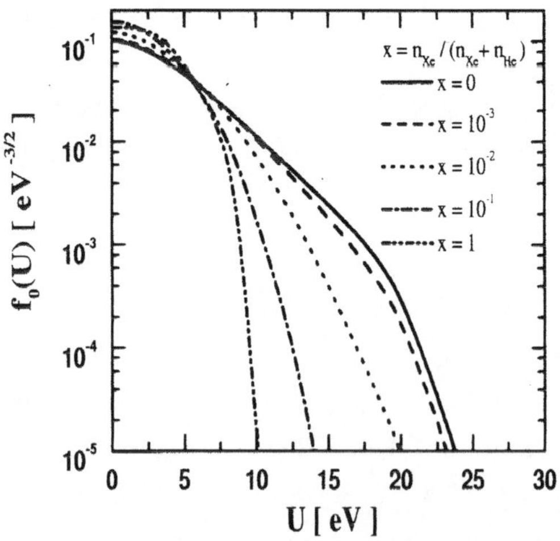

Figure 2: Calculated electron velocity distribution function in He/Xe-mixture with $E/p_0 = 2.8 V/cm Torr$

The curve starts increasing from the value of pure Helium with increasing Xenon add, is passing through a maximum and decreases to the value of pure Xenon. For the given conditions $E/p_0 = 2.8 V/cm\ Torr$ the ionization coefficient reaches a maximum at 0.6 % Xe and is at this point of nearly 3 orders of magnitude higher than in pure Helium and 6 orders of magnitude higher than in pure Xenon. The location and the value of the maximum depends beside of x on the reduced electrical field strength. The Townsend's ionization coefficient can be calculated from the relation [2]

$$\frac{\alpha}{p_0} = \frac{x\, z_{0\infty}^{Xe} + (1-x)\, z_{0\infty}^{He}}{v_d},$$

where $z_{0\infty}^{(\nu)} = \left(\frac{2e_0}{m_e}\right)^{\frac{1}{2}} \int_{U_i}^{\infty} n_g Q_i^{(\nu)} U f(U,x) dU$ is the collision rate for direct ionization, ν : Xe, He; Q_i ionization cross-section, v_d drift velocity, $f(U,x)$ energy distribution function, $n_g = 3.54 * 10^{16}$ $[cm^{-3}\ Torr^{-1}]$. The distribution function $f(U,x)$ was determined from the Boltzman-equation considering elastical and inelastical collisions. Fig. 2 shows the transition of the electron distribution function from pure Helium to pure Xenon. Even a small addition of Xenon has drastical consequences for the course of the distribution function. The influence of Xenon on the electron velocity distribution function can only be neglected for adds $x \leq 10^{-5}$. A comparison of the calculated electron energy distribtuion functions under homogeneous field conditions with measured distribution functions in the axis of the positive column of a glow discharge for equal reduced field strengths shows as to be seen in Fig. 3 a satisfactory agreement. The validity of such a comparison was shown in recent investigations about consequences of radial inhomogenity in the cylindrical column plasma to the distribution function in the axis [3],[4]. The collisional cross-sections can be found in [5]. The dependence of the mean energy, the mobility and the drift velocity of the electrons on the mixture of the nobel gases is shown in Fig. 4. Quiet similar results can be found in a Neon/Xenon mixture and will be presented in the poster.

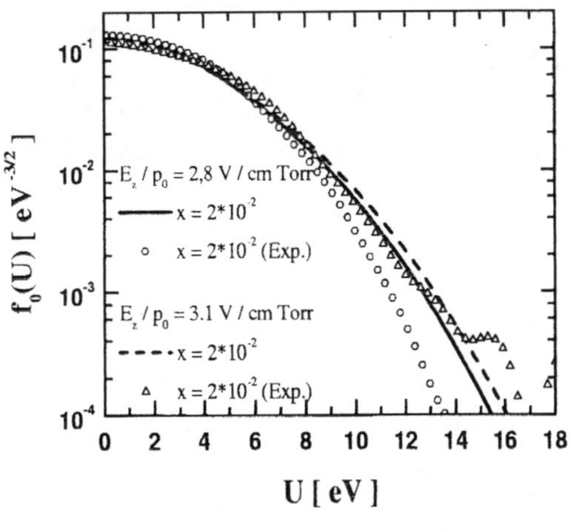

Figure 3: Calculated and measured electron velocity distribution function in He/Xe-mixture

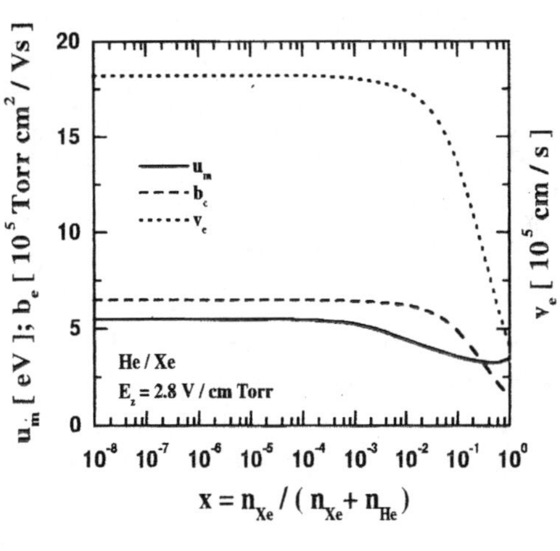

Figure 4: Mean energy, mobility and velocity of the electrons

This work was funded by the Sonderforschungsbereich 198 'Kinetik partiell ionisierter Plasmen', Project A1 of the Deutsche Forschungsgemeinschaft.

References

[1] J.P. Bœuf, C. Punset, A. Hirech and H. Doyeux, J. Phys. IV FRANCE **7**, C4-3 (1997)

[2] S. Pfau and R. Winkler, Beiträge aus der Plamaphysik **18** (1978) 131

[3] S. Pfau, M. Otte, J. Rohmann, D. Uhrlandt, R. Winkler, XIII-th ESCAMPIG, Poprad/Slovakia **20E** (1996) 89

[4] S. Pfau, J. Rohmann, D. Uhrlandt and R. Winkler, Contrib. Plasma Phys. **36** (1996) 4, 449-469

[5] M. Hayashi, private communications (1996)

F 05

A TWO- DIMENSIONAL SIMULATION OF PULSED DISCHARGE FOR A COLOR PLASMA DISPLAY

Yukio Murakami[1], Hiroshi Murakami[1], Hideomi Matuzaki[1], and Kunihide Tachibana[2]
[1]NHK Science & Technical Labs., 1-10-11 Kinuta, Setagaya-ku, Tokyo 157-8510, Japan
[2]Kyoto University, Yoshidahonmachi, Sakyo, Kyoto 606-01, Japan

INTRODUCTION

A 42-inch-diagonal, high-resolution dc PDP was developed for an HDTV receiver in 1997[1]. About one hundred 42-inch HDTV PDP receivers were later produced for the enjoyment of HDTV programs from the Winter Olympiad in Nagano in Feb. 1998.

Still, however, improvement are required to the luminance and luminous efficiency of the PDP and the cost of the receiver must also be reduced for the spread of HDTV PDP receivers to many homes. Higher efficiency, in particular, is essential regardless of what type of PDP(dc or ac) is used for the small cell-pitch panel of the HDTV display. In this study, two-dimensional spatio-temporal densities of particles were calculated in a pulsed discharge cell filled with a mixture of He and Xe gas as a first step towards this improvement of the efficiency. This paper describes the modeling for the calculation, basic equations, and the results.

MODELING AND CALCULATING METHOD

In the computer simulation, a two-dimensional fluid model with local field approximation was employed for twelve species: the electron, four types of ion (He^+, He_2^+, Xe^+, Xe_2^+), and seven selected excited species, similarly to those of one-dimensional simulation[2]. Fig.1 shows the atomic and molecular energy levels and the processes related to the calculation. The model took the imprisonment of resonant radiation and secondary electron emissions due to ions, metastable atoms and photons into consideration.

The X- and Y-axes in the spatial coordinates correspond to the directions from the cathode to the anode and parallel to the cathode, respectively. The differential continuity equations described in the next section were solved by the finite difference method. The time step was a few picoseconds, which is not enough for an electron to pass through a single mesh on the X- and Y-axes.

BASIC EQUATIONS

Twelve continuity equations for particle i (i=1~12: electron, ions, and excited species) are given by

$$\frac{\partial N_i}{\partial t} + \frac{\partial \phi_{ix}}{\partial x} + \frac{\partial \phi_{iy}}{\partial y} = Gi\,(x,y,t) - Li\,(x,y,t)\,, \quad (1)$$

where N_i is the densities of the species, ϕ_{ix}, ϕ_{iy} are X- and Y-axis components of species flux, G_i, and L_i is the production and loss terms for reactions between species i and other species. Flux ϕ_{ix} and ϕ_{iy} are given by

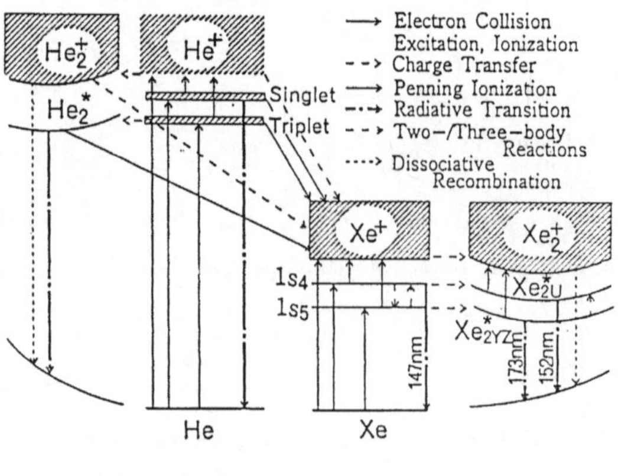

Fig.1 Energy levels and processes in He-Xe

$$\phi_{ix} = N_i \cdot v_{ix} - D_i \frac{\partial N_i}{\partial x}\,,\quad \phi_{iy} = N_i \cdot v_{iy} - D_i \frac{\partial N_i}{\partial y}\,, \quad (2)$$

where v_{ix}, v_{iy} are X- and Y-axis components of the drift velocity, and D_i is the diffusion coefficient.

The densities of an electron, ions, and excited species were calculated according to equation (1)(2) under such boundary conditions as secondary electron emission from the cathode and the circuitry condition. The electric field was calculated with Poisson's equation. The ionization and excitation coefficients for electron collision, taken from the reference[3] or calculated by solving Boltzmann's steady-state equation, were adopted for the instantaneous value of the electric field at the position in the calculation. Other reaction rate coefficients were taken from published papers.

Table 1 Condition of the calculation

Gas composition	He (90%)-Xe (10%)
pressure	27 kPa (200 Torr)
Cathode width	0.40 mm
Anode width	0.18 mm
Gap length	0.35 mm
Electrode length	0.10 mm
Bias voltage	150 V
Pulse voltage	165 V
period	4 μs
duration	1 μs
Resistor	1 MΩ

The same calculation was repeated at every step until the end of one sustain pulse period, and the second calculation was made on the initial condition of the densities at that time. The calculation was repeated until a steady state was reached.

RESULTS

Table 1 shows the conditions of the calculation. Fig.2 shows the spatio-temporal behavior of the resonance state of Xe($1s_4$), which emits effective 147nm-VUV rays for the phosphor. The results for the spatio-temporal distribution of Xe($1s_4$) and discharge current waveforms agreed qualitatively with the experimental results[4].

On the basis of the simulation method described above, we will next study the mechanism of various kinds of PDP discharge and seek to improve the efficiency.

REFERENCES

[1]T.Yamamoto et al., SID '97 Dig. Tech. Papers, pp.217-220 (1997)
[2]Y.Murakami et al., Proc. EuroDisplay '93, pp.555-558 (1993)
[3]K.Takahashi, K.Tachibana, Trans. IEE Jpn. 111-A.3, pp.182-191(1991)
[4]K.Tachibana et al., Appl.Phys.Lett., 65, 3, p.935 (1994)

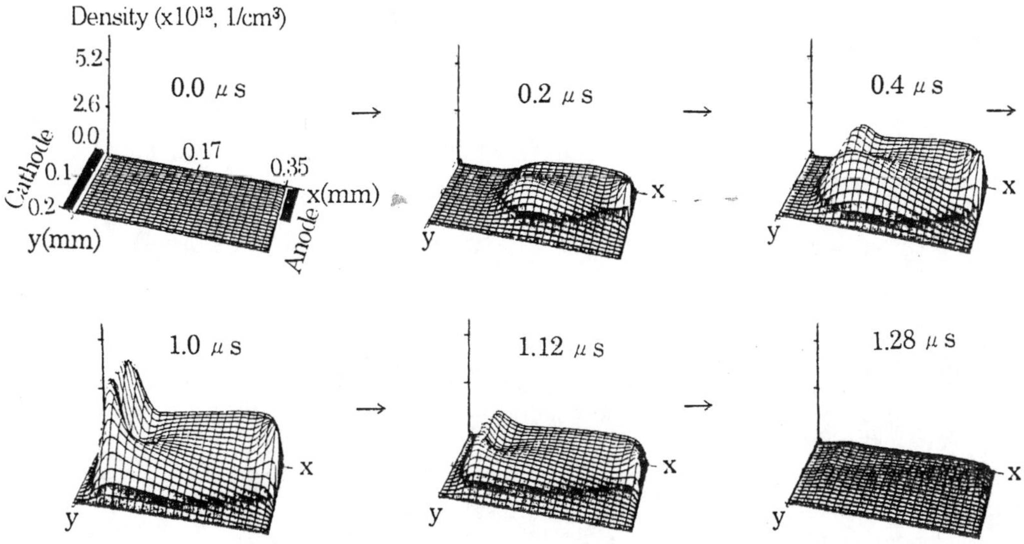

Fig.2 Spatio-temporal behavior of the Xe($1s_4$) atom

Novel Concepts

G 01

FULLERENE C_{60} AS A NEW EMITTING ADDITIVE FOR A LOW PRESSURE GAS DISCHARGE LIGHT SOURCE

V.M.Milenin, N.A.Timofeev, S.V.Kidalov*, S.V.Kozyrev*, A.Ya.Vul*
St.Petersburg State University, Ulianovskaya 1, Petrodvorets,
St.Petersburg, 198904 RUSSIA
*Ioffe Physico-Technical Institute RAS, Polytechnicheskaya 26,
St.Petersburg, 194021 RUSSIA

INTRODUCTION

Creation of new light sources by searching new additives (atomic or molecular), which would be able to replace mercury in luminescent lamps and which would be environmentally harmless, is of great interest, and all leading light companies all over the world are constantly carried out investigations in this field. This paper deals with the possibility of using a recently discovered substance - a molecular cluster C_{60} which is referred to as fullerene (or Buckminster fullerene) - as an emitting additive of a gas discharge light source.

PRELIMINARY REMARKS

Several factors, concerning physical and chemical properties of fullerene molecules [1,2], lead to the hypothesis that fullerene can be a suitable additive for a low pressure gas discharge light source.

They are as follows:

- 1. a low ionization potential 7.6 eV;
- 2. theoretically predicted optical transitions between excited levels in a range 200-260 nm;
- 3. molecules C_{60} do not dissociate during the collisions with fast electrons (up to energy 80-100 eV) or the absorption of UV photons (with energy up to 30 eV);
- 4. fullerene is stable to heating up to 900-950 C;
- 5. the pressure 10-100 mTorr of C_{60} vapour is reached at 600-700 C;
- 6. fullerene molecules have practically no ecological effects;
- 7. it is not aggressive to other materials.

The points 1-3 makes it possible to draw a conclusion that under some discharge conditions fullerene molecules, being added to a buffer gas (as mercury, being added to argon in luminescent lamps), can form emission properties, ionization balance in the plasma of a discharge and absorb the major part of applied electric power. Similar picture takes place in mercury luminescent lamps and ensures their high efficacy.

EXPERIMENTAL RESULTS

The possibility of using fullerene molecules as the emitting additive in a gas discharge light source was verified on electric discharges in rare gases He, Ne, Ar and Xe. The discharge chambers were made from transparent quartz glass allowing measurement in a spectral range 200-800 nm with the wall heating up to 700 °C. Two tipes of discharge were studied: DC discharge and HF inductive discharge at 60 Mhz. The rare gas pressure varied within 1-30 Torr and the DC discharge current density was about 0.01 A/sq.cm; the HF discharge power was high enough for the optical emission to be observed by with the naked eye. Fullerenes were introduced into the discharge either as C_{60} powder or fullerene-containing soot.

The plasma emission spectrum was measured in the range 200-800 nm at various temperatures of the discharge chamber, various gas pressures and discharge current.

The measurements have shown that the heating in the presence of fullerene-containing material leads to sweeping in the emission spectra for both DC discharge and HF discharge. At near-room temperature only noble gas atom lines are present. Increasing temperature and appearance of fullerene vapour "inhibits" these and produces new lines and bands in the blue-green and UV spectral regions. The lines observed can be identified as C_2, CN, CO and H lines. These dashes are presented in a fullerene sample and can hardly be eliminated by previous heating up to 150 °C. The higher temperature is not used, because it is possible the pre-oxidation of C_{60} [3] and following destruction of the pre-oxyded fullerene. There is also a band in the range 260 nm which can be attributed, as analysis gives, to the fullerene vapour in the plasma (or may be to some derivatives of fullerene molecules and atoms and molecules presented in the plasma) [4]. This band is situated near by the absorption peak (257 nm) of fullerene [5] and has a well-resolved molecular structure with the characteristic gaps which are close to the internal molecular vibrations of fullerene [1].

CONCLUSIONS

The first observation of fullerene emission at 260 nm in plasma conditions is presented in the paper. This radiation is near the mercusy resonance line 254 nm that makes it siutable to use this band instead of mercury line for the excitation of phosphors provided that the intensity at 260 nm is sufficiently high.

REFERENCES

[1] A.Eletsky, B.Smirnov: Usp.Fiz.Nauk, 163(1993) 33.
[2] S.Kozyrev, V.Rotkin: Sov.Phys.Semic., 27(1993) 777.
[3] H.Werner et al: Int.Workshop "Fullerenes and Atomic Clusters", St. Petersburg, June 19-24 (1995) 33.
[4] N.Bashlov et al: JETF Lett., 63(1996) 713.
[5] S.Dai et al: J.Chem.Phys., 101(1994) 4470.

G 02

HIGH BRIGHTNESS FULL COLOR LEDs

Yoshinobu Suehiro, Takashi Sato, Koji Uchida, Shigeru Yamazaki
R&D Center, Iwasaki Electric Co., Ltd
1-20, Fujimi-cho, Gyoda-City, Saitama, 361-0021, JAPAN

1. INTRODUCTION

Two types of structures are known for full color LEDs(Light Emitting Diode) with red, green, and blue chips corresponding to the three primary colors. [1],[2] Each of these structures for full color LEDs has it own merit. However both have limitations in principle in mixing different colors of light and radiating them at a high luminous intensity at the same time and require a minimum distance to observe different light colors as mixed.

Our newly developed Multi-Layer type full color LED has overcome the above-mentioned restrictions, and as a full color LEDs it has achieved good results both in mixing different colors of light and in radiating them at high luminous intensity.

2. DESIGN

As a prototype of a Multi-Layer type full color LED, a schematic front view and a schematic cross sectional view are shown in Fig. 1 and Fig. 2. This prototype has red, green, blue LEDs, spacer(clear color), and optical surfaces which correspond to the chip of each color LED. Each color LED is assembled with green, blue and

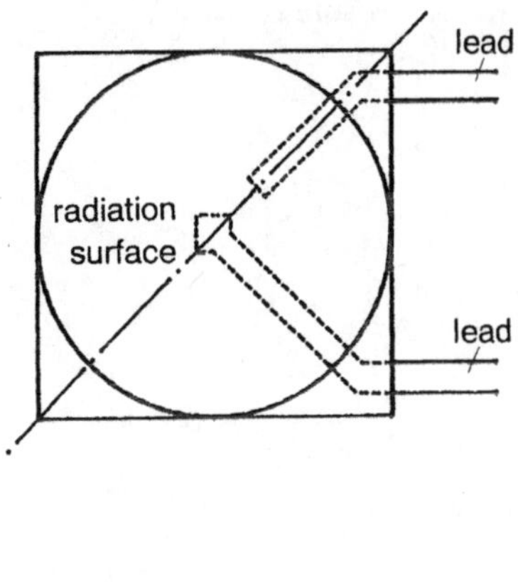

Fig. 1 : Schematic front view of Multi-Layer type full color LED

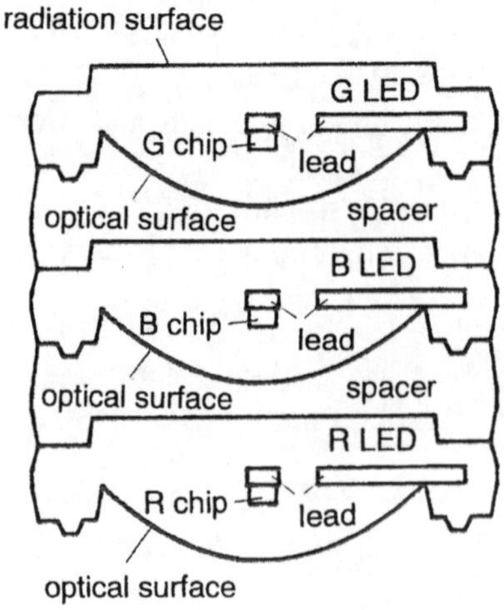

Fig. 2 : Schematic cross sectional view of Multi-Layer type full color LED

red LEDs in the upper, middle and bottom layer respectively. And spacers are put between each LED, using an interposed adhesive of a light transmissible material.

This prototype has three optical surfaces which correspond to each chip and reflect the light emitted by the chip, and has a radiation surface which outputs these reflected rays. The three optical surfaces which correspond to each chip are formed by a thin film and have approximately 2π strad, with a wide solid angle to the chip. Optical surfaces in the upper layer of the green chip and middle layer of the blue chip are dichroic mirrors. Therefore, the dichroic mirror in the upper layer will reflect light with a wavelength range of green color and transmit rays of light with wavelength range of blue and red. The dichroic mirror in the middle layer reflects light with a wavelength range of blue color and transmits light with wavelength range of red.The optical surface in the bottom layer of the red chip is formed with Ag thin film.

The radiation surface which is to output these reflected rays is perpendicular to the axis of symmetry of the three optical surfaces of each chip, and has a flat shape of approximately the same diameter as that of the path of reflected light of each optical surface at the radiation surface.

3. RESULT

Luminous intensity of the prototype's green, blue and red LEDs when lit individually is 6.6cd, 1.2cd, 2.8cd at 20mA per chip at a temperature of 25°C, respectively.

The LED's ability of constantly mixing each luminous wavelength independent of the distance is ascertained in accordance with the principle. The character of their color is visible white (x=0.29,y=0.22) and values of 10cd have been achieved, using the prototype green, blue and red at 20mA per chip.

4. CONCLUSIONS

The newly developed Multi-Layer type full color LED is capable of radiating different luminous wavelength light through each chip to the same optical path with a high level of efficiency.

Excellent luminous intensity has been demonstrated by the prototype. Value of 10cd white light has been achieved using Multi-layered blue, green and red chips, at a current of 20mA per chip.

5. REFERENCES

[1] T.Kunisata et al.,"Development of A RGB Multi-Color LED Display", *TECHNICAL REPORT OF IEICE EID*92-91 (1993-01)

[2] S.Nakamura,"LED Full Color Display ",*J.IEICE*,vol.78,no.7,pp683-688,1995

[3] Y.Suehiro et al ,"Reflector Type LED", *TECHNICAL REPORT OF IEICE EID*94-114 (1995-01)

G 03

Characteristics of xenon radiation of xenon-neon mixture operated with pulsed discharge

Masaharu Aono, Hisayoshi Kurokawa, and Masafumi Jinno
Ehime University, 3 Bunkyo-cho, Matsuyama, Ehime, 790-8577, JAPAN

1 INTRODUCTION

A small current low pressure xenon discharge lamp is independent of peripheral temperature though a mercury discharge lamp is much influenced. This is a great advantage of a small current low pressure xenon discharge lamp. However it has a serious disadvantage of low luminance. Many researchers reported on characteristics of xenon radiation, however, high luminance low pressure xenon discharge lamp has never been obtained[1-3]. In this study, the characteristics of radiation of xenon-neon mixture operated with pulsed discharge are measured experimentally as the fundamental study on a high intensity low pressure xenon discharge lamp.

2 EXPERIMENTAL SETUP

The inner diameter of lamps are 7mm. The distance between electrodes is 55mm. Xenon-neon mixture is filled in each lamp at 13.3kPa. There are three combinations of partial pressure ratio of xenon and neon as follows; (a) Ne:Xe = 99:1, (b) Ne:Xe = 90:10, and (c) Xe 100%. Only half part in length of each lamp is coated with a phosphor NP-107 ($BaMg_2Al_{16}O_{27}$:Eu). Since this phosphor is excited by only vacuum ultraviolet (VUV) radiation and since the phosphor has no afterglow, the radiation waveform of this phosphor is similar to that of the VUV radiation of xenon. Each lamp was operated by pulsed discharge. Pulse frequency was varied from 10kHz to 50kHz, and duty ratio was varied from 10% to 30%. The radiation of xenon, neon and the phosphor were observed and recorded under each condition.

3 EXPERIMENTAL RESULTS AND DISCUSSION

The radiation waveforms of xenon atom (828nm), neon atom (703.2nm) and the phosphor (451nm) are shown in Fig.1 with voltage and current waveforms. The radiation of neon is detected only when the mixing ratio of xenon is less than 1%. The radiation of xenon is strong during discharge current is flowing, and it decreases rapidly after the current is cut off. The phosphor also emits fluorescence during the current flowing. Though once the fluorescence decreases after the current is cut off, it increases again to the peak, and then it decreases slowly. The afterglows both of xenon and phosphor become strong as the partial pressure of xenon increases.

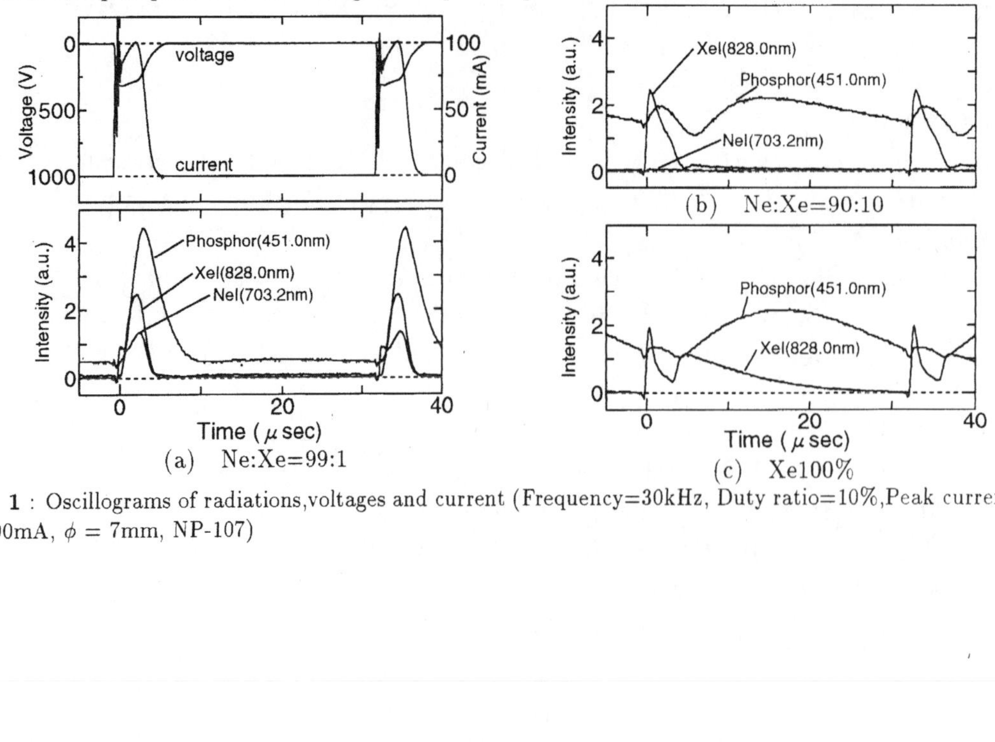

Fig. 1 : Oscillograms of radiations,voltages and current (Frequency=30kHz, Duty ratio=10%,Peak current=100mA, ϕ = 7mm, NP-107)

Here, let us discuss about mechanism of xenon radiation. The excitations of atoms and molecules during one cycle of discharge are separated into four periods and the radiation characteristics of xenon are accounted according to the Fujimoto's theory[4] as follows;

[I] **Xe Partial Pressure ≥ 10%** — Xenon molecular radiation is strong.

(i) **First Rising Period** — In this period, both electron temperature T_e and density n_e are high. Then a plasma is in the ionizing-phase. The xenon atom is ionized by the following processes; $e + Xe \rightarrow Xe^* + e$, and $e + Xe^* \rightarrow Xe^+ + 2e$, where Xe^* is an excited xenon atom.

(ii) **First Relaxation Period** — In this period, ionization plasma and recombination plasma exist simultaneously. T_e is moderate and n_e is high. The population of exited xenon atoms is minimum, then the radiation of xenon decreases and is weak.

(iii) **Second Rising Period (Afterglow)** — The plasma is in recombining-phase. Both T_e and n_e are low. Xenon ions are produced by the following process; $Xe^m + Xe^m \rightarrow Xe^+ + Xe + e$, where Xe^m is a metastable xenon atom produced during current flowing. VUV radiations are emitted by the following processes[1] ;

$$Xe^+ + e \rightarrow Xe^* \ , \quad Xe^* \rightarrow Xe + h\nu(147\mathrm{nm}), \tag{1}$$

$$Xe^m + 2Xe \rightarrow Xe_2^* + Xe \ , \quad Xe_2^* \rightarrow Xe_2 + h\nu(172\mathrm{nm}). \tag{2}$$

(iv) **Second Relaxation Period (Afterglow)** — In this period, both T_e and n_e are lower than the previous period. The radiation of xenon decreases as the population of metastable of xenon Xe^m. Since the transitions generating the radiations of 828nm and 147nm are sequential, the waveform of 147nm radiation is supposed to be similar to that of 828nm. Then, the time constant of the 147nm radiation is obtained as 7.63×10^{-6}[s] from the waveform of xenon radiation (828nm).
That of the 172nm radiation is also obtained as 1.90×10^{-5}[s] from the phosphor waveform. The decrease of radiation of xenon atom (147nm) is faster than that of xenon excimer (172nm), since the radiation of xenon atoms follows the decrease of $(n_{Xe^m})^2$ and since that of xenon excimers follows the decrease of n_{Xe^m}.

[II] **Partial pressure of xenon is very low** — Atomic radiation is dominant and molecular radiation is faint.

(i) **First Rising Period and First Relaxation Period** — Neon is ionized, since the electron temperature is high.

(ii) **Second Rising Period (Afterglow)** — In this period, atomic resonance 147nm is dominant due to radiation recombination and radiation cascade. Therefor, the phosphor is excited mainly by 147nm radiation in the afterglow. Since density of Xe^m is low, the 172nm radiation is weak.

(iii) **Second Relaxation Period (Afterglow)** — Since both densities of Xe^m and Xe are low, the 172nm radiation (generated by the process (2)) is low. However, since the decomposition of Xe^m by elastic collision between Xe^m and Ne is also rare, the weak 172nm radiation continues so long.

4 CONCLUSION

The radiation of a low pressure xenon discharge lamp operated with pulsed discharge has two peaks. The first peak of radiation is mainly radiation of xenon atoms. The radiation of xenon excimers is strong and dominant in the second peak of radiation, i.e. afterglow. For a fluorescent lamps excited by xenon molecular radiations, high efficacy is obtained by pulsed discharge of higher partial pressure of xenon. On the other hand, lower partial pressure of xenon is required in order to use xenon atomic resonance.

5 REFERENCES

[1] P.Millet, *et al* : *J. Chem. Phys.*, **69**(1), pp.92 (1978)
[2] A. Barbet, *et al* : *J. Phys. B*, **8**(10), pp.1785 (1975)
[3] T. Sakurai, *et al* : *1991 Convention Records of I.E.I.J.*, 56, pp.83 (1991)
[4] A. Hirabayashi, *et al* : *Phys. Rev. A*, **A37**(1), pp.77 (1988)

G 04

BARIUM GLOW DISCHARGES FOR LIGHTING APPLICATIONS

J. E. Lawler, Heidi M. Anderson, and J. J. Curry
University of Wisconsin, 1150 University Avenue, Madison, WI 53703, USA

1 INTRODUCTION

Radio frequency (rf) excitation methods and ceramic lamp envelopes have expanded the accessible parameter space, including temperature, pressure, and chemical dose, for discharge lamp operation. We are studying barium (Ba) as a possible new atomic radiator in a glow discharge lamp. Barium has a number of characteristics that indicate it may be a good atomic radiator in a glow discharge. 1) At practical operating temperatures Ba has a vapor pressure in the mTorr range (9 mT at 600°C). 2) The excitation energy of the primary radiating level in Ba is less than half of the ionization energy. The high efficiency of the traditional atomic radiators, Hg and Na, is due in part to the existence of low resonance levels which inhibit ionizing collisions between pairs of excited atoms. 3) The wavelength of the resonance transition is suitable for direct use as a visible light source. At 553 nm, this radiation lies at the peak of the eye sensitivity curve. For these reasons, a Ba glow discharge may prove to be useful in a novel, efficient light source. As a first step in developing a Ba light source we are measuring the efficiency of Ba-Ar direct current (dc) positive column discharges over a range of parameter space. We are also exploring the properties of Ba radio frequency (rf) inductively coupled plasma (ICP) discharges.

2 EXPERIMENT

High temperature Ba vapor is corrosive to many materials, including some of the envelope materials, electrode metals, and envelope seals used in lighting products. For example, Ba rapidly reacts with fused silica (SiO_2) to form an opaque solid. For preliminary studies we have developed some novel discharge systems for radiometric measurements on a Ba dc positive column discharge and emission and absorption spectroscopy on a Ba ICP discharge. We have also found that sapphire (Al_2O_3) is resistant to chemical attack by the Ba vapor and thus may have applications as a Ba light source envelope material. Stainless steel, which resists attack by Ba vapor, is used as the electrode material in our dc discharge system. Materials issues, although formidable, do not appear to be insolvable.

Our dc discharge system operates in an oven so that we have independent control of Ba vapor pressure and other parameters. The dc discharge system consists of a fused silica vacuum envelope in the shape of a tee and a second fused silica tube, which serves as a sacrificial inner liner, with a small hole at the viewing arm of the tee. The Ba is introduced as solid pieces of metal, and the discharge region of the device is heated externally with a clamshell oven to wall temperatures ranging from 600°C to 730°C, to vaporize the Ba. A static fill of Ar is present as a buffer gas, at pressures of 0.1 Torr to 5.0 Torr. Input power ranges from 1.5 to 7 Watts. The light from the positive column is sent into a 0.275 m spectrometer and detected with a photo diode array (PDA). The entire visible spectrum is observed simultaneously. The PDA is then calibrated with a standard tungsten-quartz-halogen lamp so we can determine the visible power radiated by the Ba dc positive column discharge.

The cylindrical inductively-coupled discharge consists of a 5 cm diameter by 10 cm long fused-silica cell surrounded by a six-turn radio-frequency coil. When thermal conduction from the cell is limited, excess heat from the discharge is sufficient to produce and maintain suitable vapor pressures of Ba. A discharge is initially ignited in a rare-gas buffer, but at operating temperatures the plasma is dominated by Ba. Optical access is maintained through two windows which are recessed from the discharge. We are able to vary the rare gas (He, Ne, Ar, Kr, or Xe) buffer pressure.

3 RESULTS

The radiometric efficiency of the dc Ba positive column is strongly dependent on the oven temperature (and thus on the Ba vapor pressure) and on the buffer gas pressure. We have observed an efficacy of at least 92 lumens/Watt, and expect to do better as we explore more of parameter space. The dc discharge, under conditions observed thus far, is dominated by neutral Ba lines and a few strong Ba ion lines. With the 553 nm green resonance radiation, strong ion lines in the blue (455 nm) and red (650 nm) portions of the spectrum and a number of additional red lines (and a few blue ones), a white light is obtained for some discharge parameters. By "white" we mean that it is a line spectrum that appears white to the eye, but does not contain any continuum radiation. Under other conditions, the discharge can appear very green.

The emission spectrum of the Ba rf discharge depends upon the buffer gas and coupled rf power. Under some conditions strong Ba ion emission is balanced with the Ba neutral 553 nm emission to produce white light. Emission spectroscopy is supplemented with absorption spectroscopy of the Ba ion and Ba neutral ground state densities.

We were pleasantly surprised to discover that white light was obtainable from a Ba discharge, as we expected the 553 nm resonance radiation to dominate.

The possibility of using ion radiation in a practical light source should be given some consideration. Although extra energy must be expended to make an ion from a neutral atom before the ion is excited, the loss of efficiency due to the ionization energy may be minimal if the ion can be repeatedly excited and radiate many times before diffusing to the wall.

4 ACKNOWLEDGEMENTS

This work was supported by the National Science Foundation.

H

Electronic & Other Power Supply

Electronic Ballast
for pulsed operation of Dielectric Barrier Discharges

Peter Schwarz-Kiene, Wolfgang Heering
Lichttechnisches Institut (LTI), Universität Karlsruhe
Kaiserstraße 12, D-76128 Karlsruhe

1 INTRODUCTION

Dielectric barrier discharges are well known as selective incoherent radiation sources. The most important difference to other light sources is the capacitively coupled energy transfer across the dielectric barrier. Therefore the lamp has to be operated at high voltage and high frequency, usually driven by a sine or square wave power converter. Typical applications are UV sources for surface modification, waste water and flue gas treatment as well as mercury free substitutes for standard fluorescent lamps.

Due to a significant increase in radiation efficiency, the square wave operation - also named pulsed operation - is preferred in some applications. Here the spatial density and shape of the microdischarges are changed compared to the sinusoidal operation [1,3]. In this paper an electronic ballast for pulsed operation of dielectric barrier discharges (DBD) is presented. The special features are the variable slew rate (dU/dt), the variety of wave forms and the high feasible bandwidth. The typical losses of the ballast are calculated for the different operating modes and wave form parameters

2 TECHNICAL DESIGN

The circuit arrangement described is a transformer coupled square wave power source optimized for capacitve loads like DBD. It can be controlled in different modes. One possibility is a controlled AC current source, the other is an AC voltage source. Combinations of both modes are possible.

2.1 Inverter details

The main components of the ballast are a switched power stage built as a full bridge, a resonant tank circuit with a following pulse transformer and a digital control unit. The circuit is shown in figure 1. A combination of microcontroller and complex programmable logic device (CPLD) controls the frequency, the phase shift and the dead-time of the power stage. This is designed as a phase shift full bridge [4]. The duty cycle of the output voltage can be controlled by phase shifting the first half bridge A with respect to the second half bridge B. The resonant tank circuit as an energy storage allows a current-source mode with variable slew rate of the output voltage and the quasi resonant switching of all power transistors. The high voltage pulse transformer converts the low level inverter output voltage of approximately 500 V to the typical DBD voltages of about 3 to 20 kV peak.

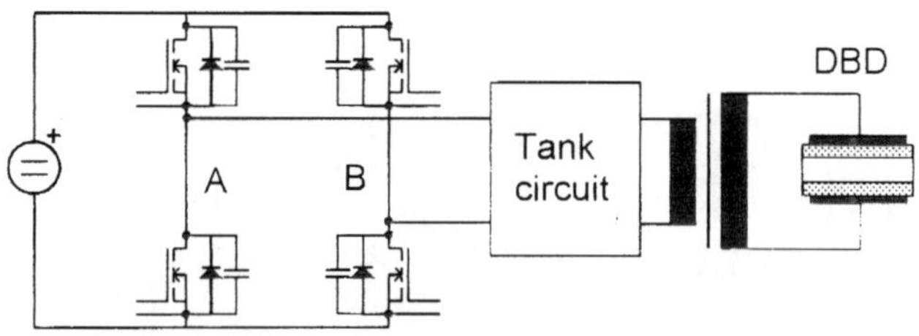

Fig. 1 Circuit arrangement

The high power prototype is designed for a maximum reactive power of about 22 kVA at switching frequencies up to 300 kHz. The gate drive circuit is optimized for the control of large power MOSFETs with typical gate charges up to 500 nC. The control signals are transferred via optical fiber. This configuration allows frequencies from almost DC to approximately 1 MHz.

2.2 Operational modes

The aforementioned ballast operates in different modes. In the current-source mode the slew rate of the lamp-voltage V_L and the lamp-current I_L is defined by the state of the tank circuit. The operation mode with a duty cycle of 100 % is feasible for operation frequencies above 10 kHz. It is presented in figure 2. The higher slew rate (typ. > 100 V/ns) causes higher peaking (overshot) of the lamp-voltage due to higher excitation of the transformers resonance frequency. It also increases electromagnetic interference (EMI). At a reduced duty cycle of e.g. 20 % as shown in figure 3, the operating frequency could be decreased to approximately 2 kHz without an increase of the magnetic flux of the transformer. Therefore the transformer could be optimized for high frequency operation with lower stray-field and higher resonance-frequency, although the operating frequency could be quite low. The slew rate of both the rising and falling edge are controllable independently.

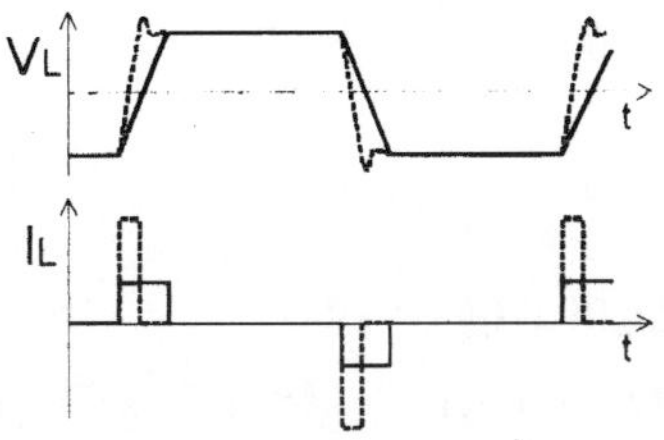

Fig. 2 current mode, DC=100%

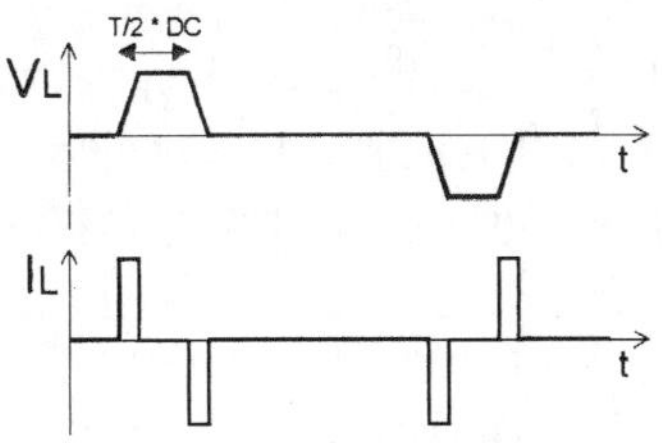

Fig. 3 current mode, DC=20%

The second operation mode is a combination of a current and a voltage source. The typical wave shape is an asymmetrical pulse with a different slew rate at the rising and the falling edge. This mode is shown in figure 4. The high slew rate (typ. >>100 V/ns) at the rising edge of each pulse is caused by turning on one of the switches of half bridge A. In this case the bridge acts as a switched voltage source. The second half bridge B is switched as in the current source mode and therefore the falling edge is controlled by the tank current. In some application the higher slew rate of the rising edge could cause higher lamp efficiency, while the lower slew rate of the falling edge reduces switching losses and EMI.

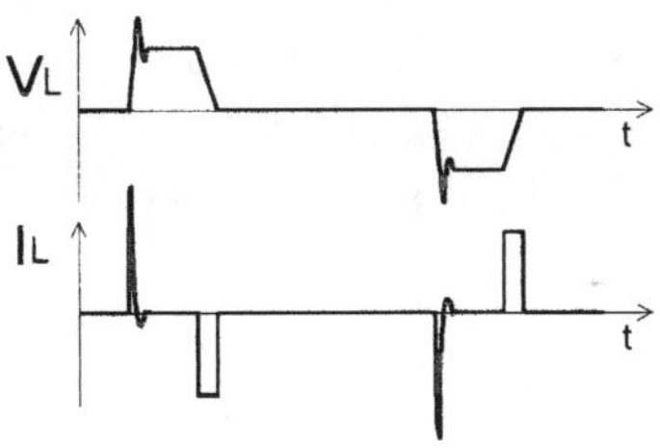

Fig. 4 Mixed mode

A third operation is given by the modification of the first or second mode as burst mode. Here a fixed number of pulses is applied to the lamp as one sequence followed by a fixed deadtime.

3 CONCLUSIONS

The efficiency of dielectric barrier discharges is linked with external operating conditions. To optimise the radiation efficiency it's fundamental to match the system consisting of the electronic ballast and the discharge lamp. Therefore the wave form parameters have to be optimised with respect to the efficiency of both. The electronic ballast presented in this paper allows to investigate the system efficiency in dependence of the wave form parameters slew rate, operation frequency and peak voltage.

4 REFERENCES

[1] R. Kling, S. Gasteiger and M. Neiger; Experimental Investigations of Microdischarges in Excimer Barrier Discharge Lamps, ICPIG Toulouse, 1997

[2] J. Kunze, N. Fröhleke, H. Grotstollen; Resonant Power Supply for Barrier Discharge UV-Excimer Sources, IEEE, New York, 1992, 2 vol. xxix+2071 pp.

[3] F. Vollkommer, L. Hitzschke; Durchbruch bei der Erzeugung von Excimer-Strahlung; Phys. Bl. 53 (1997) Nr. 9

[4] P. Schwarz, D. Flury, W. Heering; Entwicklung eines Mikrocontrollersystems zur Steuerung und Regelung von elektronischen Vorschaltgeräten, Lichttechnisches Institut, Univ. Karlsruhe, Jahresbericht 1996/1997, p. 117-119

H 02

TRANSFORMER IN A NOVEL MODIFIED HALF-BRIDGE INVERTER FOR FLUORESCENT LAMP BALLASTS

Hirofumi Matsuo[1)] and Keiichi Shimizu[1),2)]

1) Nagasaki University, Nagasaki, 852-8521 Japan

2) Toshiba Lighting & Technology Corp., Yokosuka, 237-8510 Japan

1. INTRODUCTION

Recently, in order to simplify the circuit configuration and to realize low cost, combined type inverters for fluorescent lamps have been reported [1-8], in which the functions of the active filter and inverter are combined. These inverters have achieved successfully the high power factor and low harmonic distortion of the input current and low output crest factor. which satisfy the maximum limits of IEC 1000-3-2 Class C. However, the power efficiency of these inverters is not so high, compared with that of the active filter method, in which the active filter and inverter is cascaded. Furthermore, it is often observed in practical use of these combined type inverters that the inrush current flows when the ac switch is turned-on.

The purpose of this paper is to present a novel modified half-bridge type inverter for fluorescent lamps, in which not only the high power factor, low harmonic distortion and low crest factor are obtained but also the sufficiently high power efficiency is achieved and the inrush current is removed [8]. The relationship between the THD and parameters of the leakage and exciting inductance in the transformer is analysed.

2. ANALYSIS

Fig. 1 shows a new modified half-bridge inverter for fluorescent lamps. The capacitor C_1 has large capacitance to keep the crest factor low. C_2, which is conventionally used as one of two voltage-dividing capacitors with large capacitance in the usual half-bridge inverter, is replaced by that with extremely small capacitance. Therefore, C_2 works to achieve the high power factor and low harmonics. Also, C_2 prevents the inruch current to flow through e_{ac}, ac switch which is omitted in Fig. 1, diode bridge rectifier, C_1 and C_2. The transformer T has a relatively large leakage inductance between the primary and secondary windings.

To simplify the analysis, the followings are assumed: (1) Each circuit element is ideal and has no power loss, (2) The capacitance of C_1 is large enough to be replaced by a dc voltage source E_{C1}, (3) One period of e_{ac} is so large compared with that of the switching of the main switches T_{r1} and T_{r2}, that e_{ac} is replaced by the constant dc voltage source E_{ac}, (4) The diode D_B is used instead of the diode rectifier of D_{B1}, D_{B2}, D_{B3} and D_{B4}, (5) The ideal switches S_1 and S_2 are used instead of the antiparallel connection of T_{r1} and D_1, and that of T_{r2} and D_2, respectively, (6) The noise filter of L_F, C_{F1} and C_{F2} is neglected because it does not affect the operation of the inverter.

Taking account of the above assumptions, the circuit of Fig. 1 is divided into three states. The circuit equations are derived in each state. Considering these circuit equa-

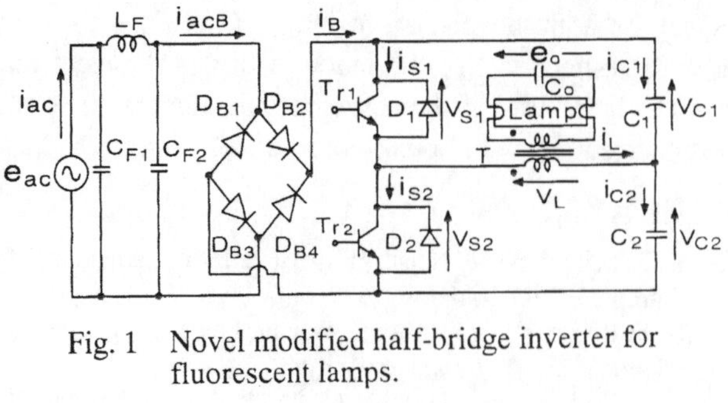

Fig. 1 Novel modified half-bridge inverter for fluorescent lamps.

tions, state transition, the energy-balance relationship between the input and output energies of the inverter during one switching period T_S [8], and the energy-balance relationship between the charged and discharged energies in E_{C1} during a half period of e_{ac}, E_{C1} is determined, and consequently the voltages and currents of the circuit elements are calculated [7].

Here, we will examine L_ℓ and L_e of T. Figs. 2 (a) and (b) show E_O and THD, respectively, for the changes of L_ℓ, taking f_S as a parameter, and E_O and THD for the changes of L_e are shown in Figs. 3 (a) and (b), respectively. It is seen in Figs. 2 and 3 that, determining L_ℓ = 0.36mH, L_e = 0.33mH and f_S = 52kHz on the point of the symbol ○ , E_O of about 300V and THD of about 10% are obtained. Therefore, the circuit parameters and variables are determined as shown in Table 1. Fig. 4 shows the observed and calculated waveforms of e_{ac} and i_{ac}. The power factor is 97% and THD is about 11%. The maximum limits of the harmonic current in IEC 1000-3-2 Class C is sufficiently satisfied. The crest factor is 1.56 and the power efficiency is 93%.

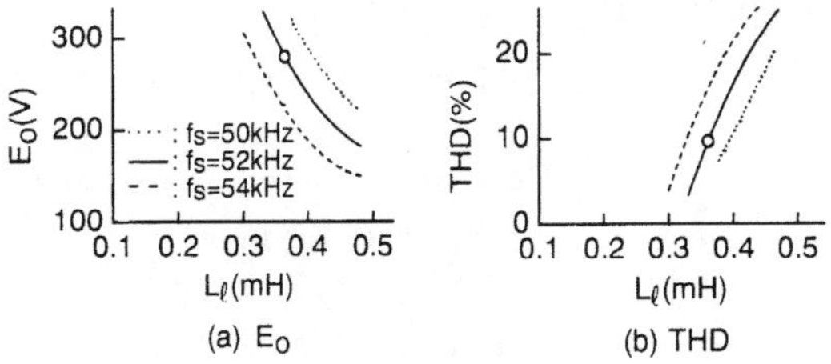

Fig. 2 For the change of L_ℓ, taking f_S as a parameter.

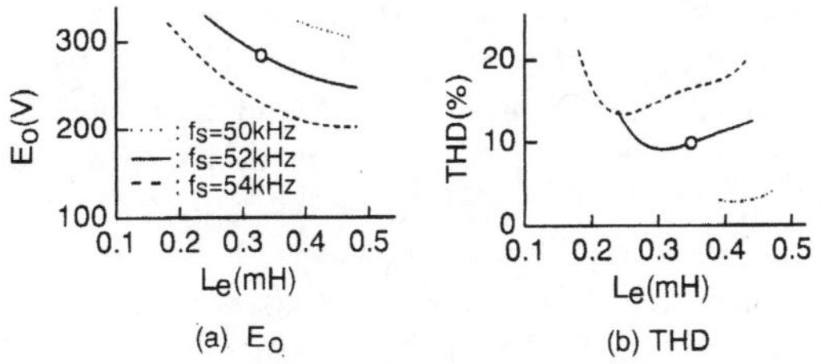

Fig. 3 For the change of L_e, taking f_S as a parameter.

Table 3 Circuit parameters and variables.

e_{ac}=200V	C_1=270μF
E_o=293V	C_2=0.027μF
f_{ac}=60Hz	C_o=0.0057μF
f_s=52kHz	L_e=0.33mH
R=1200Ω	L_ℓ=0.36mH

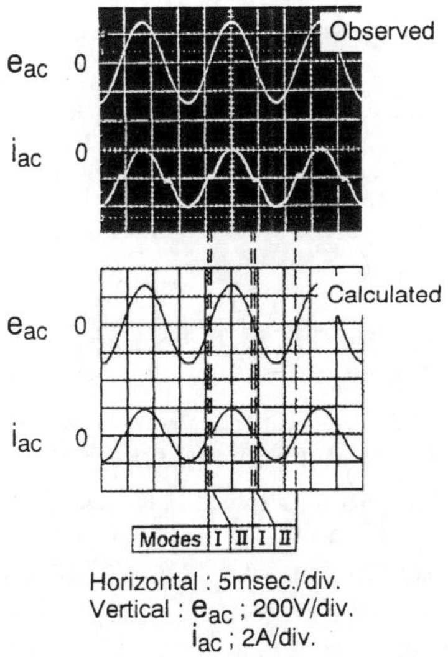

Fig. 4 Observed and calculated waveforms of e_{ac} and i_{ac}.

3. CONCLUSION

The leakage and exciting inductance of the transformer in the proposed modified half-bridge inverter for fluorescent lamp ballasts are examined. As a result, the power factor of 97%, THD of 11%, crest factor of 1.56 and power efficiency of 93% are obtained.

4. ACKNOWLEDGEMENT

This work was supported in part by the research foundation of Illuminating Engineering Institute of Japan.

REFERENCE

[1] W.Chen et al., Proc. of IEEE APEC (1996).

[2] C. S. Moo et al., ibid (1996).

[3] Eguchi et al., Technical Report of IEI of Japan, LS96-18 (1996).

[4] J. Qian et al., Proc. of IEEE IAS Annual Meeting (1997).

[5] Y. Kato et al., Technical Report of IEICE of Japan, EE97-77 (1998).

[6] T. Yamaguchi et al., ibid, EE 97-78(1998).

[7] T. Kitajima et al., ibid, EE97-79 (1998).

[8] Matsuo et al., IEICE of Japan Trans. Communications, E75-B, 11 (1992).

H 03

A DIMMING CIRCUIT IN SELF-EXCITED ELECTRONIC BALLAST FOR ELECTRODELESS FLUORESCENT LAMPS

Takashi Fukue[1], Kazurou Harada[1], Yoshiyuki Ishihara[1], Toshiyuki Todaka[1] and Futoshi Okamoto[2]
[1]Doshisha University, Kyotanabe, Kyoto 610-0321, Japan
[2]Matsushita Electric Works, Ltd., Kadoma, Osaka 571-8686, Japan

1 INTRODUCTION

The electrodeless fluorescent lamp was put to practical use in 1990 and ever since the high-frequency electronic ballasts for the lamps have been developed in point of miniaturization and high efficiency [1-3]. Above all, the self-excited electronic ballasts have the advantages of size, efficiency and cost. However it is difficult to control the output power, that is, to dim. The dimming method using an intermittent oscillation has been reported in the separately-excited electronic ballast for the lamps [4], but it has not been realized in self-excited electronic ballast for the electrodeless lamps yet.

This paper presents a novel dimming circuit using intermittent oscillation in the self-excited electronic ballast for electrodeless fluorescent lamps operated at 2.65MHz [5].

2 CIRCUIT CONFIGURATION

Fig. 1 shows the proposed dimming circuit using intermittent oscillation. A resonant half-bridge inverter of zero-voltage-switching is used for the electronic ballast. The starting and dimming circuit consists of resistance, capacitance and auxiliary switch Q3 for intermittent oscillation. There is no voltage-breakover-type device (DIAC, Zener Diode, etc.) for the starting, so that this circuit realizes intermittent oscillation without flicker.

3 OPERATION OF THE PROPOSED CIRCUIT

Fig.2 shows each state of the proposed starting and dimming circuit.

Starting of oscillation (Fig.2 (a))

DC power is applied to the circuit by the power switch, and then voltage E is divided by resistance $R1$, $R2$ and voltage of $R1$ is divided by capacitance Cdc, Cp. The electronic balllast oscillates as soon as voltage of Cp, namely, gate-source voltage of switch Q1 is increased to

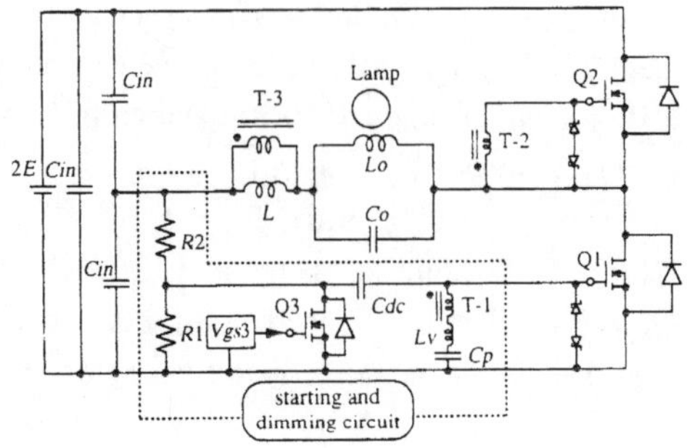

Fig.1 A novel dimming circuit using intermittent oscillation in the self-excited electronic ballast.

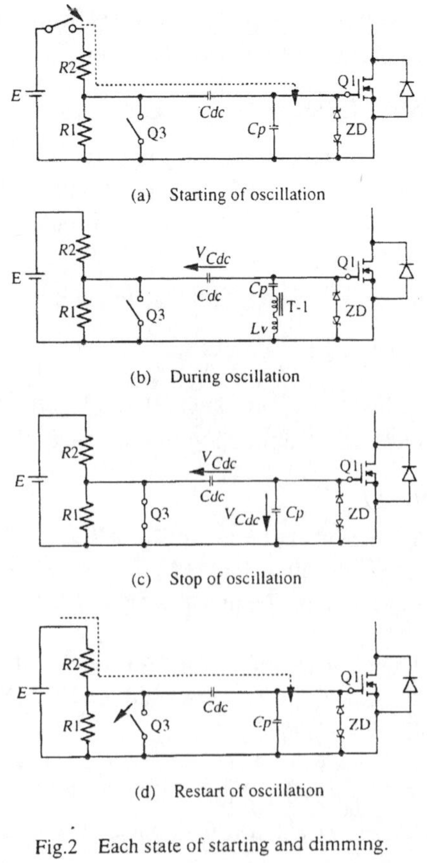

(a) Starting of oscillation

(b) During oscillation

(c) Stop of oscillation

(d) Restart of oscillation

Fig.2 Each state of starting and dimming.

threshold voltage of Q1.

During oscillation (Fig.2 (b))

Oscillation keeps up during off state of the auxiliary switch Q3. Capacitance *Cdc* blocks DC voltage so that the transformer operates without saturation.

Stop of oscillation (Fig.2 (c))

Oscillation of the electronic ballast is instantaneously stopped by switching on of Q3 because negative DC voltage of *Cdc* is applied to gate-source of Q1. Capacitance *Cp* prevents the short of electric charge in *Cdc* and the saturation of transformer.

Restart of oscillation (Fig.2 (d))

Oscillation is restarted by switching off of Q3. The interval between the switching of Q3 and the restart of oscillation is sufficiently short compared with the existing circuit using voltage-breakover-type device. This interval depends on time constant determined by mainly $R2$ and Cp ($\because Cp \ll Cdc$).

4 EXPERIMENTAL RESULTS

The proposed circuit is verified by the experiment. Fig.3 shows dimming rate vs. duty ratio of Q3 for dimming. Fig.4 shows lower limit of dimming rate vs. dimming signal frequency. Where, dimming rate is gived by

Dimming rate = (Illuminance at dimming)／(Maximum illuminance) × 100.

From Fig.3, it is found that illuminance of the electrodeless fluorescent lamp decreases in proportion to duty ratio of Q3 for dimming in each dimming signal frequency.

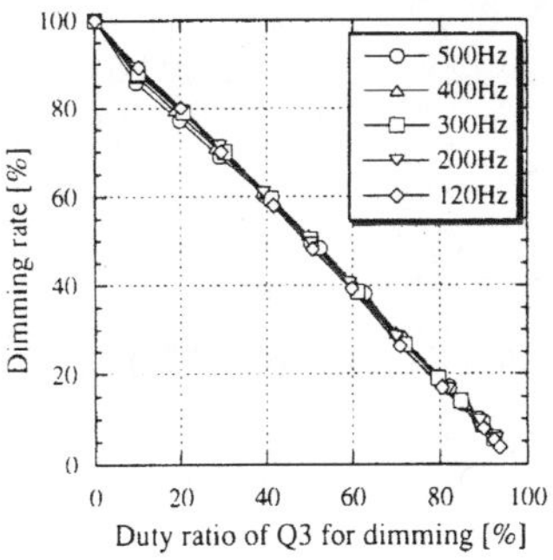

Fig.3 Dimming rate vs. duty ratio of Q3 for dimming.

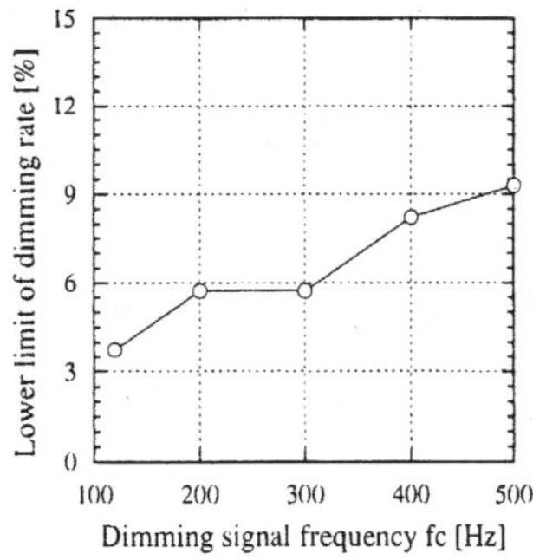

Fig.4 Lower limit of dimming rate vs. dimming signal frequency fc.

5 CONCLUSIONS

The novel dimming circuit using intermittent oscillation in the self-excited electronic ballast for electrodeless fluorescent lamps has been presented. In the experiment at 2.65MHz, the proposed circuit controls illuminance of the electrodeless fluorescent lamp from 100% to under 5% without flicker.

6 REFERENCES

[1] Sayed-Amr El-Hamamsy : "Design of High-Efficiency RF Class-D Power Amplifier," IEEE Transactions on Power Electronics, Vol.9, No.3, pp.297-308 (1994)

[2] T. Hosotani, K. Harada, Y. Ishihara, T. Todaka and F. Okamoto : "10MHz Zero-Voltage-Switched Resonant Inverter for Electrodeless Fluorescent Lamp," J. Illum. Engng. Inst. Jpn., Vol.79, No.11 (1995)

[3] Louis Robert Nerone : "Design of a 2.5MHz, Soft-Switching, Class-D Converter for Electrodeless Lighting," IEEE Transactions on Power Electronics, Vol.12, No.3, pp.507-516 (1997)

[4] T. Nakano, F. Okamoto and T. Mizumoto : "A study of light control method for the electrodeless fluorescent lamp," Proceedings of Annual Conference of the Illuminating Engineering Institute of Japan, p.20 (1993)

[5] T. Fukue, K. Harada, Y. Ishihara, T. Todaka and F. Okamoto : "A study of dimming method in electronic ballast for electrodeless fluorescent lamp," Record of the Kansai-Section Joint Convention of Institutes of Electrical Engineering, Japan, G4-25 (1997)

H 04

CHEMICAL EQUILIBRIUM IN CLUSTER LIGHT SOURCE.

B.M.Smirnov
Institute for High Temperatures, Russian Academy of Sciences, Izhorskaya 13/19, Moscow 127412, Russia

Cluster lamps are based on discharges of a high pressure where hot clusters are effective radiators. Experimental investigations of cluster lamps [1-3] respect to a microwave discharge plasma of a high pressure with clusters of different heat-resistant metals. Our theoretical studies [4,5] relate to arc of a high pressure of a rare buffer gas with tungsten clusters and below we deal with such a system. It is of importance the regeneration of tungsten in this discharge. It means [1] existence of a gaseous compound at low temperatures and its transformation in clusters at more high temperatures. For definiteness, we take this compound as WX_k, where X is the halogen atom. Thus, tungsten exists in the form of this gaseous compound near walls of the discharge tube and its clusters are formed in a region of a more high temperature. Hence, tungsten is not consumed on walls during operation of this discharge. Note that we assume the excess of a halogen X_2 compared to tungsten which are admixtures to a buffer gas.

The analysis of the chemical equilibrium in this system gives that the gaseous compound exists under the condition $E-kD/2 > \varepsilon_W$, where E is the total binding energy of atoms in the compound WX_k, D is the dissociation energy of the molecule X_2 and ε_W is the atom binding energy for bulk tungsten. Existence of tungsten clusters requires the criterion $\varepsilon_X = E/k < \varepsilon_W$. One can see that these criteria are compatible for $k > 1$. Because of the gaseous state, the molecule WX_k has the completed structure, so that in reality $k = 4, 6$. One can estimate the average binding energy per one halogen molecule $\varepsilon_X \sim 2 - 3eV$. Because $\varepsilon_W \approx 8eV$ [5], the above criteria are valid for this case.

From the chemical equilibrium of WX_k one can estimate a typical temperature T_1 when it is decomposed in atoms, and from the chemical equilibrium of clusters one can found a typical temperature T_2 when clusters are transformed in atoms

$$T_1 = \frac{\varepsilon_X}{\ln(N_o/[X])}, \quad T_2 = \frac{\varepsilon_W}{\ln(N_o/[W])}$$

Here $[X]$ denotes the total number density of free and bound atoms X, N_o is a typical atomic value, and $[W]$ is the total number density of free and bound tungsten atoms. Evidently, clusters exist in the temperature range

$$T_1 < T < T_2$$

Because of the halogen excess $[X] >> [W]$, the possibility of existence of clusters in the system under consideration respects to the criterion $\varepsilon_X < \varepsilon_W$. On the basis of the above data one can estimate $T_1 = 2000 - 2500\,K$, and for typical values $[W] \sim 10^{14} cm^{-3}$ [4,5] we get $T_2 = 3300 - 3800K$ [5]. Thus, in arc under consideration a cold discharge region near walls (the gaseous region) where a gaseous chemical compound is formed, and a neighboring discharge region (the cluster region) where clusters are generated. In the case of large currents, clusters are decayed in a central region (the vapor region), so that an atomic vapor is formed there.

One can estimate a typical cluster size by comparison of a transport time during which clusters are found in a cluster region and a typical time of the cluster growth. But a typical transport time is proportional $n^{2/3}$ (n is a number of cluster atoms) if it is determined by diffusion of clusters in a buffer gas, and a typical time of a cluster growth $\sim n^{-2/3}$ for large clusters. Hence, clusters grow faster than can go out the cluster region, so that an instability arises which leads to accumulation of clusters in this region. The instability takes place also if a transport time is determined by drift of large charged clusters under action of discharge electric fields. It stops only when clusters vary parameters of the discharge.

The instability under consideration promotes to transformation of an atomic vapor and a gaseous component of tungsten in clusters, and these clusters are located in the cluster region of arc. The development of this instability causes a redistribution of clusters on size and improves radiative parameters of this system. Probably, this effect takes place in $Na - Tl - I$ arc lamps of a high pressure. This cluster instability is of interest also for separation of heat-resistant elements in plasmachemistry processes.

References

1.Weber, B. and Scholl, R., *J.Illum.Eng.Soc.*, Summer 1992, p.93; *J.Appl.Phys.* **74**, 607(1993)

2. Scholl, R. and Weber, B., *in: "Physics and Chemistry of Finite Systems: from Clusters to Crystals".* (Edited by P.Jena, S.N.Khana, B.K.Rao). (Kluwer Academic Press, Dodrecht,1992) vol.II.p.1275.

3.Scholl, R. and Natour, G., *in: "Phenomena in Ionized Gases".*(Edited by K.H.Becker, W.E.Carr, E.E.Kunhardt). (AIP Press, Woodbury, 1996) p.373

4.Smirnov, B.M., *Phys.Scripta* **51,** 380(1995); **53,** 608(1996)

5.Smirnov, B.M., *Phys.Uspekhi* **40,** 1169(1997)

H 05

A New Electronic Starter for Miniature Metal Halide Lamps With Reactor Ballast

Zhu Ding-chang, Huangfu Bing-yan,
Department of Illuminating Engineering,
Fudan University, Shanghai, 200433, China

1. Introduction

Metal halide lamps (MHL) have the characteristics of high efficacy and good color quality. Many customers therefore have chosen to use MHL instead of HPM and HPS lamps.

It is necessary to use a starter to ignite the MHL when a reactor was used for ballasting, especially for the miniature MHL where no auxiliary electrode exists. Many previous research works shown that the aging life of the miniature MHL is not simply a matter of the quality of the arc tube and that, in particular, igniting conditions have a major impact on the miniature MHL performance.

2. Abnormal Phenomena When Lighting Miniature MHL

With the 220V/50-60Hz-line supply, using reactor for ballasting MHL is more effective and economical, but unsuitable igniter would cause the failure of miniature MHL lighting. The major abnormal phenomena can be summarized as follows.

a). The lamps extinguish immediately after breakdown.

b). The lamps "hang up" glow discharge state and cause the serious lose of the electrode materials consequently effect lamp life.

c). The aging life is quite short, typically about several hundreds hours.

d). The ballast insulation was damaged after the failure of lamps.

As the arc chamber volume become smaller, the relative concentration of impurity gases increases and causes a high breakdown and re-ignition voltage during the warm-up period. So the functions of suitable igniter for miniature MHL should be:

a). A pulse energy (peak voltage and width) sufficient to start the lamp must be generated.

b). Pulse phase appears nearby the phase of zero current.

c). Pulse energy output should be reduced or ceased in certain time after the lamp starts.

d). Pulse generation must be cut-off when the lamp will no longer start.

3. Circuit Characteristics of the New Igniter and Lighting Result

To have the functions of a) and b) mentioned above, the pulse potential should be bigger than 4 kV, the pulse width at 90 percent of peak voltage is more than 2 μs if only one pulse appears in each half cycle, and the pulse position is within ± 10 electric degree of the peak open circuit voltage. Furthermore, it is important that the instantaneous lamp voltage drop after breakdown must excess 280 volts to avoid the glow discharge state, that means a certain value of extra voltage has to be provided by a storage component in the igniter to overlap on the ballast supplying voltage. Because of the difficulty to reach high amplitude and wide duration at same time, the practical way is to produce several narrow pulse per half cycle with needed amplitude and pulse position, but the sum of the individual pulse width should be more than 2 μs. Table 1 shown the ignition parameters of this new igniter. To have the function c), some countermeasures had been taken in this new igniter and the pulse numbers will drop from (3-6) per half cycle to 1 per two cycle after 20 seconds of the lamp ignition, and the pulse amplitude reduces to half of the original value too. In the new igniter, a heat sensitive element is also available to have the function d). If excessive current throughout the lamp, the temperature of the igniter rises and damages sensitive element, finally the igniter stops work. With this new igniter, the aging life of miniature MHL is greatly improved, typically the burning life of 70 w MHL is more than 5 thousand hours.

Table 1. The Pulse Parameters of the New Igniter

amplitude (kV)	phase (degree)	extra voltage (V)	width (μs)	numbers per half cycle	total duration (μs)
6.0±0.3	70~120 (250~290)	50 ± 10	1.2 ± 0.1	3~ 6	(3.6~ 7.2)± 0.6

4. Conclusion

By employing the correct components in the igniter to ensure appropriate igniting conditions needed for miniature MHL, the practical burning life is greatly improved more than 5 times with the same quality of the lamp tube.

It is clearer that igniter parameters will extremely influence the aging life especially for lighting the miniature metal halide lamps.

Reference:

[1] W.W Byszewski et al: HID Starting: Glow Discharge and Transition to the Thermionic Arc. JIES. summer 1991. P3-P9

[2] Shunichi Sakaki et al: Built-in Starter with Pulse Cutout for HID Lamps. J.Light & Vis.Env (Japan). Vol.20. No.2. 1996.

[3] American National Standard. ANSI C 78.1381-1989.

H 06

Modelling of fluorescent lamps for power supply design

Ted Vos[1], Frans Ligthart[1†], Marcel Hendrix[1], Theo Stommen[1], Uwe Chittka[2], Joost v.d. Mullen[3], Willem v.d. Bosch[3]
[1]Philips Lighting B.V., P.O.Box 80020, 5600 JM, Eindhoven, The Netherlands
[2]Philips Research Labs, Weisshausstr. 2, D-52066, Aachen, Germany
[3]Eindhoven University of Technology, P.O.Box 513, 5600 MB Eindhoven, The Netherlands
[†]deceased March 29, 1998

1 INTRODUCTION

The aim of the present work is to create a fluorescent lamp model that describes the electrical behaviour of a fluorescent lamp and that can be implemented in the SPICE circuit simulation program. This would make it possible to simulate the lamp and its power supply simultaneously. The objective is reduction of development times for electronic gear.

2 THE DEVELOPED MODEL

2.1 The discharge model

A 3-level collisional radiative model was developed which takes into account the dominating processes that occur in a fluorescent lamp. Included processes are: collisional (de)excitation and ionisation, ambipolar diffusion and radiative decay of mercury states. Together with the energy balance for the electron gas and equations for the electron mobility this model proved to produce quite satisfactory results.

2.2 Implementation into a circuit simulator

A breakthrough was realised by fully implementing the physical model into a SPICE circuit simulation program. The set of differential equations can be solved using the XSPICE * extension. The normal circuit equations are solved in SPICE directly. A fluorescent lamp can thus be inserted from the component library like any other component. The parameters for a lamp can be altered to describe a specific lamp, or to investigate the effect of lamp parameters on electrical behaviour. Output includes physical parameters, e.g. particle densities.

3 COMPARISON BETWEEN EXPERIMENT AND MODEL

3.1 Experimental

By means of a near-resonant LCR circuit, fast positive or negative discrete changes can be forced on the high-frequency lamp current. In Figure 1 the time development of lamp current and lamp voltage are depicted after a step down in lamp current, for a TLD 36W lamp operated at room temperature. After the step is applied the Lissajous figures turn in a counter-clockwise direction towards a higher lamp voltage.

3.2 Results of SPICE simulation

In Figure 2 the results are depicted for a SPICE simulation with the same circuit and lamp parameters as used in the experiment. The comparison with the experiment is quite good.

4 CONCLUSIONS

- The implementation of a fluorescent lamp discharge model into a SPICE simulator was achieved.
- A quite good agreement between experiments and modelling results was obtained.

* Developed by Georgia Tech Research Corp.

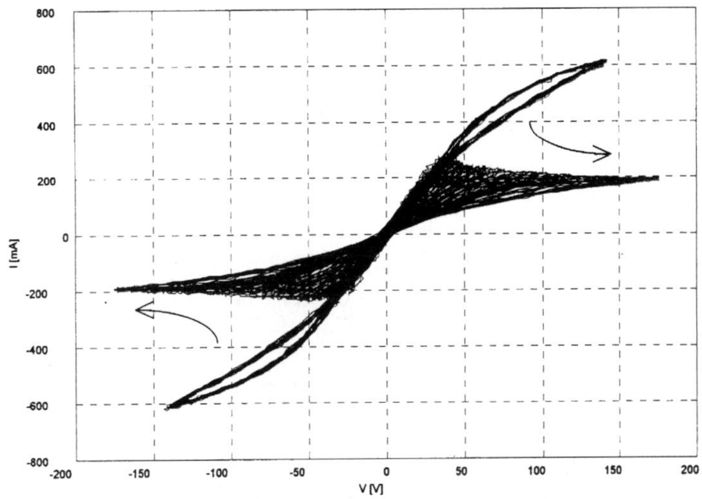

Figure 1: Experiment: time development of Lissajous curves after a current step-down for a TLD 36 W lamp operated at room temperature; I_{begin} *= 455 mA,* I_{end} *= 155 mA, f = 97 kHz.*

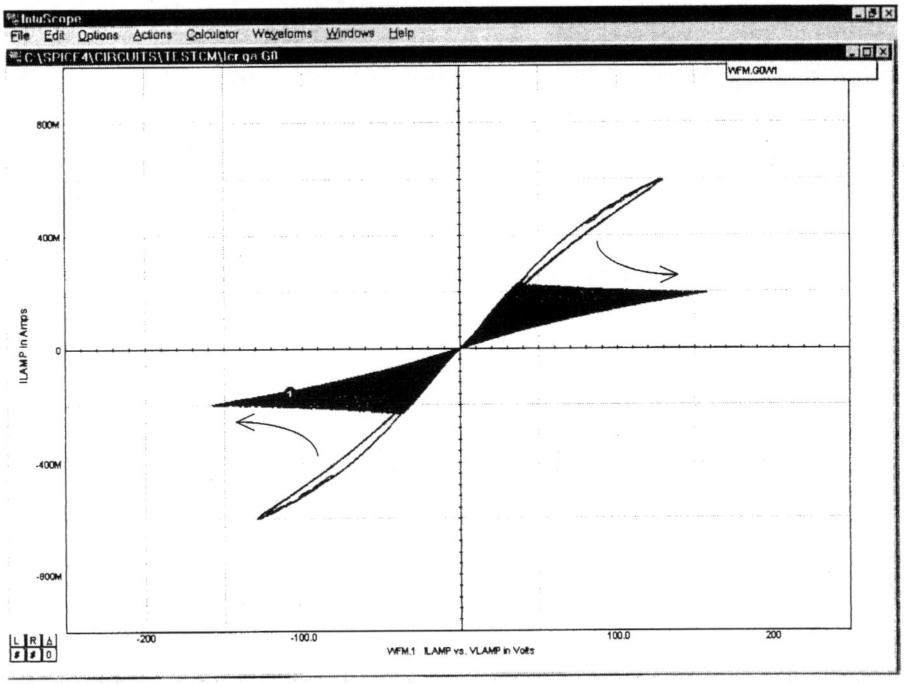

Figure 2: SPICE simulator output: time development of Lissajous curves after a current step-down; for the same conditions as in Figure 1.

H 07

High Frequency Power Supply Systems for Multiple-operating of Fluorescent Lamps

Kazuhiko TANIGUCHI [1)] Hiroshi MORITA [1)] Takashi URAYAMA [2)]

1) Kinden Co. R & D Center of Kyoto.
3-1-1 Saganakadai Kizu-cho, Souraku District, Kyoto Pref. 619-0223 Japan

2) Ritsumeikan Univ. Dept. of Electrical & Electronic Engineering.
1-1-1 Noji-higashi Kusatsu-shi, Shiga Pref. 525-8577 Japan

Introduction

Electronic ballasts, one of the application of power electronics technology, are contributing remarkable improvement of the efficacy and performance in lighting system. Most of the electric and electrical equipment have been designed to adapted with 50/60 Hz commercial power supply. Progress of power electronics, however, makes possible to produce the variety of suppies, such as the DC and high frequency with various waveform.

This work is part of the study for the multiple-operating of fluorescent lamps under the high-frequency power supply.

Block diagram of electronic ballast

Fig. 1 shows the standard block diagram of the electronic ballasts for discharge lamps. The first block consists of an AC/DC converter by which the line frequency power converts to the DC power. This block often contains a pre-regulator . The function of a pre-regulator is to generate a fixed DC voltage from the mains supply while drawing a sinusoidal current which is in phase with the supply voltage. The second block is DC-to-high frequency converter and the high frequency power is supplied to the lamp through the third block .

The third block has the function of lamp starting and lamp current limiting.

DC indoor distribution system

The popularized electronic ballasts are composed of three different circuit blocks shown in Fig. 1. Eaeh lamp then requires individual ballast. If DC power can be supplied from centralized DC power supply through distribution line, the first part of the circuit block can be removed from individual electronic ballast. In this case, reduction of size of the ballast and unit cost can be expected.

High frequency indoor distribution system

As the next step of the simplification of the system, the high frequency distribution system is considerable. In this system, the high frequency power is fed directly from the centralized high frequency power supply to the lamp circuit through the distribution line . In this case, individual lamp requires only very simple ballast which has the function of starting and current limiting, consequently further circuit simplification as well as the smaller size and light weight of the unit can be realized.

Frequency characteristics of line conductor

The high frequency distribution system is expected to be able to reduce the unit cost and increase the system reliability. However, electrical loss in the distribution line accompanies.

Minimization of this loss is very important in order to popularize the multiple lamp operating system. The frequency characteristics of line conductor's resistance and inductance affect the performance of the high frequency operating system. The high frequency impedance of the conductor is varies by the shape and arrangement of the conductor as well as the direction of current flow through the conducting wire.

Measurements were made for following three cases. Tested conductor is composed three conducting wires in parallel arrangement and is insulated polyvinyl chloride. (VVF 1.6 mm $\phi \times 3$, 20m length), this wire is widely used for the indoor distribution line in Japan.

Fig. 2 show the measured results for following cases.

case A ; only one conductor

case B ; connect center and outer wires in series, the direction of current through wires is reverse with each other

case C ; connect outer two conductors in series, the direction of current through wires is same as case B

As is clear from Fig. 2(a), resistance of the conductor increases with increasing of frequency in any case, In case A, however, the increasing rate is bigger than the case B and C. Impedance values those in case B and C are one order smaller than the case A as shown in Fig. 2(b). Therefore, in order to reduce the electrical losses of multiple lamp operation system with high frequency, wire diameter, the arrangement of wires in conductor and the direction of current flow must be optimized.

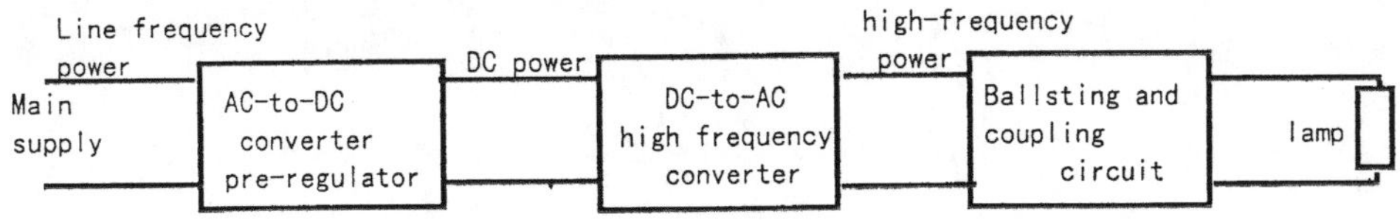

Fig. 1 Block diagram of electronic ballast

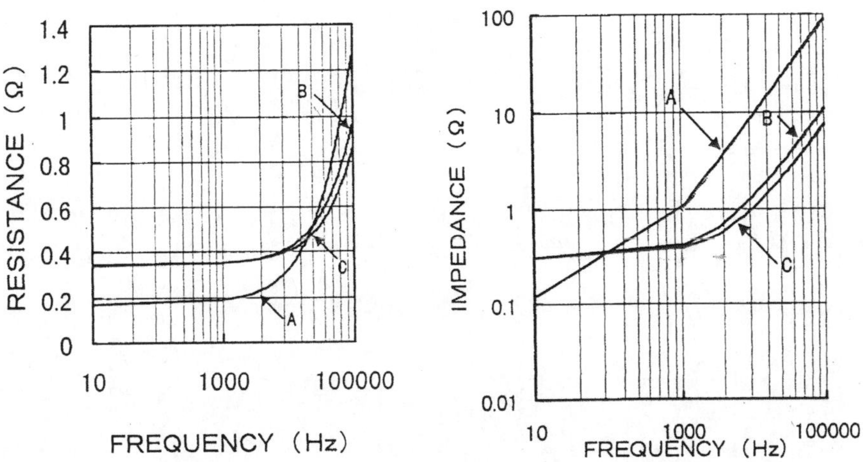

(a) conductor resistance (b) conductor impedance

Fig. 2 frequency characteristics of line conductors

I

Non Lighting Applications

I01

A Bioluminescence application for Illumination

Naoyoshi Nameda[1] and Shinichi Itoh[2]

[1] Laboratory of Biophysics, University of Muenster,
Robert-Koch-Str. 45, D-48129, Germany
(on leave from Kagoshima University, Japan)

[2] Kyoto University of Education, Fukakusa, Fushimi, 612, Kyoto, Japan

1 INTRODUCTION

A chemical reaction produces an amount of energy in comparison with electric energy produced in a battery or an electric cell. If light is emitted directly from the chemical reaction, a brighter or longer durable light suppose to be substituted of a lamp energized by a battery or a electric cell. Firefly luminescence is known as a high quantum efficiency, because it emits only visible light. Furthermore, the bioluminescence materials are neither harmful for the human being nor ecologically harmful for an environment. As this bioluminescence materials, Luciferin(Lu) and its enzyme Luciferase are well known. Recently, these materials are produced artificially by colitis germs installed in the gene of a firefly at the Kikkoman Corp. This report presents the experimental results and shows an actual application.

2 METHOD OF EXPERIMENT

Luciferin, Luciferase and ATP which is necessary for the reaction of the energy source. are materials for the bioluminescence. In this experiment, 1g of each powder was dissolved in 5oocc purified water. When the light was desired, the solutions were mixed together into a one vessel. Consequently, the light began to emit. For the variation of temperature, the vessel containing the mixture was dipped in a temperature controlled external water vessel. The external vessel was covered with heat insulation material. Luminance of the surface of the mixed solution was measured by a LS-3 Topcon Spectral Luminance Meter.

3 RESULTS AND DISCUSSION

Time dependence (in minute) of Luminance values and the spectrum (in nm) are shown in Figure 1 and 2 respectively. The surrounding water temperature was changed from 0 ° C to 50 ° C in 10 degree increments. However, at temperatures lower than 10 ° C, and above 40 ° C, they did not emit light. The chemical reaction procedure can be estimated from analysis of the spectrum variation. The excited state Lu shows strong luminescence during the process of decomposition of dioxetane part in intermediate Lu and the removal of CO_2 from the tetragonal part. We performed a semi-empirical MO calculation for the intermediate substance of Lu by using computer simulation software MOPAC93. The results indicate that a dioxetane structure in the intermediate of Lu plays an important role in the emission of light. The energy gap between the high occupied molecular orbital and low occupied molecular orbital shows the reasonable energy range to be between 500nm and 600nm.

In this experiment, the luminance value was not sufficient for actual illumination purpose, however, it is known that higher concentration of the solution shows more light emission. Even this luminance is sufficient in a dark place such as a dinner table lamp. One of the applications is a bright wine glass filled with the solution. The feature of the light is freely changing the shape in dependence on shape of the vessel.

4 ACKNOWLEDGEMENTS

I thank Dr. Kajiyama at Kikkoman Corp. for his preparation of the luminescence samples and Dr. Yukiko Itojima for her advice on the chemical experiment.

5 REFERENCE

(1) S. Itoh and N. Nameda, Molecular orbital calculations for Dioxetan as a part of the intermediate of Firefly luciferin, J. Light and Vis. Env. Vol.21, No.1, 1996, PP.22-24

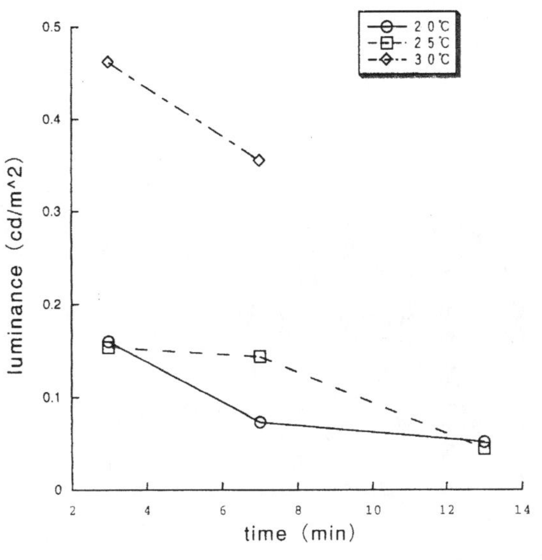

Fig.1 Luminosity versus time at some temperatures.

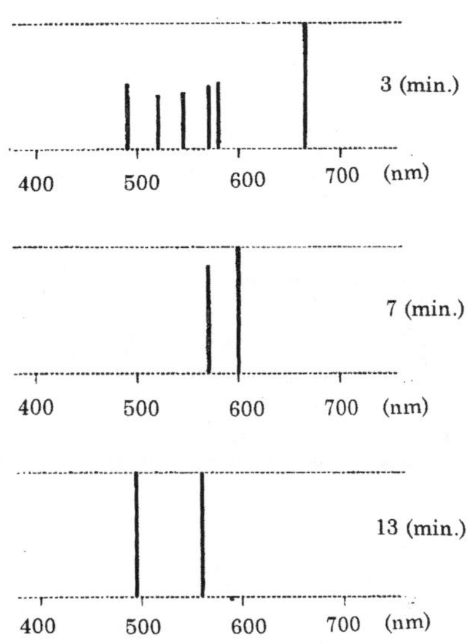

Fig.2 Time dependence (in minute) of the spectrum (in nm) at 20 ℃ .

I 02

DIRECTLY MODULATED INFRARED RADIATION SOURCE FOR OPTICAL GAS ANALYSERS

J. Scherzer, S. Grob, Heraeus Noblelight GmbH, Reinhard-Heraeus-Ring 7
63801 Kleinostheim

1. Introduction

Absorption spectroscopy is a universal measuring method for direct, continuous and quantitative analysis of one or several known gases in a gas mixture. This method uses the property of multiatomic gases in order to absorb radiation in the infrared sector of the spectrum. Gas analysers which are based on the absorption principle are equiped with infrared lamps, the radiation of which is modulated mechanically by means of a rotating shutter disc (mechanical chopper). This kind of modulation requires a highly sophisticated precision engineering. Direct modulation of the infra-red radiation is achieved by periodical change of the power input to an infrared lamp. Sufficient thermical speed of the infrared lamp is an essential requirement.

2. Structure and properties of a directly modulated IR-radiation source

2.1 Geometrical structure of the radiation source

The general structure of the lamp is shown in Fig. 1 + 2. In Fig. 2 the lamp is represented with view of the radiation window (1). Two C-fibre-strips are arranged along the focus line of two spherical reflectors (3). The sectional drawing (Fig. 1) demonstrates that the metallic lamp housing is composed of two parts. In the inferior part (4) of the housing the power supply cables (5) are fixed vacuum-tightly and connected with C-fibre-strips. The inferior part of the housing also serves as supporting base for the reflector unit (3). An opening (6) for the evacuation, indispensable for this kind of lamp, is also situated in the inferior part of the casing. The upper part of the housing (7) which is vacuum-tightly connected with the inferior part includes the radiation window (1) made of CaF_2. This window with an excellent transmission over a wide IR-spectral range, is vacuum-tightly linked with the upper part of the housing.

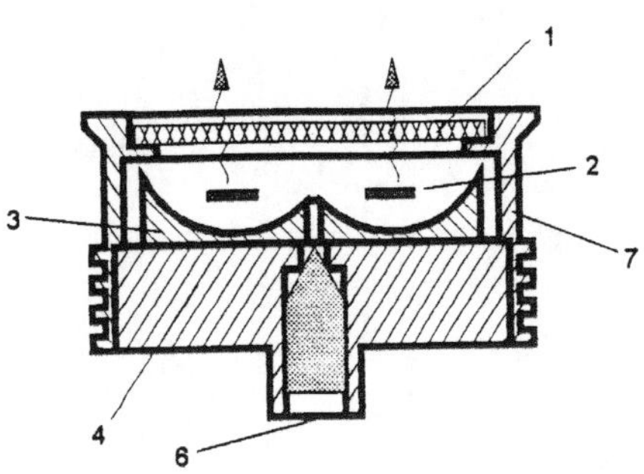

Fig.1

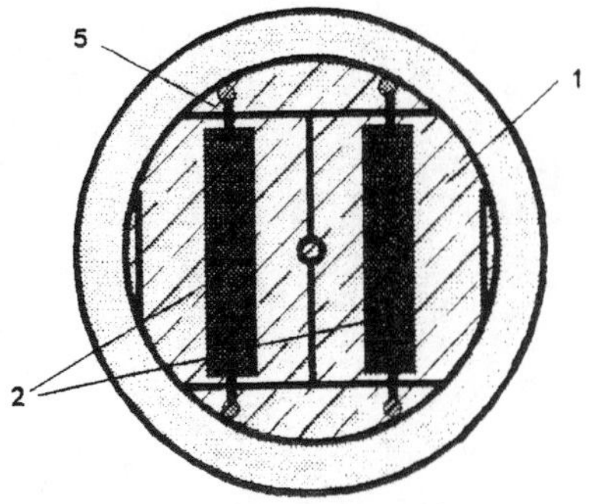

Fig.2

2.2 Electrical and physical data

Modulation of the radiation power is achieved by modulation of the power input (quartz-stabilised) and can be chosen within the range of 2 - 6 cycles per second. Particularly relevant known quantities for modulated infrared lamps for analysis concerned are depth of modulation and degree of modulation. The degree of modulation is defined by:

$$m = \frac{\text{maximal radiation power - minimal radiation power}}{\text{minimal radiation power}} \times 100\%$$

The depth of modulation results from the difference between maximal and minimal radiation power.

<u>electrical and physical data of the lamp:</u>

	antiphase	in-phase
voltage	ca. 7 V	ca. 20 V
average power with 5 Hz	ca. 6,5 W	ca. 7,5 W
optional modulation frequency	2-6 Hz	2-6 Hz
max. strips/ribbon temperature	700-800 °C	700-800 °C
modulation depth (temperature lift)	200-300 °C	200-300 °C
total emissivity of the radiation strips	ca. 0,9	ca. 0,9
diameter of the radiation window	ca. 20 mm	ca. 20 mm

3. Conclusions

A directly modulated infrared-radiator has been developed for NDIR-spectroscopy. Since the application of this kind of lamp in analysers does not require any mechanical modulation facilities, production costs can be saved. The application of C- fibre-strips with fast thermical reaction achieves sufficient modulation depths at relevant wavelengths (e.g. at CO/4,6 µm) so that measuring signals are produced when applying this kind of lamp in NDIR gas analysers which are of the same scale as with the use of conventional lamp modulation units.

4. References

[1] Staab, *Industrielle Gasanalyse,* Oldenbourg Verlag München Wien 1994

[2] Heinz Gatzmanga und Dieter Baschant, *Direktmodulierbare Infrarotstrahlungsquellen für optische Konzentrationsmeßgeräte,* Fachaufsatz in tm-Technisches Messen 63 (1996), Oldenbourg Verlag

I 03

ACOUSTIC RESONANCES IN PULSED VORTEX-WATER-WALL HIGH-PRESSURE-ARGON-ARC LAMPS

D. Kouroussis, R. Bonert, F. P. Dawson
Department of Electrical and Computer Engineering,
University of Toronto, Toronto, Canada M5S 3G4
Tel. #: 416-978-8540, Fax #: 416-971-2325
e-mail: dawson@ecf.utoronto.ca

1 INTRODUCTION

Vortex-water-wall-high-pressure-argon-arc lamps are used to generate intense light and are capable of operating at radiative efficiencies greater than 50% [1]. 200 kW lamps are being considered for use in a pulsed dc mode of operation in the next generation of rapid thermal processing applications [2]. One of the major issues to address is the avoidance of acoustic resonances during pulsed conditions. Acoustic resonances distort the positive column and as a result disturb the positional and temporal optical flux density at the surface of materials being processed [3]. This paper describes the power supply and the conditions under which the lamp is operated. Experimental results and conditions conducive to the generation of acoustic resonances are described.

2 OPERATING REQUIREMENTS

Fig. 1 shows a diagram of the rapid thermal processing system. Measurements of the reflected radiation are made when the lamp is pulsed to the high current state and during the interpulse period when the lamp is operated at a low current level as shown in Fig. 2. These two measurements are then used to concurrently calculate, with the aid of Stefan Boltzman's radiation law, the hemispherical emissivity and surface temperature of the processed surface. The interpulse period is designed to be much shorter than the thermal response time of the processed material.

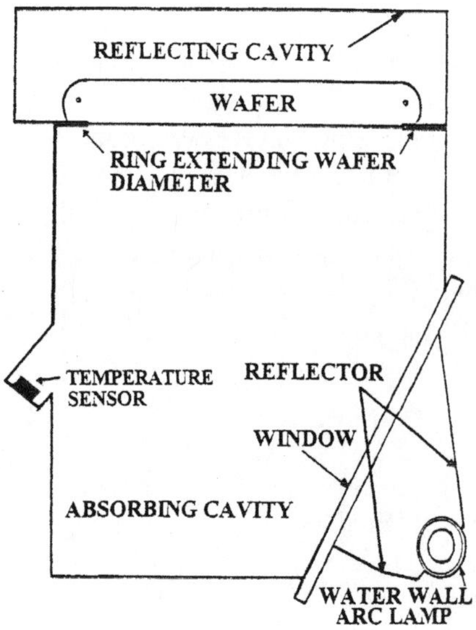

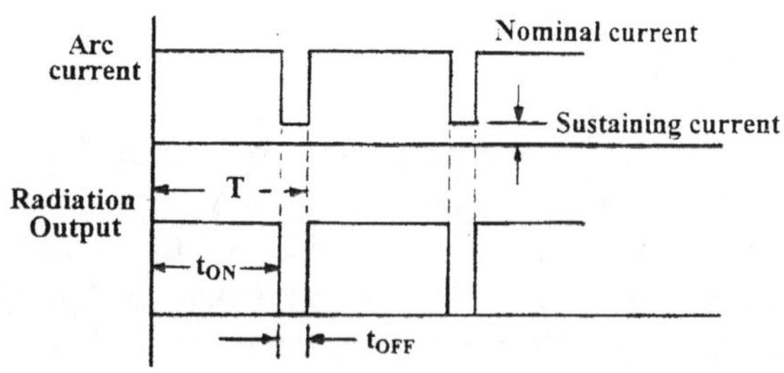

- T << thermal response time of the wafer
- $t_{min} < t_{off} < t_{max}$ where
  - t_{min} is dictated by the time taken to calculate temperature and emmisivity
  - t_{max} << thermal response time of the wafer

Fig. 1 Rapid thermal processing system   Fig. 2 Pulsed power timing diagram

3 POWER SUPPLY

Fig. 3 shows a diagram of the pulsed source. A power supply consisting of an ignition supply and sustaining supply connected in series is placed in parallel to the pulsed source. The ignition supply is shorted out once an arc is sustained. The pulser is able to provide square wave current pulses. A capacitor at the output is designed to limit the voltage overshoot which occurs when the lamp is pulsed from the sustaining current level to the rated current level.

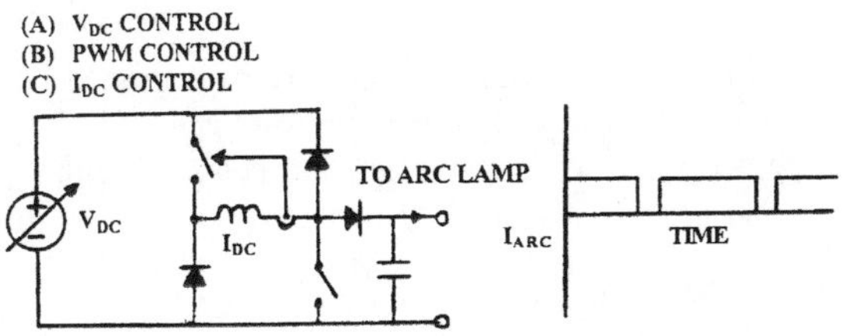

Fig. 3 Pulsed power supply

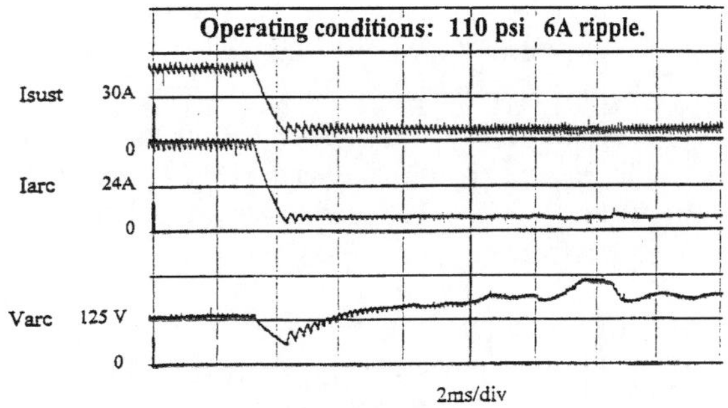

Fig. 4 Experimental waveforms showing the onset of an acoustic resonance

4 EXPERIMENTAL RESULTS

Fig. 4 shows the voltage and current waveforms for the pulsed arc lamp. The arc voltage, after being pulsed to a higher current level, exhibits oscillatory behaviour. This oscillation can be attributed to an acoustic resonance mode. The magnitude and duration of the oscillation increase with pressure and with the magnitude of the ripple current. The oscillation magnitude increases significantly above a pressure of 7 atm. and if the ripple current exceeds 6 amperes. The acoustic mode was visible over one half of the interelectrode gap (150 mm interelectrode gap) and took over 100 ms to decay. Three nodal points were observed.

5 CONCLUSIONS

A pulsed dc source has been used to determine the conditions for the onset of acoustic resonance modes in a vortex-water-wall-high-pressure-argon-arc lamp. The acoustic mode appears to be well damped if the pressure is below 7 atm. and the ripple current in the dc source is below 6 amperes.

6 REFERENCES

[1] D. Camm, R.A. Nodwell, 1974, US Patent Application No. 478872

[2] D. Camm, B. Lojek, "High power arc lamp RTP system for high temperature annealing applications", 2nd International Rapid Thermal Processing Conference, 1994, Monterrey California, pp 259-262

[3] H.L. Witting, "Acoustic resonances in cylindrical high pressure arc discharges", Journal of Applied Physics, 49(5), 1978, p 2680

I 04

KINETIC STUDY OF ($ArNe^+$) and ($ArHe^+$) HETERONUCLEAR IONS USING FLASH X-RAY EXCITATION

Laurent Hure[1], Eric Robert[1], Christophe Cachoncinlle[1] and Jean Michel Pouvesle[1]
Jochen Wieser[2] and Andreas Ulrich[2]

[1] GREMI, CNRS/Université d'Orléans, B.P. 6759, 45067 Orléans Cedex 02, FRANCE
[2] Physikdepartment E12, Technische Universität München, James Franck Str. 1, D-85747 Garching, GERMANY

ABSTRACT Flash X-ray excitation of binary mixtures of Ar with He and Ne is reported. Fluorescence at 143, 146 and 216, 224 nm, respectively, has been observed in high pressure gases (1 to 10 bar). Two hypotheses for the population of the upper state of $ArNe^+$ emission are proposed and discussed.

1 INTRODUCTION

In the past few years, we performed flash X-ray excitation of high pressure rare gas (Rg) targets [1]. The Rg are known to emit intense and broadband fluorescence continua in the UV-VUV spectral region. In the course of these studies of the molecular ion continua [2] we performed excitation of different binary mixtures. X-ray excitation of mixtures containing Rg(x) and Rg(y) can lead to fluorescence of heteronuclear ions (Rg^+(x)Rg(y)). The precursor states of these fluorescences have been identified a long time ago [3]. The kinetic mechanisms of the light emitting process is being studied because of the potential interest of these heteronuclear species as intense light sources in the UV-VUV.

In this work, we present a kinetic study of argon-neon and argon-helium mixtures excited by an energetic flash X-ray source. The fluorescence of $ArHe^+$ using flash X-ray excitation has been obtained for the first time. A long-lived emission of the $ArNe^+$ ion has been observed for different pressures and neon admixtures. A simple kinetic model including only key reactions enables us to reproduce the experimentally observed time dependence of the light emission quite properly and to discuss the population paths leading to the $ArNe^+$ states.

2 EXPERIMENTAL SET UP

A detailed description of the X-ray excitation source, gas handling system and data acquisition has been published in Ref. [1]. The excitation source consists of a powerful flash X-ray device. It emits X-ray pulses of 25 ns (FWHM) duration at a repetition rate from single shot up to 50 Hz. Each pulse corresponds roughly to 10^{10} X-ray photons with energies from 5 to 50 keV (mainly consisting of W-Lα characteristic lines).

The gas handling system consists of a high pressure cell and a mixture unit. Two gases can be mixed, one at flow rates from 10 to 1000, the second from 0.1 to 5 cm^3/min. Fluorescence is detected using a VUV spectrometer and a photomultiplier tube. The phototube signals are recorded using a fast transient analyzer.

3 RESULTS

Besides the emission of the second and molecular ion continua of Ar intense emission of $ArNe^+$ at 216 and 224 nm and of $ArHe^+$ at 143 and 146 nm are observed. Time integrated spectra of a mixture of 1.3 bar Ar containing 1% of Ne and a mixture of 3 bar Ar containing 1.8% of He are shown in Figure 1. Flash X-ray excitation enables this fluorescence to be studied at high pressure up to 10 bar, which is not easily obtained using other more conventional excitation methods.

Fig. 1: Time integrated spectra of mixtures of Ar/Ne (100/1) at 1.3 bar and Ar/He (100/1.8) at 3 bar.

Intense production of the heteronuclear ions fluorescence requires rather low admixtures of light Rg (He or Ne). This observation is a first indication of energy transfer reactions from highly excited argon states (Ar^{2+}, Ar^{+*}) being important (first ionization potentials of He and Ne are 24.59 and 21.56 eV, respectively), in particular for the case of helium which is almost transparent to X-ray photons emitted by the flash X-ray device. A long-lived $ArNe^+$ fluorescence (from a few tens to 200 ns duration depending on the gas mixture) has been observed, in contrast to the $ArHe^+$ emission which always follows the excitation pulse (25 ns FWHM).

Table 1 Key reactions for ($ArNe^+$) emission

	reaction	reaction rate
k_1	$Ne^+ + 2\,Ar \rightarrow Ne^+Ar + Ar$	10^{-33} $cm^6.s^{-1}$
k_2	$Ne^+ + Ar \rightarrow Ar^+ + Ne$	$2\ 10^{-13}$ $cm^3.s^{-1}$
k_3	$(Ar_n^{m+})^* + Ne \rightarrow Ne^+ + Ar_n^{(m-1)+}$	$5\ 10^{-12}$ $cm^3.s^{-1}$
k_4	$Ne^+Ar \rightarrow Ar^+Ne + h\nu$	10^8 s^{-1}
k_5	$Ne^+Ar + Ar \rightarrow$ products	$5\ 10^{-13}$ $cm^3.s^{-1}$
k_6	$(Ar_n^{m+})^* \rightarrow$ products + $h\nu$	$3\ 10^7$ s^{-1}
k_7	$Ne^+ + Ne + Ar \rightarrow Ne_2^+ + Ar$	$4\ 10^{-30}$ $cm^6.s^{-1}$

Two interpretations for the long lifetime of $ArNe^+$ are proposed and discussed below.

A simple kinetic model has been tested to fit the experimental data. The reactions and rates used for modeling are listed in table 1. Comparison between the simulated and experimental temporal behaviour of the $ArNe^+$ emission is shown in Fig. 2. The rather good agreement for different pressures and Ne admixtures indicates that the key reactions have been taken into account. The reaction rates are close to those proposed by Tischler et al. [4] for He. In our case, k_3 and k_5 have been lowered. In this scheme, a rather long lifetime of the $(Ar_n^{m+})^*$ species in combination with the low value of k_5 causes the long-lived fluorescence of $ArNe^+$. The almost four orders of magnitude difference in the rate constants k_1 and k_7 (table 1) might be questionable. Nevertheless a two orders of magnitude difference between measured and calculated rate constants was pointed out for rate k_7 (table 1) in the case of helium admixtures [4] instead of neon.

An alternative hypothesis can also explain the experimental observations. It is based on the assumption that the state correlated with the $Ne^+(^2P_{3/2})$ + Ar asymptote is metastable, with a natural lifetime of about 100 ns. This second hypothesis is also consistent with the short lived emission of the $ArHe^+$ ion for which no spin splitting exists for the upper state of the transition. Another point can be brought forward in favor of this second interpretation. A single long lived state would be in better agreement with the experimentally observed exponential decay at low pressures (Fig. 2, 1.3 bar Ar, 13 mbar Ne).

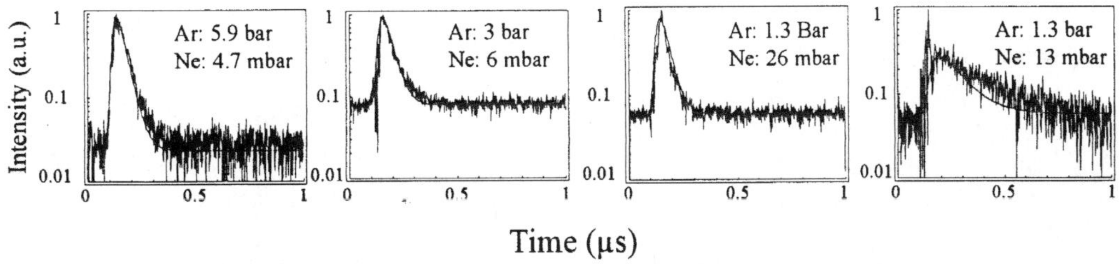

Fig. 2: Experimental and modeled time evolution of $ArNe^+$ fluorescence at various Ar and Ne pressures.

ACKNOWLEDGMENTS

Support of this work by INTAS 96-0351, MENRT and TMR-FAPS is gratefully acknowledged.

REFERENCES

[1] E. Robert, A. Khacef, C. Cachoncinlle and J.M. Pouvesle, Opt. Comm. **117**, 179 (1995).
[2] E. Robert, C. Cachoncinlle, A. Khacef, R. Viladrosa and J.M. Pouvesle, ICPIG **III**, 32 (1997).
[3] Y. Tanaka, K. Yoshino and D.E. Freeman, The Journal of Chemical Physics **62**, 4484 (1975).
[4] H. Tischler, P. Delaporte, B. Fontaine and M.L. Sentis, IEEE JQE **1**, 877 (1995).

I 05

Existing State and Future Prospect of Lighting Education of Light Sources and its Related Technologies in University

Tetsuo YAMAYA*[1], Masato OKI*[2], and Tatsuo ITOH*[3]

*1 Department of Electrical Engineering, College of Industrial Technology, NIHON University,
1-2-1, Izumi-cho, Narashino-shi, Chiba-ken, 275-8575 JAPAN
Telephone : +81-474-74-2379, Facsimile : +81-474-74-2399
E-mail Address : yamaya@ee.cit.nihon-u.ac.jp

*2 Department of Architecture, Ashikaga Institute of Technology,
268-1, Ohmae-cho, Ashikaga-shi, Tochigi-ken, 326-8558 JAPAN
Telephone : +81-284-62-0605, Facsimile : +81-284-62-4235
E-mail Address : oki@ashitech.ac.jp

*3 TATSUO ITOO LIGHTING DESIGN LAB.
C-408, 1-3-18, Shibuya, Shibuya-ku, Tokyo, 150-0002 JAPAN
Telephone : +81-3-3407-3524, Facsimile : +81-3-3407-3525
E-mail Address : itooldl@cb.mbn.or.jp

1. INTRODUCTION

Lighting is one of important factor for composing comfortable visual environment. Science and its related technology of light sources, that is incandescent lamp, fluorescent lamp, HID lamp, EL lamp and so on have been remarkably progressed during the 19th and 20th centuries. Now, development of new light sources and its application are expected towards to the 21st Century.
However, education method of light sources in university is considered to keeping a traditional stance. For future educational trends, it is indispensable to develop a new educational method for university students. In this paper, existing state and future prospect of lighting education relating to light sources and its related technologies in university are descried

2. EXISTING STATE OF LIGHTING EDUCATION

Artificial lighting is related not only electrical engineering but also other engineering related technologies of various fields. Contents of lecture in universities were surveyed each field of electrical engineering, architecture and environmental design to know how educate light source and its related fields. Hours of lecture relating lighting was investigated and analyzed by syllabus of each lecture which are delivered in several university. **Figure 1** shows hours of lecture relating lighting in university. As the results of this investigation following are cleared.

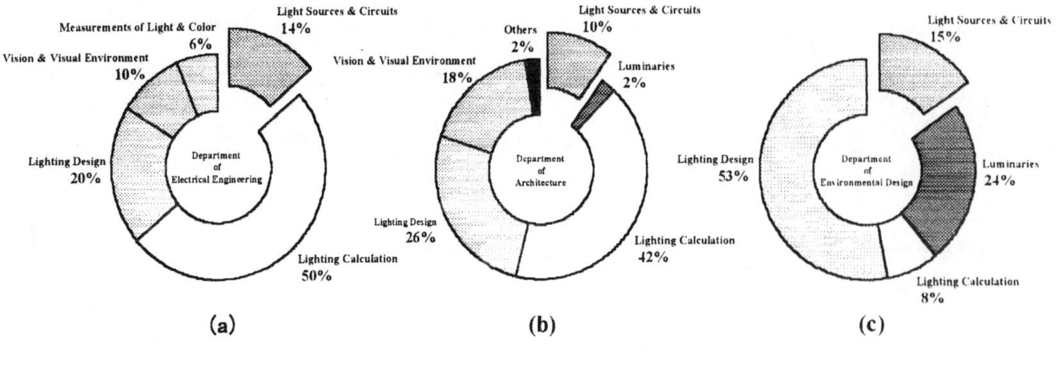

Figure 1 : Existing state of lighting education in university.

As to approaching "a grey plasma source," it is strait forward in accordance with the Plank and radiation transfer laws : the highest spectral output hv_{max} (in eV) and the Black Body temperature T (eV) are correlated as hv_{max} (eV) = 2.82 T (eV) [2]. To get the maximum output in the middle of the UVC area (hv_{max} =5.6 eV or 220 nm), the Black body temperature must be 2 eV (T°=23.000° K). A respective black body discharge emits ca. 30% of its energy in photons with energies over 7 eV (below 180 nm), which cannot be used here anyway and only damage a quartz port. "A grey plasma" of the same temperature does not emit these photons and could have the UVC efficiency up to 60% of all radiated energy. Such conditions have been calculated and reached in a shock wave around high-current pulsed discharges [1, 2]. So it could be done with a plasma-focus as well. Respective calculations with the mean free path of radiation, the Rosseland absorption coefficient, etc. [2] and some previous works listed in [1] became our base to find optimal electrical, material, geometrical and gas parameters for two pilot UVC systems with plasma-focus elements:

A). PF-CS (Fig.2) with a pulse energy from 10 to 100 J, a pulse peak power in the UVC - up to 2 MW/cm^2, with the UVC content of up to 50%, the repetition rate of 1 Hz, a quartz port size of 1 to 3 cm. The size of its power-control unit HxLxW - (270x420x440 mm); a purged or a sealed lamp.

B). PF-IS with a pulse energy from 100 to 1000 J, a radiator with quarts port of 200 mm (Fig 2), a pulse repetition rate of 1 to 10 Hz, depending on a pulse energy. The total output - up to 5 kw, Power supply/control unit has a size of HxLxW 1100x500x600.

Fig. 2: Photo of PF-CS control unit with a small lamp and with a large lamp from PF-IS system.

4. PROSPECTIVE USE AND CONCLUSIONS.

There are sufficient data to build commercial pulsed UVC systems with plasma-focus elements which could be more efficient than flash-lamps for "spot" applications in the UVC curing, in disinfecting food packages or medical waste, in the UV -Oxidation of concentrated hazardous products, for an optical bonding in opto-electronic industry, for a lab analysis as a pulsed UVC source with a calibrated temperature from 20.000° K to 40.000° K, for CVD (Cold Vapor Deposition) with UVC light for the semiconductor industry, etc. An interested and potent industrial partner is required to finance the engineering of its commercial version and its market introduction.

5. REFERENCES.

[1] Alex Wekhof, *Plasma-focus as a powerful pulsed VUV-UVC source,* WEK-TEC, Heilbronn, Germany, 1998.

[2] Ya. B. Zel'dovich, Yu. P. Raizer, *Physics of Shock Waves and High-Temperature Hydrodinamic Phenomena*, Academic Press, New York and London, 1966.

(1) Hours of lectures for light source and related lighting circuit are only 10 to 15 % of the total lighting lecture.

(2) Lecture for lighting materials and display devises are almost not include in the lighting lecture.

(3) Lecture of structure of lamp, electrical circuit of discharge lamp and light characteristics of light sources and light materials are limited at very fundamental level in electrical engineering department and architectural department.

(4) Experimental and excise system in studio is used for supporting the lecture relating light sources at environment design department.

(5) Tendency of contents of lecture are mostly so called soft field such as theory of lighting, artificial lighting planning and daylighting planning.

3. FUTURE PROSPECT OF LIGHTING EDUCATION [1],[2]

As nowadays lighting lecture is carried out by mostly traditional methods such as text book and blackboard. So several educational tools for lighting education should be developed urgently. Development of an individual self-education system using multimedia such as Internet and personal computer etc. is indispensable. Therefore, a new system of lighting education applying CMI and CAI is proposed and a practical educational system is developed by the authors. Outline of the system is shown in **Figure 2**. Using this new system, students can be studied repeatedly by individual self-education methods.

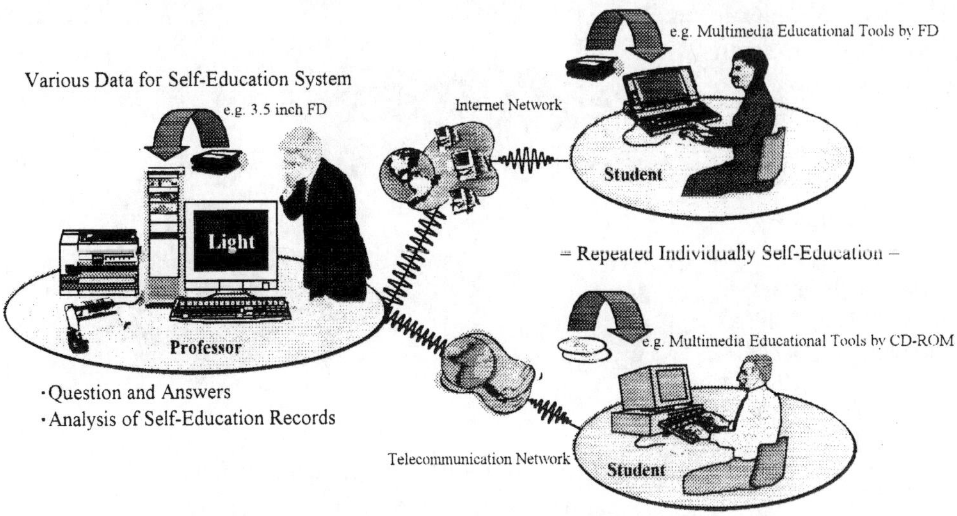

Figure 2 : Outline of the new system of lighting education.

4. CONCLUSION

Results of the survey, it is clear that hours of lighting lecture in universities for so called hard fields of lighting such as light source, lighting electric circuit etc. is too small. Therefore, as urgent developing theme of lighting education, development of supplemental educational method by visual tools is to be urgently expected. On this view point, the authors is proposed a new lighting educational system and outline of the system is described in this paper.

5. ACKNOWLEDGMENTS

A part of this study has been carried out by the grants for encouragement of research from The Research & Education Grant-in-Aid of The Illuminating Engineering Institute of Japan.

6. REFERENCES

[1] Moritsuna HIRAHARA and Tetsuo YAMAYA : "Development of Multimedia CAI System for Illuminating Education", *Proceedings of The Light Wave Forum '97 of The Illuminating Engineering Institute of Japan*, pp. 41-46, (1997)

[2] Tetsuo YAMAYA, Minoru ISOMURA, Yoshihiko OHTANI, Masato OKI and Tatsuo ITOH : "On a Proposition of The Multimedia CAI System for Lighting Education", *Proceedings of The 1998 Annual Conference of The Illuminating Engineering Institute of Japan*, (1998)

I 06

A PULSED UVC SOURCE WITH A PLASMA-FOCUS ELEMENT.

Alex Wekhof
WEK-TEC, Frankfurter Str. 85 C, 74072 Heilbronn, Germany

1. INTRODUCTION.

A plasma-focus has long been used as an intense pulsed X-rays source and has advantages of a high brightness and of a pointed geometry. These advantages so far have not been commercially realized in a plasma-focus device for the UVC (180-280 nm) spectral range because of numerous obstacles. Major ones are a damage to a quartz window with discharge by-products and with electrode materials, an electrode erosion and a poor matching of a discharge with its PFN (Pulse Forming Network). This work was undertaken to solve above problems for two pilot UVC systems.

2. A PLASMA-FOCUS AS A PULSED SOURCE OF THE UVC CONTINUUM.

A typical plasma-focus has a coaxial geometry and a very low impedance [1]. A coaxial discharge is originated by an electrical breakdown at a cold cathode and is propelled by electromagnetic forces to the anode, where for a very short time it is squeezed into a dense plasma hemisphere (Fig.1). A high density electron beam funnels from this plasma to the center of the anode. This configuration emits an intense radiation in a broad spectral region (up to 80% of a deposited energy) by the end of the first half-period, while up to 20% of a deposited energy goes to a plasma jet departing from the anode area at the end of the first half-period.

If there is a gas of a sufficient density and a composition around a plasma-focus, then at the end of the first half-period a plasma-focus forms a shock wave. Its front can recapture some of the radiation and kinetic energy from the plasma-focus and get heated up, becoming a dense plasma layer. It radiates differently than a plasma-focus itself. This absorption-re-emission removes a direct dependance of all emitted radiation on the discharge impedance and on PFN, which in turn allows to remove many technical obstacles for building "a grey plasma" pulsed source of a pre-determined electron temperature and a spectral output [1].

Indeed, since parameters of a shock wave depend more on a physical geometry of a plasma focus and on a total deposited energy rather than directly on a discharge current, than longer and lower current pulses could be used instead of short and high current pulses usually generated in a plasma-focus. It makes possible to reduce a number of current periods and respective high reverse voltages. The latter is highly desirable since otherwise it degrades capacitors, blocks a normal use of an SCR (Silicon Control Rectifier) switch and of a switching mode power supply for charging capacitors. It also reduces an electrical load on electrodes to an extend that available materials could be used without getting much of its sputtering and evaporation. The latter otherwise builds on a quartz port a dense film which blocks the UVC light. The price to get rid of it is to deal with a shock wave instead - which is possible to handle by conventional means [1].

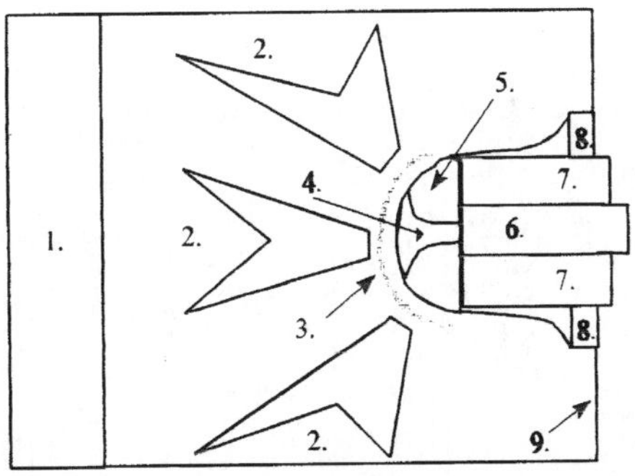

Fig. 1: Schematic of a plasma-focus lamp: 1.- quartz port; 2- UVC and other light; 3- shock wave; 4-electron beam; 5- plasma-focus; 6- anode; 7- insulator, 8- cathode; 9- enclosure

I 07

MODELING OF CATHODE HEATING IN VORTEX-WATER-WALL-HIGH-PRESSURE-ARGON-ARC LAMPS

W. Hu[1], F. P. Dawson[1], M.S. Benilov[2]
[1]Department of Electrical and Computer Engineering,
University of Toronto, Toronto, Canada M5S 3G4
Tel. #: 416-978-8540, Fax #: 416-971-2325
e-mail: dawson@ecf.utoronto.ca
[2]Departamento de Fisica, Universidade da Madeira, Funchal, Portugal

1 INTRODUCTION

Vortex-water-wall-high-pressure-argon-arc lamps are used to generate intense light and are capable of operating at radiative efficiencies greater than 50% [1]. 200 kW lamps similar to the one shown in Fig. 1 are being considered for use in a pulsed dc mode of operation in the next generation of rapid thermal processing applications [2]. One of the major issues to address is a quantitative understanding of factors related to the improvement in electrode lifetimes. This can be achieved by developing better models which can predict the cathode temperature profile. This paper builds on the cathode modeling work initiated by Benilov and Marotta [3]. The method of calculating the heat flux density, current density and cathode temperature profile are described.

2 MODELING OF CATHODE TEMPERATURE PROFILE

2.1 Heat flux density calculation

High pressure arcs operating above atmospheric pressure and at high current levels can be assumed to have a cathode layer which is collisionless. The use of refractory cathode materials such as tungsten ensure that the vapour pressure of the evaporated cathode material is insignificant in comparison to the background pressure of argon. The space charge sheath potential distribution and the surface electric field are calculated using Poisson's equation and the Bohm velocity criterion at the sheath edge. The sheath/presheath interfacial ion density is modeled using a correlation formula. The emission current is assumed to be thermionic and includes a surface field dependent Shottky correction factor. The electron density, electron temperature and Ar^+, Ar^{++}, Ar^{+++} ion densities in the bulk plasma, for a fixed value of cathode voltage drop, are calculated using Saha's equation, the charge neutrality equation and Dalton's partial pressure law/equation of state. The species contributing to current flow include thermionically emitted electrons, ions and backdiffusing electrons. The heat flux density as a function of tungsten cathode temperature for different cathode voltage drops is shown in Fig. 2. An ion temperature of 10K ^{0}K and a total pressure of 1 atm. are assumed.

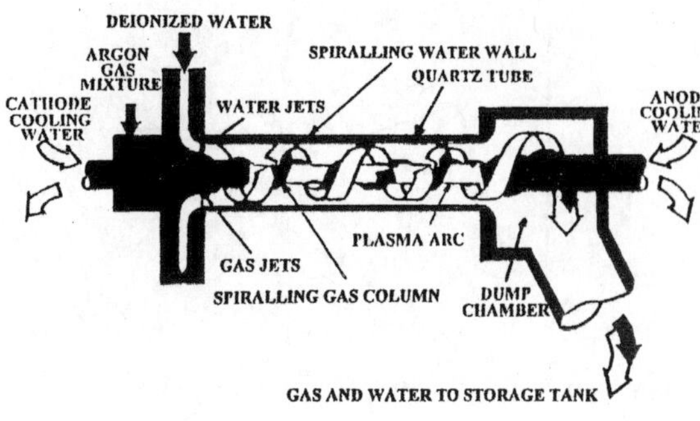

Fig. 1 Vortex arc lamp

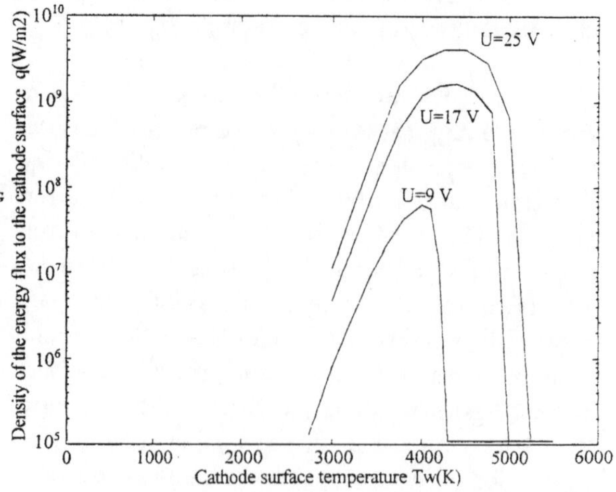

Fig. 2 Heat flux density as a function of cathode temperature for different cathode fall voltages

2.2 Cathode temperature, current density and heat flux density distributions

PHOENICS a computational fluid flow program was used to calculate the axisymmetric temperature, current density and heat flux density profiles at the cathode's surface for a preassumed cathode voltage drop. The thermal conductivity, heat capacity and electrical conductivity of the cathode were assumed to be nonlinear functions of temperature. The electrode was first preheated with a uniform heat flux density extending over a radius of 0.5 mm. Switchover to the calculated values was initiated after the surface temperature stabilized. A simple underrelaxation scheme was used to to update the the heat flux density profile after each iteration. No solutions were obtained for less than 20V. Figures 3 and 4 show the steady state heat flux density and current density respectively for an assumed cathode voltage drop of 25 V. It is clear that a valid solution requires the existence of a minimum and a maximum in the heat flux density distribution. The location of the maximum coincides with the radius of the arc spot and is approximately 1 mm for this example.

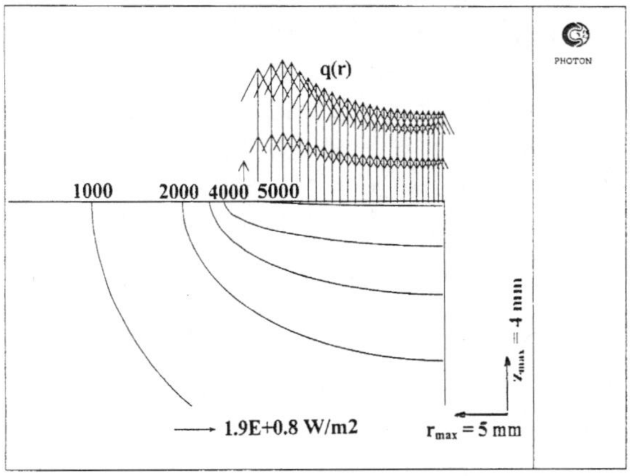

Fig. 3 Calculated heat flux distribution

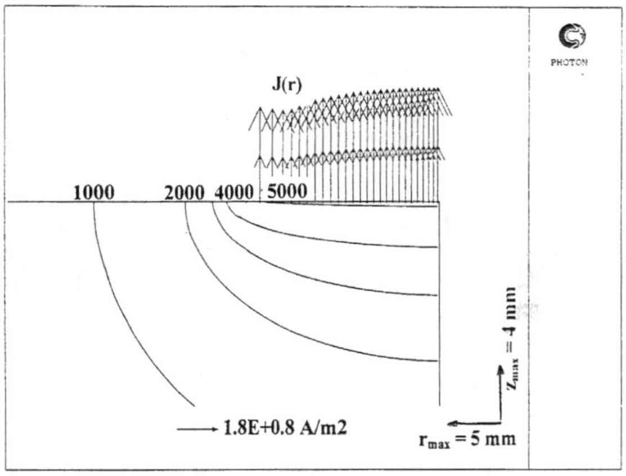

Fig. 4 Calculated total current density distribution

3 CONCLUSIONS

A computational fluid flow program has been used to calculate the current density , heat flux density and temperature distribution at the cathode surface. Benilov's and Marotta's model of the cathode region for atmospheric arcs was used to estimate the heat flux density at the cathode surface. Arc spots occur if the cathode potential drop is greater than approximately 20 volts.

4 REFERENCES

[1] D. Camm, R.A. Nodwell, 1974, US Patent Application No. 478872

[2] D. Camm, B. Lojek, "High power arc lamp RTP system for high temperature annealing applications", 2nd International Rapid Thermal Processing Conference, 1994, Monterrey California, pp 259-262

[3] M.S. Benilov, A. Marotta, "A model of the cathode region of atmospheric pressure arcs", J. Phys. D: Appl. Phys. 28 (1995), pp 1869-1882

I 08

MICROWAVE SLOT ANTENNA TYPE ULTRAVIOLET LIGHT SOURCE FOR FLUID TREATMENT

D. Korzec[1], D. Boonyawan[2], and J. Engemann[1]

[1] *fmt*, University of Wuppertal, Obere Lichtenplatzer Str. 336, 42287 Wuppertal, Germany

[2]FNRF, Chiang Mai University, Chiang Mai 50200, Thailand

1. INTRODUCTION

(i) For many technological applications, especially for treatment of gases, liquids, or continuous substrates such as wires, fibers or threads, UV-sources are favorable radiating coaxially into a cylindrical volume. *(ii)* For water purification, sterilization or biological wastes neutralization mercury lamps are used, which is a health and environmental hazard. Because many standards and measurement techniques are based on the Hg wavelength 253.7 nm, a discharge with very similar emission wavelength would be very advantageous. *(iii)* Low cost high efficiency discharges with high flexibility are needed to successfully introduce innovative UV-technologies. Especially promising are microwave discharges [1]. These three requirements are fulfilled by a a novel microwave UV-lamp concept presented in this work.

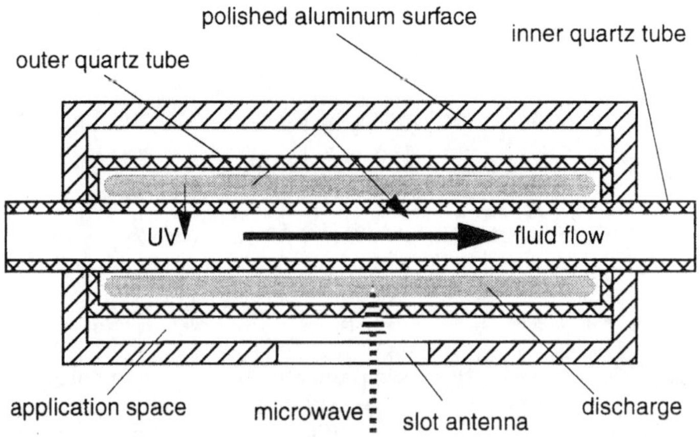

Figure 1

Operation principle of the microwave slot antenna plasma source for UV treatment of fluids.
High efficiency is achieved due to: microwave distribution, surface wave excitation, UV-reflection, UV-nonabsorbing coolant and selective UV-spectrum.

2. OPERATION PRINCIPLE

A novel concept of UV-lamp based on the SLAN type plasma source [2] has been developed. A main part of the lamp is a double quartz tube (see Fig. 1). The treated fluid is flowing along the source axis through the inner tube. Discharge is ignited in the toroidal space between the inner and the outer tube. The microwaves propagate in the application space between the outer quartz tube and the metal casing. To avoid the UV-absorption and to allow efficient passive cooling, the application space is hermetically closed and filled with helium. The microwaves are introduced into the application space through the slot antenna positioned parallel to the source axis. The outer wall of the application space is made of aluminum and polished for ultraviolet reflection.

3. UV GENERATION

The discharge is ignited in a gas mixture of sulfur hexafluoride, argon and helium. The inspiration for the gas composition was a commercial mixture (Messer-Griesheim) for excimer lasers containing 0.2% fluorine, 10% argon, 40% helium and 49.8% neon. For pressures below

1 mbar all tested gas compositions show the emission with many lines and bands. For pressures of more than 100 mbar, an enhancement of few selected emission lines was possible. For wave lengths 252 nm and 288 nm the emission intensities are on an order of magnitude stronger than for other wavelengths. The molecular bands in the emission background can be suppressed by optimization of the argon-to-helium ratio. An atomic silicon in ground state was identified as a source of such selective spectrum (see Fig. 2). The atomic fluorine from the sulfur hexafluoride is "etching" the quartz wall. The volatile product of such reaction - silicon tetrafluoride - is transported into the bulk of the discharge, set free and excited for the atomic emission. This explanation is in agreement with a dynamic behavior of the discharge. Immediately after cutting the gas supply a bright, pink discharge is observed. After about 15 minutes it transforms into a dark, ink blue looking "silicon" discharge. Concurrently the emission intensities of silicon lines increase.

Figure 2

Optical emission specturm for the gas mixture of 0.2% SF_6+15% Ar + 84,8% He at pressure of 100 mbar and 1 kW microwave power.
The lines depicted in the figure correspond to the atomic silicon present in the discharge due to the chemical reactions of fluorin with the quartz wall.

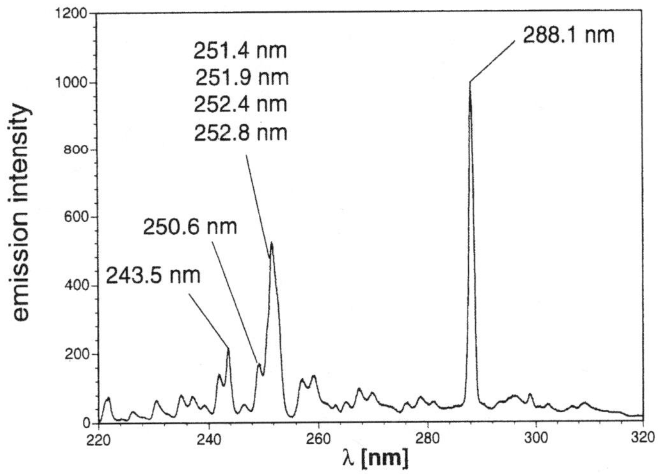

4. CONCLUSION

New microwave excited ultraviolet lamp is presented, which is especially suitable for treatment of fluids. The gas mixture for high intensity ultraviolet radiation of 252 nm from atomic silicon has been optimized. It allows the replacement of 253,7 nm mercury discharges.

5. ACKNOWLEDGMENTS

The authors like to thank Karl-Heinz Weggässer for constructing and machining the source. The research of D. Boonyawan in the University of Wuppertal was supported by International Atomic Energy Agency (IAEA) in Vienna.

6. REFERENCES

[1] Kumagai, H. and M. Obara (1990). Microwave discharge-pumped excimer lamp, *Review of Laser Engineering* **18**, pp. 456-473.

[2] D. Korzec, F. Werner, R. Winter, and J. Engemann (1996). Scaling of microwave slot antenna (SLAN): A concept for efficient plasma generation, *Plasma Sources Science and Technology* **5**, pp. 216–234.

K

Materials

K 01

New Materials and New Structure for Hot Cathodes

M.Hamada, A.Takeishi, M.Takahashi, M.Yodogawa, H.Harada
ELECTRONICS DEVICE BUSINESS GROUP ,TDK CORP.
15-7 Higashi-Owada 2-chome, Ichikawa, Chiba 272-0026

1. Introduction

We have developed new structure of hot cathode which consists of a small ceramic container filled with ceramic granules to store heat and to generate arc discharge without preheating. Moreover, we have developed new materials for it which consist of compounds of BaO, Ta_2O_5 and ZrO_2 and which are covered with TaC thin layer to realize more than 10 thousand hours of life using the above structure. This hot cathode has many superior characteristics as following.
1)Hot cathode without preheating circuit. 2)High luminance in thin tube.
3)Longer life than those of usual cathodes. 4)Higher discharge efficiency to realize saving energy.
5)Simple fabricating process without decomposing emitter materials.

2. New hot cathode materials and new structure

The hot cathodes usually consist of alkaline metal oxides and are sputtered by ion collisions during lightening to reach the end of the life. We have been researching for new hot cathode materials to overcome the above disadvantage. In 1986 we confirmed arc discharge in the ceramic cathodes made of semiconducting $BaTiO_3$.[1] In 1989 we have succeeded in continuous arc discharge using small ceramic containers filled with ceramic granules as shown in Fig.1 as cathodes.[2]

The diameter of the granules is about 100μm and the contacts and the spaces among them realize a suitable electronic conduction and storing of heat to cause the transition from glow to arc discharge immediately after lightening by partial heating as shown Fig.2. As preheating circuit is not required and the cathode structure is simple and compact, it is possible to use the cathodes even in thin fluorescent lamps and in incandescent lamp bulbs of the same size as usual ones. In these cases high energy efficiency is realized to save energy.

The materials consist of a few compounds of BaO, Ta_2O_5 and ZrO_2. Most of these are semiconductive and have strong durability against sputtering by ion collisions. As they are covered with TaC thin layer which is formed during firing process and has also strong durability against sputtering, their life has been considerably improved. Moreover, as thermal decomposition is not required, the fabricating process of tubes is simple. An example of application and typical dimension of this new cathode are shown in Fig.3. Suitable dimensions of containers and the diameters of granules are determined by lamp currents.

Average life time of hot cathodes are shown in Table 1. The life of hot cathodes is defined by arc discharge time. In both 30 mA and 240 mA of lamp currents the life of more than 10 thousand hours were confirmed in the new cathodes. The cathode fall voltages of these cathodes were about 20 % higher than those of usual ones as shown in Table 2. In these cases the cathode fall voltage was defined as the value measured at 1.5 mm of the distance between cathodes.

3. Summary

We have developed original structure of hot cathode which consists of small container filled with granules to store heat which is made of new ceramic emitter materials. It is applicable to thin fluorescent tube to realize high energy efficiency and long life.

4. References

[1] M. Hamada *et al.* EUROPEAN PATENT APPLICATION Publication number EP0249196
[2] M. Hamada *et al.* EUROPEAN PATENT APPLICATION Publication number EP0643416

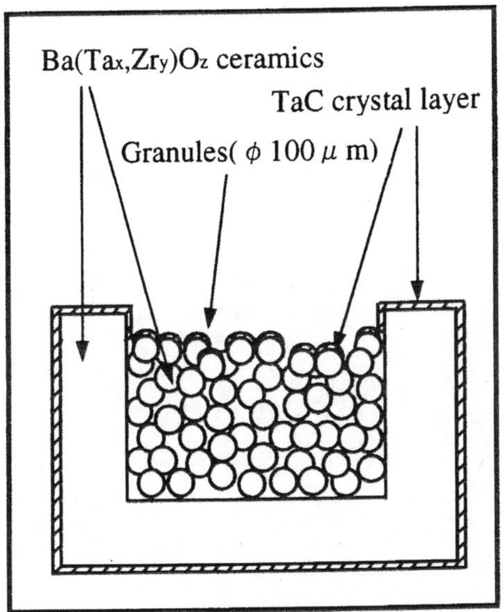

Figure 1 New hot cathode structure

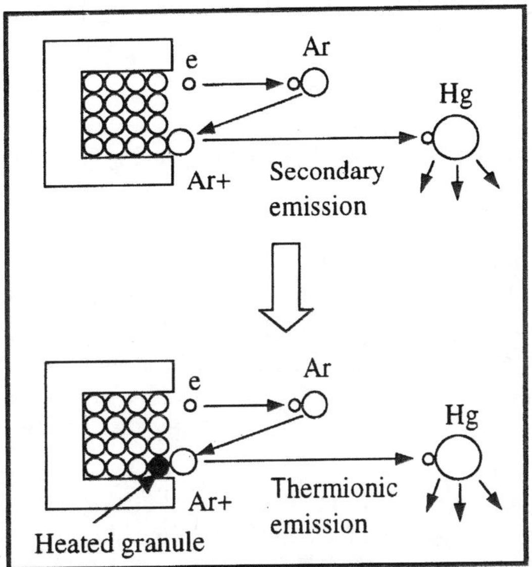

Figure 2 Arc discharge mechanism

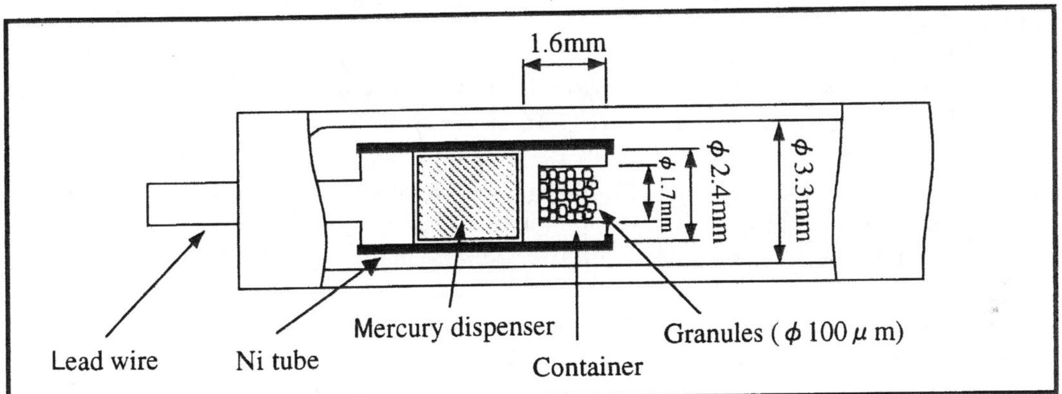

Figure 3 An example of application on lamp of new hot cathode

Table 1 Life time of hot cathode

	Type A	Type B
Lamp		
Inner gas	Ar, Hg	Ar, Hg
Ar pressure	9.3kPa	2.7kPa
Inner diameter	3.3mm	6.9mm
Effective length	80mm	80mm
Lamp current	30mA	240mA
Frequency	30kHz	30kHz
Cathode		
Outer diameter	2.4mm	3.0mm
Inner diameter	1.7mm	2.3mm
Hight	1.6mm	1.6mm
Granule size	100 μ m	100 μ m
Average life	12,000 hr	10,000 hr

Table 2 Cathode fall voltage

Measuring conditions	
Effective length	1.5mm
Inner gas	Ar, Hg
Ar pressure	2.7kPa
Lamp current	30mA
Cathode fall voltage	19V
Reference Conventional hot cathode	16V

K 02

PINCH PROTECTION OF QUARTZ GLASS BURNERS BY CHROMIZING

Maarten Steinmann
Philips Lighting B.V., P.O.Box 80020, 5600 JM Eindhoven, The Netherlands

1. INTRODUCTION

High thermally loaded quartz burners suffer from oxidation of the molybdenum feedthrough parts during lamp burning, provided no outer bulb is present. Oxidation of the molybdenum parts affects lamp life strongly beyond a critical quartz pinch temperature of 350 °C (IEC norm). Prolonged oxidation either leads to leaky quartz pinches or to electrical failure due to melt down of the molybdenum foil near the welding spot.

Chromium coating of molybdenum feedthrough parts is one of the best protective agents against corrosion and oxidation [1]. If applied, permissible pinch temperatures are far beyond the present critical temperature of 350 °C.

2. TECHNICAL DETAILS

2.1 Mechanism

Oxidation of molybdenum feedthrough parts undermines lamp life in two ways: (i) small amounts of MoO_3 already build up a high internal mechanical load causing pinch cracking upon exceeding a critical value and (ii) oxidation of the Mo-foil increases the electrical resistance, R, of the foil. Upon exceeding a critical value of R, the Mo-foil melts, ending up with electrical failure of the feedthrough. Both failure mechanisms occur in Halogen and HID burner concepts. In practice, Halogen burners mainly fail due to pinch cracking and leaky pinches, according to the first mechanism, whereas HID burners mainly fail due to electrical breakdown, according to the second mechanism and promoted by high run-up lamp current. The time scale of both mechanisms is ruled by the same rate law i.e. parabolic time dependence of MoO_3 formation. The latter can be measured isothermally. From the rate constant at different temperatures always the same lamp life-temperature dependency is derived, irrespective of burner type, see figure 1. Upon applying chromium the lamp life can be dramatically increased, as is shown by isothermal furnace tests at different temperatures. Furnace testing offers a powerful tool for a quick assessment of protection effectiveness. Nevertheless, additional lamp burning tests are necessary to verify the protection effectiveness under typical burning conditions i.e. non-isothermal and involving electrical heat dissipation.

2.2. Halogen burners

High wattage Halogen lamps (e.g. 1kW SSTV) are used for studio and stage lighting. Application presently tends to further miniaturisation of luminary size and to eliminate forced cooling.

Consequently pinch temperature of burners applied in such luminaries often exceeds 450 °C hence causing poor lamp life. To check life time performance, burners are tested under 'hot-pot' condition i.e. reduced filament voltage while establishing an average pinch temperature of 480 °C. Table 1 shows the effect of chromizing under 'hot-pot' burning condition. Upon applying chromium, lamp life is increased by a factor of 10-15 while burners without protection already fail after 100 hours.

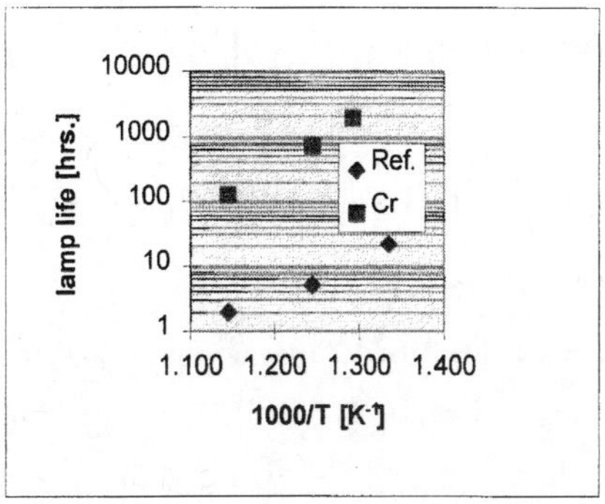

Fig 1: Arrhenius plot of lamp life

2.3 HID burners

Medium pressure Hg quartz discharge lamps (HOK 3 kW UVC lamps) are presently used for disinfection of water. Future application requires long lamp life. For realising aimed lamp life, pinch protection is necessary given the extreme burning conditions i.e. high thermal loading of the quartz pinch, and high humidity of the ambient atmosphere. The application of chromium contributes significantly to lamp life. Aimed lamp life target (50%/ 2500 hrs./480 °C) can be realised easily, see table 1.

TABLE 1. Effect of pinch protection by chromizing

50% failure point [hrs.]	furnace test (480 °C)	'hot pot' lamp burning
SSTV 1 kW ref.	20	100
SSTV 1 kW + Cr	5100	1400
HOK 25/120 ref.	7.5	350
HOK 25/120 + Cr	3020	4000

3. CONCLUSIONS

In case of burning in free air, high thermally loaded quartz burners suffer from oxidation of the molybdenum feedthrough parts which seriously undermines lamp life. Pinch protection by chromizing the molybdenum feedthrough parts strongly retards the oxidation process of these parts.

In this way it has been shown possible to realize the required long lamp lifes even under conditions of extreme pinch temperatures.

4. REFERENCES

[1] R.Meijer; *Metal Foil Lead-in Conductor for Electric Lamp*; patent US 3,105,867; 1963

K 03

ANNEALING BEHAVIOR OF EMISSION BANDS INDUCED BY PILE IRRADIATION IN FUSED SILICA

Shoichi Nasu[1],Shinya Ishida[1],Akio Koshino[1],Ryouichi Yamamoto[1],
Hidehito Nannto[1],Takaaki Tanifuji[2],Kenji Noda[2],and Nobu Kuzuu[3]
[1]Kanazawa Institute of Technology, Ohgigaoka 7-1, Nonoichi, Ishikawa, 921 Japan
[2]Japan Atomic Energy Research Institute, Tokai, Ibaraki, 319-11 Japan
[3]Fukui University, Bunnkyou 3-9-1,910 Japan

1 INTRODUCTION

Vitreous silica has numerous applications such as lamp envelops, optical parts for short wave lengths, and windows for laser fusion reactors, since it is resistant to intense electromagnetic waves or particle beams. In the light of this background, our research group has been studying the effects of irradiation on optical properties in various vitreous silica using various irradiation facilities. In this Symposium, irradiation effects of the diamagnetic point defects of the 4.2 eV and the 3.2 eV emission bands of fused silica having the intrinsic B2β band and the 4.4 eV and the 1.9 eV emission bands induced by pile irradiation are reported, since they have still been less clarified, as compared to the paramagnetic E′ center[1].

2. EXPERIMENTAL

The samples investigated were commercially available fused quartz plates of GE214, GE144, GE124 and GE012 with mirror polished surfaces These samples were irradiated with JRR-1 in JAERI. All optical measurements were made at room temperature. The optical absorption(OA) spectra in the visible and ultraviolet region (6.2~1.5 eV) were obtained with a Hitachi U-3000 double beam spectrophotometer. The continuous wave photoluminescence(PL) and PL excitation(PLE) spectra were obtained with a Hitachi F-3010 fluorescence spectrophotometer. For the measurement of 1.9 eV PL, a filter cutting down below 3.55 eV (above 350 nm) was used to reduce the effect of the twice wavelength(~580 nm, ~2.2 eV) due to the intence 4.4eV PL.

3.EXPERIMENTAL RESULTS

We noticed the following results from the experimental results.

1. The E′ center (5.8eV) and the B2α band were observed after pile irradiation.
2. The OA spectrum at 5.8 eV was decomposed into six Gaussian bands[2].
3. The intensity of PL near the intrinsic 4.2 eV of the B2β band increased, and the peak position of the band was shifted to 4.37 eV [2,3] after irradiation. The PL spectrum was also decomposed into two Gaussian bands (the intrinsic and irradiation induced bands) [2].
4. The intensity of 3.2 eV PL of the B2β band decreased drastically and any shfts of the peak position were not observed after pile irradiation and after annealing as shown in Fig.1.
5. The 1.9 eV PL band was observed after pile irradiation. The intensity of this band depends on the OH content in the samples as shown in Fig.2.
6. The intensity of 1.9eV PL band incresed to 450℃ and then decreased and was not observed at 700℃.

4.ACKNOWLEDGEMENTS

We are grateful to president H.Murakami (Souei Tsuusho Inc) for financial support.

5.REFERENCES

[1] D.L.Griscom, J.Ceram.Soc.Jpn. 99 (1991) 923.

[2]S.Nasu,K.Nishiwaki,S.Ishida,A.Koshino,R.Yamamoto,H.Nanto,T.Tanifuji,K.Noda and N.Kuzuu, Radiation Effects and Defects in Solids, in press

[3] R.Boscaino, M.Cannas, F.M.Gelardi, and M.Leone, Phys.Rev. B54 6194 (1996)

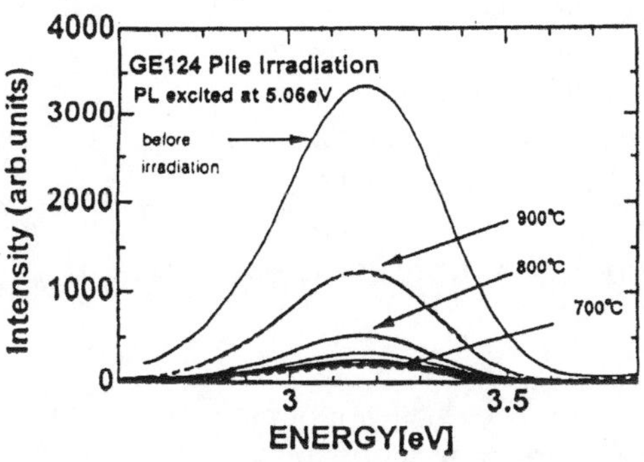

Fig. 1: Annealing behavior of the 3.2 eVPL spectra in the pile irradiated GE124

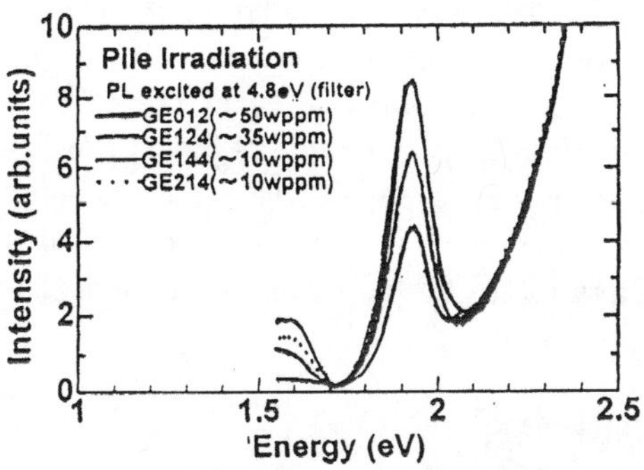

Fig.2: OH content dependence of the 1.9 eV PL in the pile irradiated GE124

K 04

WALL BLACKENING IN METAL HALIDE LAMPS CONTAINING RARE-EARTH BROMIDES

W.van Erk
Philips Lighting BV, P.O. box 80020,
5600 JM Eindhoven, The Netherlands

1 INTRODUCTION

Wall blackening in high pressure metal halide lamps is caused by the deposition of tungsten on the wall of the arc tube. It is not deposited as a uniform film but in the form of small crystals. For metal halide lamps containing scandium iodide or at least one rare-earth iodide crystal growth theory has been used to describe wall blackening phenomena [1]. In these lamps wall blackening cannot be avoided because under the conditions prevailing close to the bulb wall gaseous tungsten iodides are not stable enough. The situation can be different in metal halide lamps containing rare-earth bromides. In this contribution it will be shown why wall blackening can be strongly suppressed in lamps containing rare-earth bromides. These salt fillings are used in compact metal halide lamps for special applications like studio and stage lighting and LCD projection.

2 SUPERSATURATION

When a lamp is ignited for the first time its wall is still virginal. The first phase of the blackening process is nucleation, the formation of stable crystals on the wall. In the second phase these stable crystals grow. The rate of both processes is strongly dependent on the supersaturation, the driving force of the grow process [2]. The supersaturation β is defined as the difference in free enthalpy between tungsten in the gas phase and tungsten in the crystalline phase divided by the thermal energy kT. The value of the supersaturation can be obtained from thermodynamic calculations assuming local chemical equilibrium. [1, 3].

In metal halide lamps the tungsten in the gas phase mainly comes from the hot electrode tip, temperature 3000 K or more, where the arc attaches. Close to the relatively cold arc tube wall , typically about 1200 K, the dominant tungsten species in the gas phase are tungsten halides like WX_3 and WX_4 , where X= Br and I. In rare-earth metal iodide lamps the equilibrium pressure of the tungsten iodides is low, resulting in relatively high supersaturations (β= 18 - 25). Tungsten bromides however, are much more stable at temperatures of 1000 to 1500 K. It will be shown that in rare-earth metal bromide lamps the value of β near the wall can be decreased by increasing the excess of bromine in the gas phase, i.e. by raising the $HgBr_2$ pressure. A low supersaturation means a low nucleation rate, not many tungsten crystals on the wall, and a low growth rate of the crystals. Both effects have a positive impact on the lumen maintenance of the lamps.

3 NUCLEATION

Previously we have shown that in metal iodide lamps both the supersaturation as well as the nucleation rate are highest at the bulb wall. Nuclei can be formed either in the gas phase: homogeneous nucleation, or at the wall: heterogeneous nucleation. For both mechanisms holds that the nucleation rate J depends very strongly on β . It will be shown that there must be a critical value of the supersaturation, β^*, below which the nucleation rate is practically zero. If the classical formula for homogeneous nucleation is used and is assumed that the surface energy σ is 1.8 J/m^2, than at 1200 K $\beta^* = 10$ [4]. That this value of β^* is not small is due to the relatively high surface energy of tungsten. In the range $0 \preceq \beta \prec \beta^*$ the formation of new stable crystals will hardly occur.

Steady state burning: Assuming no condensation of tungsten and a summed tungsten pressure at the wall equal to its value at the hot electrode tip, β can be calculated. For rare-earth metal iodide lamps this gives $\beta = 18$ to 25 and β is not strongly dependent on the excess iodine pressure [1]. Most likely $\beta > \beta^*$ and nuclei will be formed during steady state burning. For lamps filled with rare-earth metal bromides the situation is different. Upon increasing the mercury bromide pressure β will decrease. As an example the results of calculations and life time experiments of lamps filled with $DyBr_3$ and CsBr will be shown. Lamps for which the calculated value of $\beta = 6$ showed an excellent lumen maintenance, whereas lamps without excess $HgBr_2$ and $\beta = 16$ had a lumen maintenance of 54 % at 750 hour.

Run up from a cold start to steady burning: During the first stage of run up the wall is very cold and there is hardly any halide in the gas phase. This means that the supersaturation is very high then, about 300, and that nucleation will be very fast. However, at 100 C the amount of mercury halides in the gas phase is already sufficient for the formation of tungsten halides and this reduces β dramatically. Due to further increase of the wall temperature β will decrease even more. This not only gives a strong reduction of the nucleation rate but also causes that just formed small nuclei may become subcritical and will disappear again. Only the crystals which grow fast enough during run up will survive.

For fillings with a low supersaturation the combined effect gives only some blackening of the wall near the electrodes (crystals formed during run up) and hardly any elsewhere. Blackening will be related to the number of (cold) starts and not to the burning time.

4 CONCLUSIONS

By using rare-earth metal bromides instead of iodides and by using an optimised excess of mercury bromide in the gas phase very compact metal halide lamps with an excellent lumen maintenance can be designed. To achieve this, it is not necessary to create a situation where the gas phase at the bulb wall is unsaturated with respect to tungsten. If the supersaturation is below a certain critical value new tungsten crystals will only be formed in the run up phase of the lamp and not during steady state burning. Lumen maintenance will be related to the number of starts and not to the burning time.

5 REFERENCES

[1] W. van Erk, P.H.L.M. Cobben and P. Bennema, Proc. Symposium on High Temperature Light Sources, New Orleans 1993, Proc. vol. 93-16, Electrochem. Soc. , 94 (1993)

[2] "Crystal Growth", ed. B.R. Pamplin, Pergamon Press, Oxford (1980)

[3] E. Schnedler, Proc. Symposium on High Temperature Light Sources, Toronto 1985, Proc. vol. 85-2, Electrochem. Soc.,95 (1985)

[4] "Crystal Growth, An Introduction" ed. P. Hartman, North Holland Publishing Company, Amsterdam (1973)

K 05

PHOTOLUMINESCENT MATERIALS BASED ON EU^{2+} DOPED ALKALINE EARTH HALIDE PHOSPHATES

Dagmar Nötzold
Institute of Physical Chemistry, Ernst-Moritz-Arndt-University Greifswald,
Soldtmannstraße 23, D - 17489 Greifswald, Germany

1 INTRODUCTION

Synthetic alkaline earth chloridephosphates doped by divalent europium are efficient photoluminescent materials. They are used as blue component in threeband fluorescent lamps and are characterized by high quantum efficiencies [1,2]. We have investigated these materials to study the relations between structural and optical properties in dependence on the chemical composition of the polycrystalline materials [3-7].

2 TECHNICAL DETAILS

Mixed crystals of the composition

$Sr_{5-x-y}Ca_xEu_y(PO_4)_3Cl$	with	$0 \le x \le 4.99$	and	$0 \le y \le 0.1$,	
$Sr_{5-x-y}Ba_xEu_y(PO_4)_3Cl$	with	$0 \le x \le 4.99$	and	$0 \le y \le 0.1$,	
$Sr_{5-y}Eu_y(PO_4)_3Cl_{1-x}Br_x$	with	$0 \le x \le 1$	and	$0 \le y \le 0.05$,	and
$Sr_{5-y}Eu_y(PO_4)_3Cl_{1-x}F_x$	with	$0 \le x \le 1$	and	$0 \le y \le 0.05$	

were prepared by solid state reaction by firing the starting mixture at 700 and 1050°C in reducing atmosphere (N_2 / H_2).
The crystal structure, optical properties at room temperature, and the emission spectra of the polycrystalline materials at liquid nitrogen temperature were investigated. The emission of the samples was excited with 39424 cm^{-1} radiation (253.65 nm).
The structure parameters were refined by the Rietveld method [8,9].

3 CONCLUSIONS

The X-ray diffractograms of the mixed crystals show the characteristic reflections of the hexagonal apatite structure. The space group of the materials is $P6_3/m$.
There are two crystallographically different sites for the cations in the unit cell, the 4f and the 6h Wyckoff positions. Eu^{2+} ions occupy preferentially the 6h cation positions.
The luminescence of the samples arises from transitions between $4f^65d^1$ and $4f^7$ configurations of the Eu^{2+} ion. It depends strongly on the composition of the polycrystalline materials, and there are significant differences in the optical properties of the Eu^{2+} ion between the four systems given above. The emission band of the starting material, $Sr_{4.99}Eu_{0.01}(PO_4)_3Cl$, appears in the spectral region between 19000 and 25000 cm^{-1} with a band width at half height of 1728 cm^{-1}. The maximum lies at 22463 cm^{-1} (Fig. 1). Mixed crystals of the composition $Sr_{5-y}Eu_y(PO_4)_3Cl_{1-x}Br_x$ show the most intensive emission (Fig. 1). The optical properties are discussed in connection with the composition and the structure parameters of the polycrystalline materials.

4 REFERENCES

[1] F.C. Palilla and B.E. O'Reilly, J. Electrochem. Soc. **115** , 1076 (1968).
[2] M. Nakamoto, T. Nishimura and H. Kohmoto, *Three-Narrow-Band de Luxe Fluorescent*

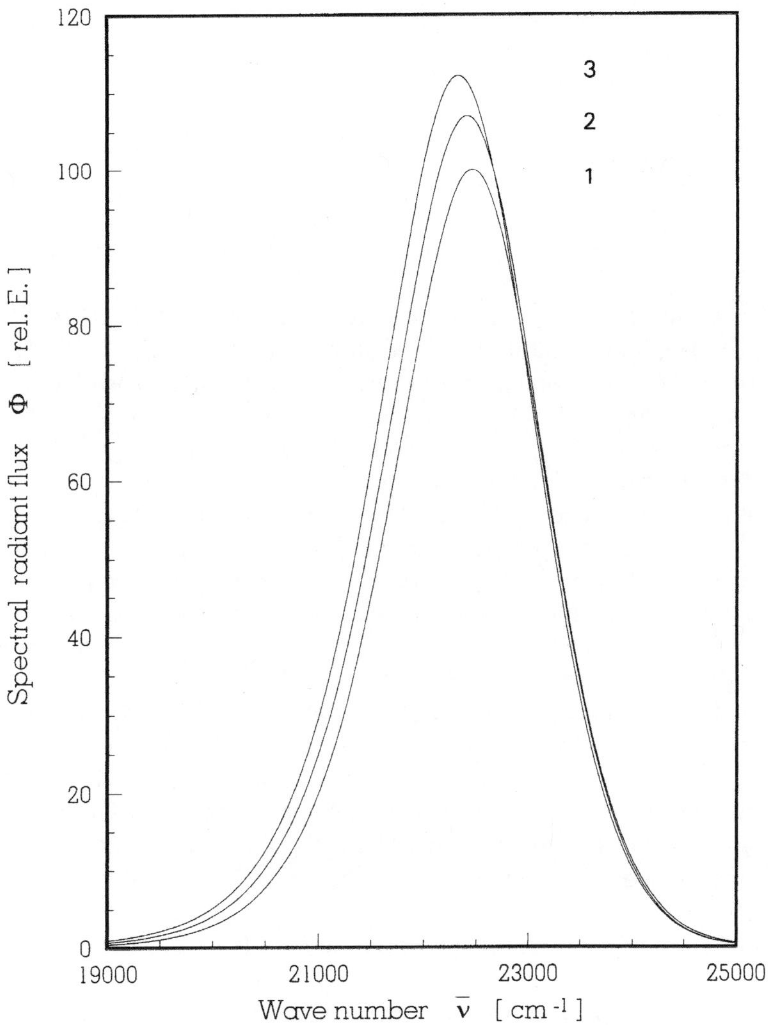

Fig. 1: Emission spectra at 293 K in dependence on the composition for samples $Sr_{4.99}Eu_{0.01}(PO_4)_3X$ with X: 1. Cl ; 2. $Cl_{0.5}$ $Br_{0.5}$; 3. Br.

Lamp with Newly Developed Phosphors, Toshiba Corporation, Prospectus No. 15 (Narva/ED Ue 1133), 1982.

[3] D. Nötzold, G. Herzog and I. Henke, Z. Anorg. Allg. Chem. **613** , 127 (1992).

[4] D. Nötzold, H. Wulff and G. Herzog, phys. stat. sol. (b) **191** , 21 (1995).

[5] D. Nötzold and H. Wulff, phys. stat. sol. (a) **158** , 303 (1996).

[6] D. Nötzold and H. Wulff, phys. stat. sol. (a) **160** , 227 (1997).

[7] D. Nötzold and H. Wulff, phys. stat. sol. (b) **207** , 271 (1998).

[8] H. M. Rietveld, J. Appl. Cryst. **2** , 65 (1969).

[9] J. Schneider, Acta Cryst. **A43** , Suppl. C, 295 (1987).

K 06

VACUUM ULTRAVIOLET EXCITATION SPECTRA OF POWDER PHOSPHORS CONTAINING ISOLATED MO_4 AND LINKED MO_4 TYPE TETRAHEDRAL IONS

Shinkichi Tanimizu[1], Teruki Suzuki[2], Masatoshi Shiiki[2], and Choichiro Okazaki[2]

[1] Consultant, 2-13-32 Nishimoto-cho, Kokubunji, Tokyo 1850023, Japan
[2] Central Research Laboratory, Hitachi Ltd., Kokubunji, Tokyo 1850014, Japan

The well-known phosphors applied in plasma display panels(PDP) are $BaMgAl_{10}O_{17}$:Eu (blue), Zn_2SiO_4:Mn(green), $BaAl_{12}O_{19}$:Mn(green), Y_2O_3:Eu(red) and (Y,Gd)BO_3:Eu(red). These phosphors were historically selected from vacuum ultraviolet(VUV) excitation spectra in the range of 100 - 200nm [1]. PDP phosphors are excited by one of the Xe resonance lines (147 nm, 8.43 eV) and/or the Xe_2 molecular band (peaked at 170 nm, 7.3 eV), which cover mostly the intrinsic absorption region of oxygen-dominated lattice phosphors.

Searching for a criterion to find good host-lattices, we considered the electronic states of an isolated MO_4 ion as the simplest case and tried to interpret the VUV excitation spectra of various phosphors. The VUV spectra were measured on 1m Seya mount monochromator. The values of the transition energy from $2t_2$ to $2a_1$ vary such that BO_4 > PO_4 > SiO_4 > AlO_4 in order of decreasing energy (Table 1). The $2t_2 \rightarrow 2a_1$ transition of PO_4 lies in the region of 8.3 - 9.1 eV for a mean distance of P-O, 1.55 ± 0.03 Å, and the oscillator strength of the transition is 0.69, so a combination of PO_4 (absorber) and VO_4 (emitter) can be selected as a good candidate. This indeed has proved to be the case for Y(P,V)O_4 :Tm(blue) and (Y,Gd)(P,V)O_4 :Eu(red). Figures 1 and 2 show the experimental results. The same transition of BO_4 lies at 10 eV, and YBO_3 contains only BO_4 ions. However, the VUV band for YBO_3 :Eu appeared at 7.5 eV. This discrepancy might be caused by two kinds of BO_4, one of which is very distorted [2].

There exists a number of phosphors containing linked MO_4 ions in which each tetrahedra shares vertices (Figs. 3b and 3c). The efficiencies of the phosphors consisting of these host-lattices were rather poor due to the occurrence of the energy dissipation through oxygen 2p levels. For example, TbP_5O_{14}, one of the ultra-phosphates, is not efficient in the VUV region [3], while (La,Ce)PO_4 :Tb, well-known as a lamp phosphor, is efficient. Additionally, $Sr_2P_2O_7$:Eu and $Sr_4Al_{14}O_{25}$:Eu are not efficient, while $Sr_3Mg(SiO_4)_2$:Eu is efficient (Fig. 4).

An exception can be pointed out for the phosphors activated with Mn^{2+}. Zn_2SiO_4 has a rather unusual structure, since both Zn sites and the Si site are tetrahedrally surrounded by oxygens with Zn and Si tetrahedra sharing corners [4]. The high efficiency of Zn_2SiO_4 :Mn suggests that the VUV levels of Mn^{2+} even located near the surface are highly advantageous for an inner-ion relaxation or for a charge transfer transition. Similar phenomena were observed in the spectra of Mn^{2+} for Ca-halophosphate:Sb,Mn (Fig. 5b) and Zn-borate :Mn.

In conclusion, the host-lattices containing isolated MO_4 ions are better than those containing linked MO_4 ions, and a system of PO_4-VO_4 with a zircon structure was confirmed to be a good candidate. In particular, carefully prepared (Y,Gd)(P,V)O_4 :Eu is superior to (Y,Gd)BO_3 : Eu in both chromaticity and efficiency. However, there are some good candidates among the host-lattices containing linked MO_4 ions in the case of Mn^{2+} activated phosphors.

References

[1] C.R.Ronda, J. Luminescence 72-74 (1997) 49.
[2] G.Chadeyron, M.El-Ghozzi, R.Mahiou, A.Arbus, and J.C.Cousseins, J.Solid State Chem. 128 (1997) 261.
[3] T.Fukuzawa and S.Tanimizu, J. Luminescence 16 (1978) 447.
[4] A.S.Povarennykh, "Crystal Chemical Classification of Minerals" Vol.1 (Plenum, 1972) 384.

Table 1. Mean M-O distance and $2t_2 \rightarrow 2a_1$ transition energy

Oxy-anion	Mean M-O distance	$2t_2 \rightarrow 2a_1$ transition
BO_4^{5-}	1.43 Å	10.2 eV
ClO_4^{-}	1.46	10.3
SO_4^{2-}	1.51	9.5
PO_4^{3-}	1.55	8.7
SiO_4^{4-}	1.61	8.1
AlO_4^{5-}	1.70	8.0
GeO_4^{4-}	1.80	7.0

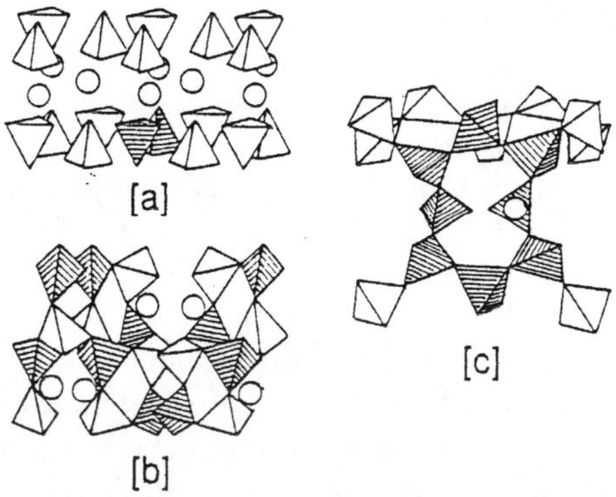

Fig. 3 Schematic drawing of isolated MO_4[a] and linked MO_4[b][c].

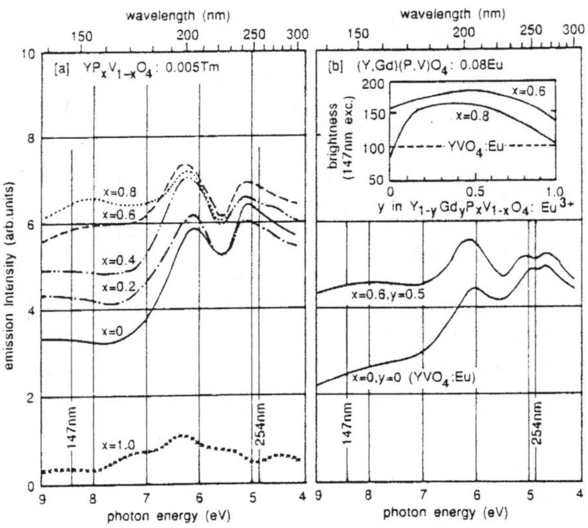

Fig. 1 Excitation spectra of $YP_xV_{1-x}O_4$: Tm [a] and $Y_{1-y}Gd_yP_xV_{1-x}O_4$: Eu [b].

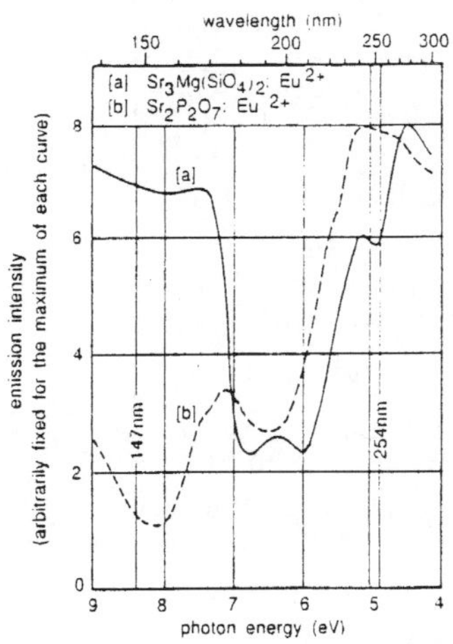

Fig. 4 Excitation spectra of $Sr_3Mg(SiO_4)_2$: Eu and $Sr_2P_2O_7$: Eu.

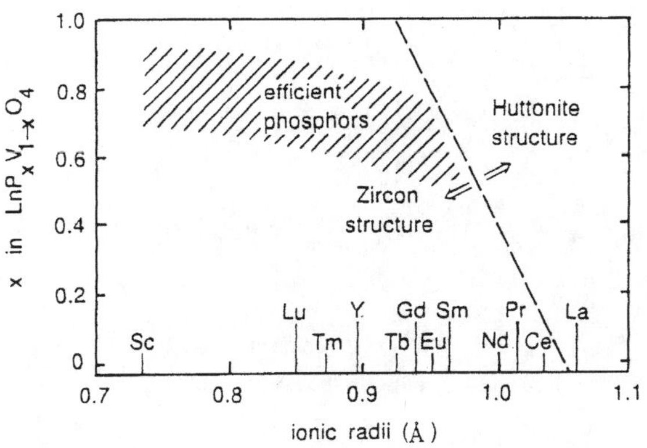

Fig. 2 Efficient phosphor region for a system of PO_4-VO_4 under 147 nm excitation.

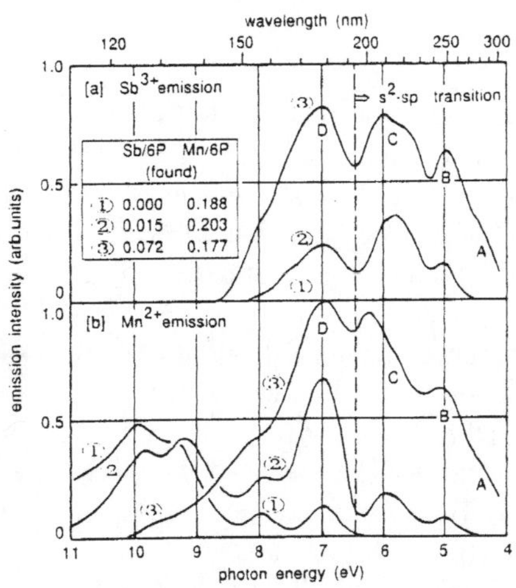

Fig. 5 Excitation spectra of $Ca_{10}(PO_4)_6(F,Cl)_2$: Sb, Mn.

K 07

SUBSTRUCTURE OF ARC SPOTS AT THERMIONIC CATHODES

Burkhard Jüttner and Heinz Pursch

Humboldt-University, Institute of Physics, Hausvogteiplatz 5/7, D-10117 Berlin, Germany

1 INTRODUCTION

Thermionic lamp cathodes at high pressure exhibit contractions of the discharge at the cathode when the current falls below a certain value. Generally it is assumed that the result is enhanced local heating compensating the effect of the reduced power input on the thermionic emission. The present paper studies the substructure of such footpoints in a model discharge.

2 EXPERIMENTAL DETAILS

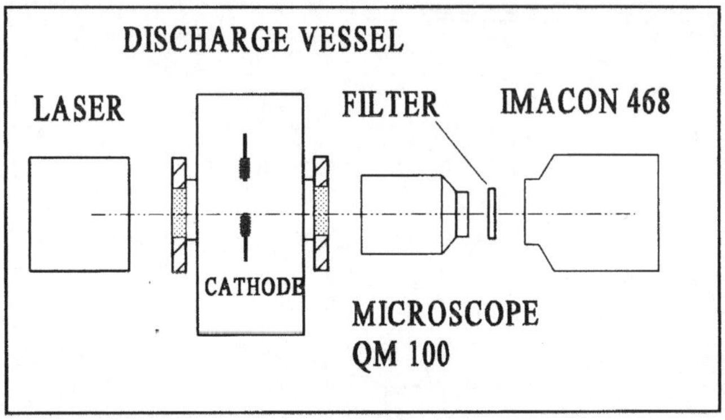

Fig. 1: Schematic of the imaging system

Fig. 1 gives a schematic of the device. Electrodes from high pressure lamps have been used, consisting of a W-tip surrounded by a W-spiral and covered with activation paste containing barium. They are mounted in a clean UHV-discharge vessel, filled with argon at 1-2 bar. The discharge current is 0.5-3 A. The cathode is viewed by the long-distance microscope QM 100 and imaged by seven channels of the high speed camera IMACON 468. A filter serves to remove the argon light. In this way, emission pictures of the footpoint are obtained with spatial resolution < 10 μm and time resolution < 1 μs. Illuminating the cathode from the backside with a pulsed dye laser produces absorption pictures with the same system. The pulse duration is 0.5 ns, the wavelength is set to 508 nm.

3 RESULTS

Figs. 2, 3 show the footpoint at a current of 1 A. An internal irregular structure is evident, consisting of several elements with a size of 15-20 μm. The number of such elements is increasing with the current, as shown in fig. 4. Each of them conducts a current of 0.1-0.2 A. When the footpoint is situated at the cathode contour, absorption pictures can be obtained. Fig. 5 explains the measuring principle. (1) One channel of the camera is exposed 5 μs to laser and footpoint. Because of the short time, the light emitted from the footpoint cannot be seen. The picture reveals an absorption structure at the right hand side of the cathode (arrowed). (2) Another channel is exposed to

Fig. 2: Emission picture. Exposure 500 μs, Current 1 A.

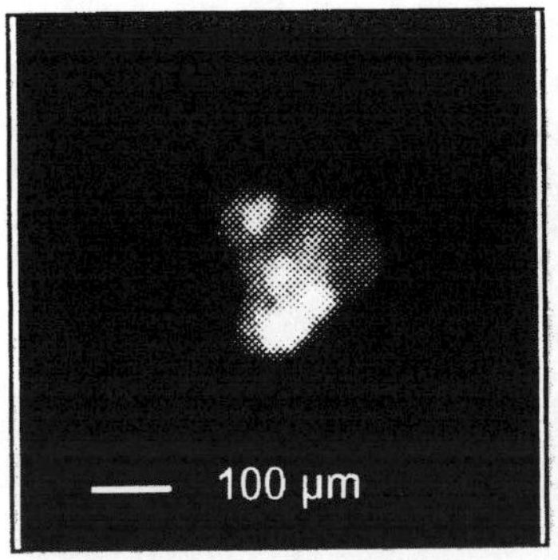

Fig. 3: Detail from fig. 2.

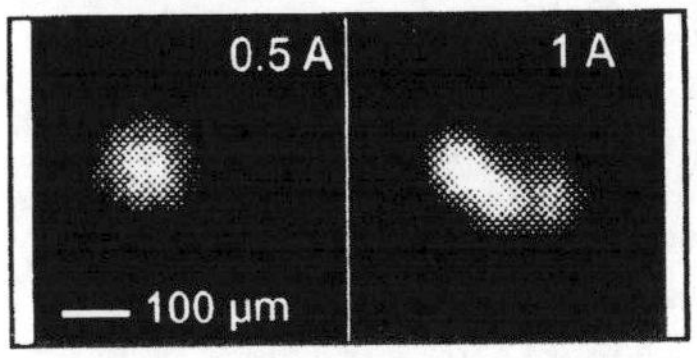

Fig. 4: Emission pictures at 0.5 and 1 A. Exposure 500 μs.

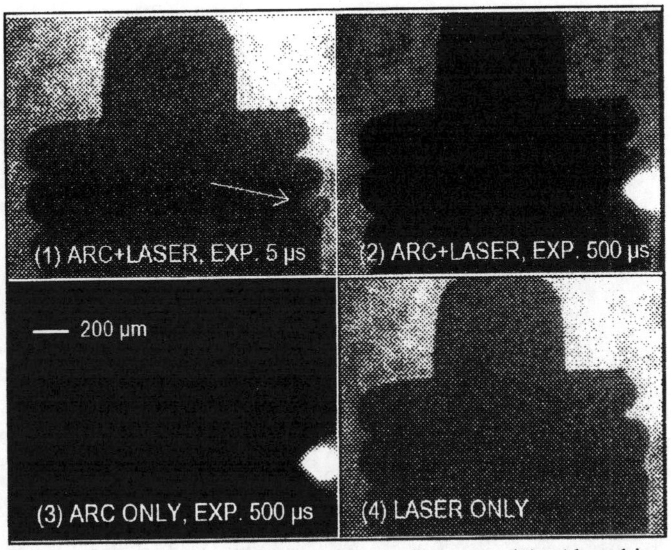

Fig. 5: Principle of absorption pictures. (1) Absorbing structure, (2) spot position, (3) spot alone, (4) reference.

laser and footpoint light with 500 μs exposure time. This picture indicates the position of the footpoint by the emitted light. (3) The remaining 5 channels are not synchronized to the laser pulse. Instead, they depict emission pictures at various times. (4) Finally, a reference picture is taken without arc. Subtracting picture (1) from (4), the absorbing region appears bright. This is shown in fig. 6. Evaluating a number of such events we find various sizes of 10-20 μm, and also a substructure resembling that of the emission pictures.

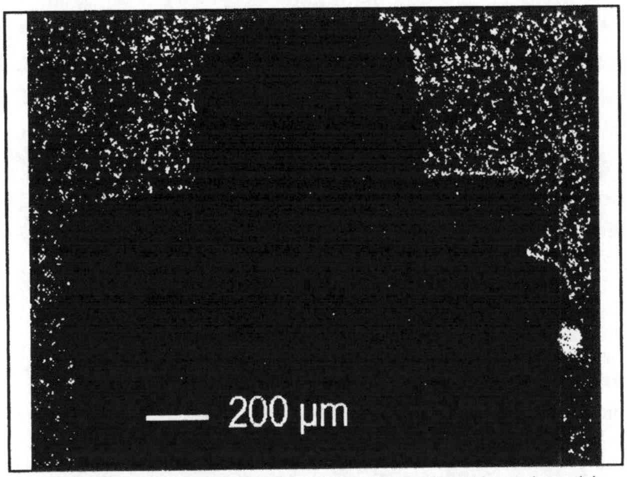

Fig. 6: Processed absorption picture. The absorbing structure appears as white region. The current is 1 A.

4 CONCLUSIONS

Contracted footpoints at thermionic cathodes exhibit irregular substructures with a size < 20 μm and are associated with absorbing regions in front of the cathode. These structures can be interpreted as vapor producing microspots as in the case with cold cathodes. Thus, both the emission and the erosion mechanism will basically be changed.

5 ACKNOWLEDGEMENTS

The work is funded by the BMBF, grant 13N7108/1.

K 08

Enhancement of Rare Earth Metals by Chemical Vapour Transport in Metal Halide Lamps

K. Hilpert[1] and U. Niemann[2]

[1]Institute for Materials in Energy Systems, Research Centre Jülich (KFA), P.O. Box 1913, D-52425 Jülich, Germany

[2]Philips GmbH Research Laboratories, P.O. Box 500145, D-52085 Aachen, Germany

1 INTRODUCTION

The arc tube of metal halide lamps contains a mixture of molten salts made up of metal iodides with, in some cases, admixtures of metal bromides as lamp filling. The salt melt is present as condensed phase at the "coldest spot" of the inner wall of the discharge vessel. During operation of the lamp the metal iodides and bromides in part vaporize. The iodides and bromides of the rare earth metals, especially those of thulium, holmium and dysprosium, are particularly important for the light generation since these metals display many lines in the visible spectrum when excited in the high temperature arc column of the lamp.

The intensity of the light emission increases if the concentration of the rare earth metals in the column is raised. This is possible by the use of chemical vapour transport reactions due to the formation of gaseous complexes containing rare earth metals. Such complexes are for example formed between the alkali metal halides AX and the rare earth metal halides LnX_3. The advantage of the rare earth metals concerning the light emission contrasts with the disadvantage of their corrosive reactions with the material of the arc tube (quartz glass, polycrystalline alumina). The corrosive attack can be reduced if the thermodynamic activity of LnX_3 in the salt melt is lowered.

2 TECHNICAL DETAILS

2.1 Experimental results

Vaporization studies and thermodynamic modelling were used to elucidate the potential for the enhancement of the rare earth metal concentrations in the gas phase by chemical vapour transport and for the reduction of the LnX_3 thermodynamic activity in the melt. The systems $NaI-DyI_3$, $NaBr-DyBr_3$, $CsI-DyI_3$, and $CsBr-DyBr_3$ were considered as an example. The results reported here are part of our systematic studies of high temperature lamp chemistry [1].

The vaporization measurements were carried out by Knudsen effusion mass spectrometry in the temperature range between 600 and 1300 K covering the complete concentration range of the four systems considered. The gaseous species $DyX_3(g)$, $(DyX_3)_2(g)$, $AX(g)$, $(AX)_2(g)$, $ADyX_4(g)$, and $A_2DyX_5(g)$ (A = Na, Cs; X = Br, I) could be identified in the vapour and their thermochemical data concerning the dissociation were determined. The hetero complexes $ADyX_4(g)$ and $A_2DyX_5(g)$ turned out to be particularly important since they are mainly responsible for the observed enhancement of Dy in the vapour phase. Moreover, the thermodynamic activities of NaX and DyX_3 (X = Br, I) resulted for the various phases of the systems. The activities and stabilities of the gaseous species were incorporated into a data base and used for thermochemical computations.

2.2 Thermochemical discussion

The Dy enhancement at different temperatures resulted for the complete concentration range of the four systems considered (fig. 1). The enhancement is discussed with respect to the demand for a small thermodynamic activity of DyX_3 in the melt in order to decrease corrosion effects. Optimal compositions meeting the two demands for high Dy enhancement in the gas phase and low activity of DyX_3 in the melt are reported. The highest Dy enhancement and the lowest thermodynamic activity of DyX_3 was observed in the CsI-DyI_3 system. The possible enhancement is higher and the thermodynamic activity is significantly lower than that for the NaI-DyI_3 system preferably applied so far.

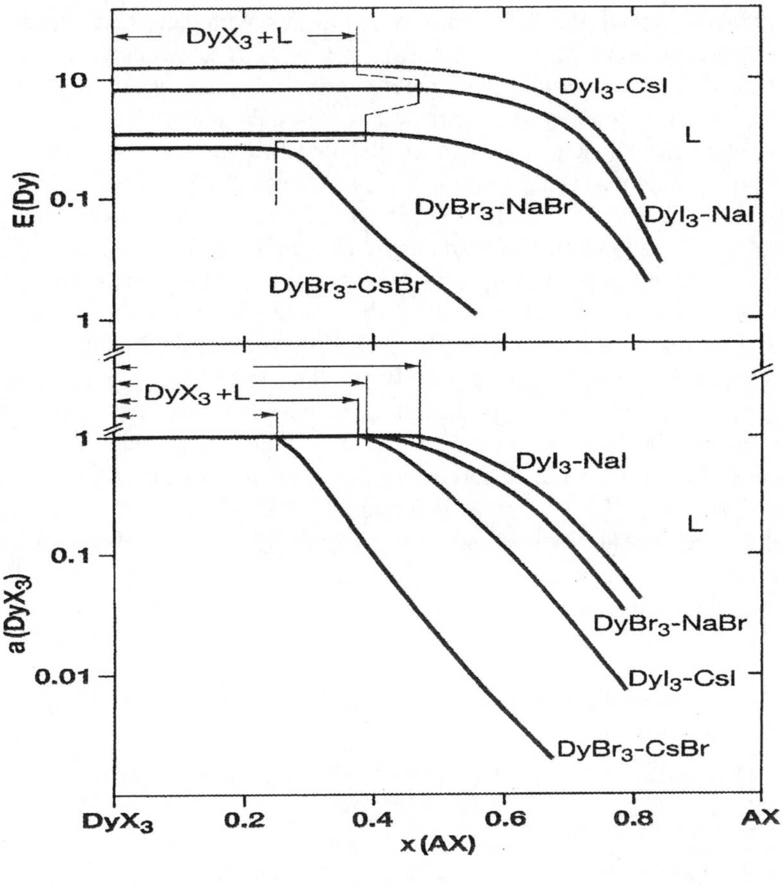

Fig. 1: Enhancement E(Dy) and Thermodynamic Activity a(DyX_3) of DyX_3 at 1050 K for the AX-DyX_3 Systems (A = Na, Cs; X = Br, I)

3 CONCLUSIONS

Lamp fillings containing CsX-DyX_3 mixtures proved to be superior to NaX-DyX_3 (X = Br, I) systems with respect to Dy enhancement and lowering of the DyX_3 activity. NaI should, therefore, be replaced by CsI, if the presence of Na is not required to decrease the colour temperature of the light emitted. A further advantage incidental to the application of Cs is that it does not penetrate the quartz glass of the discharge vessel unlike Na.

4 REFERENCES

[1] K. Hilpert and U. Niemann, Thermochimica Acta **299,** 49–57 (1997).

K 09

DYNAMICS OF QUARTZ WALL ABLATION DURING A PULSED DISCHARGE IN XENON FLASH LAMP

Sergei Raikov
Institute of Molecular and Atomic Physics, National Academy of Sciences of Belarus,
70 Scorina av., 220072 Minsk, Belarus

1 INTRODUCTION

An ablation of transparent wall of a flash lamp plays together with electrode erosion one of the major roles in plasma evolution during a high-current pulsed electric discharge in gases at lowered pressures. Besides, induced opacity and deterioration of lamp wall are in direct dependence on sputtering and volatilization of constructional materials, mainly, quartz tube and metallic electrodes. The electrode erosion is studied much more full and in detail. Moreover, frequently for simplicity of quantitative consideration and modelling of these processes erosion of a material of electrodes specially is taken into account only. As to an ablation of lamp wall, dynamics of evaporation and dissociation of SiO_2 and especially the influence of silicon species on gas plasma parameters are investigated during the discharge pulse basically qualitatively, and the available quantitative results are received, mainly, by accumulation of the appropriate information for rather large number of pulses. In order to obtain needed experimental data new spectroscopic diagnostics must be applied, as the currently available techniques are not suited to monitor the dynamical behavior of this plasma. In a study of such systems, one has to cope with a combination of short time scales, high selectivity, and low and high densities of the active species. The most sensitive absorption method – intracavity laser spectroscopy (ICLS) is of great value in such measurements. The ICLS is compared favourably with a commonly used laser-induced fluorescence, due to the possibility of diagnostics of strongly luminous object. Besides, it allows to record simultaneously a large number of spectral line profiles by a broadband backlight continuum much shorter than plasma lifetime.

2 TECHNICAL DETAILS

Measurements were carried out using the ICL spectrometer shown schematically in Fig.

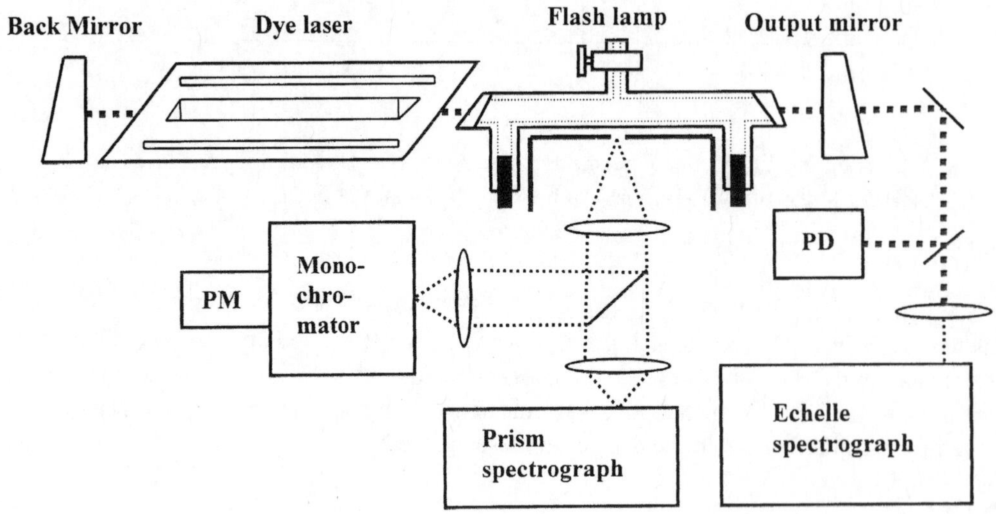

A tunable multimode flash lamp-pumped dye laser with variable pulse duration (1-10 μs) radiates a smooth (spectral structureless) broadband (10-15 nm) spectrum in the visible range. A quartz flash lamp with quartz windows specially designed for laser and optical probing of plasma of high-current pulsed gas discharge at different pressures was located in the nonselective laser cavity. The quartz tube is the same as for one of the types of a standard flash lamp for laser pumping: 16 cm in length, 1.1 cm in inner and 1.5 cm in outer diameters. It is filled with high-purity xenon at the pressure ranged between 1.3 and 8 kPa. Close to a rectangular electric pulse with variable duration (up to 1.4 ms) was applied to the lamp electrodes with discharge energy E = 0.5-1 kJ. A compact (1377 mm focal) high resolution (1 pm) echelle grating spectrograph with double dispersion was constructed and applied in the ICLS measurements. The effective path length of laser radiation through the flash lamp reached 1 km for the 10-μs laser pulse. The minimally determined value of Nf (where N – density of the absorbing species, and f – oscillator strength) was about 10^7 cm^{-3}. Simultaneously conventional optical analyses of pulsed plasma have been performed with a prism spectrograph and a 0.5-m grating monochromator.

The pure xenon plasma without seen presence of silicon was possible to realize only at the lowest discharge energies, i.e. at $E \leq 0.5$ kJ. Thus are recorded practically all Xe absorption lines from the excited electronic configurations: 6s, 6s', 6p (8.3-10 eV) – in the beginning (20-70 μs) and in the end (1400 μs) of a discharge; Xe^+: 5d, 5d', 5d", 6s, 6s', 6s", 6p, 6p' (11.8-16.5 eV) - during almost all discharge pulse (40-1400 μs); and Xe^{++}: 5d', 6s (16.5-18 eV) – in the middle part of a pulse (400-1000 μs). The populating of the Xe^+ exited levels repeats the electric pulse temporal shape. There are also absorption lines of gas components of a various nature: (i) volatilized from constituent elements of a lamp and residual gases – N, N^+, C, C^+, H, O; and (ii) present in xenon as impurities – He, Ne, Ar, Kr. Even in case of absence of seen quartz ablation some of these gas components and there compounds are chemically active to SiO_2 and form reducing atmosphere, in which a decomposition of heated superficial quartz layers occur.

At increase of discharge energy (E>0.5 kJ) the dynamics of plasma drastically varies. In addition to the mentioned above spectral lines practically all absorption lines of Si from the excited configurations are recorded: 3d, 4s, 4p (4.9-6 eV); Si^+: 4s', 4p (10-14.6 eV); and Si^{++}: 4s, 5d, 5f, 6h (19.7-30.1 eV). Silicon species appear with a large delay from the discharge onset (400-500 μs). It is connected with a necessity of rather large expenditure of energy to heat, melt, evaporate and dissociate SiO_2. Besides, the sputtered silicon species in the beginning are located close to a tube surface, and in process of growth of their pressure they are mixing with gas environment. Si and Si^+ species are detected down to the end of the discharge and in afterglow. In turn Si^{++} ions exist only in the middle of a pulse (500-1100 μs), and their density compared with Si^+ is insignificant. In a maximum the value of N for Si^+ at the excited levels reaches $5 \cdot 10^{11}$ cm^{-3}. Simultaneously different impurities present in quartz (Na, K, Ca) are detected in plasma. The above described dynamic picture of SiO_2 ablation and silicon species appearance in plasma is accompanied by sharp decrease of species densities at the excited levels of all gas components. At the end of the discharge (>1100 μs) the energetic threshold of quartz evaporation is not reached, therefore the excited levels of gas species are repeatedly populated almost up to starting values (at ~1300 μs). The dynamics of electron density and plasma temperature as well as the absolute mass of ablated SiO_2 during one discharge pulse have been found from spectral data.

3 CONCLUSIONS

Melting and cooling of superficial quartz layers and also complex and simultaneous action on a tube surface of ablation products and gas components result in formation of irreversible blooms and unhardening of flash lamp wall even during a single pulse, especially under forced operating conditions.

K 10

EUROPIUM DOPED ALKALINE RARE EARTH ORTHOPHOSPHATE PHOSPHORS

Matthias Kloss, Annemarie Rohmann, Ute Sasum, Lothar Schwarz, Detlef Haberland

Institute of Physical Chemistry, Ernst-Moritz-Arndt-Universität Greifswald,
Soldtmannstr. 23, D-17489 Greifswald, Germany

According to earlier results of optical investigations on cerium and terbium bidoped double phosphates of the type $K_3RE(PO_4)_2$ [1] we studied europium doped systems ($M_3RE_{1-x}Eu_x(PO_4)_2$, M = Rb; K, Na; RE = La, Gd) as an approach to find new red emitting phosphors. A further aim of the investigations was to study the influence of gadolinium on the europium luminescence centres in these solid solutions [2].

The compounds were prepared by solid state reactions as crystaline powders. and the structure verified by x-ray diffraction pattern and FT-IR spectra, respectively. Thermal analysis and REM photographs are used to confirm the formation of the double phosphates. The unit cells of rubidium and potassium compounds are monoclinic (SG $P2_1/m$, Z = 2) like $K_3Nd(PO_4)_2$ [3] and those of sodium compounds orthorhombic (SG $Pbc2_1$, Z = 24) like $Na_3Nd(PO_4)_2$ [4].

The materials were tested mainly by low pressure mercury excitation for application in lamps. As a refinement of the results about the structure of the doped double phosphates the $^5D_0$-$^7F_0$ transition is a sensitive indicator of the situation around the luminescence center, it was determined from the emission spectra of trivalent europium in this matrices (see Fig.1). Therefore VUV measurements of reflection spectra [5] were carried out to determine as well as the absorption edge of the basic lattices as some of the characteristic 4f transitions of trivalent lanthanide ions. These results and high resolved europium(III) emission and excitation spectra in the UV/VIS range at room temperature and at 77 K are used to characterize the luminescence centers and to determine the ground term multiplet ^7F_J in the different lattices.

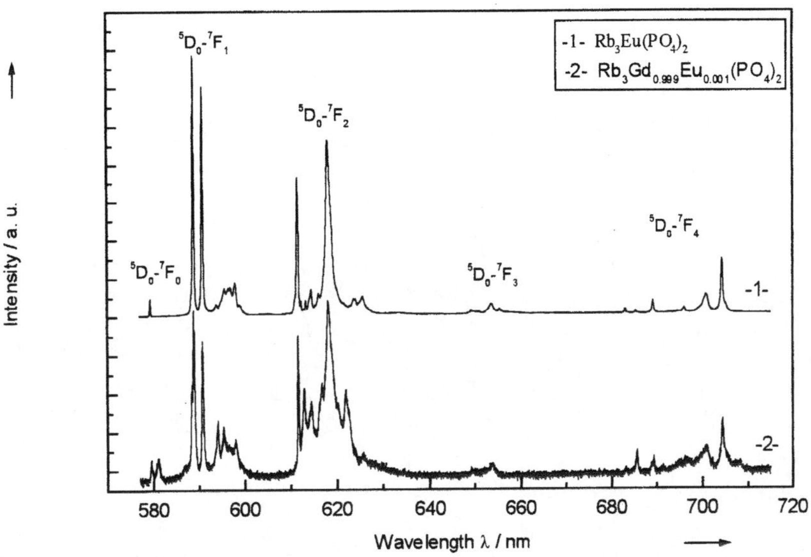

Fig. 1 Emission Spectra of Rubidium Rare Earth Double Phosphates at 77 K

The intension is to compare of emission centres via the spectra of $Na_3RE_{1-x}Eu_x(PO_4)_2$, $K_3RE_{1-x}Eu_x(PO_4)_2$ and $Rb_3RE_{1-x}Eu_x(PO_4)_2$. The optical properties are discussed in connnection with a term scheme proposal deduced from the observed $^5D_0$-^7F_J and $^5D_1$-^7F_J series.

References:

1 B.Finke, L.Schwarz, P.Gürtler, M. Kraas, M. Joppien, J. Becker: J. Lumin. 60/61 (1994) 975-978
see also B. Finke, L.Schwarz: Radiation and Defects in Solids 135 (1995) 407-411

2 M.Kloss, B. Finke, L.Schwarz, D.Haberland: J.Lumin.72-74 (1997) 684-686

3 Y.P. Hong, S.R. Chinn: Mater.Res. Bull. 11 (1976) 421-428

4 R. Salmon, C. Parent, M. Vlasse, G. Le Flem: Mat. Res. Bull. 13 (1978) 439-444

5 L.Schwarz, B.Finke, M.Kloss: DESY Jahresbericht Hamburg II (1995) 131-132

SCANDIUM SILICATE AND TUNGSTEN SILICIDE EQUILIBRIA

Richard Snellgrove
OSRAM SYLVANIA INC., Beverly, MA 01915, U.S.A.

1 INTRODUCTION

An important factor affecting the maintenance of MH lamps dosed with Na-Sc chemistry is the reaction of the scandium component of the fill with the "quartz" arc tube walls. Not only does this reaction have consequences for lamp performance but the SiI_4 generated can lead to electrode meltback. These reactions are often cited but little has been published regarding the details of what controls the degree of scandium silicate formation. This study is a first step in that direction and the further investigation into the details of the reaction mechanism.

2 TECHNICAL DETAILS

2.1 Scandium silicate formation

Scandium silicate crystals are formed by means of vapor phase transport of silicon, scandium, and oxygen to the growing surface. This is indicated by both the microstructure of the crystals and the fact that they will grow on materials other than silica, such as tungsten. The formation reaction

$$4ScI_3(g) + 7SiO_2 \rightarrow 2Sc_2Si_2O_7 + 3SiI_4(g) \quad (1)$$

was studied by means of furnace experiments in which ScI_3 and other reagents, such as SiI_4, I_2, and Hg, were heated for various temperatures and times in sealed silica tubes. Amounts of reaction products were determined by microscopic estimation of crystal volumes and/or chemical analysis, and corrected for dissociation or combination at the temperature of the run.

Figure 1. Reaction (1) reversal (900°C)

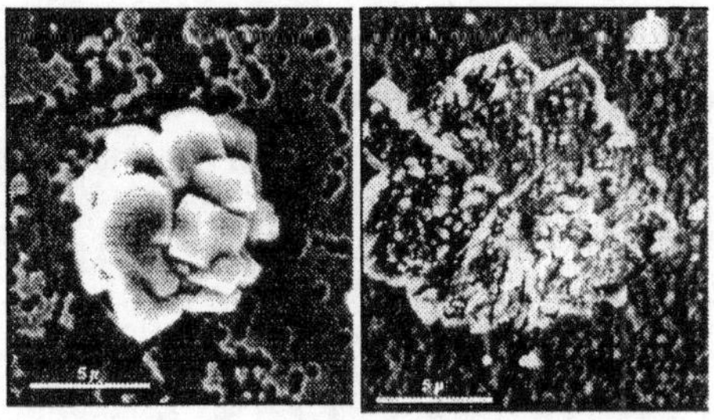

Previous studies [1] have suggested that reaction 1 might approach equilibrium but is not readily reversible. The present study confirms that no reaction occurs when just SiI_4 is added to tubes containing scandium silicate (left side Figure 1), but if ScI_3 is also present, then reaction (1) is rapidly reversed and scandium silicate converted back into silica, some of which is transported away and some remains as relict outlines of scandium silicate (right side of Figure 1). This confirms a complex vapor phase mechanism for the reaction involving scandium containing intermediates, and also suggests that equilibration should occur readily at 900 °C.

To test equilibration and calculate the equilibrium constant a series of experiments were carried out. A number of variables were tested to demonstrate equilibrium rather than kinetic control of the extent of reaction. The reaction is quite rapid, requiring just a few hours to come to equilibrium, but thereafter is independent of time. The equilibrium constant is independent of iodine, ScI_3, and Hg pressure within the accuracy of the analytical data. If SiI_4 greater than the equilibrium amount is added, formation of scandium silicate is completely blocked.

The effect of iodine, mercury, temperature, and pressure are best understood if reaction 1 is written in terms of the dominant species at arc tube wall temperatures:

$$4ScI_3(g) + 7SiO_2 \rightarrow 2Sc_2Si_2O_7 + 3SiI_2(g) + 6I(g) \quad (2)$$

The degree of reaction (but not the constant) will be suppressed by iodine but increased by mercury or any other agent that reacts with free iodine. This was experimentally confirmed.

Reaction 2 is also decreased by pressure (less gas), increased by temperature (highly endothermic), and decreased by a condensed phase of ScI_3. These latter effects are illustrated in Table 1.

Table 1. Calculated fraction of ScI_3 reacted (Shaded values indicate condensed ScI_3).

P (kPa) \ T (K)	1000	1100	1200	1300	1400
1	0.028	0.089	0.221	0.478	0.789
10	0.002	0.035	0.079	0.166	0.322
101	0.000	0.002	0.018	0.063	0.112

The value for the equilibrium constant agrees well with that calculated from literature values after making two corrections: The enthalpy of formation of scandium silicate was lowered by 14 kJ at 900 °C based on recent heat of solution measurements [2], and fused silica values were used rather than those for the stable crystalline form.

2.2 Tungsten silicide formation

If a tungsten electrode is present, silicon iodides will react further to form tungsten silicide on the electrode and free iodine:

$$SiI_4(g) + 5/3W = 1/3W_5Si_3 + 2I_2(g) \tag{3}$$

This reaction was also studied experimentally. It is much more sluggish than reaction 1 and requires several hundred hours to approach equilibrium at temperatures around 1000 °C. Equilibration was confirmed by approaching from both directions. SiI_4 was added directly or generated by reaction 1. The equilibrium constant (1 atm. standard state) for different reactants is given in Table 2. Contrary to the trend in K_3, formation of the silicide, W_5Si_3, is strongly favored for high temperature regions of electrodes because the I_2 pressure is greatly reduced by dissociation.

Table 2. Experimental constant for 3

Inputs	T	K_3
$W+ScI_3+SiO_2$	920	246
$W_5Si_3+I_2$	1040	143
$W+SiI_4$	1060	33
$W+ScI_3+SiO_2$	1100	4

3 CONCLUSIONS

- Scandium silicate crystals grow from the vapor, which supplies the scandium, silicon, and oxygen through complex intermediates.
- Scandium silicate formation can be reversed by reaction with SiI_4 if ScI_3 is also present.
- Scandium silicate formation comes to equilibrium within a few hours at 900 °C with an equilibrium constant of 0.003 (1 atm. standard state)
- The silicon iodides produced react with tungsten to form W_5Si_3, and this reaction will come to equilibrium in several hundred hours at 900 °C.
- The two reactions are coupled together and will shift in response to changes in temperature and iodine pressure.

4 REFERENCES

[1] E. Fischer and R. Kaiser, Proc. Symposium on Science and Technology of High Temperature Light Sources, Toronto, 1985, 85-2, Electrochem. Soc., 218

[2] A. Navrotsky, Private communication.

K 12

PHOTON CASCADE EMISSION OF Pr^{3+} IN OXIDE HOSTS

William Beers[1] and Alok Srivastava[2]
[1]GE Lighting, NELA Park, Cleveland, OH 44112
[2]GE Corporate Research and Development Center, 1 Research Circle, Niskayuna, NY 12309

1 INTRODUCTION

The Photon Cascade Emission (PCE) of Pr^{3+} generates two photons per ultraviolet (UV) photon absorbed [1]. A quantum yield of 140% for 185nm excitation was achieved at room temperature in Pr^{3+} activated YF_3. Additional fluoride host lattices that exhibit PCE have been identified [2]. Since fluoride hosts are unstable in conventional mercury discharges, we focused our work on obtaining PCE in oxides.

2 TECHNICAL DETAILS

2.1 Background

The photon cascade emission process (PCE) can be understood on the basis of the energy level scheme of the Pr^{3+} ion (Fig. 1). If the lowest energy component of the Pr^{3+} 4f5d configuration is located above the $^1S_0$ state, then nonradiative transitions may populate the $^1S_0$ state following $4f^2$ ($^3H_4$) $\rightarrow$ 4f5d excitation. The first photon emitted corresponds to the $^1S_0 \rightarrow {}^1I_6$ transition. Following the population of the $^3P_0$ state through $^1I_6 \rightarrow {}^3P_0$ nonradiative decay, radiative decay of the $^3P_0$ level results in the emission of the second photon. If the 4f5d configuration is not isolated from the $^1S_0$ state, excitation into the 4f5d state results in broad-band ultraviolet luminescence due to the 4f5d $\rightarrow 4f^2$ transitions and PCE is not observed. In general, weak crystal field results in the high-energy position of the Pr^{3+} 4f5d configuration. In our search for appropriate oxide hosts, we considered a point charge model of oxygen atoms around the activator and assumed any covalency contributions to the crystal field were negligible second order effects. In a point charge model, the magnitude of the crystal field parameter falls off as $1/R^5$ (R is the average Pr^{3+} - O^{2-} distance). Therefore, longer Pr^{3+} - O^{2-} distance are desirable. We postulated that if the coordination number around Pr^{3+} was 10 or greater, that the 5d crystal field splitting would be sufficiently small to generate PCE.

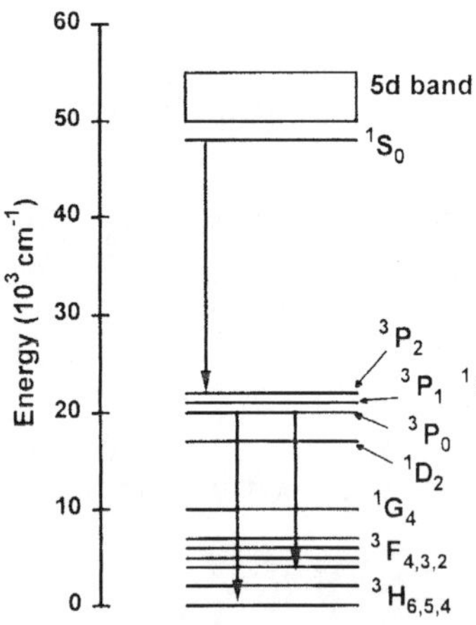

Fig 1. Pr^{3+} energy level diagram

2.2 $Sr_{0.99}Pr_{0.01}Al_{11.99}Mg_{0.01}O_{19}$

The structure of $SrAl_{12}O_{19}$ is magnetoplumbite. The Pr^{3+} can occupy a Sr^{2+} site and have the charge compensated by a Mg^{2+} on the Al^{3+} site. This places 6 oxygens at a distance of 2.75 Å and an additional 6 oxygens at a distance of 2.785 Å. [3] The emission spectra of $Sr(Mg,Al)_{12}O_{19}:Pr^{3+}$ is compared with $YF_3:Pr^{3+}$ in Figure 2. The simultaneous occurrence of the 405nm ($^1S_0 \rightarrow {}^1I_6$) and 485nm ($^3P_0 \rightarrow {}^3H_4$) radiation confirms the PCE. Transitions to the $^1G_4$ produce a UV photon as well as an IR photon. Since in $SrAl_{12}O_{19}$ the $^1S_0 \rightarrow {}^1G_4$ dominates while in YF_3 the $^1S_0 \rightarrow {}^1I_6$ is dominating, the overall visible quantum efficiency is lower in $SrAl_{12}O_{19}$ than YF_3. The conditions for maximizing the $^1S_0 \rightarrow {}^1I_6$ transition probability are symmetry related and can be examined by the Judd-Ofelt theory [1].

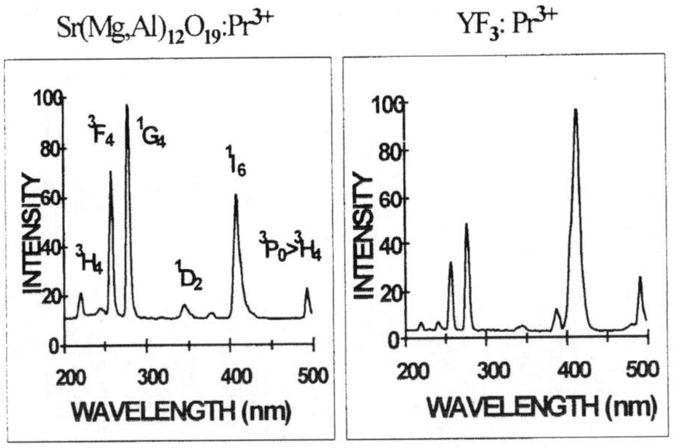

Fig 2. Emission spectra 185nm excitation

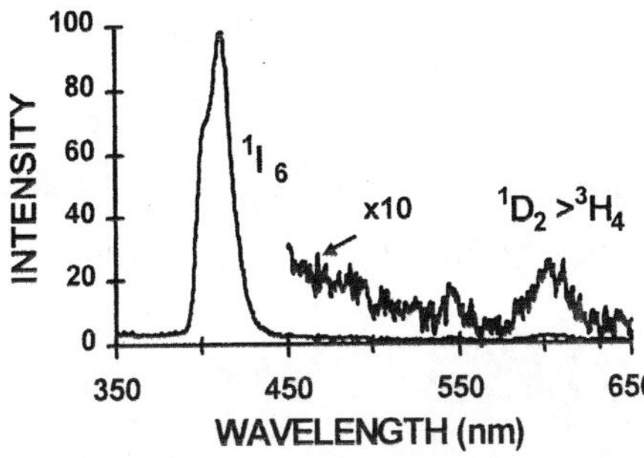

Fig 3. Emission spectra of $La_{0.99}Pr_{0.01}MgB_5O_{10}$

2.3 $La_{0.99}Pr_{0.01}MgB_5O_{10}$

The crystal structure of $LaMgB_5O_{10}$ contains edge-sharing LaO_{10} polyhedra that form linear zigzag chains along the b-axis [4]. The average La^{3+} - O^{2-} distance is 2.61 Å. The emission at 410nm (Fig 3) is indicative of photon cascade emission, however we do not observe the $^3P_0 \rightarrow$ $^3H_4$ transition. Upon careful examination, the $^1D_2 \rightarrow {}^3H_4$ transition is observed confirming that photon cascading, albeit inefficient, is occurring. Therefore, the $^3P_0$ state must be decaying nonradiatively to the $^1D_2$. We attribute this to multiphonon relaxation [5].

3 CONCLUSIONS

Photon cascade emission from Pr^{3+} is possible in oxide hosts as long as the crystal field affects are small. Although this condition is necessary, efficient PCE also requires correct symmetry and low vibrational frequencies. Details of the investigation are published [5].

4 ACKNOWLEDGEMENTS

This work received partial financial support from the NIST ATP program (NIST 70 NANB 3H1372)

5 REFERENCES

[1] W. W. Piper, J. A. DeLuca, and F. S. Ham, J. Lum., **8**, 344 (1974).

[2] J. L. Sommerdijk, A. Bril, and A. W. de Jager, *ibid.*, **8**, 341 (1974); J. L. Sommerdijk, A. Bril, and A. W. de Jager, *ibid.*, **9**, 288 (1974).

[3] A. J. Lindop, C. Matthews and D. W. Goodwin, Acta Cryst. B, **31**, 2940 (1975).

[4] B. Saubat, C. Fouassier, P. Hagenmuller, and J. C. Bourcet, Mater. Res. Bull., **16**, 193 (1981).

[5] A. M. Srivastava and W. W. Beers, J. Lum., **71**, 285 (1997); A. M. Srivastava, D.A. Doughty, and W. W. Beers, J. Electrochem. Soc., **143**, 4113 (1996); A. M. Srivastava, D.A. Doughty, and W. W. Beers, J. Electrochem. Soc., **144**, L190 (1997).

K 13

Quartz Corrosion Reactions in the Discharge Vessel of Metal Halide Lamps

T. Karwath[1], W. van Erk[2], and K. Hilpert[1]

[1]Institute for Materials in Energy Systems, Research Centre Jülich (KFA), P.O. Box 1913, D-52425 Jülich, Germany

[2]Philips Lighting B.V., P.O. Box 80020, 5600 JM Eindhoven, The Netherlands

1 INTRODUCTION

Metal halide lamps contain a salt melt essentially made of metal iodides and/or bromides which in part vaporise under operating conditions. The salt melt is present as condensed phase at the cold spot of the inner wall of the discharge vessel. Important constituents of the melt are the rare earth metal halides LnX_3 (X = Br, I) such as DyX_3, HoX_3, and TmX_3 because rare earth metals are efficient radiators in the visible range of the emission spectrum. In metal halide lamps with a quartz glass vessel typical temperatures for the coldest spot in the vessel are 1000 to 1300 K. During prolonged burning the quartz glass wall of the discharge vessel becomes opaque due to a corrosion reaction caused by the filling. Yamane and Suenaga [1] have shown that in DyI_3-TlI and NaI-ScI_3 lamps with high wall loadings SiO_2 is transported from hotter to colder sites on the wall.

A mechanism for this corrosion process has been proposed by van Erk and Rietveld [2,3]. They claimed that a transport mechanism involving silicon oxy-iodides might be the most effective one. This statement was based on thermodynamic calculations using estimated data for SiOI and $SiOI_2$, compounds the existence of which not had been proven. The aim of this study is to find experimental evidence for the proposed model.

2 TECHNICAL DETAILS

2.1 Experimental set-up

Annealing experiments with evacuated sealed quartz ampoules containing TmI_3 or $TmBr_3$ as well as vaporisation studies using Knudsen effusion mass spectrometry were carried out at temperatures between 800 and 1700 K in order to check the proposed corrosion mechanisms and to get a detailed insight into the chemical interactions with the quartz at the inner surface of the discharge vessel.

One and two compartment Knudsen cells as well as Knudsen cells with a gas inlet system were used in order to simulate the conditions in the discharge vessel. The corrosion of the quartz glass after the annealing experiments and the vaporisation measurements was studied by X Ray diffraction, XRD, and Scanning Electron Microscopy.

2.2 Results and discussion

The gaseous species formed by the reaction of quartz glass with solid TmI_3 or $TmBr_3$ as well as the vapours of TmI_3 or $TmBr_3$ and SiI_4 or $SiBr_4$ were identified. Interesting new results on the corrosion reactions and the chemical vapour transport of SiO_2 were obtained. The vapour species SiOI, $SiOI_2$, and $SiOBr_2$, predicted in Refs. [2, 3] for example were identified for the first time by the vaporisation studies.

Furthermore thermodynamic data were deduced for the gaseous species and incorporated into a data bank. Subsequent thermochemical calculations have proven that quartz transport in the presence of primarily formed SiI_4 up to about 900 K simultaneously takes place via the reactions

$$SiO_2 + SiI_4(g) \leftrightarrow 2\,SiOI_2(g) \quad (1)$$
$$SiO_2 + SiI_4(g) \leftrightarrow 2\,SiO(g) + 4\,I(g) \quad (2),$$

whereas at higher temperatures it is dominated by the reaction following eqn. (2). In the presence of $SiBr_4$, however, the transport of quartz is governed at all temperatures by the reaction

$$SiO_2 + SiBr_4(g)\,) \leftrightarrow 2\,SiOBr_2(g) \quad (3).$$

The XRD measurements showed the formation of thulium silicate compounds as reaction products and indicate the formation of a solid solution at the surface of the quartz glass.

3 CONCLUSIONS

Knudsen effusion mass spectrometry using one and two compartment cells as well as a cell with a gas inlet system have been successfully used to study the various corrosion reactions which occur in metal halide lamps with a quartz glass wall.

The vapour species SiOI, $SiOI_2$, and $SiOBr_2$ have been clearly identified in the gas phase. Thermochemical calculations were carried out to get more insight into the mechanisms of quartz transport in the presence of either SiI_4 or $SiBr_4$.

4 REFERENCES

[1] M. Yamane and M. Suenaga, J. Appl. Phys. **54,** 107-111 (1983).

[2] W. van Erk and T. Rietveld, Proc. Sympos. Sci. Tech. High Temp. Light Sources 85-2, 57-71 (1985).

[3] W. van Erk and T. Rietveld, Philips J. Res. **42,** 102-118 (1987).

K 14

CATHODE FLUORESCENCE AS AN INTRINSIC PROBE OF CATHODE PROCESSING AND PERFORMANCE

A.Buckley, S.Dejmanee, G.J.Wilson and R.Devonshire

High Temperature Science Laboratories, Chemistry Department,
University of Sheffield, Sheffield, S3 7HF, UK.

As part of a research programme on lamp and vacuum tube cathode science and technology we have been developing a range of diagnostics for cathode processing and performance. One novel approach we use for investigating triple oxide cathodes involves doping the cathode matrix with a fluorescent rare earth ion and monitoring the fluorescence characteristics (*spectrum; lifetime; and intensity*) at various stages of cathode processing and cathode operation. This *in situ* technique is essentially monitoring the local electronic environments of fluorescent ions located at or near the surface of the cathode. The motivation for the study was twofold: firstly, to evaluate a new spatially resolving cathode surface diagnostic and, secondly, to relate the results to measurements of local barrier height (*work function*) using an STM modulation technique developed in the HTSL.

Fluorescent cathode materials are prepared by co-precipitation of a carbonate mixture from a solution of barium, strontium, calcium and rare earth nitrates. The concentration of the rare earth component has been varied in the range down to 0.005% by mass. The dried co-precipitated material is strongly fluorescent.

In the experiments reported here a Tb^{3+} doped material is applied to 10V, triple coiled D_2 lamp filaments as a suspension either in ethanol or in organic binder materials. When dried it is converted to an oxide matrix using a range of heating regimes. The characteristics of the Tb^{3+} fluorescence band at 543nm are followed throughout the processing and at selected cathode operating times. The Tb^{3+} fluorescence is excited by modulating the uv output of an Ar^{+} laser and is detected with a photomultiplier tube after dichroic filters have selected the desired band. The signal is recovered with a EG&G 7260 lock-in amplifier. The emission lifetime is obtained by scanning the excitation modulation frequency whilst recording the phase and amplitude of the signal.

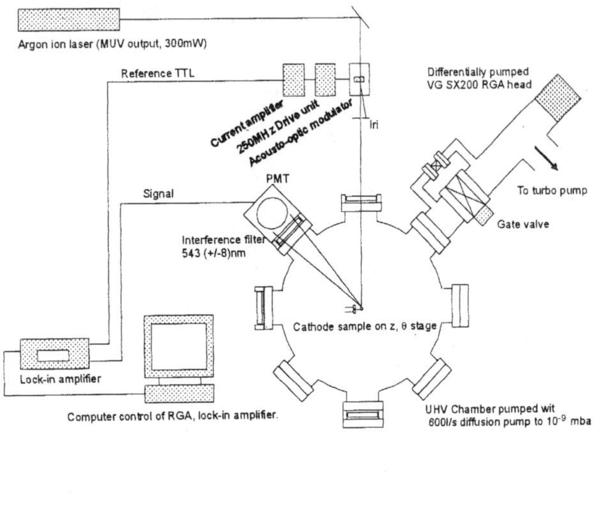

Figure 1.
Processing chamber and diagnostics
(Gas analysis can be performed during processing and correlated with fluorescence lifetime measurements)

The experimental configuration is shown in Figure 1. The use of a UHV grade vacuum chamber makes it possible to monitor the pressure and composition of gases evolved during the carbonate to oxide conversion in addition to the measurement of cathode temperature and fluorescence.

Figure 2 below shows the results of a study of the cathode breakdown process for a 0.02% Tb^{3+} doped triple oxide carbonate mixture which was heated to a final temperature of 1200K in a stepped regime. In the intervals between each of the heating steps the fluorescence lifetime was measured. The evolution of CO_2 and the fluorescence lifetime, τ, are both shown as a function of process time. The fluorescence technique reveals a very clear processing end-point. The lifetime falls dramatically from 100s μs to a value near 10 μs as the conversion process nears completion. The reduction in lifetime may indicate the formation of a more dense energy level structure in the matrix leading to an increase in non-radiative decay channels and an effective quenching of the fluorescence.

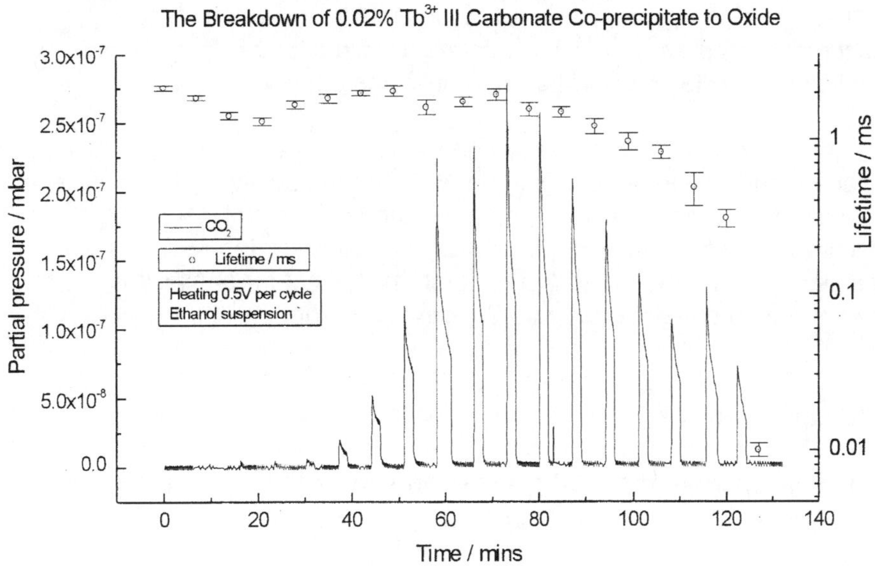

Figure 2. Monitoring of a cathode breakdown process.

We have observed also that the fluorescence signal which characterises the processed material survives cathode operation in a discharge. This observation has led us to follow the emission as a function of lamp life and also to perform experiments which seek to correlate cathode degradation mechanisms with the fluorescence parameters. The use of laser excitation makes it possible for us to perform spatially resolved measurements over the cathode's surface, again as a function both of the processing and of operation as an electrode. The work is currently being extended to include other rare earth ions.

K 15

THE APPLICATION OF ATOMIC FORCE MICROSCOPY TO THE STUDY OD WALL REACTIONS IN HALOGEN LAMPS AND METAL HALIDE DISCHARGE LAMPS

S.P. Oliver, J.C.McGourlay and R. Devonshire

High Temperature Science Laboratories, Chemistry Department, University of Sheffield, Sheffield, S3 7HF, UK.

The assessment of the changing envelope topography in an operating lamp is a useful indicator of the processes involved in wall/atmosphere interactions. The use of AFM (Atomic Force Microscopy) has allowed us to make a detailed study of this interface. Use of the Burleigh ARIS 3300 Personal AFM yields resolution of at least 1nm in the x and y directions and 0.1nm in the z direction. With resolution at this level extremely subtle changes in wall topography can be detected. Wall degradation can be monitored after very short operation times (minutes) using AFM, thus offering significant reductions in the time necessary to investigate wall reactions compared to other possible diagnostic methods such as light transmission which require lamp operation of thousands of hours.

Lamps operated for various lengths of time have been studied in detail. The contrast between wall topography in lamp regions of differing temperature, i.e. top, side and base, is studied in each case. Tungsten halogen and metal halide discharge (MHD) lamps have been studied in this way. Considering the fundamental differences between these two types of lamps it is not surprising that we have observed contrasting envelope topography development.

Gross wall erosion and silica transport is observed in MHD lamps. Wall silica is removed from the hot upper regions (1250K) and deposited in the cooler base regions (900K) [1,2]. The hot region etching is localised in pits, typically 50-400nm deep and up to 5μm across after 100 hours operation. The base regions show large raised deposition features with typical heights of 400-600nm following 100 hours operation.

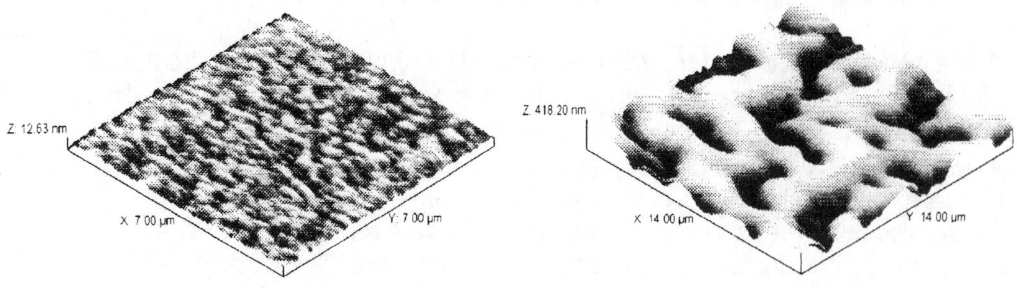

AFM images of inner arc tube wall; unoperated lamp (left), upper region of lamp operated for 100 hours (right).

The tungsten halogen lamps studied have a lower wall temperature range (<800K) and a much simpler chemical dosant makeup (comprising HBr plus inert gas) than metal halide discharge lamps. Despite this generally less reactive regime, the AFM technique has detected subtle changes in the lamps' interior walls after operation. These changes are characteristic of roughening and etching processes similar in some respects to those seen in MHD lamps though at a much reduced magnitude (pit depth 10nm after a few hundred hours). The AFM is also proving useful for the imaging of non-glassy wall deposits. The filament regenerative cycle by the action of HBr in tungsten halogen lamps is well documented [3,4]. Through observation of the amount of tungsten deposited on the lamp wall and the positioning within the lamp of this deposit, it is possible to monitor the efficiency of the regenerative cycle across the whole of the lamp and throughout its operational lifetime.

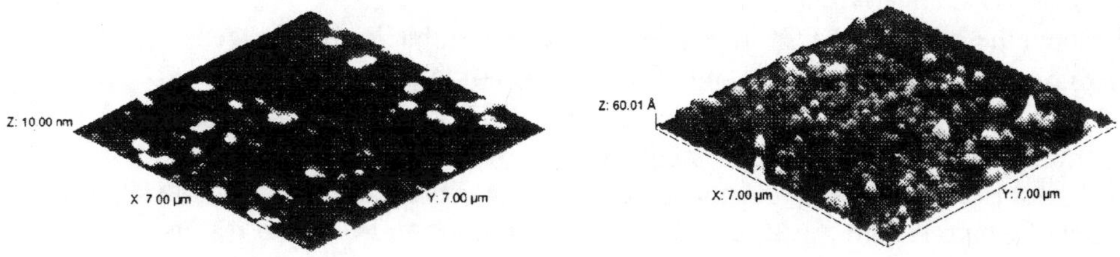

AFM images of inner Tungsten-halogen lamp wall, left, unoperated lamp, right, upper region of lamp operated for 280 hours. Below are the surface height histograms for the two images.

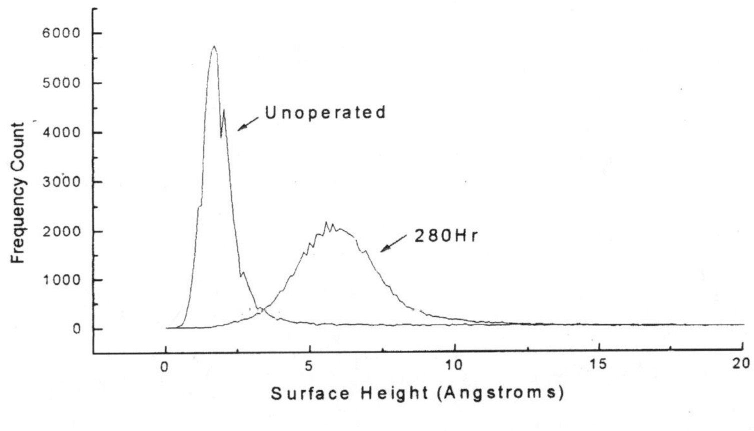

References

[1] M. Yamane and M. Suenaga J. App. Phys., 54(1983)107.
[2] W. Van Erk and T. Rietveld High Temp Lamp Chem I., 1(1985)57.
[3] G.M. Neumann J. Less Common Metals, 35(1974)51.
[4] F.J. Harvey Metall. Trans. A, 7A(1976)1167.

K 16

AN OXYGEN DISPENSER FOR IMPROVING THE LUMINOUS FLUX OF METAL HALIDE LAMPS

Claudio Boffito[1] and Ben J. de Maagt[2]
[1] SAES Getters S.p.A, Viale Italia 77, 20020 Lainate (MI), Italy
[2] Philips Lighting B.V., P.O. Box 80020, Eindhoven, The Netherlands

1. INTRODUCTION

The luminous flux of many types of HID lamps may be reduced by carbon deposits on the outer surface of the arc tube as a result of cracking of organic contaminants present in the outer envelope. The paper presents the characteristics of a silver-oxide (Ag_2O) oxygen dispensing device meant to remove the carbon thus improving the luminous flux. The dispenser was tested in metal halide lamps of the HPI type.

2. MATERIAL CHARACTERISATION AND CONFIGURATION

The source of oxygen has been studied taking into account the specific constraints posed by the application, which can be summarised as follows:

- ability to withstand the baking-exhaust process with no or negligible loss of oxygen,
- quick release of oxygen, after tipping-off, in the early period of lamp operation,
- ability to release a "calibrated" dope of oxygen (i.e. sufficient to convert carbonaceous contaminants, avoiding any deleterious oxidising effect on other lamp components and getter).

The investigation has been conducted by considering, as possible candidates, various metal oxides with suitable O_2 equilibrium pressure and dissociation kinetics. Based on results of tests, carried out by simulating possible operating conditions, silver oxide resulted to be the most convenient choice, mainly thanks to the quick release of oxygen upon heating.

The oxygen dispenser has been configured as a metal container, holding a small dose of silver oxide. A sketch, with dimensions, is shown in Fig. 1.

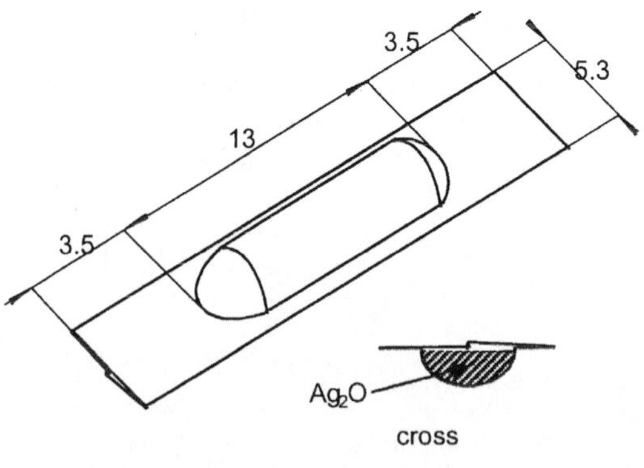

Fig. 1: Sketch of oxygen dispenser (dimensions in mm)

The average filling weight is of about 38 mg, capable to release about 2 Ncm^3 of oxygen, which can, in turn, convert up to 1 mg of carbon. Yield characteristics of oxygen dispensers have been investigated by heating test samples in an evacuated sealed system up to 400°C, and recording the pressure raise taking place. The samples were preliminarily submitted to a thermal treatment up to 200 °C, for about 20 min, under pumping, to simulate possible degassing processes and check thermal stability under these conditions.

A typical yield curve obtained with such a treatment is plotted in Fig. 2, where the amount of gas given off and sample temperature are shown as a function of time (preliminary baking at 200 °C is omitted).

As one can see, the total amount of gas released (about 2 Ncm^3) by raising the temperature up to 400 °C practically corresponds to the theoretical (stoichiometric) value, confirming that no or negligible loss of oxygen has occurred during the preliminary baking treatment at 200 °C. The yield curve also clearly points out the high rate of release of oxygen.

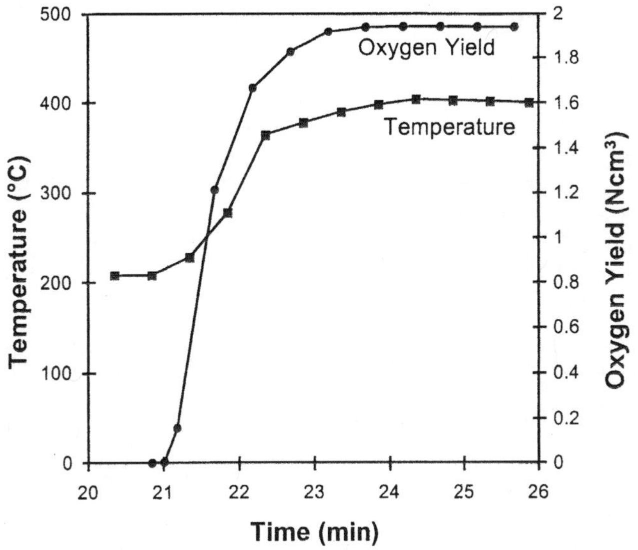

Fig. 2: Typical yield curve for oxygen dispensers

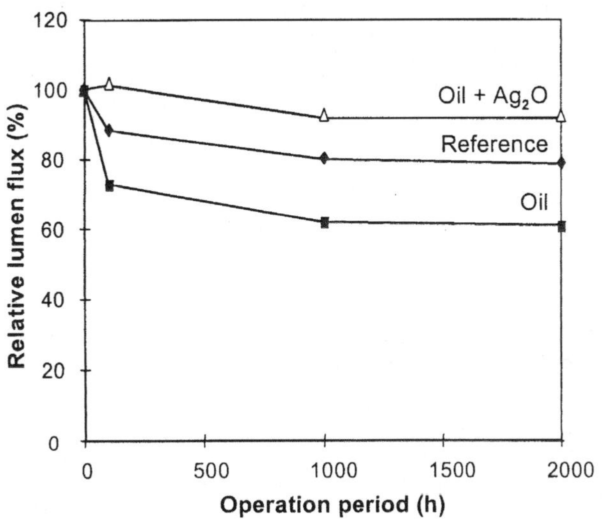

Fig. 3: Demonstration of the effect of an oxygen dispenser in HPI-T 250W lamps

3. APPLICATION IN HPI LAMPS

The oxygen dispenser was tested in the outer envelope of metal halide lamps of the type HPI-T 250W. The device was mounted in the outer envelope containing Ne/N_2 as a filling gas at a position where a temperature of 350 °C is reached during operation. No oxygen is released during the lamp manufacturing process because the process temperatures remain below 350°C.

Once being released the oxygen reacts with the carbon deposit on the hot arc tube surface to form carbondioxide and carbonmonoxide. These gases are harmless for HPI lamps and are captured anyway by the applied PH/DF Zr_2Ni getter upon further lamp operation.

Fig. 3 shows the evolution of the relative luminous flux of a set of HPI-T 250W lamps up to 2000 hours of operation. "Reference" lamps were compared with lamps deliberately contaminated with some diluted oil leading to about 20µg of carbon on the arc tube wall. The latter situation corresponds to a "huge calamity".

The results show that the oxygen dispenser is able of maintaining the luminous flux at a high level in the presence of large amounts of organics. An average luminous flux improvement of 10% and a more robust luminous flux level compared to lamps manufactured with the former standard production can be realised.

4. CONCLUSIONS

Application of a silver oxide (Ag_2O) oxygen dispenser is a simple and reliable way to improve the luminous flux of HPI lamps considerably.

L

Diagnostics

L 01

CHARACTERIZATION OF A MERCURY DISCHARGE IN NON-LTE CONDITIONS

Hatem Elloumi[1], Eckhard Kindel[2], Conrad Schimke[2], Georges Zissis[1] and Heinz Schöpp[2]
[1] Centre de Physique des Plasmas et de leurs Applications de Toulouse
Université Paul Sabatier, 118, Rte de Narbonne, 31062 Tlse Cdx, France
[2] Institut für Niedertemperatur-Plasmaphysik
17489, Robert-Blum-Str. 8-10, Greifswald, Germany

1 INTRODUCTION

In the past years, a lot of measurements in mercury arcs or metal halide discharges reveal the existence of severe departure from LTE. This deviation from equilibrium leads to complicated distributions such that the plasma cannot be correctly described by a single temperature. Indeed, the electron temperature is not equal to the temperature of the heavy particles. In this case, the only practical way of investigation is the direct measurement of the atoms density in their ground state, which normally accounts for the major part of the atomic density. Number densities of excited atoms can in principle be determined using passive emission spectroscopy.

The main part of this paper deals with the measurements of the ground state density and their use for the determination of deviation from equilibrium. Accurate and independent measurements of a such parameter are very important for a good understanding of non-equilibrium phenomena.

2 GROUND STATE DENSITY-DIAGNOSTIC TECHNIQUES

The Experiments are carried out on a set of cylindrical lamps of same internal diameter (18.2 mm) and inter-electrodes distances (72 mm). Our discharge lamps operate vertically through a current invertor (frequency equal to 500 Hz). A resistive ballast limits the arc current.

2.1 Hook method

Measurements of the anomalous dispersion in the vicinity of absorption lines of atoms has been used for the determination of oscillator strength or concentrations of the lower levels. This method, introduced by Rozhdestvenski is known as the hook method.

At known value of f_{ik} the density N_i of the lower level can be determined by measurement of the hook separation Δ using the formula below:

$$N_i = \frac{\pi K \Delta^2}{r_0 l f_{ik} \lambda_0^3} \qquad (1)$$

Where Δ is the wavelength separation between the hook extrema (Fig. 1), l the absorption length of the plasma and K the hook constant (nearly the order of interference). The experimental arrangement used for this method consists of a Mach-Zehnder interferometer, a dye laser as light source and an image processing unit.

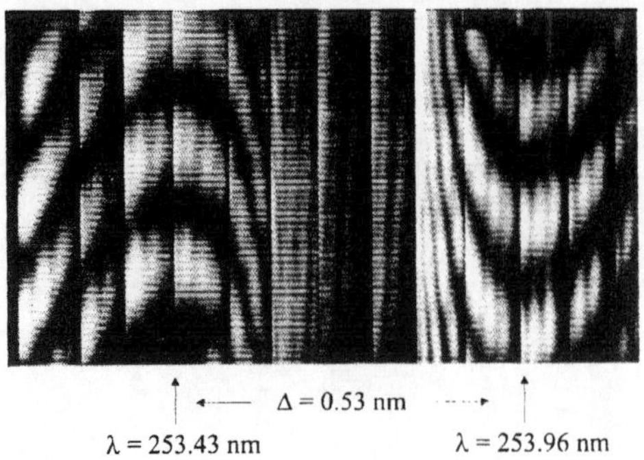

Fig. 1: Interfernce pattern

2.2. $\Delta\lambda_R$ method

In a recent paper [1] we have proposed a method allowing the determination of the ground state density in a high pressure mercury discharge from the broadening of the red wing of the 253.7 nm self reversed resonance line. We refer to this method as the $\Delta\lambda_R$ method. This method is based on the properties of the strongly self reversed resonance line 253.7 nm. The ground state density can be calculated using the formula below :

$$\alpha(\lambda_R - \lambda_0)^{3/2} = \int_{-x_0}^{x_0} n_{Hg}^{\ 2}(x)\,dx \qquad (2)$$

where λ_0 is the central wavelength corresponding to the 6^3P_1-6^1S_0 transition and α is a constant characterising the resonance line.

The experimental arrangement used for this method consists of a high resolution monochromator (THR Jobin-Yvon) with a dispersion about 0.13 nm/mm in double pass configuration.

3 RESULTS AND DISCUSSIONS

Two types of experiments were performed to illustrate the non applicability of the LTE assumptions in the medium pressure (about 0.3 bar). First a check has been made in previous investigations on whether the ground state density obtained with the $\Delta\lambda_R$ and hook methods are consistent with measurements using optically thin lines and assuming LTE in high pressure discharges (about 5 bar). For this purpose we had compared the results obtained using the optically thin line 577 nm with Abel inversion and those obtained by these two methods. A convenient agreement was found (see Refs 1 and 2).

On the other hand, Abel inversion was applied in order to obtain absolute radial ground state density. Figure 2 shows a comparison of the ground state density using hook and $\Delta\lambda_R$ methods and those assuming LTE, while keeping the arc current and the geometry constant. For the corresponding measurements that are described in the following, the density was held with less than 10% uncertainty.

These distributions illustrate, first, the fair agreement between the results obtained from the hook and $\Delta\lambda_R$ methods and second the discrepancies that exist when comparing these series of measurements with the LTE measurements, especially for the low loaded discharge (4 mg Hg). Similar results are obtained when comparing the mean values of these densities

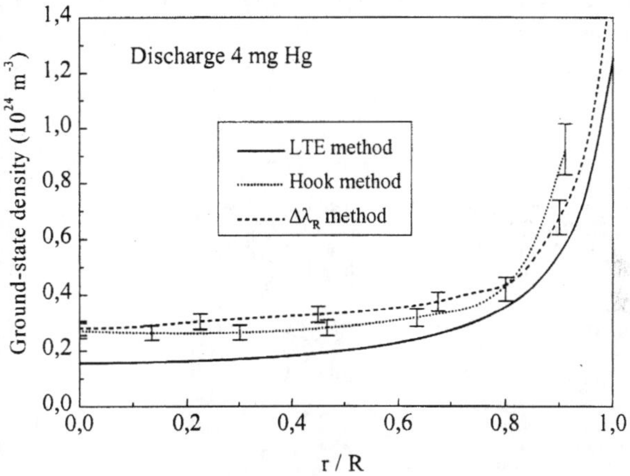

Fig. 2: The measured radial distributions of the ground-state densities;

4 CONCLUSION

In this study, we have described two methods for the determination of the ground state density. These densities thus obtained will be used for the characterisation of the mercury discharges not very far from local thermodynamic equilibrium.

5 REFERENCES

[1] H. Elloumi et al. J. Phys. D: Appl. Phys. **30** (1997) 1893-1898
[2] E. Kindel et al. J. Phys. D: Appl. Phys. **31**, (1998)

L 02

ACOUSTIC RESONANCE PHENOMENA IN HIGH PRESSURE DISCHARGE LAMPS

Sylvain Epron, Alin Petru Cojocaru, Michel Aubès, Jean-Jacques Damelincourt
CPAT, 118 route de Narbonne, 31062 Toulouse Cedex, France

1 INTRODUCTION

A discharge lamp is made by a cylindrical tube filled with a gaseous mixture. For classical applications of lighting (urban and industrial lighting) or for the applications used on the vehicles (car headlights, landing light), it is often more advantageous to supply those lamps in high frequency (20 - 100 kHz). The experience shows that, under certain conditions and for specific frequencies, may appear some undesirable acoustic resonance phenomena.

2 GENERAL RESULTS ON ACOUSTIC RESONANCE

2.1 Undesirable effects

These perturbations, who may appear above hundreds of hertz until very high frequencies, disturb the discharge channel and makes the whole structure to vibrate (see figure below). An accelerated ageing of the electrodes [1], extinction [2] or even destruction of the lamp can be caused by this arc sneaking. The range of frequency covered by the phenomena arise specific problems which must be studied.

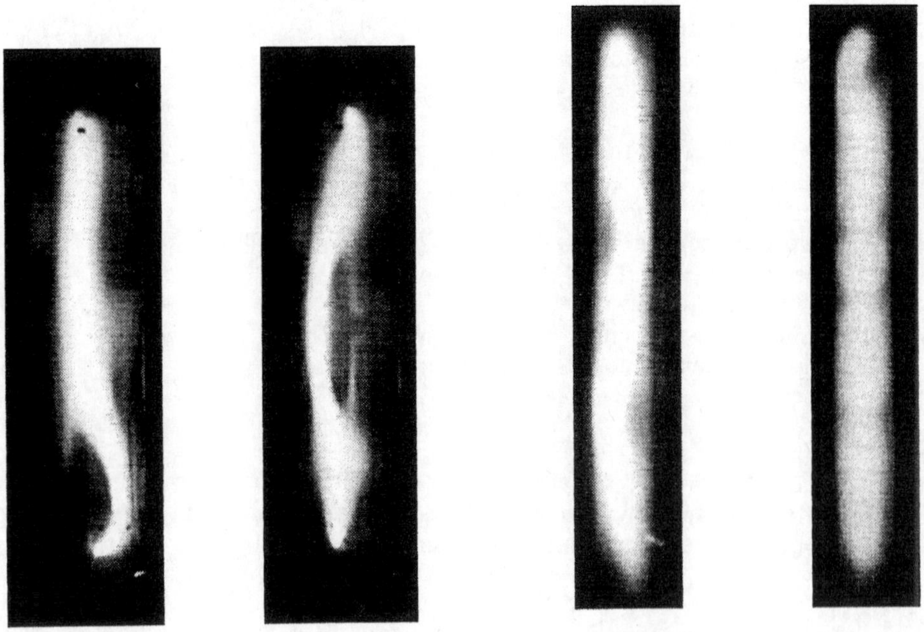

3000 Hz 7900 Hz 18600 Hz 40000 Hz

Appearance of the discharge path at the indicated high frequency
for the high pressure mercury lamp (left); for the high pressure sodium lamp (right).

2.2 Theoretical aspect

We will examine first, from a theoretical point of view, making the analogy with the phenomena observed in a resonant pipe, how can be generated the acoustical waves in a plasma. Solving the equation of the sound wave propagation for the cylindrical cavity, we give the expression for the resonance frequencies and the associated pressure distribution. Those distributions will allow to explain the particular deformation of the discharge.

2.3 Experimental results and discussions

In the second part of our presentation we show the experimental results obtained for the various lamps supplied by a current inverter working at variable frequency and duty cycle. This converter ensures the operation with alternative square wave current. The frequency may be varied between 20 - 50 kHz. This supply system allows to operate also under variable duty cycles. This is the final aspect is used for the study of excitation conditions of acoustic instabilities. We show experimentally that two parameters change significantly the value of resonance frequency:

1) the geometry of discharge tube: the resonance frequencies depends linearly of 1/2R and/or 1/L, in other words when the diameter 2R or the length L decreases, the frequency increases, making a wider range of utilisation without inconvenient instabilities.

2) the temperature of the discharge: the resonance frequencies depends equally of the average temperature of the discharge. From the moment when the lamp is switched on, those frequencies are "shifted" progressively towards high values as the average temperature increases towards his permanent value (reached on about 6 - 7 minutes).

We have also shown that the excitation conditions of acoustic oscillations depends:

1) of the gas density: the excitation of acoustic oscillations is strongly dependent of the quantity of mercury or sodium introduced in the discharge tubes. For instance, for the high pressure mercury lamps, the necessary duty cycle for the excitation of oscillations is bigger when the mass of mercury introduced in discharge tube is high.

2) of the gas pressure: like for the density, higher is the gas pressure, bigger is the necessary duty cycle for the excitation of the acoustic oscillations.

3 CONCLUSION

One part of those experimental results allowed to validate the theoretical expressions obtained, and another part to verify the theoretical predicted deformation of the discharge path. The results as a whole makes now a good support for the modelling of the acoustic phenomena.

4 REFERENCES

[1] R. A. Petr and T. R. Burkes, Appl. Phys. Lett. **36-7**, 536 (1980)

[2] J. K. Buckley, Illum. Engr. **57**, 589 (1962); **58**, 365 (1963)

L 03

MEASURING THE TEMPERATURE DISTRIBUTION AT THERMIONIC CATHODES IN AN ARGON MODEL LAMP

Daniel Nandelstädt, Jens Luhmann, Berthold Michelt and Jürgen Mentel
Department of Electrical Engineering and Information Technology,
Ruhr University Bochum, D-44780 Bochum, Germany,
✆ (+49) 234 700 3483, ✉ dnandel@aeeo.ruhr-uni-bochum.de

1 INTRODUCTION

*This work is part of investigations aiming at a „**Fundamental Characterization of Thermionic Cathodes**"*. While the physical processes in the arc column of gas discharge lamps are by now well understood there are still many unsolved problems related to the arc cathode interaction. In order to investigate thermoemitting cathodes for HID lamps systematically an experimental setup was developed in which a DC arc is operated under conditions typical for lamps. Important discharge parameters as e.g. arc current, gas species, gas pressure and electrode geometry can be varied. Electrical and optical diagnostic techniques allow the complete characterization of the plasma cathode boundary layer, so that theoretical sheath models can be verified. As a result reliable predictions of the behaviour of these cathodes will become possible, which is important for developing and optimizing lamps. As power fluxes in the cathode region are of great importance for the design of lamps this paper focuses on the measurement of the cathode temperature distribution and the power fluxes in the model lamp.

2 EXPERIMENTAL SETUP

2.1 The model lamp

An arc is operated with a DC current between 0.5 and 10 A in a quartz tube of 9 mm in diameter. Previously the chamber is evacuated by an UHV pumping system down to a pressure of 10^{-5} Pa and hereafter filled with the operating gas. The discharge is ignited by a 18 kV voltage pulse. A 90 Ω resistor in series to the discharge is used to limitate the arc current. As shown in fig.1 the electrodes are fixed in a holding device which is cooled by flowing silicon oil. Electrodes with diameters up to 4 mm can be inserted. The cathode bottom immerses into the cooling medium so that it is kept at a known, constant temperature. By stepping motors the electrode gap can be varied during operation up to a maximum of 100 mm. All measurements presented here were performed in argon at pressures ranging from $1{\cdot}10^5$ to $6{\cdot}10^5$ Pa using cylindrical tungsten electrodes of 1.5 mm in diameter.

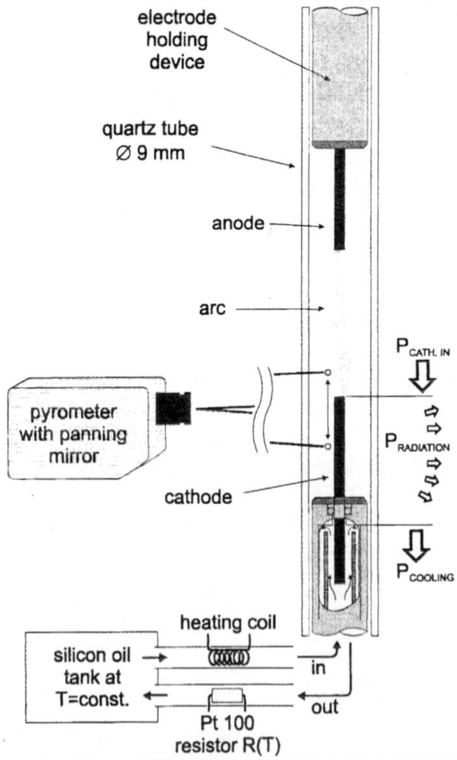

Fig.1: drawing of the model lamp and the power fluxes in the cathode region

2.2 Measuring the cathode temperature

The cathode temperature is measured with a one wavelength pyrometer working in the range of 0.7 to 1.1µm, focusing the cathode in a side on view (see fig.1). It is equipped with a panning mirror. Thereby it is possible to measure temperature distributions along the cathode body.

2.3 The cathode energy balance

Assuming that convection and Joule heating play only a minor role, the cathode power balance can be reduced to the following equation: $P_{CATHODE,IN} = P_{RADIATION} + P_{COOLING}$ (see fig.1). $P_{RADIATION}$ is the power irradiated from the cathode surface. It can be calculated from the temperature distribution. The power output $P_{COOLING}$ from the cathode to the oil is ascertained by applying a calorimetric method. It is based on adjusting the temperature of the outflowing silicon oil without burning arc to the same level as in the case of lamp operation. For this purpose a heating coil in front of the cathode is integrated in the cooling circuit (see fig.1). The power $P_{CATHODE,IN}$ denotes the flux from the arc into the cathode, which complies with the product of arc current and cathode fall. So measuring the power fluxes offers another posibillity to determine the cathode fall additionally to that described in the authors other LS-8 paper „Experimental Determination of the Cathode Fall in a High Pressure Low Current Argon Arc."

3 EXPERIMENTAL RESULTS

Fig.2 shows axial cathode temperature distributions for different cathode lenghts and arc currents. Position zero marks the cathode plasma boundary. Below this position no temperature can be measured by the pyrometer as the focus spot is still in the argon plasma. During all measurements the arc attached to the cathode in the diffuse, spotless mode. Other operating conditions like gas pressure, arc length and electrode cooling were kept at constant values. It can be seen from the measurements, that all temperature profiles are of a linearly decreasing shape. A higher arc current or a longer cathode rod result in an increasing maximum temperature. Comparing $P_{COOLING}$ from measurements with numerical simulations given in the LS-8 paper of *Boetticher et.al.*, on the basis of the measured temperature profiles, approximately 20 W is found in both cases.

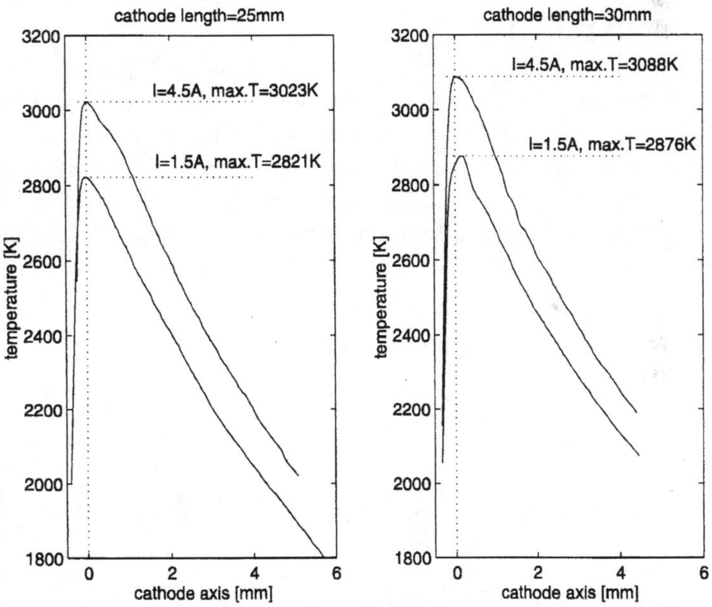

Fig.2: temperature distributions of 1.5 mm Ø tungsten cathodes, $p=2.5\cdot10^5$ Pa, $T_{COOLING}$=20°C, L_{ARC}=10 mm

4 OUTLOOK

In further experiments the influence of gas pressure, cathode diameter, cooling temperature and cathode position on the temperature profile shall be determined and the cathode energy balance shall be exactly verified.

5 ACKNOWLEDGEMENTS

This work is supported by the German Federal Ministry of Education and Research (BMBF). Reference number **13N6895**.

L 04

EXPERIMENTAL DETERMINATION OF THE CATHODE FALL IN A HIGH PRESSURE LOW CURRENT ARGON ARC

Jens Luhmann, Daniel Nandelstädt, Berthold Michelt and Jürgen Mentel
Department of Electrical Engineering and Information Technology,
Ruhr University Bochum, D-44780 Bochum, Germany,
✆ (+49) 234 700 3482, ✉ jluhmann@aeeo.ruhr-uni-bochum.de

1 INTRODUCTION

*This work is part of investigations aiming at a "**Fundamental Characterization of Thermionic Cathodes**"*. The experimental setup for the arc discharge operated in a model lamp and some other results is to be found in the contributed papers submitted to LS-8 "Measuring the Temperature Distribution at Thermionic Cathodes in an Argon Model Lamp".

To verify and further develop models of cathode boundary layer, measured values of the cathode fall are important. In this paper a method to determine the cathode fall from measurements of the plasma potential by a Langmuir probe is introduced. First results are presented.

2 THE EXPERIMENT

It is only possible to measure the cathode fall together with the voltage drop of other parts of the arc. The Langmuir probe is a diagnostic tool for measuring the local plasma potential by analysing the I-U-characteristic of the probe. Figure 1 shows the experimental setup. Two independent probes are used. Both probes are inserted into the plasma, but only one of them is attached to an increasing voltage ramp supply ranging from 0 to 60 V. The probe-current is measured by a combined DC- and AC-current-measurement-system as a function of its potential. In this case the cathode is used as reference potential.

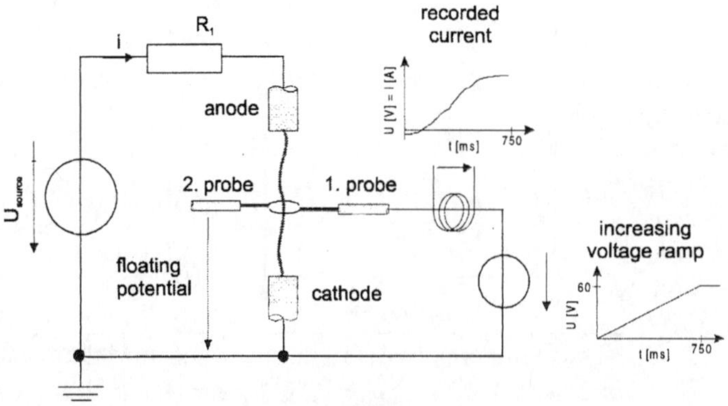

Fig. 1: experimental setup for measuring the local plasma potential by a Langmuir probe

To determine the cathode voltage drop, the plasma potential U_{PP} is taken with different distance $L_{probe\text{-}cathode}$ between probe and cathode. By plotting the plasma potential U_{PP} as a function of the distance $L_{probe\text{-}cathode}$, the cathode voltage drop is received as Y-axial section, see fig. 3, [1 & 2].

2.1 The probes

The probes consisted of cylindrical bended pure tungsten 0.5mm in diameter, fig. 2. Both probes lie in the same layer. By using two half circular probes, it is possible to get a contact between the cold probes and the hot discharge and measure the current voltage characteristic curve for a single probe.

2.2 First results

First measurements were performed with pure tungsten and Y_2O_3 activated lamp electrodes at a pressure of $1.65 \cdot 10^5$ Pa and a current ranging from 2 to 5 amps.

The cathode voltage drop turned out to be nearly independent of the current at this current ranging if the same electrode is used. However, the voltage drop of the Y_2O_3 activated electrode is about 3 volt lower than that of the pure tungsten electrode.

Concerning the Y_2O_3 activated cathode it is noticeable, that the measured voltage drop is about 2 Volt lower for 5 Amps than for 2 to 4 Amps. The reason is a weak dependence of the cathode fall on the current.

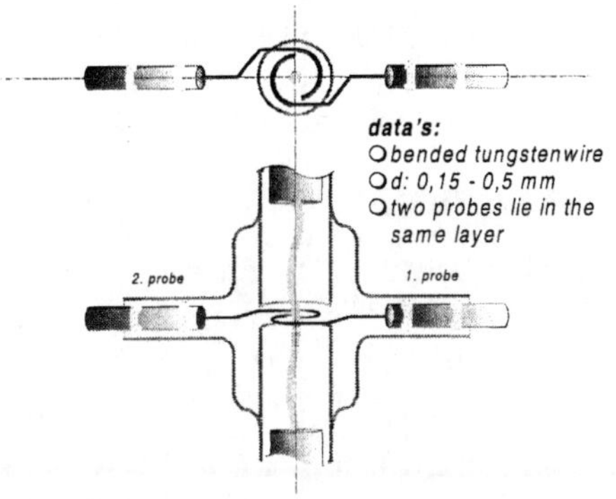

Fig. 2: Setup of the two used half circular probes

3 CONCLUSION

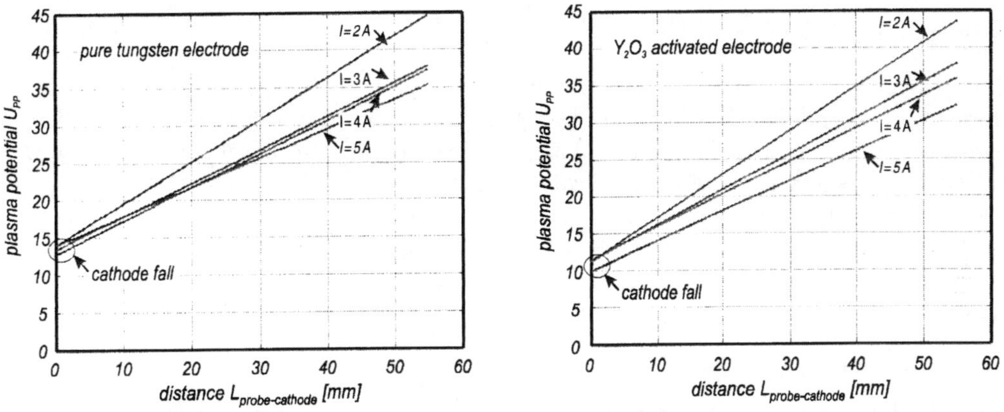

Fig. 3: Comparison of the voltage drop between probe and cathode for a pure tungsten and a Y_2O_3 activated cathode

A comparison of the voltage drop of a pure tungsten and a Y_2O_3 activated electrode measured in a high pressure argon arc has been shown. In future, the dependence of parameters like pressure, material of electrodes, and influence of the different mode of current transfer will be analysed.

4 REFERENCES

[1] J.D. Swift, M.J.R. Schwar; *"Electrical Probes for Plasma Diagnostics"*, Iliffe Books LTD, London, 1970

[2] P.M. Chung, L. Talbot, K.J. Touryan, *"Electric Probes in Stationary and Flowing Plasmas: Theory and Application"*, Springer-Verlag, ISBN 3-540-06800-7, 1975

5 ACKNOWLEDGEMENTS

Financial support supplied by the German Federal Ministry of Education and Research (BMBF) is gratefully acknowledged by the authors. Reference number 13N6895.

L 05

INVESTIGATION OF THE ATTACHMENT OF AN ARGON ARC TO A THERMIONIC TUNGSTEN CATHODE

Daniel Nandelstädt, Jens Luhmann, Berthold Michelt and Jürgen Mentel
Department of Electrical Engineering and Information Technology,
Ruhr University Bochum, D-44780 Bochum, Germany,
✆ (+49) 234 700 3483, ✉ dnandel@aeeo.ruhr-uni-bochum.de

1 INTRODUCTION

*This work is part of investigations aiming at a „**Fundamental Characterization of Thermionic Cathodes**"*. The experimental setup for the arc discharge operated in a model lamp and some other results is to be found in the contributed papers submitted to LS-8 „Measuring the Temperature Distribution at Thermionic Cathodes in an Argon Model Lamp" and „Experimental Determination of the Cathode Fall in a High Pressure Low Current Argon Arc".

From earlier experiments with HID lamps [1] it is known that the attachment of a high pressure arc to a thermionic cathode can operate in different modes. Until now this phenomenon could neither be measured reproducibly nor be explained in a satisfying way. Hence this effect still represents a problem for lamp manufacturers. As different cathode modes can be observed in the presented argon model lamp too, this discharge arrangement with ist numerous variable parameters offers a good chance to better characterize the different modes of attachment, determine their conditions of existence and understand the transition from one to another mode.

2 EXPERIMENTAL SETUP

All measurements presented here were performed in the argon model lamp at pressures ranging from $1 \cdot 10^5$ to $6 \cdot 10^5$ Pa using cylindrical tungsten electrodes of 1.5 mm in diameter. For optical investigations of the cathodic arc attachment a long focal length microscope (QUESTAR QM100) is used. In combination with a CCD camera and a PC framegrabber card this allows taking time resolved pictures of the arc cathode boundary region in a side on view. The cathode body temperature distribution can be measured with a resolution of 0.2 mm as described in one of the authors other LS-8 papers. Arc current and total lamp voltage are captured by a digitizing oscilloscope.

3 EXPERIMENTAL RESULTS

3.1 Different modes of arc cathode attachment

In agreement with former experimental results [1,2] two different thermionic cathode modes can be observed: Thermionic **spot mode** and thermionic **spotless mode**. In fig.1a the spotless mode is shown for the case of a pure tungsten cathode of 1.5mm in diameter and a current of 4A. The arc root attaches to the cathode front in a diffuse and obviously uniform mode. This leads to a nearly constant temperature of the whole cathode front surface which amounts to about 3000K. In fig.1b the same cathode with a lower current of only 3A is pictured. In this case the thermionic spot mode has established so that the arc is constricted in front of the cathode and forms a bright cathode spot of about 200µm in diameter. This mode of operation comes along with a much higher cathode temperature inside the spot which is in the order of magnitude of the

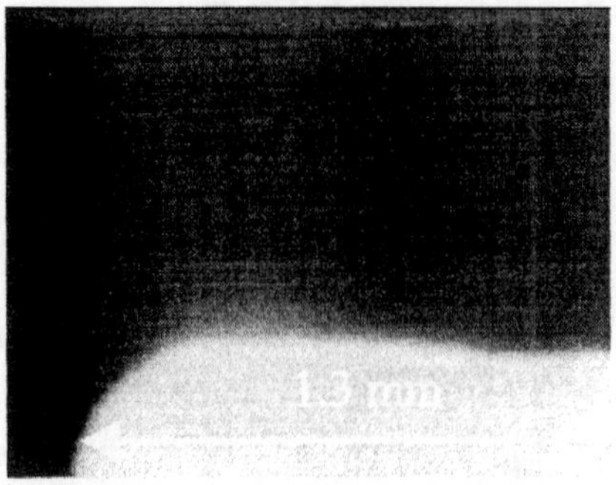

Fig.1a: thermionic spotless mode on a 1.5 mm tungsten cathode at I=4A

melting point of tungsten. However the global cathode temperature outside the spot is explicitly lower than in the former case even if the arc current is increased to the same magnitude. In which mode the discharge is working is determined by the parameters current, pressure, electrode geometry and electrode cooling.

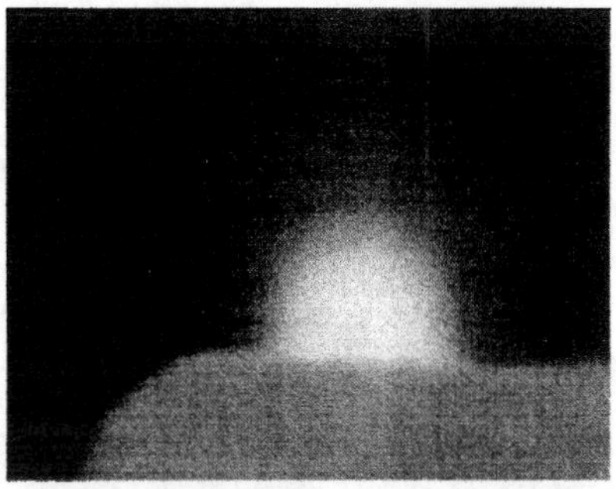

Fig.1b: thermionic spot mode on a 1.5 mm tungsten cathode at I=3A

3.2 Transitions between thermionic cathode modes

It was found that by applying a slowly increasing current ramp (steepness 0.04A/s) to the discharge the system changes at a current I_{C1} from the spot mode to the spotless mode. The transition occurs suddenly without a preceding enlargement of the spot size. Electrical measurements show that this mode jump is accomplied by a small voltage step of +0.5V. By decreasing the current the arc enters the spot mode again and the arc voltage returns to the former lower level. This second transition appears at a critical current I_{C2} which is somewhat lower than I_{C1} so that the current voltage characteristic forms a hysteresis. Repeating this cycle allows the precise determination of individual values for I_{C1} and I_{C2} which are statistically spread over a small current range. In fig.2 the two critical transition currents are plotted as cumulative frequency distributions for 25 cycles. The transition from spot to spotless mode due to an increasing current here occurs at a mean current I_{C1}=3.81 A. While the current is again decreased the mean current for the transition back to the spot mode amounts to I_{C2}=3.47A.

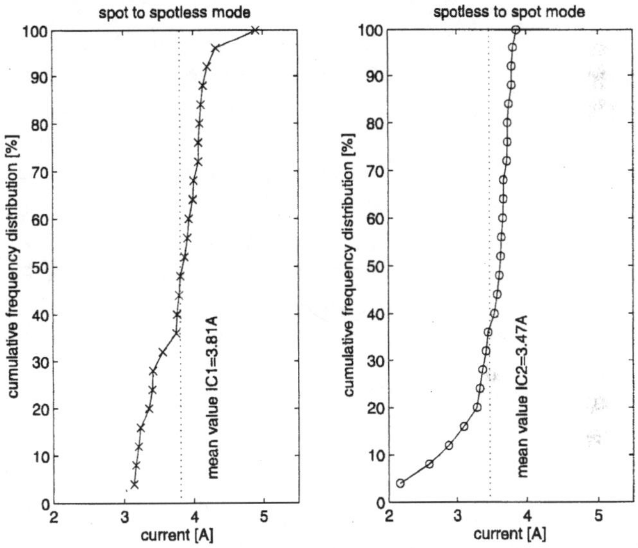

Fig.2: transition currents for 25 cycles at a 25 mm long, 1.5 mm diameter tungsten cathode, p= $3.8 \cdot 10^5$ Pa

4 OUTLOOK

In order to specify the conditions of existence for the different thermionic cathode modes more detailed as in this report further measurements under variation of the parameters current, cathode length and diameter are to be made. Moreover the investigation of real lamp electrodes will show the relevance of these experiments for the development and design of HID lamps.

5 REFERENCES

[1] W.Thouret, W.Weizel and P.Günther, Zs. für Physik **130**, 621 (1951)

[2] S. Anders and A.Anders, IEEE Trans. on Plasma Science **19**, 20 (1991)

6 ACKNOWLEDGEMENTS

This work is supported by the German Federal Ministry of Education and Research (BMBF). Reference number **13N6895**.

L 06

SPECTROSCOPIC DIAGNOSTICS OF A LOW PRESSURE DEUTERIUM DISCHARGE

A.Buckley, G.J.Wilson, and R.Devonshire

High Temperature Science Laboratories, Chemistry Department, University of Sheffield, Sheffield, S3 7HF, UK.

At the LS:7 Symposium we reported both experimental [1] and theoretical / modelling [2] work on the low pressure deuterium (D_2) gas discharge. This work has continued in a number of directions. Here we report on spectroscopic investigations whose purpose is to measure both the atomic (D; and H, see below) and molecular (D_2) gas temperature distributions and electric field distributions in an operating, conventional, D_2 lamp structure. The results are being compared with the predictions of our computer model of the discharge which is now well developed. The techniques used in this present work are line shape analysis of atomic emission spectra and vibrational-rotational coherent anti-Stokes Raman scattering (CARS) of D_2.

Figure 1 : Deuterium Lamp Structure

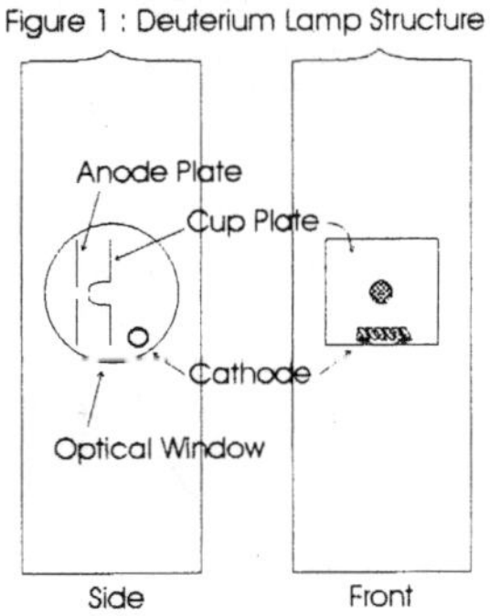

The structure of the D_2 discharge lamp that was the basis of the present studies is shown in Figure 1. Essentially, it is a standard commercial lamp structure with the shielding around both the cathode and anode removed to allow access for the spectroscopic diagnostic measurements. The cylindrical lamp envelope is adapted to take optical windows to give undistorted access both to the discharge axis and to the axis perpendicular to it. The windows remove spatial distortions introduced by the curvature of the cylindrical lamp envelope. The windows made it possible to determine the temperature distributions in the discharge in the cathode, anode and cup regions of this modified lamp structure.

To perform the atomic emission technique an image of the discharge is created on the entrance slit of a Spex 1.25m spectrometer with 1 to 1 magnification. At each spatial location (achieved by moving the lamp on precision translation stages) the emission profiles of the first five lines in the Balmer series are scanned with better than 0.1 cm^{-1} resolution. The lines shapes are then simulated using an HTSL Systems Ltd. spectral fitting program to obtain both the best fit temperature and the local electric field which best describe the spatially averaged emission. The Balmer α line is the main basis of the temperature mapping technique as it has the smallest Stark broadening contribution of the Balmer series of lines; the most accurate values are extracted by extrapolation of parameters obtained for a series of Balmer lines. The transverse spatial resolution of the detection system is defined in one dimension by the monochromator entrance slit width (20 μm and below), and in the other, perpendicular, dimension by an entrance slit shutter (1 mm). The depth of the image is not well defined in this configuration and may be considered to be track like in nature.

The detection system comprises a cooled photomultiplier and a photon counter. Figure 2 overpage shows a map of the intensity distribution of the D_α line viewed normally to the discharge's axis. It shows also the temperature profile along three tracks indicated in the intensity map.

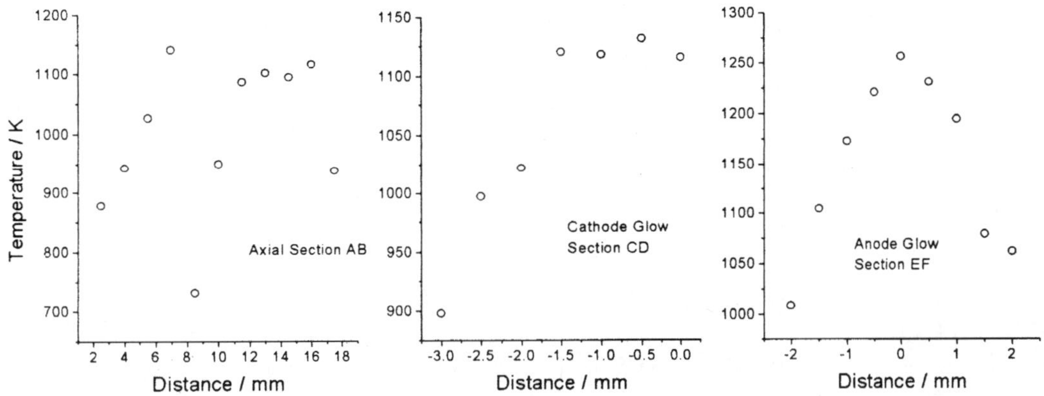

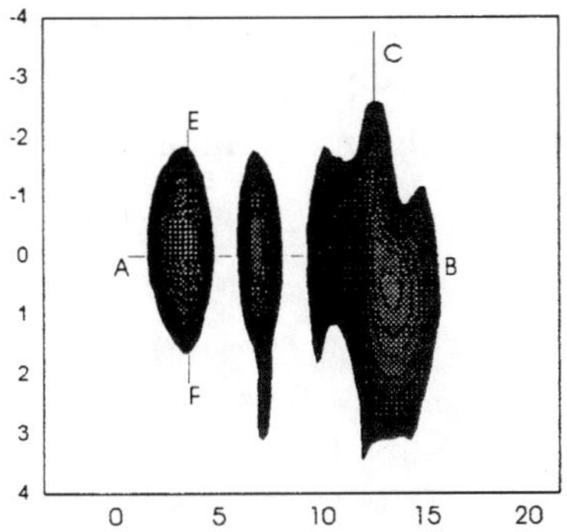

In contrast to the atomic emission technique the CARS technique is capable of giving high spatial resolution in three dimensions. For the folded-BOXCARS phase matching configuration used in these experiments the signal generation volume is a small cigar shaped volume (50 μm x 50 μm x 1mm). The scanning CARS spectrum of D_2 is obtained for several ro-vibrational levels in the fundamental and where the temperature is sufficiently high, in the region of the first hot band. In general the reported temperatures are rotational temperatures which may be assumed with some confidence to be equal to the translational temperatures at the same location.

The experimental CARS spectra are then compared with CARS spectra simulated using an HTSL Systems Ltd. spectral fitting program to yield best fit values of the respective ro-vibrational temperature. An example of an experimental spectrum and its computer simulation is shown in Figure 3. The spectrum was taken from the cathode glow region of the discharge and yields a rotational temperature of 699 K.

The work is allowing us to make direct comparisons of the atomic and molecular temperatures at the same point in space (recognising, as mentioned above that the emission temperatures are somewhat confused). The results are providing critical tests of the 1-D / 2-D theoretical model of the discharge and guiding its further development.

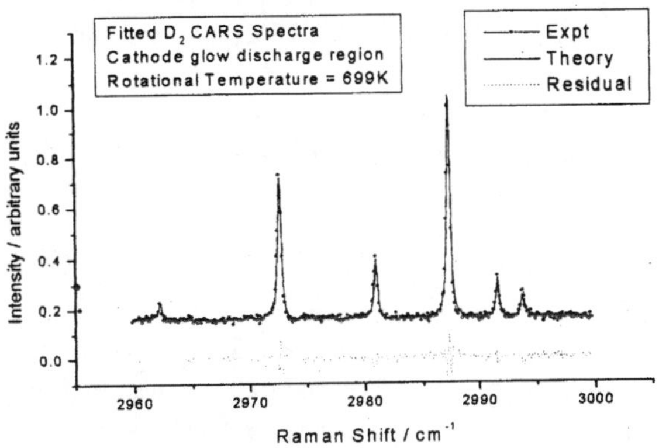

L 07

DETERMINATION OF ABSOLUTE DENSITIES OF ADDITIVES IN HID LAMPS

Eckhard Kindel, Manfred Kettlitz, Conrad Schimke, and Heinz Schöpp
Institut für Niedertemperatur-Plasmaphysik e.V. an der Ernst-Moritz-Arndt-Universität Greifswald, Robert-Blum-Straße 8-10, D-17489 Greifswald, Germany

1 INTRODUCTION

The description of the plasma state and the calculation of relevant properties is based on the knowledge of particle densities and the radial temperature distribution. Often only the starting parameters of discharges with different atoms are known and there is a lack of information about the densities in the arc. Exemplary a Hg-TlI lamp was used for the experiments. For such lamps partial pressures have been determined from optically thin [1] as well as optically thick Tl lines [2] by emission spectroscopy. In this work population densities are determined by using the interferometric hook method and the arc temperature profile by emission spectroscopy.

2 EXPERIMENT

The measurements are performed in a vertically mounted cylindrically quartz lamp without an envelope. The discharge tube has an internal diameter of 18 mm, a volume of 5 cm^3 and a 27 mm electrode (tungsten) separation. It is filled with 30 mg mercury and 0.98 mg thallium iodide. The ac power at 50 Hz is varied between 125 W and 300 W. To measure the absolute ground state densities in mercury and thallium the interferometric hook method [3] for the resonance transitions 6^3P_1-6^1S_0 ($\lambda = 253.7$ nm) for mercury and $7^2S_{1/2}$-$6^2P_{1/2}$ ($\lambda = 377.6$ nm) for thallium is used. A dye laser as background light source is applied. The laser beam traverses the lamp side-on through the axis to obtain the mean value of the densities over the line of sight.

In order to get the radial profile of the densities the radial temperature profile must be known. The absolute side-on radiance of the optically thin mercury lines (λ = 577.0 / 579.1 nm) are measured for various axial distances by a CCD camera system mounted to a 0.5m-spectrometer. Carrying out an Abel inversion the emission coefficient and by further processing the radial temperature profile [4] is obtained. The radial ground state density profiles for mercury as well as for thallium are received by taking into consideration the ideal gas law and neglecting in a first approximation the formation of molecules (HgI_2, TlI) near the wall. For a comparison the thallium density profiles are measured from the optically thin line at 655 nm ($8^2P_{3/2}$-$7^2S_{1/2}$).

3 RESULTS

The temperatures in the arc for various input powers are shown in figure 1. In the arc centre they are between 5400 and 5600 K. For 125 W the highest arc temperature is observed and the profile shows a stronger contraction than the other ones.

The radial profiles for the ground state densities of thallium and mercury are shown in figure 2. Those for mercury do not vary very much with the input power because all Hg is evaporated. The ground state densities of thallium are 100 times smaller than those of mercury. They are between 1.2 and $7 \cdot 10^{22}$ m^{-3} at the arc axis. For the lowest input power a significant decrease of the Tl density is found because the whole amount of the TlI is probably not evaporated.

The measured mean values for mercury are for the three higher input powers $8.9 \cdot 10^{24}$ m^{-3} and $7.8 \cdot 10^{24}$ m^{-3} for 125 W. For thallium the values are $2.38 \cdot 10^{22}$ m^{-3} for 125 W and $1.03 \cdot 10^{23}$ m^{-3} for the higher powers. The mixing ratio of Tl and Hg varies between 0.012 for 300 W and 0.003 for 125 W. A comparison between the Tl densities obtained by the hook method and emission spectroscopy shows a good agreement.

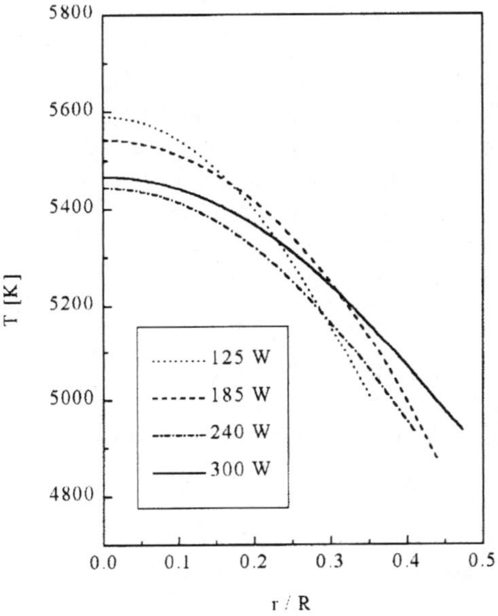

Fig. 1: Arc temperature profiles of a Hg-TlI-discharge for various power values along the reduced radius.

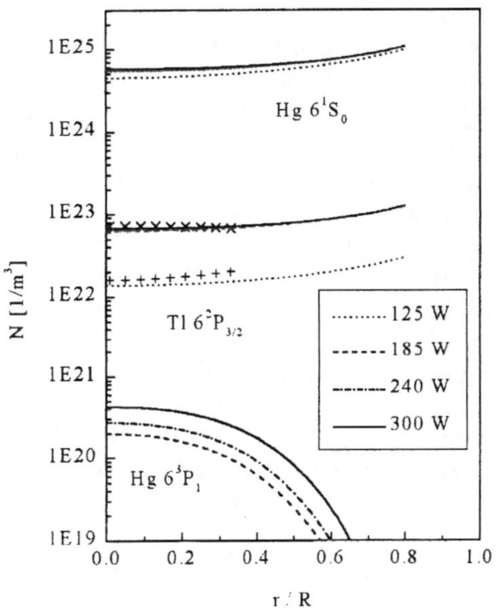

Fig. 2: Densities of the 6^1S_0 and 6^3P_1 levels of Hg and the $6^2P_{3/2}$ level of Tl along the reduced radius for various powers. All curves belong to the hook method except the values for Tl from the Tl 655 nm line for 125 W (+) and 300 W (x).

4 CONCLUSION

It is shown that the hook method is a useful tool for the determination of absolute densities of additives in high pressure discharges.

5 REFERENCES

[1] D. Karabourniotis, S. Couris, J. J. Damelincourt and M. Aubes, IEEE Trans. Plasma Sci. **PS-14**, no. 4, 325 (1986).

[2] W. Funk and H.-G. Kloss, Beitr. Plasmaphys. **13**, 101 (1973).

[3] E. Kindel, M. Kettlitz, C. Schimke and H. Schöpp, J. Phys. D: Appl. Phys. **31**, 1352 (1998).

[4] H. Hartel and H. Schöpp, J. Phys. D: Appl. Phys. **29**, 2881 (1996).

L 08 Dependence on the Operating Frequency of Cathode Fall Volatage

Tadao Uetsuki, Noriyuki Taguchi and Wataru Noda

Matsushita Electric Works, Ltd.
1048, Kadoma, Osaka 571-8686, Japan

Introducttion

It has been reported by Watanabe that the cathode fall voltage at high frequency (HF) operation is lower than at low frequency (LF)[*1]. The cathode temperature at HF is lower than at LF because of disappearing the heating power in anode cycle at HF.

The current from the cathode is shown as

$$I \propto T^2 \exp(-\Phi / kT) \exp(0.40\sqrt{E}/T)$$

This formula shows electric field of the cathode fall increases with the cathode temperature decreasing. This is inconsistent with the result of Watanabe[*1]. We have tried an analysis looking of the peak value of the cathode fall voltage (CFV). If the peak value of the CFV depends on operating frequency, Crest Factor (CF) should be critical for HF operation. This is because higher peak current causes higher peak of CFV, which will influence on the lamp life by spattering. To eliminate this phenomenon by heating the cathode, we checked how much power should be input to the cathode.

The purpose of this paper is to report the followings by measuring the CFV.

(1) The relationship between the peak of CFV and operating frequency

(2) The relationship between CF and CFV

(3) The possibility of reducing the peak value of CFV by heating the cathode.

We used two methods to measure the CVF, one is the direct method and the other is indirect method. The former is to check the lamp voltage of the negative glow lamp and the latter is to check the plasma potential of negative glow of fluorescent lamp.

Experiment

Firstly we checked the relationship between CFV(and cathode temp.) and frequency of the negative glow lamp shown in Fig.1, whose diameter is 35 mm and argon pressure is 400 Pa by the direct method. The circuit is shown in Fig.2. The Impedance Z is 75 ohm. We measured them at 60Hz (LF) and at 50kHz(HF) operations.

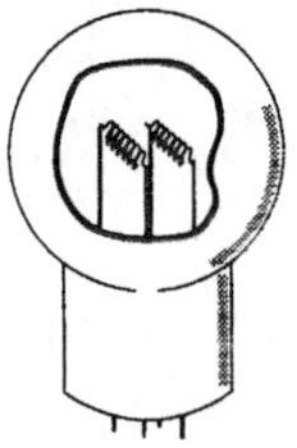

Fig.1

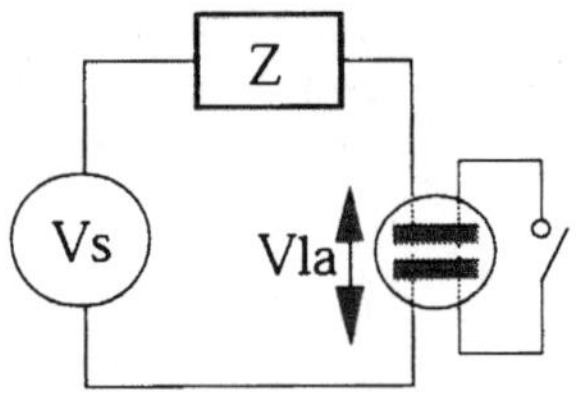

Fig.2

Vs	LF	HF
Vla(rms)	14.4 V	11.8 V
Vla(peak)	17.8 V	19.0 V
Temp.	1040 ℃	990 ℃

Table 1

The results (shown in Table 1) show the followings.

(1) The cathode temperature becomes lower, as operating frequency becomes higher.

(2) The peak of lamp voltage becomes higher, as operating frequency becomes higher.

(3) The r.m.s. of lamp voltage becomes lower as operating frequency becomes higher.

Secondly, we checked the relation between CFV(and cathode temp.) and cathode heating ratio by the direct method. We measured them at 50kHz(HF) by using the same lamp and the circuit shown in Fig.3.

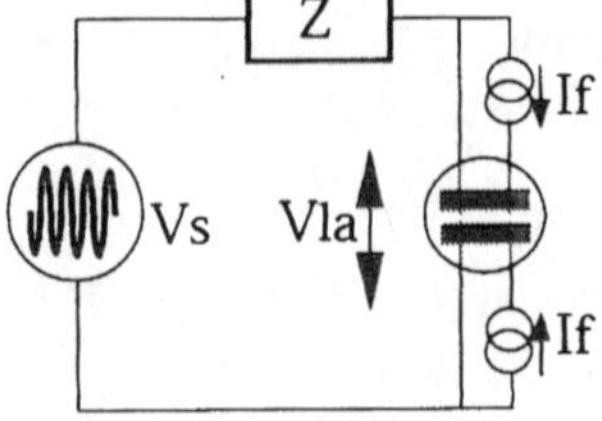

Fig.3

The results (shown in Table 2) show the followings.

(1) The peak of CFV becomes lower, as the cathode temperature becomes higher by heating cathode at HF.

(2) The peak of CVF becomes same level as at LF, when the cathode temperature becomes same level as at LF.

If	0 A	0.10 A	0.15 A
Vla(rms)	11.8 V	11.4 V	11.2 V
Vla(peak)	19.0 V	18.0 V	17.2 V
Temp.	990 ℃	1030 ℃	1040 ℃

Table 2

Thirdly we checked the relation between CFV of fluorescent lamp and cathode heating ratio by the indirect method, shown in Fig.4. The operating frequency was 50kHz.

The results (shown in Table 3) show that the peak of CFV becomes lower as the cathode temperature becomes higher by heating cathode.

This result does not contradict the result of negative glow lamp measured by the direct method.

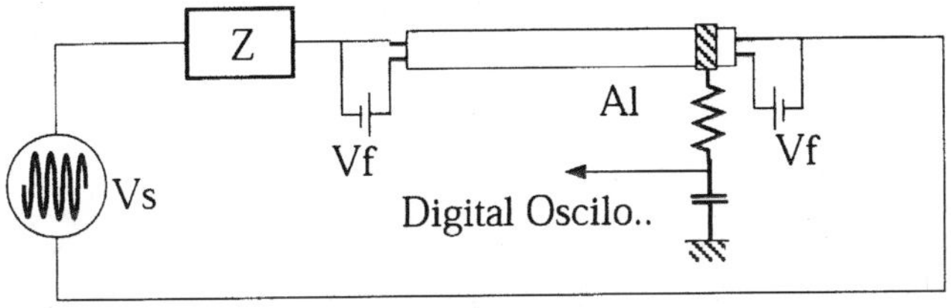

Fig.4

Vf	0 V	4V
Vc(peak)	19.0 V	10.0 V

Table 3

Lastly we checked the relation between the peak of CVF and CF by the indirectly method shown in Fig.4. The measurement results (shown in Table 4) show the followings.

(1) When the cathode is not heated, the peak of CFV becomes higher as CF becomes higher.

(2) When the cathode is heated, the peak of CFV does not vary so much if CF becomes higher. This peak value of CVF is much lower than that without heating cathode.

CREST FACTOR	1.4	1.8	2.1
Vc (w/o heat)	15 V	16.5 V	17 V
Vc (w/ heat)	11 V	10.5 V	11 V

Table 4

Discussion

The results getting from two methods do not contradict each other. So we can get to the followings regarding the relation among the operating frequency, the cathode temperature, CVF (Cathode fall voltage), and CF (Crest Factor).

(1) As operating frequency becomes higher, the cathode temperature becomes lower and the peak of CVF becomes higher

(2) When the cathode is not heated, the peak of CFV becomes higher as CF becomes higher.

(3) When the peak of CFV becomes higher, it can be made lower by heating cathode,

(1) is ascribed to the increase of the cathode fall electric field (Schottky Effect) caused by the decreasing cathode temperature because of no heating power to the electrode during the anode cycle at HF. (2) is also ascribed to the increase of the cathode fall electric field, because the cathode temperature cannot follow the increasing ratio of circuit current.

This phenomenon influences the lamp life by ion spattering to the cathode. But (3) shows that heating the cathode by external power source can decrease the peak of CFV for both situations, (1) and (2). And we confirmed that the power for heating the cathode is small enough for practical circuit from these measurement.

Conclusion

* The peak of CFV increase, when the operating frequency becomes higher (50kHz) or when CF becomes higher.

* The peak of CFV can be decreased by heating cathode externally. We found out the peak of CFV could be decreased to the level where the spattering to the cathode did not influence, if CF=2.1. And we also found out the external heating power was no problem for actual circuit from these measurement.

Refference

*1 Y.Watanabe, S.Murayama, Jpn. J. Appl. Phys. 32 Part1, No.8, pp.3593-3600(1993)

L 09

THE ELECTRON DENSITY AND TEMPERATURE IN THE PHILIPS QL-LAMP

J. Jonkers, J.A.M. van der Mullen, J. van Dijk and D.C. Schram
Department of Applied Physics, Eindhoven University of Technology
PO Box 513, 5600 MB Eindhoven, the Netherlands

1. INTRODUCTION

There is an increasing interest in electrodeless lamps since reactions between the electrodes and the plasma are avoided which leads to a considerably longer life time and less restrictions to the choice of filling than conventional discharge lamps. One special class of electrodeless lamps is that formed by the inductive fluorescent lamps. For some years commercial lamps of this type have been available: the "QL-lamp" of Philips since 1992 and the "Genura" of General Electric since 1994 [1]. Recently the third major lighting company Osram Sylvania announced the "Endura".

Within the Eindhoven University of Technology a 2D model is being developed which accurately describes the QL-lamp. The electron density and temperature are key-parameters in the description of the plasma. Moreover it appears that due to the relatively high gas temperature (550 to 800 K, depending on the filling pressure) and electron density the mercury ground state atoms are seriously depleted in the most active zone of the discharge.

2. METHODS

The electron density and temperature in the QL-lamp are obtained using two different methods:

1. from the argon $4s^3P_2$ metastable density and the gas temperature which were obtained from radially resolved diode laser absorption measurements [2] and
2. from the electron particle and energy balances.

Each method results in a set two equations with two unknowns (n_e and T_e) which can be solved iteratively. Results are presented for three different argon filling pressures (33, 66 and 133 Pa) and fixed RF power (80 W).

2.1 Absorption measurements

The electron temperature is estimated from the ratio of the argon metastable and ground state densities. A Collisional Radiative Model (CRM) [3,4] is used to correct for the non-equilibrium population. The electron density is obtained from the heavy particle heat balance.

2.2 Size-stabilised plasma theory

A good estimation for the electron temperature can be obtained from the electron particle balance:

$$n_e n_1^{Hg} S_{CRM}^{Hg} \approx \frac{D_a}{\Lambda_{ne}^2} n_e$$

The CRM which is used to calculate the ionisation coefficient S_{CRM}^{Hg} is treated in a different contribution [4]. The gradient length in the electron density profile Λ_{ne}, which dictates the electron temperature, is estimated from the position of the maximum in the gas temperature.
In a similar way the electron energy balance yields a value for the electron density.

3. RESULTS

The results are depicted in Figure 1. The agreement between the temperatures as found by both methods is striking: the differences are much smaller than the error bars (due to various assumptions and inaccuracies in the measurements and the model). Also the electron densities agree fairly well with each other.

For increasing argon filling pressure the losses of charged particles are reduced which results in a lower electron temperature and a higher electron density. It should be noted that apart from the fact that a higher argon pressure obstructs the diffusion, it is also found that the center of the most active zone is situated more to the outside so that (in our simple model) the characteristic size of the plasma Λ_{ne} is larger. Both effects (lower diffusion coefficient D_a and larger Λ_{ne}) benefit the residence time of the charged particles, so that the electron temperature is lower and the density higher.

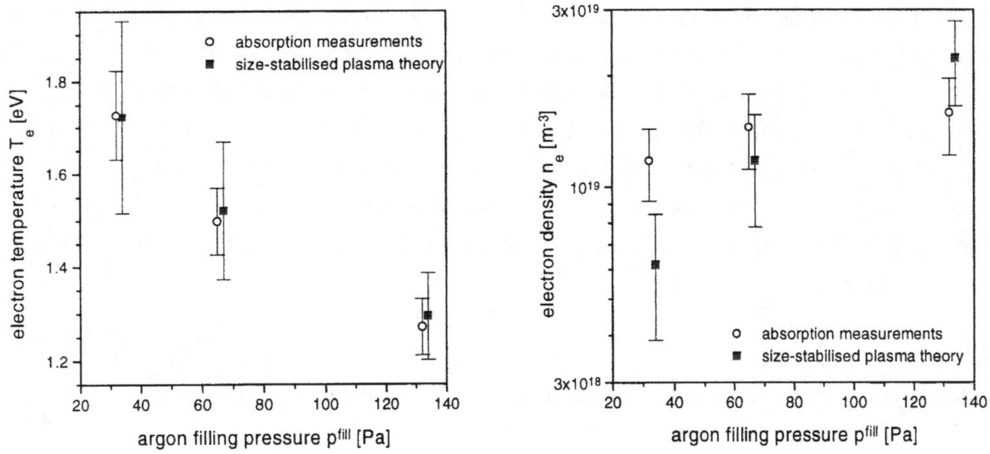

Figure 1: The electron temperature and density as estimated from the two different methods. The error bars stem from the estimated inaccuracies of the measurements, the used assumptions and the CR model. The points at the same argon filling pressure are plotted next to each other, so that the error bars do not interfere.

The combination of the relatively high electron density and gas temperature causes that the density of mercury atoms in the most active part of the lamp (5.7 at 33 and 2.6×10^{19} m^{-3} at 133 Pa) is significantly lower than near the outer wall of lamp (around 9.0×10^{19} m^{-3}).

4. REFERENCES

[1] D.O. Wharmby and S.-A. El-Hamamsy, Proceedings of the 7th International Symposium on the Science and Technology of Light Sources, Kyoto Japan, 27 (1995).

[2] J. Jonkers, M. Bakker and J.A.M. van der Mullen, J. Phys. D: Appl. Phys. **30** 1928 (1997).

[3] D.A. Benoy, J.A.M. van der Mullen and D.C. Schram, J. Quant. Spectrosc. Radiat. Transf. **46** 195 (1991).

[4] J. van Dijk, H.P.W. van der Heijden, A. Hartgers, J. Jonkers and J.A.M. van der Mullen, *"Towards a general Collisional Radiative Model"*, this conference.

L 10

Spectroscopic diagnosis of a high-pressure metal iodide(ScI_3, NaI, Hg) HID lamp plasma

Makoto Kubo [1], Hiroaki Taniguchi [1]*, Shigeyuki Mori [2]
Tomiki Inagaki [2], Kunihide Tachibana [1]

[1] Kyoto University, Dept. of Electronic Science & Engineering
Yoshida-honmachi, Sakyo-ku, Kyoto, 606-8501, Japan
[2] Japan Storage Battery Co. Ltd., Lighting Equipment Div.
Nishinosyo, Kisshoin, Minami-ku, Kyoto, 601-8501, Japan

1. INTRODUCTION

Metal iodides (ScI_3, NaI) are added to a high pressure mercury (Hg) discharge lamp in order to improve the spectral luminous efficiency as well as the color rendering properties of the lamp. The knowledge of the spatial distributions of these metal atoms and ions in an arc plasma is useful for understanding the discharge physics and also for the comparison with a computer simulation for developing a new spectral distribution lamp. In this proceeding paper we describe only a part of the measuring results of the radial distribution of ground-state Sc atoms and ions in a lamp plasma by using laser induced fluorescence (LIF) method. We have also measured by the emission spectroscopy method [1].

2. EXPERIMENTAL

The testing metal iodide discharge lamp was operated at AC 60Hz and 250W as the average power. The lamp plasma which was the object of diagnosis was generated in a cylindrical quartz luminous tube. It had an inner diameter of 16mm, a discharge length of 32mm, and was filled with 40mg of Hg, 30Torr of argon gas and the iodides of Sc and Na. The molar ratio of these iodides was NaI / ScI_3 = 7.0 / 1.0. The laser diagnostic system consists of a Nd:YAG-laser and a dye-laser which was pumped by YAG-laser. Both laser lights were mixed in a crystal to generate the tunable sum frequency. The LIF measurement was done at three points in the luminous tube along the vertical axis (z=0: at the center of the discharge length and at the points of $\pm$9mm from z=0) and at about 0.5 mm intervals on radial. In this experiment a boxcar integrator was used for reducing the background emission from the plasma of which gate-width was set at 50nsec. Laser pulse width was 15nsec (FWHM) and the laser operation (10Hz) was synchronized with the AC lamp power supply (60Hz).

3. RESULTS AND DISCUSSION

The radial distribution of ground state Sc atoms (upper part) and ions (lower part) in the lamp plasma measured by LIF is shown in Fig.1,which have been taken at the 90^0 phase angle of the discharge current (the input power maximum). The radial profiles in Fig.1 are distorted to some extent due to the self-absorption along the line of sight of LIF signals. Compensation for the effect has been tried by the other absorption measurement combined with the Abel inversion technique to get more symmetric profiles. The arc temperature is shown in Fig.2, which has been derived from the Boltzmann plot of population densities measured by emission line intensities. The highest temperature position lies at the center axis as usually seen in arc plasmas. The concave profile of Sc atoms Fig.1 can be explained by the neutral atom density decrease due to the ideal gas law under constant pressure condition and the temperature dependence of ionization degree given by Saha's equation. The decrease of Sc ion density, however, is much less because the ionization degree increases toward the center. The dependence of radial profiles on the axial height z (z=0 shows the center of electrode length) may be attributed to the convective flow of Sc atoms and the thermo-chemical balance between source iodide molecules and decomposed species. Detailed arguments will be given at the conference.

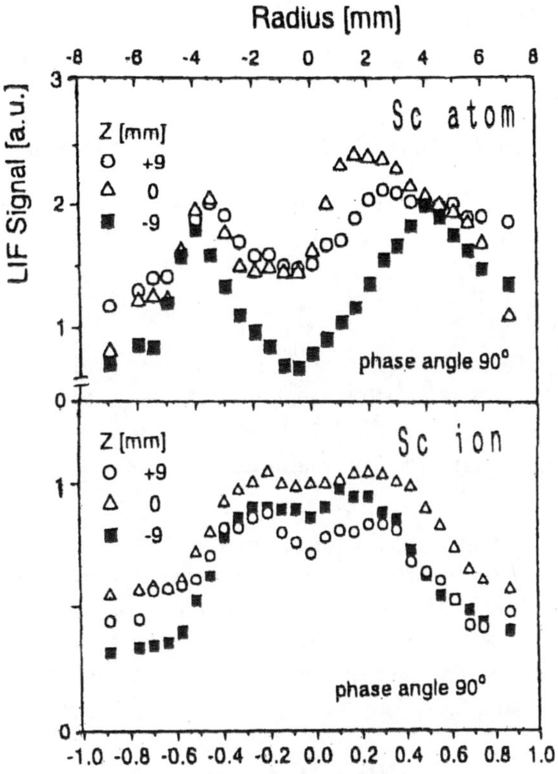

Fig.1 The radial distribution of ground state Sc atoms and ions

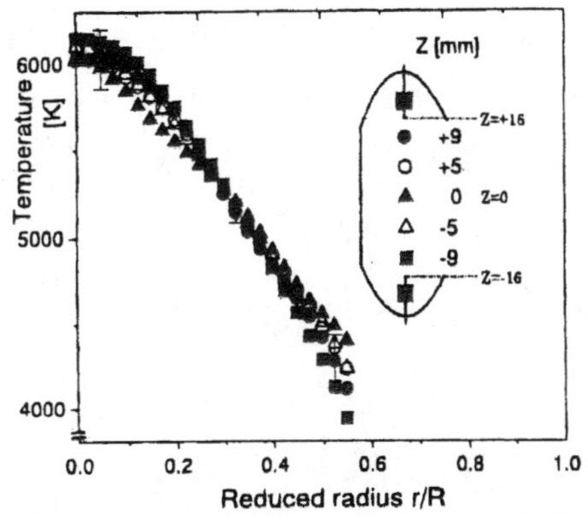

Fig 2. The radial distribution of arc temperature

4. ACKNOWLEDGMENTS

Authors wish to thank Mr.Y.Hukumoto for experimental assistance on LIF works.

5. REFERENCES

[1]H.Taniguchi, M.Kubo, K.Tachibana, S.Mori, T.Takasu., Lux pacifca'97 (1997) F-45

----

* After 1st of April in this year, the position is following.
USHIO INC., 1194, Sazuchi, Bessho-cho, Himeji 671-0224, Japan

L 11

THE LABORATORY FOR UV AND VUV RADIOMETRY OF THE PHYSIKALISCH-TECHNISCHE BUNDESANSTALT AT BESSY I

J. Hollandt, H. Rabus, M. Richter, R. Thornagel, and G. Ulm

Physikalisch-Technische Bundesanstalt
Abbestraße 2-12, D-10587 Berlin

1 INTRODUCTION

The Physikalisch-Technische Bundesanstalt (PTB), the German national institute of metrology, operates its sections 'UV and VUV Radiometry' and 'X-Ray Radiometry' at the Berlin electron storage rings BESSY I and BESSY II. At BESSY I six experimental stations on four beamlines provide synchrotron radiation for radiometric applications in the broad spectral range from 400 nm down to 0.1 nm [1].

This paper describes the calibration techniques and the two experimental stations at BESSY I employed for the calibration of radiation sources and of radiation detectors in the spectral range from 40 nm to 400 nm, which are the major activities of the laboratory for 'UV and VUV Radiometry'. In the year 2000, after the anticipated shut down of BESSY I, these instrumentations will be transferred to the new radiometry laboratory of PTB at the storage ring BESSY II, that is currently under commisioning.

2 CALIBRATION OF UV AND VUV RADIATION SOURCES

On the basis of the fundamental Schwinger equation the spectral radiant power Φ_λ of the synchrotron radiation of BESSY I can be calculated from storage ring parameters and geometrical quantities thus making BESSY I a primary radiation source standard. In the UV spectral range the overall relative uncertainty of the spectral radiant power, $\Delta\Phi_\lambda/\Phi_\lambda$, of BESSY I is 0.04% ($1\sigma$) [2].

The radiometric characterization of a radiation source is performed by comparing the unknown spectral radiant power of the source with the calculable spectral radiant power of BESSY I. In the spectral range from 40 nm to 400 nm, the instrumentation applied for this procedure consists of: a flux-limiting aperture stop; a toroidal mirror, which images either the tangent point of the storage ring or the source under investigation into the entrance slit of a 1-m, 15° McPherson type monochromator; and a detector system (Fig. 1) [3]. The complete instrumentation can be rotated

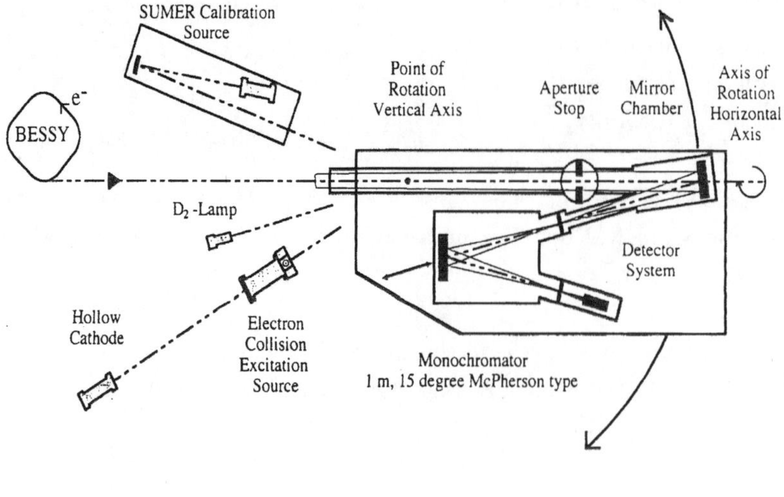

Fig.1: Schematic top view of the instrumentation for the characterization of UV and VUV radiation sources. The synchrotron radiation beam line and various sources under investigation are shown.

around a horizontal axis which coincides with the optical axis. Thus the highly polarized synchrotron radiation can be detected with the optical plane parallel and perpendicular to the electron orbit plane. From these measurements the polarization property of the instrument is determined and its effect on the flux comparison is properly considered.

The synchrotron radiation spectrum can be optimized to the requirements of a specific calibration by changing the energy of the stored electrons from 800 MeV (nominal energy) to 340 MeV or 850 MeV. Furthermore, the radiant power of the synchrotron radiation can be adjusted by variation of the number of stored electrons over twelve orders of magnitude to match the radiant power of the source under investigation. The instrumentation allows to perform calibrations of spectral radiance and spectral radiant intensity with a relative uncertainty of 2% (1σ).

3 CALIBRATION OF UV AND VUV RADIATION DETECTORS

The calibration of radiation detectors is achieved by comparison with the synchrotron radiation cryogenic electrical substitution radiometer, SYRES, as the primary detector standard. SYRES is a thermal detector based on the equivalence of radiant and electrical heating, that allows for the absolute determination of radiant power in the 1 μW to 10 μW range with a relative measurement uncertainty of below 0.2% (1σ) [4].

The instrumentation used for detector calibration in the 40 nm to 400 nm spectral range consists of a monochromator in combination with additional optics (Fig. 2) that produce small-bandwidth radiation of high spectral purity at a fixed focal point. For detectors, such as e.g. photodiodes, that can be operated at radiant power levels of 1 μW to 10 μW, the spectral responsivity s is determined by measuring the same monochromatized synchrotron radiation beam alternately with SYRES and the dectector under investigation.

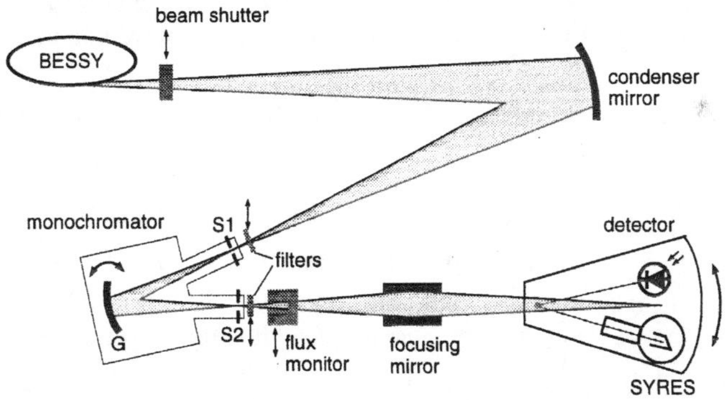

Fig.2: Schematic top view of the instrumentation for the calibration of UV and VUV detectors with the primary detector standard SYRES.

When photon counting detectors, which would be heavily overloaded with a μW radiant power input, are calibrated, the photon flux is adjusted between the measurements with the two detectors by reducing the number of stored electrons [5]. With this procedure, the spectral responsivity can be determined with a relative uncertainty of about 3% (1σ), whereas in the case of a direct comparison the relative uncertainty is about 0.7% (1σ).

4 REFERENCES

[1] G. Ulm, B. Wende, Rev. Sci. Instrum. **66**, 2244 (1995).
[2] H. Rabus, F. Scholze, R. Thornagel, G. Ulm, Nucl. Instr. and Meth. A **377**, 209 (1996).
[3] J. Hollandt, W. Jans, M. Kühne, F. Lindenlauf, B. Wende, Rev. Sci. Instrum. **63**, 1278 (1992).
[4] H. Rabus, V. Persch, G. Ulm, Appl. Opt. **36**, 5421 (1997).
[5] P. Verhoeve et al., Proc. LTD-7, S. Cooper (Ed.), MPI f. Physik, Munich, 97 (1997).

L 12

PHOTOCELL ENHANCED TECHNIQUE FOR MEASURING ELECTRODE STARTING TEMPERATURE OF FLUORESCENT LAMPS

E. E. Hammer, FIEEE, FIES
GE Lighting, Nela Park, Cleveland, Ohio, 44112

1 INTRODUCTION

Past experience, supported by extensive measurements and consistent with discharge theory, indicate that a minimum electrode temperature of 700°C is needed to minimize the E mix sputtering during starting so that the rated lamp life performance can be reliably achieved. To further differentiate the criteria, it must be demonstrated that essentially no asymmetry occurs in the arc current waveshape during the glow-to arc transition time because asymmetry in the arc current signature indicates the electrode temperature is not hot enough to supply the needed electrons for both halves of the cathode-anode cycle. The testing tool needed, therefore, had to be capable of reliably repeating the estimated electrode temperature and be able to capture in real time the visual profile of the glow-to arc current waveform signature.

2 MEASUREMENTS

Figure 1 illustrates the starting signature of an F32WT8 lamp with a programmed electronic rapid start type ballast having the proper preheating conditions. As is evident by the signature, this starting profile is good because the glow to arc transition time is relatively fast with no evidence of asymmetry or partial rectification in the arc current waveshape. This indicates that the electrodes were sufficiently hot and that there was minimum sputtering during the start. The photocell trace shows the correct time to record the Rh value so that the Rh/Rc ratio can be calculated. In this case, that ratio was approximately 5.0 which is an excellent value because it should give greater than 20,000 cycle starts on short cycle burning as well as rated lamp life on standard burning cycles. The top two graphs in this Figure 1 show the circulating cathode current and cathode voltage measurements associated with this start so that the Rh/Rc ratio can be accurately calculated.

This photo-enhanced technique can also be used for preheat type starting scenarios. Using a 13W CFL hèx type lamp on an electromagnetic type 230V ballast, the starting time was approximately 3.0 seconds. The related photocell trace showed that this particular start had five bimetal switch openings before the lamp started. The starting signatures revealed that the previous bimetal openings did not start the lamp because the electrodes were still not hot enough. It was only when the RH/RC ratio reached approximately 4.49 that the lamp started. If more preheat was supplied by the ballast and the digital scope via the photocell detector. With the voltage and current leads connected as shown, the input power is applied and the resultant starting waveshapes recorded by the high speed digital for the starting scenario, less sputtering damage would have occurred and therefore, better lamp life performance could have been achieved. However, the overall evaluation of the starting signature was still considered satisfactory.

3 CONCLUSIONS

This report discussed a photocell enhanced technique to evaluate the lamp-ballast compatibility relationship during the starting/operating scenario of fluorescent lamps. The technique, which fundamentally uses the dynamic Rh/Rc methodology, can be used for preheat as well as programmed or rapid start type circuitry using either low frequency electromagnetic ballasts or high frequency electronic ballasts. Although not discussed in this report, the technique is also useful for evaluating instant start type ballasts as well. Considerable discussion of this basic testing technique has occurred at both ANSI [USA] and IEC [European] committee meetings and hopefully will be incorporated into those standards at some future date.

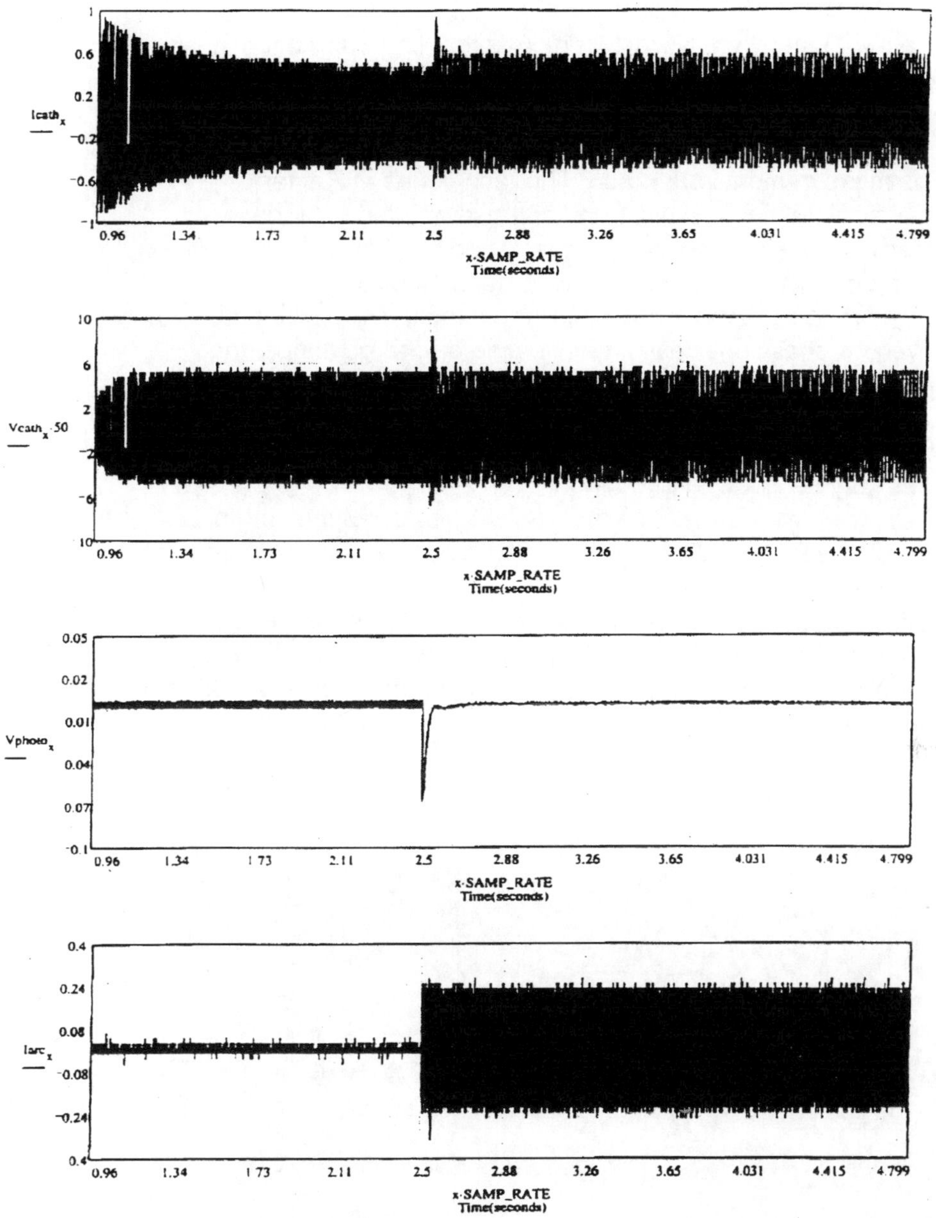

Figure 1

Good Electronic Signature for F32WT8 Lamp with Programmable Electronic Rapid Start Ballast having Proper Preheat Conditions

L 13

RAMAN MICROSCOPY AS A NON-DESTRUCTIVE DIAGNOSTIC TOOL FOR LAMP CHEMISTRY

S.P.Oliver and R.Devonshire

High Temperature Science Laboratories, Chemistry Department, University of Sheffield, Sheffield, S3 7HF, UK.

An inherent problem with many analytical techniques that find application in lamp chemistry (e.g. Scanning Electron Microscopy/Energy Dispersive X-ray Spectroscopy, Mass Spectroscopy) is the need to fracture the lamp in environmentally controlled conditions and maintain the fragments in these same conditions until the measurements are made. The techniques of Raman spectroscopy have the potential to avoid many of these problems. Modern Raman techniques such as scanning Raman microscopy can provide a wealth of information in a non-destructive manner. We have been exploiting these techniques for a wide range of lamp problems. For example, we have been investigating the vapour phase species in metal halide discharge lamps, the problem of dose migration down electrode seals, and the diagnostics of new generation PCA metal halide lamps which have translucent envelopes. The measurements can be performed on lamps with even their outer envelopes intact and thus processes may be followed throughout normal lamp life. This present paper reports results we have obtained on the first of these problems; metal halide compounds have very large Raman scattering cross sections and are very favourable species to study using the technique.

A high quality Raman microscope allows the user to target and analyse features to a resolution of 1μm and use of a long throw objective makes analysis possible through a lamp envelope. Indeed it is possible to analyse individually the inner surface of an outer envelope, the outer surface of an arc tube, arc tube inner wall deposit and the electrodes (or filament) without destroying the lamp. Raman resonances down to ≈25cm^{-1} can be detected using good quality optics.

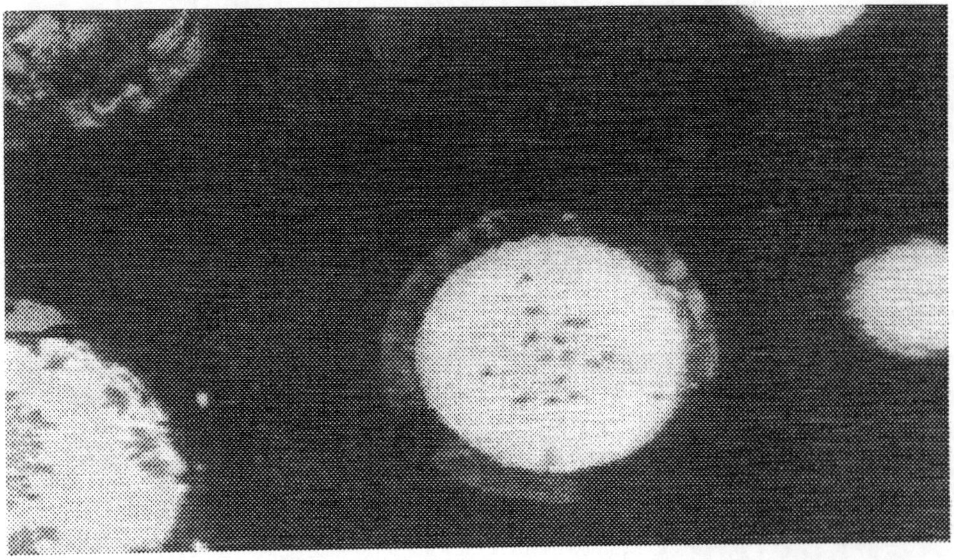

CCD camera white light image of centre of cold spot region (see below), showing 'halo' deposit surrounding mercury spot, image extends 300x150μm.

In the study we were interested to observe the species which are present in the vapour phase of an operating metal halide lamp. To achieve this goal we developed a method of condensing the active vapour phase species has been developed. The creation of an artificial cold spot at a chosen point on the lamp wall immediately prior to switching the lamp off causes the vapour phase species to condense rapidly at that point. The condensate at the generated cold spot is visible to the eye, roughly circular in shape and approximately 3mm in diameter. Several distinct features can be seen within this region using the microscope. The centre consists of mercury droplets, some of which are surrounded by a 'halo' of coloured material. Smaller areas of other coloured deposits appear elsewhere. Raman analysis of these cold spot regions yields a wealth of data.

An extensive catalogue of Raman spectra of individual lamp species is being compiled to enable as complete and as detailed as possible an interpretation of these spectra. The catalogue comprises spectra of all the lamp materials present in a given lamp and also the individual dose components. In addition, species believed possibly to be created in the lamp environment are also being catalogued.

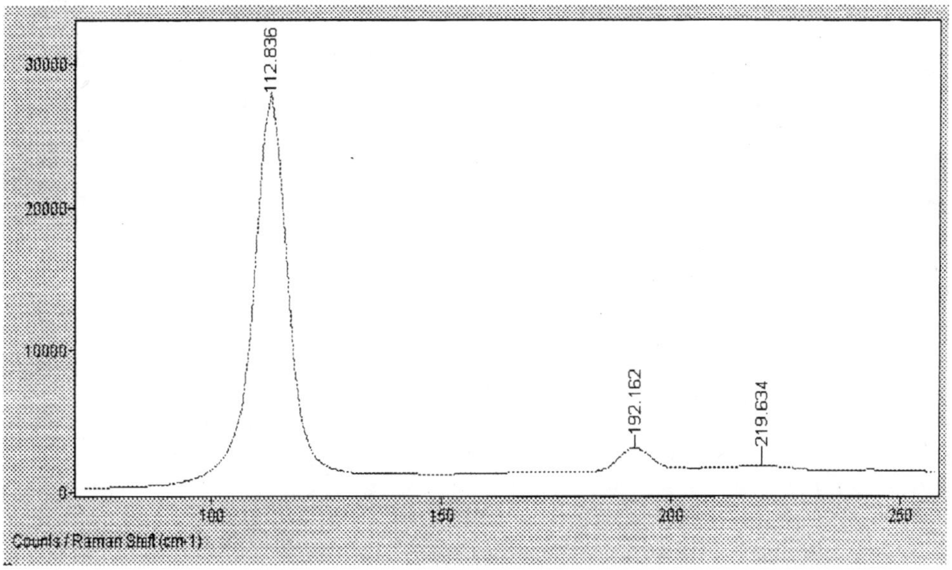

Raman spectra of 'halo' material. The observed resonances correspond to HgI.

Other capabilities of Raman microscopy that we have yet to explore include the ability to analyse envelope material, for example, in respect of the extent of crystallinity present in otherwise glassy materials. This avenue has considerable potential importance in lamp research and demands further investigation.

We are planning to establish in the HTSL a comprehensive modern Raman analysis facility based on multiple excitation sources, microscope and translation stages, etc., to address a wide range of problems in lamp science and technology.

L 14

Deuterium lamps

Spectroscopic measurements of electron densities and gas temperatures

Dr. A. Carstens, Dr. E. Arnold, Dr. K.-J. Dietz, Dr. E. Smolka

Heraeus Noblelight GmbH, Reinhard-Heraeus-Ring 7, 63801 Kleinostheim, Germany

Prof. Dr. H.-P. Popp

Lichttechnisches Institut (LTI), Kaiserstraße 12, 76131 Karlsruhe, Germany

Introduction

Deuterium lamps stand out by a high stability of intensity and by a line-free continuum in the wavelength region from 160 to 350 nm. To obtain the necessary radiation density, the plasma is intensified by a molybdenum constriction with 1 mm or 0.5 mm in diameter. This UV standard light source is used in research, industry and medicine for transmission, absorption, and fluorescence measurements.

Plasma parameters, which are important for the estimate of the thermal load of the constriction, are still unknown for the majority of deuterium lamp types. In our research the electron densities and the gas temperature in the plasma centre of commercial deuterium lamp model D800 are determined by varying the current density from 38 A/cm^2 to 306 A/cm^2. Plasma parameters are obtained by comparing measured and calculated Balmer line profiles.

Experimental Setup

The line shape of the four Balmer lines: D_α, D_β, D_γ and D_δ are measured. Due to small half-widths of < 0,05 nm, we are forced to use a highly resolving experimental setup. The measurements are made by using a Czerny-Turner monochromator with a focal length of 3.4 m, a 13 cm wide grating of 2160 lines/mm and an entrance slit and exit slit of 25 µm widths.

Analysis of the Measurements

The comparison of the measured and calculated line profiles shows the electron density and the gas temperature. An additional verification secure the simultaneous consideration of four line profiles to determine two parameters. To achieve the calculated line profiles, we convolute the Doppler profile with the Stark profile. The result is folded with the combined apparatus profile and fine structure. The apparatus profile is determined by measuring a He-Ne laser line. The Stark-broadened part of the line profiles, caused by statistically distributed electric microfields of free electrons and ions, yield the electron density. The Stark data of Griem's improved theory [1] are used. The gas temperature is determined by analysing Doppler broadening. The Doppler effect mainly influences the D_α line, whereas the influence of Stark broadening is strongest for the D_δ line.

Conclusions

Diagnostic results in the plasma spot by different current densities are summarised as shown in table 1:

Current densities [A/cm²]	Electron density [cm⁻³]	Gas temperature [K]
38	$1.6 \times 10^{13} \pm 4.0 \times 10^{12}$	1800 ± 180
76	$2.5 \times 10^{13} \pm 7.0 \times 10^{12}$	3000 ± 300
153	$3.2 \times 10^{13} \pm 8.0 \times 10^{12}$	3700 ± 370
306	$5.0 \times 10^{13} \pm 1.3 \times 10^{13}$	6800 ± 680

Table 1: Plasmadiagnostic results of deuterium lamps type D800 varying the constriction diameter (0.5 mm and 1 mm) and the current (300 mA and 600 mA)

An increase in the current density by a factor of 4 roughly increases electron density by a factor of 2. Measurements along the diameter of the discharge show a strong decrease of the gas temperature from 1800 K to 1300 K by the lowest current density of 38 A/cm^2.

References

[1] H. G. Griem, *Spectral line broadening by plasma*, Academic, New York, 1974

L 15

A TEMPERATURE AND EMISSIVITY DIAGNOSTIC FOR FLUORESCENT LAMP ELECTRODES

Richard C. Garner
OSRAM SYLVANIA Development Inc., 71 Cherry Hill Dr., Beverly, MA 01915 USA

1 INTRODUCTION.

Since fluorescent lamp life is typically limited by the life of its electrodes, knowledge of electrode temperature and emissivity provides valuable information on lamp life. High electrode temperature during steady-state, and large overshoots of temperature during starting, cause large losses of emitter material. Emissivity and temperature are connected to other discharge parameters which influence electrode life, such as cathode fall and ion current impinging on electrode.

We have developed a diagnostic with which we determine temperature and emissivity at all positions along an electrode simultaneously. The diagnostic uses a combination of blackbody emission and hot/cold resistance ratio. Unlike measurements based on emission alone, a knowledge of emissivity is not necessary to obtain temperature. In fact, a calibrated optical system provides emissivity. And, unlike measurements based on hot/cold resistance ratio alone, the spatial nonuniformity of the temperature during lamp operation does not limit the accuracy of the results. In fact, using a linear array detector the spatial profile of the temperature can be determined.

2 DESCRIPTION OF DIAGNOSTIC

Temperature and emissivity are determined using measurements of both blackbody emission and hot/cold coil resistance ratio. First, blackbody emission power from all positions along the electrode is measured when the lamp is on. Then, a functional relationship is obtained between blackbody emission power and hot/cold resistance ratio (i.e., temperature) when the lamp is off and d.c. current is applied to the coil. The temperature of the hot spot is then determined by interpolating the value from this functional relationship using the previously measured blackbody emission power. In addition, emissivity is determined from the ratio of the blackbody emission power from the hot spot to the blackbody emission power from a calibrated blackbody source at the hot spot temperature just determined.

The electrode temperature is assumed to be uniform during application of d.c. current to the coil. In reality, the temperature is near uniform in most of the central region but decreases near the ends where the electrode is held by thick posts. We attempt to correct for this with the aid of an electrode thermal energy balance model.

A 128 pixel, InGaAs linear array detector and an interference filter are used to measure the blackbody emission at 1.55 μm. The entire length of the electrode is imaged, with ½ magnification, to the active area of the linear array. The spatial resolution, referenced to the electrode, is less than 100 μm. It is limited by spherical aberrations of the imaging lens system. The temporal resolution is 0.5 msec, limited by the linear array circuitry and control electronics.

3 EMITTER MATERIAL LOSS

A major application of the diagnostic is in estimating emitter material loss due to evaporation. We estimate such loss using a simple model in which Ba, produced at the electrode through a reaction between BaO and W, diffuses away from the electrode to the lamp wall through a neutral argon atmosphere. The equilibrium Ba pressure at the electrode is given by $\log(P_{Ba})=-17200/T+8.56$, where P_{Ba} is in Torr and T is in degrees Kelvin.[1]

For the diffusion problem each point along the electrode is treated independently. The geometry is such that each electrode position together with the glass wall constitute a pair of infinite parallel planes separated from each other by a distance equal to the lamp radius.[2] The density of Ba at the glass wall is zero. The density just in front of the electrode is given by the equilibrium Ba pressure (evaluated at the electrode temperature at the given position) divided by the temperature (i.e., an ideal gas in thermal equilibrium with the electrode).

In this model mass flux loss Γ ($kg/m^2/sec$) at each position along the electrode is given by

$$\Gamma = \frac{m}{R}\left(\frac{0.06}{P_{Ar}}\right)\left(\frac{133P_{Ba}}{kT}\right)$$

where R is the lamp radius in meters, m is the mass of Ba in kg, and P_{Ar} is the argon pressure in Torr. Total mass loss rate (kg/sec) is obtained by integrating over the entire electrode, taking into account the temperature variation, as measured.

The two most striking departures of this model from reality are 1) the geometry and 2) the material transport mechanisms leading to deposition of material back onto the electrode. Although more realistic models for these provide us with more accurate absolute material loss, this simple model still provides accurate relative material loss for the parameter regime of interest to us. The reason for this is that material loss is most strongly dependent on variations of the source function (i.e., Ba pressure/density at the electrode) and more weakly dependent on changes of geometry and transport. Geometry does not change and transport changes very little over the parameter regime of interest.

4 REFERENCES

[1] W.C. Rutledge and E.S. Rittner, J. Appl. Phys., **28**, 167 (1957).

[2] J.F. Waymouth, *Electric Discharge Lamps*, (The MIT Press, Cambridge, 1971).

L 16

Temporally and Spatially Resolved Spectroscopic Investigations of Dielectric Barrier Discharges

F. Adler, S. Müller
Institute of Low-Temperature Plasma Physics
Robert-Blum-Straße 8-10, D-17489 Greifswald, Germany

Introduction

Dielectric Barrier Discharges (DBD) are convenient sources of UV and VUV radiation for many technological applications [1]. The purpose of our work is to understand the dynamic behaviour of the DBD and, in particular, the mechanism of excimer formation and relaxation. In this work we present results of temporally and spatially resolved spectroscopic investigations in the VUV range.

Experimental Details

The investigations concerning the emission spectra, produced by the decay of excited molecules, were performed using a single discharge configuration. In order to achieve spatially resolved results a single discharge was mounted movably. Thus, different parts of the microdischarge could be focused onto the entrance slit of a monochromator, which allowed a spatial resolution of about 15 μm. The temporally resolved measurements were realized by measuring the temporal evolution of each part focused onto the entrance slit. Using time-resolved photon counting a time-resolution of 500 ps could be achieved.

The spectra of xenon and argon were analysed at different discharge parameters such as pressure (200 – 770 Torr) and sustaining voltage (sinusoidal or square wave type).

Experimental Results

The discharge structure in the spectral excimer radiation region of a single microdischarge strongly depends on the external parameters mentioned above. As can be seen from Fig. 1 the observed structure of the microdischarge in this spectral region was found to be qualitatively similar compared with that optical structure arising from the radiation of the whole spectral range (compare with [2]). With increasing pressure the microdischarge becomes more filament-like and the discharge areas at the surface spread out. Increasing applied voltage causes a further radial spreading of the discharge at the electrode areas. Furthermore, it could be observed that the radiation maximum near the anode shifts towards the anode with increasing pressure. This behaviour can be explained by the generation of repulsive electrostatic forces due to wall charges and spreading effects at the surface.

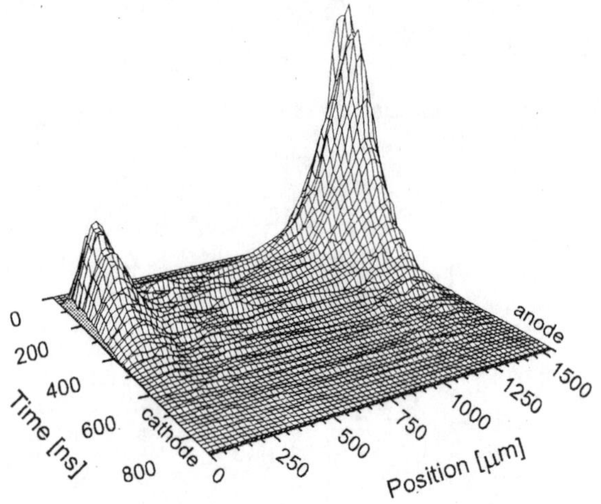

Fig. 1 Typical time-resolved discharge structure in Xe
- Wavelength: 172 nm (2. continuum)
- Discharge parameters:
400 Torr Xenon
Applied Voltage: 2.6 kV

Fig. 2 shows the temporal radiation development of the discharge related to the applied voltage and the discharge current. As seen from Fig. 2 a remarkable delay of the radiation maximum compared to the current maximum occurred. In Xe the radiation reached its maximum 200 to 250 ns after the current peak. In Ar the delay was even more, at low pressures (200 Torr) up to 1.5 µs. Moreover, the decay time of the radiation in the investigated wavelength region was rather long compared to the behaviour of the discharge current. The decay times were found to be 250 to 750 ns in Xe and 6 to 12 µs in Ar. Corresponding to the discharge development seen in streak images [2] the continuum emission starts at the anode first.

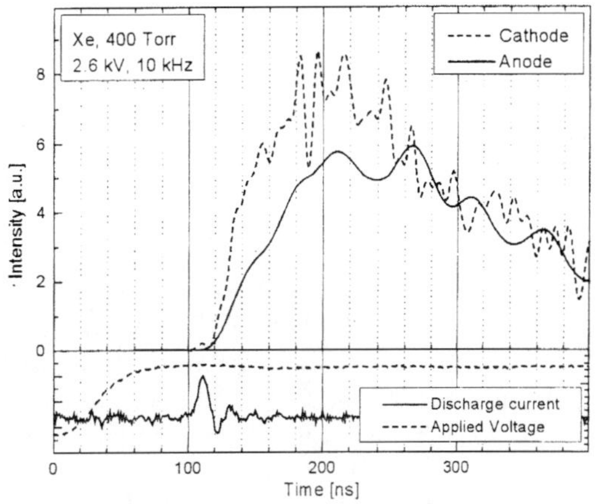

Fig. 2 Discharge current and the corresponding light emission of the 2. continuum

The temporal behaviour of the DBD is of complex nature. In Ar the decay times at the cathode were found to be different from that at the anode. At low pressures the decay times at the cathode were longer than at the anode but with increasing pressure this behaviour reversed.

In Fig. 3 the temporal behaviour of the 1. and 2. continuum is demonstrated. This figure shows the temporal development of the continuum radiation at 200 Torr Xe. Corresponding to these time curves in Fig. 4 the VUV-spectra at different times are plotted. These figures indicate that the 1. continuum starts before the 2. continuum and also reaches its maximum first. The situation described above can be explained by means of the chemical dimer path way. The formation of electronically excited rare gas dimer evolves via highly excited states.

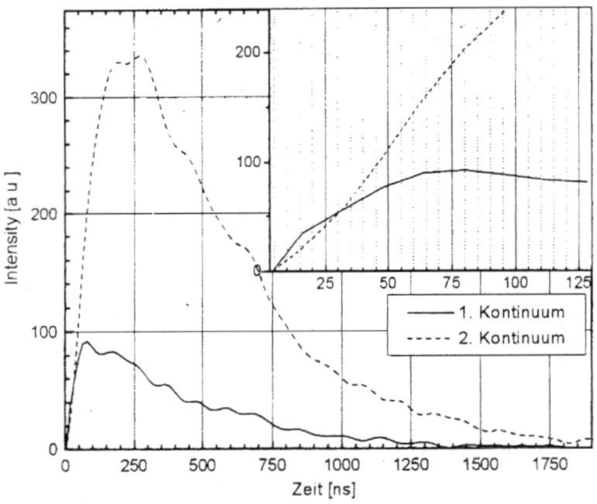

Fig. 3 Temporal behaviour of the 1. and 2. continuum

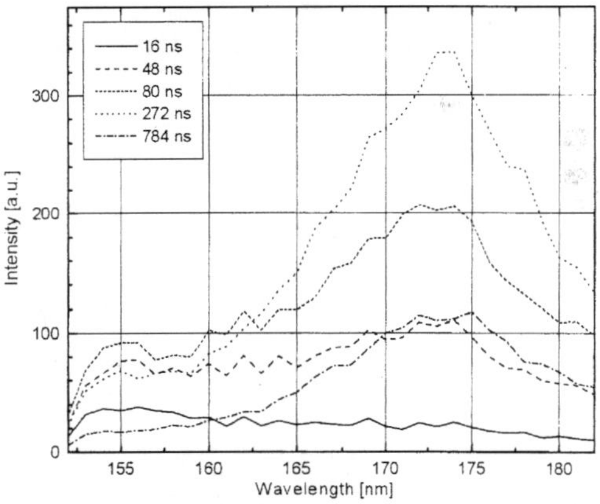

Fig. 4 Time-resolved VUV-Spectra (pressure: 200 Torr, applied voltage: 2.2 kV)

References

[1] U. Kogelschatz, B. Eliasson, W. Egli
Dielectric Barrier Discharges. Principle and Applications
Proc. XXIII ICPIC, Invited Papers (1997) C4-47.

[2] S. Müller, R.J. Zahn
Investigation of the Electrical and Optical Properties of DBD's
Proc. LS:7, Kyoto/Japan, 8/1995.

L 17 HIGH PRESSURE QUADRUPOLE MASS SPECTROMETER FOR RESIDUAL GAS ANALYSIS IN FLUORESCENT LAMPS

Massimo Righetti and Paolo Manini
SAES Getters S.p.A, viale Italia 77, 20020 Lainate (MI), Italy

1 INTRODUCTION

The detrimental effect of gaseous impurities in the filling gas of fluorescent lamps is well known. Lumen output, duration and electrical characteristics (e.g. starting voltage) may already be negatively affected by impurities concentration in the tens of ppm range [1, 2, 3].
This paper describes a new technique to measure impurity concentration down to the ppm range in a noble gas (e.g. argon) at 1-10 hPa pressure, which is the typical filling gas pressure in fluorescent lamps. With respect to alternative techniques, for example based on HID gas chromatograph [4], the presented system allows real time monitoring of gas evolution, opening the possibility to directly monitor interesting phenomena taking place inside these widespread light sources.

2.1 EXPERIMENTAL SET UP DESCRIPTION

The technique is based on the Balzers SPM200 quadrupole mass analyzer [5]. Due to modification of the ionization chamber, this instrument allows the direct analysis of gases at a maximum pressure of 10^{-2} hPa, avoiding the well known problems related with high pressure analysis. A schematic description of the experimental set-up is given in fig. 1, the differential pumping system reduce the pressure in the sample area by a factor $5x10^{-3}$. The presence of the argon purifier guarantees the quality of the filling gas for the calibration and blank tests. A non-evaporable getter pump is added to improve the background impurity level before starting the analysis.

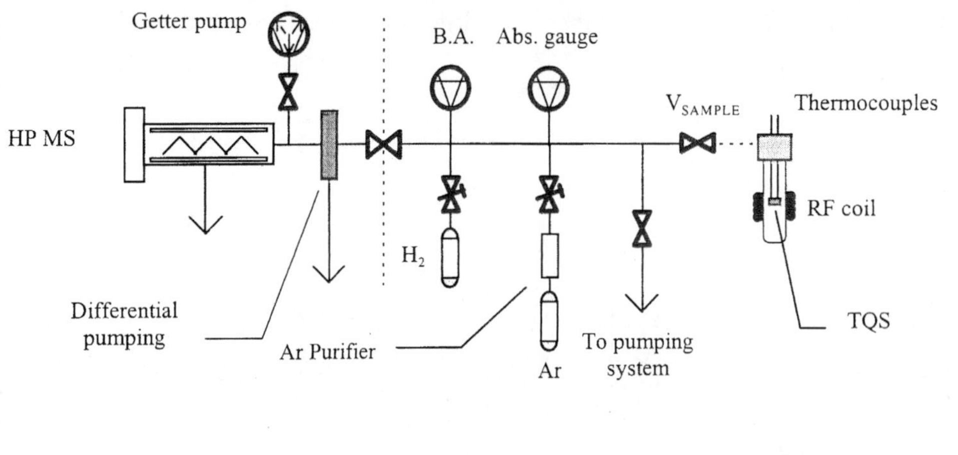

Fig. 1: UHV bench with high pressure mass spectrometer (HP-MS)

2.2 ANALYSIS PROCEDURE AND APPLICATIONS

In the following, two examples of application of the system will be given:
1) effect of the discharge on hydrogen concentration in a hydrogen doped fluorescent lamp. A standard 36W fluorescent lamp was connected to the sample area. After opening of the lamp, $3*10^{-5}$ Pa*m^3 (= 0,30 ccmbar) of hydrogen was added to the filling gas. The lamp was then ignited and kept "burning" for some minutes, at the same time the H_2 partial pressure was measured (fig. 2);
2) getter activity for hydrogen of a high yield mercury dispenser with getter capabilities, i.e. a total quality shield (TQS) [6]. The test was performed in a lamp at working condition, i.e. in presence of 2 hPa argon pressure. A TQS was mounted in a glass bulb connected to the sample area, the system was baked overnight at 120°C (background pressure stabilize in the low 10^{-6} hPa range when isolated from the vacuum pumps). The dispenser was then heated at 900 °C for 10" in 2hPa Ar atmosphere to release the mercury and activate the getter material. 10^{-3} Pa*m^3 (= 10 ccmbar) of hydrogen, much lower than the total getter capability, was subsequently introduced onto the TQS

(kept at 200 °C, temperature reached by the shield during working condition of the lamp). Evolution of the hydrogen concentration is reported in fig. 3, showing the capability of TQS to easily absorb the H_2 dose.

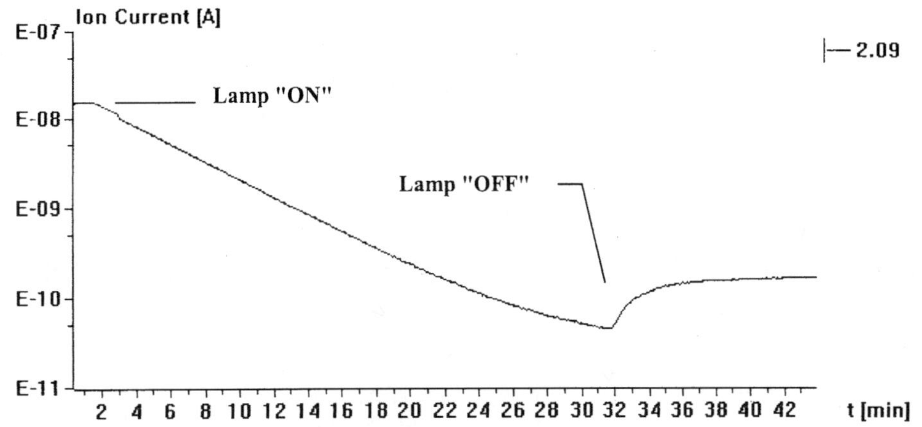

Fig. 2: trend of H_2 ion current in H_2 doped 36W fluorescent lamp during the discharge

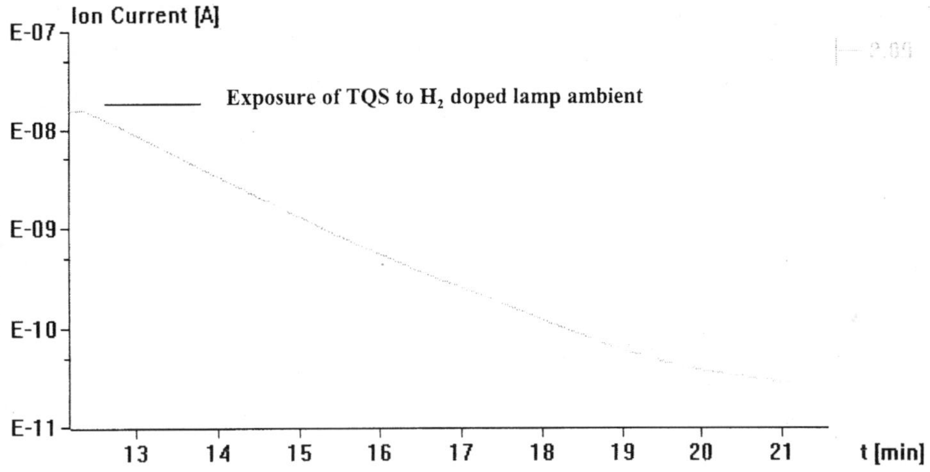

Fig. 3: decrease of H_2 partial pressure in the sample volume due to the pumping effect of TQS

3 CONCLUSIONS

A technique able to detect gas impurities down to the ppm level in a noble gas at a relatively high pressure (1-10 hPa) has been presented. The system is based on a BALZERS SPM 200 high pressure quadrupole mass analyser which allows real time monitoring of the sample gas. Two applications have been presented: 1) trend of hydrogen concentration during the discharge in a hydrogen doped fluorescent lamp and 2) pumping effect for hydrogen of a getter and mercury dispenser (TQS) in 2 hPa of argon contaminated with hydrogen. Other applications, i.e. : correlation between impurities level and electrical parameters (e.g. starting voltage), desorption from phosphors etc. , will be subjected to future investigations.

4 REFERENCES

[1] E. Rabusin "Getters for Lamps", SAES Getters S.p.A. Technical Publications (1997)
[2] R. B. Winkler et al, Annalen der Physik 7, 39 (1982)
[3] G. Bergmann, Lighting Research and Technology, Vol. 10 No. 1 (1978)
[4] M. Borghi and L. Rosai, J. Vac. Sci. Technol., Vol 12 No 5 (1975)
[5] N. Müller, G. Rettinghaus and G. Strasser, J. Vac. Sci. Technol. A 8 (3), (1990)
[6] S. Giorgi and M. Righetti (presented at LS8 Greiswald - Germany - 1998)

Modelling

M 01

Modelling of arc-electrode systems - preliminary results for high pressure xenon arc lamps

J. Wendelstorf, I. Decker, H. Wohlfahrt and G. Simon
Technische Universität Braunschweig, Postfach 3329, D-38023 Braunschweig
http://www.ifs.ing.tu-bs.de/jw/aglbphys.html

1 INTRODUCTION

Properties like lifetime and peak intensity of high pressure arc lamps depend on the electrode plasma transition of the discharge. The main high pressure lamp plasma can be described by local thermodynamic equilibrium (LTE) models [1-3], but the formation of the cathode hot spot is not fully understood. It dominates the overall lamp system for electrode stabilised discharges (short arc lamps) and it may be responsible for the lamps lifetime (all discharges).

To compute predictions for new lamp configurations, self consistent models are required. There are two approaches currently under investigation. The model of Fischer [4] is implemented into an adaptive finite element (FEM) model by the group of Neiger [5,6] and a finite difference hydrodynamic model is extended to include non-LTE boundary layers. Initial results for the second approach are sketched in this paper.

For the validation of such models, the cathode spot transition in high pressure Xenon lamps [7] is proposed.

2 DETAILS OF THE LAMP MODEL

The 2- or 3-D modelling of stationary high pressure arc columns is state-of-the-art for a number of application areas. For a 4 MPa Xe short arc lamp, Palacin [3] computed the flow field and temperature distribution in the lamp plasma and found reasonable agreement with experimental data. But pure LTE-models need boundary conditions, especially at the cathode surface. As discussed in [9], the inverse problem of deriving these boundary conditions from comparison with spectroscopic plasma temperature data will require very high experimental accuracy and the results will depend on arc configuration and current.

The cathode surface temperatures and electrical current densities have to be computed self-consistently, rather than prescribed. The approach sketched here is an iterative scheme linking the (2-D) electrode and LTE-plasma models by a locally 1-D sub-layer model. The overall computation region is selected as the part of the lamp significantly affected by the discharge details (figure 1).

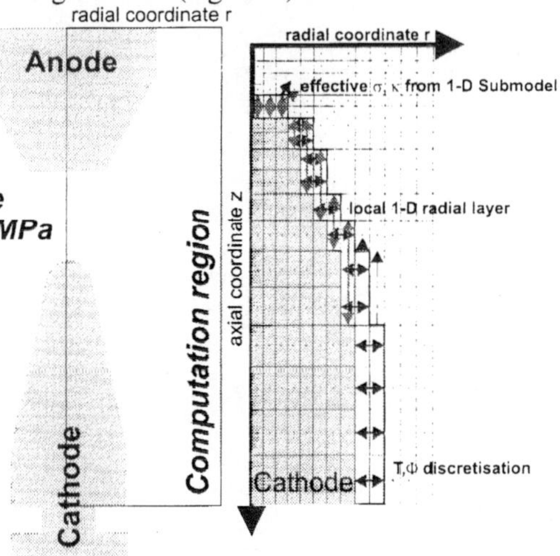

Fig. 1: Schematic diagram of the computation region (left) and the discretisation details for the boundary layer.

2.1 THE ARC ELECTRODE SYSTEM

The description of the LTE-arc plasma is composed of fluid flow, heat transport and current continuity balances formulated in cylindrical co-ordinates and solved by finite difference methods [8]. Inside of the electrodes, heat and current transport is computed. Temperature dependent transport coefficients are used [7,12-14].

For the iterative linkage of the plasma and electrode current and heat flow calculations the simple approach of conjugate heat transfer is used. This method implies the description of the locally 1-D boundary layers (anode and cathode sheath) by effective electric and thermal conductivities. These have to be derived from the locally 1-D sub-layer description (*transfer function approach*).

2.2 Lamp configuration and results

The lamp investigated here and by Hoppstock [7] is a 4 MPa Xenon short arc lamp with 2mm arc length (see figure 1). For a 1:1000 TlI doping, the cathode spot constricts at 8 A.

First results were computed by discarding any sheath model, but adding an additional heat flux at the cathode surface (electron emission cooling, ion heating and thermal radiation loss [10]), a *no-sheath transfer-function.* The plasma shows a diffuse attachment to the cathode for all currents. At 9A, temperature and flow fields are in good agreement with experimental data. The computed arc voltage is too small by the amount of the cathode sheath drop not included in the model (7V, [7]).

The ambipolar diffusion model of Lowke et.al. [10] is able to produce results for atmospheric argon arcs, but it is not applicable for the plasma regime in the lamps investigated here. A 1-D model for a collision dominated space charge sheath [11] was used for the cathode and gives arc voltages 1-2V below measured (using the *no-sheath* model for the anode). At 7.5 A, convergence becomes too slow to judge the model's ability to describe the spot transition. The temperature distribution at 9A agrees with spectroscopical data within the experimental error (see figure 2 and [7]).

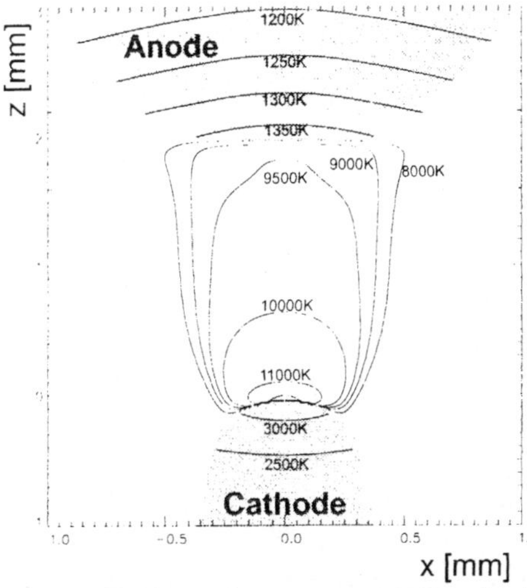

Fig. 2: Computed temperature distribution for a 9A, 4MPa Xe lamp with 1:1000 TlJ.

3 Conclusions

The integrated electrode-plasma models developed for 0.1 MPa argon arcs can be adapted for high pressure lamp situations, but different 1-D boundary layer models have to be used.

The transition of the cathode operation mode from diffuse to constricted within a small current range can not be predicted by neglecting the cathode sheath. Additionally, further physical effects may play a role for this phenomenon, e.g.

- the electron emission mode changing from thermionic to field enhanced.
- deviations of electron and heavy particle temperatures (PLTE).

4 Acknowledgements

Financial support by the BMBF under the „Forscherverbund Lampenkatoden" (Contract No. 13N7105) is gratefully acknowledged.

5 References

[1] P Y Chang and W Shyy, *Int. J. Heat Mass Transfer* **35** (1992) 1857

[2] K Charrada et.al., *J. Phys. D: Appl. Phys.* **29** (1996) 2432

[3] A C Palacin, Diss., U Karlsruhe, (1988)

[4] E Fischer, *Philips J. Res.* **42** (1987) 58

[5] H Wiesmann and M Neiger, *Proc. 11th Int. Conf. on Gas Discharges and Their Applications*, Tokyo, **2** (1995) 380

[6] H Wiesmann and M Neiger, *7th Int. Symposium on the Science and Technology of Light Sources (LS-7)*, Kyoto (1995) 215

[7] R Hoppstock, Diss., U Karlsruhe (1987)

[8] J Wendelstorf et.al., In: H Cerjak (ed.), *Mathematical Modelling of weld phenomena 3*, IOM, London (1997) 848-897

[9] J Wendelstorf et.al., *Proc. 11th Int. Conf. on Gas Discharges and Their Applications, Greifswald,* **I** (1997) 62

[10] J J Lowke et.al., *J. Phys. D: Appl. Phys.* **30** (1997) 2033

[11] M S Benilov, *J. Phys. D: Appl. Phys.* **30** (1997) 1115

[12] R Wilhelm, Diplomarbeit, U Karlsruhe, 1996

[13] P D Desai et.al. *J. Phys. Chem. Ref. Data* **13** (1984) 1069

[14] S W H Yih and C T Want, Tungsten, Plenum Press, New York (1979)

M 02

Modeling of High Intensity Discharge Lamps

Matthias Born, Hermann Giese, Thomas Krücken
Philips Research Laboratories
Weisshausstrasse 2, D-52066 Aachen, Germany
M. Neiger, Lichttechnisches Institut der Univ. Karlsruhe (TH)
Kaiserstr. 12, D-76128 Karlsruhe, Germany

June 2, 1998

1 Introduction

The software presented here calculates self consistently flow field, plasma temperatures, electric field, heat transfer, bulb and electrode temperatures in High Intensity Discharge (HID) lamps in full three dimensional geometry. Within the discharge Maxwell equations, energy equation and Navier Stokes equation for compressible fluids are solved self consistently. Using the resulting heat transfer from the plasma to the inner wall, bulb temperatures can be predicted with great accuracy. The model used to compute bulb and electrode temperatures takes into account solid state heat conduction within bulb and feedthrough, radiation from the outer wall as well as wall to wall radiation among the inner walls.

2 Physical Model

Of all light sources, HID lamps offer the highest light intensities. An HID burner always consists of a bulb, usually made of transparent quartz, two (tungsten) electrodes, and a tight electrical feedthrough. The filling consists of a buffer gas, usually mercury, and some additional substances to influence the radiation spectrum. During operation the gas pressure is typically of order 5 MPa ($\approx$50 bar). Since such a discharge has a negative electrical characteristic, these burners must be operated with a ballast which adjusts current or power input. The resulting operating voltage is then mostly determined by plasma properties. It must, however, lie within a certain range.

Main goal of the presented software is the evaluation of electrical and thermal properties of the discharge as well as temperature profiles of bulb/electrodes. Especially maximum and minimum temperatures at the inner wall of the bulb are crucial. Differences as little as 20 K can lead to significant changes of maintenance or performance of a lamp: The hottest temperature is often close to the limits the bulb material can handle, the coldest temperature determines the gas pressure and composition in the entire discharge. At the rather high pressures, natural convection within the discharge is an important heat transfer mechanism [1]. The operating voltage is also determined by temperature profiles.

The software is based on the commercial finite element fluid dynamics package FIDAP. It is structured such that the simulation of the plasma and bulb/electrode are separated. Plasma and bulb/electrode simulations are linked through the exchange of data at the inner

One usually starts with a simulation of the plasma. An estimated inner wall and electrode temperature profile must be entered as boundary condition. Before running the plasma simulation, a chemical equilibrium program is invoked to evaluate and tabulate the plasma material properties viscosity η, mass density ρ, electrical conductivity σ, specific heat c_p and thermal conductivity λ including chemical energy transport as a function of temperature for a given filling and an estimated coldest spot temperature. Optically thin radiation is treated as heat sink. Non optically thin radiation leads to a contribution to the effective thermal conductivity in the discharge. The latter can be either calculated from a theoretical model based on radiation diffusion of various mercury spectral lines [2] or from a semiempirical approach based on measurements of operating voltage and mean wall temperature [3].

The plasma simulation is by far the most complicated since it solves self consistently the electric continuity equation to evaluate the electric field, the energy equation to evaluate temperatures and heat fluxes, and the Navier Stokes equation for compressible fluids with natural convection driven by gravity to evaluate the flow field. The situation gets even more complicated since one cannot apply usual boundary conditions for the electric field: As these lamps are operated with a ballast limiting the current or power input, we impose a global condition for the total power deposited, instead.

As a result of a plasma simulation the heat flux profile onto the inner bulb wall is computed. This heat flux profile is then used as boundary condition in a bulb/electrode simulation. Here the energy equation is solved to evaluate the temperature profile. At the outer wall, radiation boundary conditions and convective heat transfer boundary conditions are imposed. At the inner wall energy is exchanged via infrared radiation within the bulb. Evaluated are temperatures in the bulb, electrode and feedthrough. Outer wall temperatures and electrode temperatures can be used to compare the calculations with measurements. The inner wall temperature profile can be used in a next step as boundary condition for a new plasma calculation, if the deviations from the temperatures used as boundary conditions in the previous plasma run are too large. In a similar way the coldest spot temperature can be used to rerun the chemical equilibrium program to update the plasma material properties. This procedure can be iterated over several steps.

The decoupling of the two simulation modes allows to study many problems, where the interaction is negligible, separately thus significantly reducing computing time. For example studying the influence of bulb parameters like wall thickness on inner wall temperature does not require to simulate the plasma for each computation. Since the exchange of the required data between different simulations is automatically handled by the software, the user does not notice any additional effort.

References

[1] H. Giese. *Theoretische Untersuchungen zur Konvektion in Quecksilber-Hochdruckgasentladungslampen.* PhD thesis, Universität Fridericiana Karlsruhe, 1996.

[2] R. Wilhelm. Berechnung des Strahlungsdiffusionsanteils der Wärmeleitfähigkeit von Hochdrucklampen-Plasmen. Technical report, Lichttechnisches Institut der Universität Karlsruhe, 1997.

[3] M. Born. Modeling of radiation heat transfer in high pressure discharge lamps. To be published.

M 03

NON-LTE MODELLING OF ARC COLUMN OF HIGH PRESSURE SODIUM DISCHARGE LAMPS

Sándor Holló

GE Lighting TUNGSRAM

Váci út 77, Budapest, H-1340 HUNGARY

INTRODUCTION

The concept of Local Thermodynamic Equilibrium (*LTE*) has been widely accepted as standard approach for describing phenomena in high pressure discharges and therefore became a key tool for modelling arc column behavior of high pressure discharge lamps [1]. However, it was known from the early days of high pressure discharge modelling that for numerous situations LTE assumption is not valid. Earlier work of Waszink [2] showed that close to the wall of a high pressure steady-state sodium arc electron temperature deviates significantly from that of the heavy particles. Same effect was found more recently in high pressure mercury arcs [3].

This work presents a consistent time dependent detailed kinetics model within the frame of which the different non-LTE mechanisms can be quantitatively studied.

THE MODEL

In the core of the model there are three sets of time dependent balance equations accompanied by radiation transfer. Cylindrical 1D symmetry is considered.

- *Balance of electrons/ions, excited and ground states.* Quasi-neutrality assumed, transport of electrons and ions are governed by ambipolar diffusion. Detailed kinetics of electron and radiation driven reactions are included.
- *Energy balance of electrons.* Key terms are heat conduction, Joule-heating, reaction heat, energy transferred to heavy particles.
- *Energy balance of heavy particles.* Ion/atom temperature is determined by energy exchange with electrons and heat conduction toward the wall.
- *Radiation.* For quantitative modelling full radiation transfer is solved for each individual transition. For quick qualitative trials diffusion approach of radiation transfer is used. Key line broadening mechanisms like Na-Na, Na-Xe, Na-Hg interactions are included, too. As a result detailed output spectra can be obtained.

The above full detailed kinetic *(Kin)* model can be reduced to a partial LTE description *(pLTE)* by assuming that all the reactions are in local equilibrium at the local electron temperature. Special care was taken to consistently derive the different transport terms coming into the electron energy balance (heat transfer via transport of excited and ionized states). Finally, having the electron temperature be equal to the heavy particle temperature reduces the system to the well known *LTE* approach.

Straightforward boundary conditions were imposed at the axis. At the wall deexcitation of the excited states and wall recombination of ions were qualitatively included. No energy transfer was considered in between electrons and the wall. Radiative boundary condition was set for the energy balance of the heavy particles, by which wall temperature could also be calculated.

The self-consistent set of equations were solved on a non-uniformly discretized mesh by fully implicit time marching.

RESULTS

The *Table* and the *Figure* illustrate some of the key conclusions found for a DC steady state case.

- Validity of *LTE* is reaffirmed for processes close to the axis in steady-state
- Two temperature *pLTE* approach is superior for regions close to the wall and for describing dynamic behavior
- For wall sheath phenomena and fine details of spectra kinetic description is needed

		Experiment	LTE model	pLTE model
Voltage	*[V]*	**93.5**	**92.1**	**92.1**
Wall T	*[K]*	**1405**	**1466**	**1416**
Lm/W	*[-]*		**138**	**152**
Radiative efficacy	*[%]*		**48**	**52**
Atom/Ion T at Axis	*[K]*		**4166**	**4153**
Electron T at Axis	*[K]*		**4166**	**4162**
Electron T at Wall	*[K]*		**1466**	**2286**

Table

Comparison of experimental data with LTE and pLTE model results. DC Na-Xe discharge.
Current: 2.8A.
Tube diameter: 2.5mm.
Arc gap: 98 mm.
Na pressure: 15 kPa

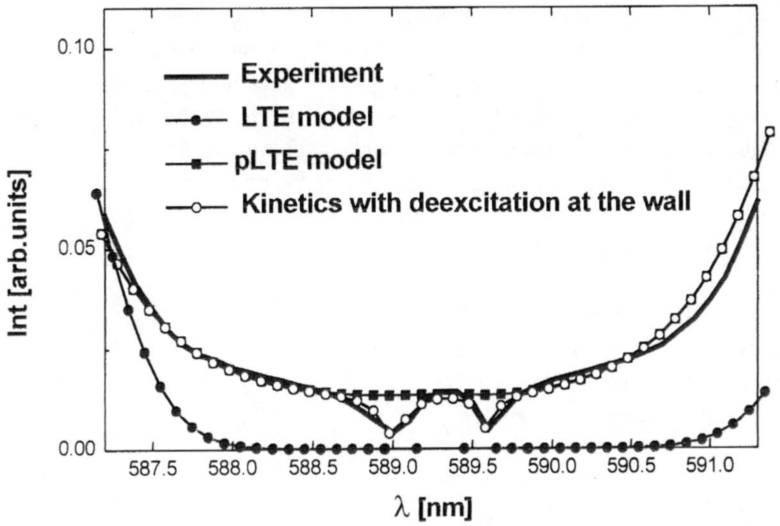

Figure

Spectra near Na D-line for the same DC case as in the Table. Fine details of D-line reabsorption can be interpreted just by the detailed kinetic approach *(Kin)*.

ACKNOWLEDGEMENTS

Valuable discussions and experimental support was greatly appreciated from J.T.Dakin, B.Nyíri, J.Sárközi, D.O.Wharmby.

REFERENCES

[1] J.T.Dakin, T.H.Rautenberg, E.M.Goldfield, *J.Appl.Phys.* **66** (1989) 4074

[2] J.H.Waszink, *J.Phys.D: Appl.Phys.* **6** (1973) 1000

[3] K.Charrada, G.Zissis, M.Aubes, *J.Phys.D: Appl.Phys* **29** (1996) 2432

M 04

A COLLISIONAL-RADIATIVE MODEL FOR HIGH-FREQUENCY DISCHARGES

N. V. Denisova[1], **A. Skudra**[2]
[1] Institute of Theoretical and Applied Mechanics, 630090, Novosibirsk, Russia
[2] Institute of Atomic Physics and Spectroscopy, University of Latvia, Raina blvd. 19, LV 1586, Riga, Latvia

1 INTRODUCTION

High-frequency electrodeless light sources (HFELS) are widely used as intensive sources of a line spectrum. In order to optimise design, filling and working regimes of HFELS, it is necessary to understand the physical processes taking place in the high-frequency (HF) discharge. Various models of electrodeless discharges have been developed to analyse their operation. The electromagnetic properties of inductively coupled (electrodeless) discharge have been considered in [1] - [3].

The present work was motivated by the need for the model which can be used in spectroscopic applications. In order to relieve the theoretical approach of the HF discharge, we prepared and investigated cylindrical HFELS, filled only with helium at various pressure. A stationary collision-radiative model of electrodeless high-frequency discharge has been developed.

2 THEORETICAL MODEL

The structure of the model consists of a set of coupled rate equations for the n=15 excited states of atoms, the energy and density balance equations for the electrons, including elastic and inelastic collisions with atoms and the coupled Maxwell's equations. This model has been described in [4] in detail. The calculations were performed in [4,5] for a discharges in noble gases Ar, Kr, Xe at pressure 0.5 - 10 Tor. Here we present a modified model for a discharge in He.

A method for a self-consistent solution for the plasma-field system in an electrodeless HF discharge is developed. It is assumed that the skin effect is significant, and a computational procedure is given to obtain the spatial distribution of the main plasma parameters and field strengths. The discharge is supported in a tube of radius R and is represented by an infinitely long cylinder placed inside a solenoid. The plasma is assumed to be homogeneous along the z axis and the parameters are assumed to depend on the radial coordinate r. The system of equations in the situation with a distinct skin effect may be formulated as follows:

$$\frac{\partial n_e}{\partial t} = D\Delta n_e + S_e n_e \qquad (1)$$

$$\frac{3}{2}\frac{\partial}{\partial t}(n_e kT_e) = div(q_{cc}) + S^+ - S^- + Q^+ - Q^- \qquad (2)$$

$$N_m[\sum_{k<m} A_{mk}\theta_{mk} + n_e(wI_m + \sum_{k\neq m} w_{mk})] - [\sum_{k>m} A_{km}\theta_{km}N_k + n_e\sum_{k\neq m} w_{km}N_k] = 0 \qquad (3)$$

$$\Delta H_z - iH_z/\delta^2 = 0; \qquad rotH_z = 4\pi\sigma E_\varphi / c \qquad (4)$$

Equation (1) describes the process of ambipolar diffusion, D being an appropriate diffusion coefficient, S_e is the electron source term due to collisional ionization. Electron energy balance is expressed by the Eq.(2). q_{cc} defines the total (conductive + convective) flux of electron heat transfer. S^+ and S^- are the source and sink terms for heating and cooling of the electron gas due to elastic and inelastic collisions with atoms, respectively. Q^+ is the energy contribution from an external field, Q^- is the work performed against the field which is determined by ambipolar diffusion. The rate equations (3), in which m is the variable index, is written in simplified form, thereby taking into account only the terms playing an important role in the typical situations. The terms $A_{mk}\theta_{mk}$ are the Einstein coefficients modified in accordance with the Holstein formula for trapping by the transverse dimension of cylindrical discharge, and θ_{mk} is an escape factor. N_m and N_k are the populations of levels m and k, respectively. Other quantities that appear in Eqs. (3) are wI_m, which is the collisional ionization rate coefficient, and w_{mk}, which is the rate coefficient for collisional transition $m \rightarrow k$. The Maxwell equations (4) are simplified according to the regime of a well-defined skin layer, δ is the depth of the skin layer and σ is the complex conductivity of a plasma. It is appropriate to make some comments regarding the numerical solution of Eqns (1), (2). From a computation point of view, it is more convenient to solve transient equations using so-called «set-in» method to obtain a steady state solution. Several calculations were performed to test the model. The intensity of the lines λ_1=587.6 nm and λ_2= 728.1 nm of RF discharge in He was calculated as a function of the discharge pressure in the interval 0.1 ÷ 10 Tor. Rather good agreement between experimental results and calculations was obtained.

3 CONCLUSIONS

Good coincidence between the experimental results and theoretical calculations leads to the conclusion that the described model permits us to obtain the various important characteristics of HF discharge.

4 REFERENCES

[1] J. H. Denneman, J. Phys. D: Appl. Phys. **23**, 67 (1992).
[2] G. G. Lister, M. Cox, Plasma Sources Sci. Technol. **1**, 67 (1992).
[3] R. B. Piejak, V. A. Godyak, B. M. Alexandrovich, Plasma Sources Sci. Technol. **1**, 179 (1992).
[4] N. V. Denisova, N. A. Preobrazhensky, Spectrochimica Acta **49B**, 185 (1994).
[5] A. E. Bulischev, N. V. Denisova, A. Skudra, Opt. Spectrosc. **67**, 788 (1989).

M 05

THEORETICAL MODELLING OF THE CATHODIC REGION OF HIGH INTENSITY DISCHARGE LAMPS

H. Schmitz, K.–U. Riemann
Inst. für Theoretische Physik, Plasma- und Atomphysik, Ruhr-Universität Bochum
Unversitätsstr. 150, 44780 Bochum, Germany

1 INTRODUCTION

This work is part of investigations aiming at a **"Fundamental Characterization of Thermionic Cathodes"**. Although High Intensity Discharge (HID) lamps have been in everyday use for many years, many details concerning the physical processes are not yet understood. This holds especially for the transition from the partial-equilibrium plasma to the cathode body. It is our aim in this contribution to make a self-consistent analysis of the cathodic region applying general principles of the plasma boundary layer theory.

2 MODEL

The transition-region is dominated by a number of different effects including for example the ion-kinetics, the electron emission from the cathode surface and the heat conduction in the cathode body. Because of this a description of the whole cathodic region is very complicated. To be able to handle this complexity the region is divided into four distinct model-zones:

Saha-plasma: This zone forms the end on the plasma-side of the Model. The Saha plasma is in partial local thermodynamic equilibrium (PLTE). This means that ions and electrons can be described by thermodynamical equations while in general the ion-temperature differs from the electron-temperature. The Saha-equation is used to obtain charged particle densities.

presheath: The presheath connects the Saha-plasma with the thin space charge sheath. It is a quasineutral layer dominated by the ion-dynamics. In this layer the ions are accelerated towards the cathode so they fulfill the Bohm-criterion [1,2] when entering the sheath. The presheath extends over a few ionization-lengths. In its description ionization, three-body recombination, and charge-exchange (CX) collisions as well as the ion-inertia are considered.

space-charge sheath: The space-charge sheath screens the plasma from the negative cathode potential over a distance of a few Debye-lengths. Because the Debye-length is small compared with all lengths of the ion dynamics, the ions fall through this region collisionless.

cathode body: To achieve a self consistent description of the cathodic region the temperature of the cathode surface must be calculated. To this end the energy flux into the cathode (incl. ion heating, electron emission cooling and radiative cooling) is combined with the heat cunduction equation of the cathode body to obtain the energy equation.

In the presented contribution we formulate equations for the four model zones and solve them self-consistently. Our main attention is directed to the description of the presheath.

The different scale lengths of ionization and of recombination on the one hand and of CX-collisions and of strong inhomogeneities on the other hand make it suggest a further division of the presheath into a collision dominated layer and a Knudsen layer.

The **collision dominated layer** that is directly adjoined to the Saha-region can be described by an hydrodynamic approach describing the motion of ions and neutrals. This approach is closely related to a model of Benilov [3] and includes ionization, recombination and CX-collisions.

The **Knudsen-layer** is located in front of the space charge sheath and has an extension of a few ion mean free lengths for charge exchange collisions. This length scale is small compared with the ionization length. Therefore the Knudsen layer can be regarded to be free of ionization and recombination effects. However strong inhomogeneities occur in this region and the need to describe the layer with a kinetic approach using the Boltzmann equation arises. The theory utilized for the Knudsen-layer is based on a model given by Biehler [4].

3 CONCLUSIONS

Using the model equations for the Saha-plasma, the presheath, the space-charge sheath and the cathode body, the physical quantities are calculated selfconsistently using a minimum number of given parameters. Values calculated are e.g. the cathode-fall voltage U_c or the cathode surface temperature T_c. These results are compared with measurements presented in papers submitted to LS-8 by Luhmann, Nandestädt, Michelt and Mentel and good agreement is achieved.

4 ACKNOWLEDGEMENTS

This work is supported by the BMBF, Project No. 13N6895

5 REFERENCES

[1] D. Bohm: The Characteristics of Electrical Discharges in Magn. Fields. Ed. by A. Guthrie and R. Wakerling, New York: McGraw-Hill 1949

[2] K.-U. Riemann, J. Phys. D: Appl. Phys. **24**, 493 (1991)

[3] M.S. Benilov, J. Phys. D: Appl. Phys. **28**, 286 (1995)

[4] S. Biehler, G. Ecker, K.-U. Riemann, Phys. Fluids **31**, 1999 (1988)

M 06

MATHEMATICAL MODELLING OF MULTIPLE REGIMES OF CURRENT TRANSFER TO HOT ARC CATHODES

Mikhail Benilov
Universidade da Madeira, Largo do Município, 9000 Funchal, Portugal

1 INTRODUCTION

Multiplicity of modes of current transfer to hot arc cathodes is an effect important for design of high intensity discharge lamps [1, 2]. This effect is explained in the present work by means of non-uniqueness of multidimensional thermal balance of a finite sample heated by a non-linear external energy flux [3].

2 THE MODEL

The model is similar to that described in [4] and is as follows. A steady-state temperature distribution is considered in a cathode in the form of a right cylinder of the height h whose cross section is not necessarily circular. The equation of heat conduction assumes the form of the Laplace equation in terms of the heat flux potential ψ. The top surface of the cathode, $z = h$, is maintained at a fixed temperature T_c by external cooling. The lateral (inactive) surface is thermally insulated. The bottom surface, $z = 0$, is current-collecting; the respective boundary condition is $\partial\psi/\partial z = -q(\psi, U)$, where q is the density of the heat flux from the plasma to the current-collecting surface. The function q is considered as a given function of the local surface temperature T and of the voltage drop across the near-cathode layer U, which is constant along the surface. The specific form of this function for a thermionic cathode is taken from [5].

3 BIFURCATION ANALYSIS

The formulated problem may have a one-dimensional solution $\psi = \psi(z)$ which describes a diffuse regime of current transfer to the cathode; an example of the current-voltage characteristic described by the one-dimensional solution is given in Fig. 1. Besides, the problem may have multidimensional solutions $\psi = \psi(x, y, z)$ which branch off from (or join) the one-dimensional that. A procedure of finding the bifurcation points is as follows [3]. After the one-dimensional solution has been determined, one should find for each U a wave number k of steady-state perturbations which can branch off at U considered. The bifurcation points are associated to values of U at which k takes the values determined by the spectrum of the homogeneous Neumann problem for the two-dimensional Helmholts equation considered in the cross-section of the cathode. As an example, three initial bifurcation points are depicted in Fig. 1 for the case when the cathode is in the form of a circular cylinder of the radius $R = 2$ mm. The solutions branching off in the first, second, and third bifurcation points (B, C, and D, respectively) describe a mode with a spot at the edge of the cathode, with two spots at the edge, and with a circular spot at the center or a ring spot at the edge.

The present theoretical approach predicts a spot mode at low currents and the diffuse mode at high currents, with a transition accompanied by hysteresis. The transition from the diffuse mode to the spot mode is accompanied by an increase of the voltage drop in the near-cathode layer. This transition occurs at the point B at which the respective wave number

$k(j_B)$ equals $k_1 = 1.841/R$. The effect of the design parameters on j_B is illustrated by Fig. 2. (For convenience, the respective temperature of the cathode surface is also presented; note that the dashed lines 1 and 5 coincide). A reduction of the front area of the cathode results in a decrease of the transition current, as well as a cathode design with an increased heat resistance or a decrease of the thermal conductivity of the cathode material. A decrease of the work function of the cathode material results in a small increase of j_B for wide cathodes and in a decrease of j_B for cathodes not too wide. An increase of the argon pressure results in an increase of j_B. A substitution of an inert gas by mercury results in a small decrease of j_B for wide cathodes and in an increase of j_B for cathodes not too wide. These conclusions conform to the available experimental data (e.g., [4] and references therein).

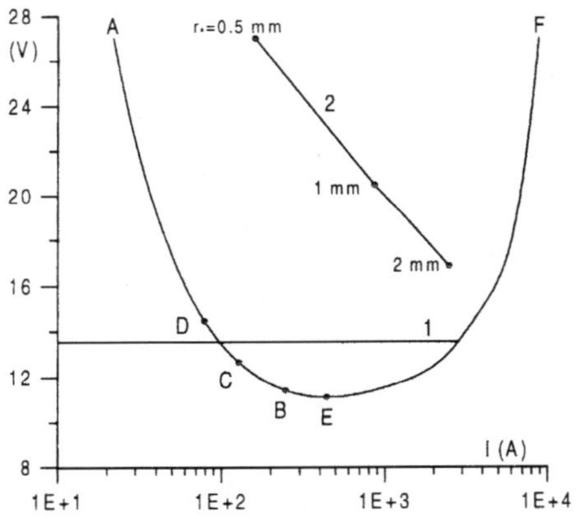

Fig. 1. Current-voltage characteristics of various modes of current transfer. Tungsten cathode, $h = 10$ mm, $R = 2$ mm, $p = 1$ atm, Ar plasma. AEF: the diffuse mode. 1: the normal-spot mode. 2: the spot on an infinite plane cathode.

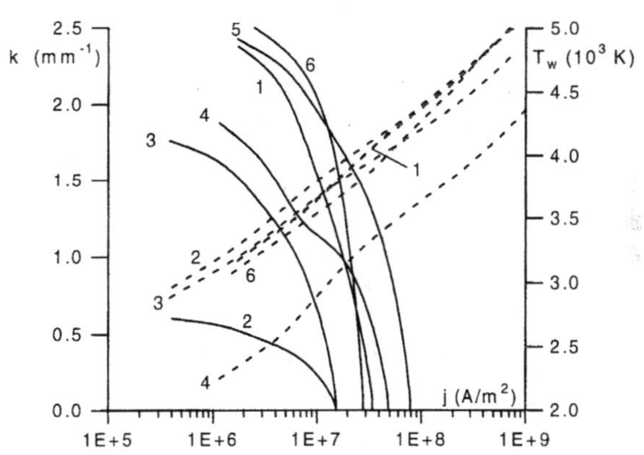

Fig. 2. The wave number (solid lines) and the temperature of the cathode surface (dashed lines). 1: tungsten cathode, $h = 10$ mm, Ar plasma, $p = 1$ atm. 2: $h = 40$ mm. 3: Zr cathode. 4: thoriated-tungsten cathode. 5: $p = 5$ atm. 6: Hg plasma.

4 ACKNOWLEDGMENTS

The work has been performed within the activities of the projects *Nonlinear, Stochastic and Complex Systems: Their Mathematical Theory and Application* of the program Praxis XXI (Portugal) and *Fundamental Characterization of Thermionic Cathodes* of BMBF (Germany) and was financially supported by FEDER, CITMA, and Heinrich-Hertz Stiftung. Part of the work has been carried out in Ruhr-Universität Bochum (Germany).

References

[1] J. F. Waymouth, J. Light and Vis. Env. **6**, 53 (1982).

[2] E. Fischer, Philips J. Res **42**, 58 (1987).

[3] M. S. Benilov and N. V. Pisannaya, Sov. Phys. - Tech. Phys. **33**, 1260 (1988).

[4] W. Neumann, *The Mechanism of the Thermoemitting Arc Cathode* (Akademie-Verlag, Berlin, 1987).

[5] M. S. Benilov and A. Marotta, J. Phys. D: Appl. Phys. **28**, 1869 (1995).

MODELLING OF HIGH PRESSURE DISCHARGE LAMPS INCLUDING ELECTRODES

Peter Flesch[1], Hartwig Wiesmann[2], and Manfred Neiger[1]

Lighting Technology Institute (LTI), University of Karlsruhe, Box 6980, 76128 Karlsruhe, FRG.
E-mail: Peter.Flesch@etec.uni-karlsruhe.de

[2]Ushio Europe B.V., Netherlands (formerly at LTI)

INTRODUCTION

The time-independent modelling of high pressure discharge lamps without dividing the discharge into special zones is the subject of this paper. We introduce the model equations first proposed by Fischer [1] and present results for mercury and xenon lamps using the finite element method.

THEORY

.1 Equations for temperature, potential and electrical conductivity

The high pressure discharge in a lamp is described by the electric potential V, the plasma temperature T_p, and the electrical conductivity σ. Additionally we calculate the temperature T_E in the electrodes. The potential V is determined by current continuity (1), the temperature T by the energy balance in the discharge (2) and in the electrodes (3). To calculate the electrical conductivity σ we use an approach first proposed by Fischer [1] which includes the effect of strong diffusion of electrons near the electrodes due to a large gradient in the concentration of electrons (4) (see also Fig. 2 for a comparison between the calculated σ and σ_{LTE}). A detailed derivation of (4) is the subject of a separate contribution to this conference.

$$\nabla j = \nabla(\sigma E) = -\nabla(\sigma \nabla V) = 0 \qquad (1)$$

$$\nabla(\lambda_p \nabla T_p) + \sigma(\nabla V)^2 - u = 0 \qquad (2)$$

$$\nabla(\lambda_E \nabla T_E) = 0 \qquad (3)$$

$$\Delta\sigma = -\frac{q^I n}{2D_{i,n}\alpha}\left(1 - \alpha\frac{\sigma}{\sigma_{LTE}}\right)\sigma \qquad (4)$$

(j = electric current density, E = electric field, λ_p & λ_E = thermal conductivity of the plasma/electrodes, u = net emission coefficient, q^I = ionization rate coefficient, n = particle density, $D_{i,n}$ = ion diffusion coefficient, α = ion density / electron density, σ_{LTE} = electrical conductivity in local thermal equilibrium (LTE)). The boundary conditions for the potential V are given by the input current I and by the condition that the electrodes are at constant potential. The temperature T of the inside walls is set constant to 1200 K. At the boundary between electrodes and discharge we consider heat conduction, cooling and heating due to the electron and ion current, and the radiative loss of energy as a boundary condition for T. As only the cathode-half of a lamp is treated in this paper, the electrical conductivity σ is set to its LTE-value at the midplane between cathode and anode. At the cathode surface the electrical conductivity σ is connected to the emission of electrons by the cathode [2].

.2 Geometry and material functions

The equations are solved using the finite element (FE) method assuming rotational symmetry. The advantage of this method is, that regions with large temperature gradients can be resolved with a finer mesh than others. This leads to a very fine mesh in front of the electrodes and a low density of mesh points in the rest of the discharge (Fig. 1). Moreover it is easily possible to consider arbitrary electrode shapes with the FE method. The material functions (λ_p, u, $D_{i,n}$, n, σ_{LTE}) depend on the temperature T_p and are calculated using the Chapman Enskog method. In the present paper we present modelling results for the cathode-half of D1-type Hg- and Xe-lamps (Fig. 1).

3 RESULTS

The potential V, the temperature T_p, and the electrical conductivity σ are computed in the discharge. The potential V can be used to calculate the electric field E and the cathode drop voltage V_c. An example of the temperature distribution T in the discharge and the cathode for Hg as filling gas is given in Fig. 1. The plasma temperature T_p and a comparison between σ and σ_{LTE} (see equation (4)) on the arc axis are shown in Fig. 2 for the same geometry as in Fig. 1 with Hg and Xe as filling gas. The input currents I are chosen so that the lamp power (in the cathode-half) is approximately the same for both fillings (Hg: 26W, Xe: 23W). For both figures the pressure p is 4 MPa and the work function is 2.6 eV. With our model we have analysed the effect of different electrode shapes, currents, work functions and gas pressures. Further results from the computations are the temperature distribution within and on the surface of the cathode, electron and ion current density distributions, electric field distribution on the cathode surface and the adjacent plasma, and different contributions to the energy balance of the cathode and the discharge.

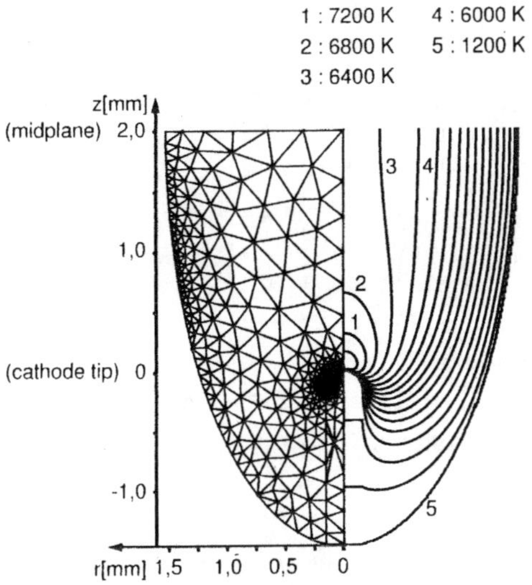

Fig. 1: Lamp geometry of a D1-type lamp (cathode-half) with the finite element mesh (left) and isothermes (right) for Hg, I = 0.75A.

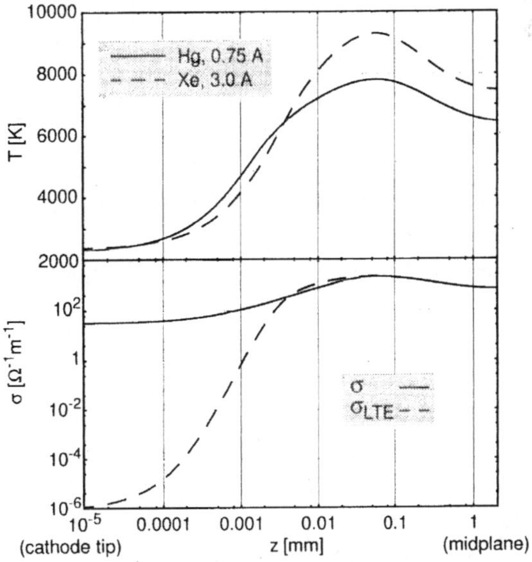

Fig. 2: Plasma temperature T_p (top) on the arc axis for Hg (I = 0.75 A) and Xe (I = 3.0 A). σ and σ_{LTE} (bottom) on the arc axis for Xe (I = 3.0A).

4 ACKNOWLEDGEMENTS

This work was supported by the BMBF, FKZ 13N7107/0 and the DFG (Graduiertenkolleg „Numerische Feldberechnung").

5 REFERENCES

[1] E. Fischer, *Modelling of low-power high-pressure discharge lamps*, Philips J. Res. 42, pp. 58-85 (1987)

[2] H. Wiesmann, *Zweidimensionale Simulation von Hochdruckgasentladungslampen unter Berücksichigung der Elektrodenbereiche mit der Finite-Elemente-Methode*, PhD dissertation, University of Karlsruhe (1997), ISBN 3-8265-2632-5

M 08

On Abnormal Intense Radiation at the Rotation of Electron Around a Dielectric Sphere

L. Sh. Grigorian†§, H. F. Khachatrian†§, P. F. Kazarian§

†Institute of Applied Problems in Physics, 375014 Yerevan, Armenia
§Department of Physics, Yerevan State University, 375049 Yerevan, Armenia. E-mail: gayane@arminco.com

The presence of matter causes a number of important electromagnetic phenomena (Vavilov-Cerenkov radiation, transition radiation, radiation at particle channeling etc.). It was shown [1] that the influence of matter on the radiation from a relativistic charge rotating in constant magnetic field (synchrotron radiation [2]) could be rather strong. In [1] the intensity of radiation from an electron uniformly rotating in vacuum with the velocity $v = r_e\omega_e$ about a dielectric sphere was calculated. The magnetic permeability of the sphere was assumed to be equal 1 and the radius of the sphere r_0 - to be smaller than the radius of electron orbit r_e, $r_0 < r_e$. It was also supposed, that the electron orbit is in the equatorial plane of the sphere. The effects connected with the absorption of radiation were ignored. In the present work we continue the analysis of results obtained in [1].

At the given $\omega = k\omega_e$ (k is the number of harmonic) the intensity of radiation is determined by the electron energy E_e, relative electric permittivity ϵ of the sphere and the ratio of the radius of sphere to that of electron orbit (r_0/r_e). Numerical calculations in the range of $1 \leq k \leq 50$ at $1 \leq E_e \leq 5MeV$, $1 < \epsilon \leq 5$ and $0 < r_0/r_e < 1$ showed that if the Cerenkov condition ($v > c/\sqrt{\epsilon}$) for the electron and the matter of sphere is satisfied, then there exist discrete values of the parameter r_0/r_e, at which anomalously strong radiation takes place on harmonics with $k \gg 1$. During one turn of electron about the sphere $n_k \gg e^2/hc$ quanta of electromagnetic field are emitted, whereas at the rotation in infinite continuous medium or in empty space the similar quantity $n_k < e^2/hc$ [2-4].

k	E_e MeV	$\epsilon = 3$ μ	$n_k(\mu)$	$\epsilon = 5$ μ	$n_k(\mu)$
20	1	6.6433228	4.13	5.2992	1.76
20	3	0.5432354	*201*	3.482	0.34
20	5	1.480803	*133*	2.596109	*133*
40	1	0.82132	9.64	1.13910742	*2260*
40	3	1.2224	0.65	0.9986	0.65
40	5	4.801	0.16	1.50036	1.45

For some values of $r_0/r_e = 1 - 0.01\mu$ (see the 2, 3 and 4-th rows of the Table) the radiation is superintensive:

$$n_k > 2\pi r_e/\lambda_k = kv/c.$$

In these cases the quantum of field is emitted sooner than the electron on the orbit has time to fly the distance λ_k (the wavelength of the k-th harmonic). Such an intense radiation is formed inside the sphere as a result of repeated reflection of Cerenkov radiation from the surface of the sphere.

The effect under consideration may be recommended to use in sources of intense radiation.

References

[1] L. Sh. Grigorian, H. F. Khachatrian, Journal of Contemporary Physics (Proc. Nat. Acad. Sci. Arm.), in press.

[2] A. A. Sokolov, I. M. Ternov, *Relativistic Electron*, Nauka, Moscow, 1983 (in Russian).

[3] V. P. Zrelov, *Vavilov-Cerenkov Radiation and It's Application in High Energy Physics*, Atomizdat, Moscow, 1968 (in Russian).

[4] L. D. Landau, E. M. Lifshitz, *Classical Theory of Fields*, Pergamon Press, London, 1979.

M 09

THREE-DIMENSIONAL SIMULATION OF EMISSION PROPERTIES FOR A GENERAL LIGHT SOURCE

Carles Pizarro, Santiago Royo, Núria Tomàs and Josep Arasa
Optics&Optometry Department, Technical University of Catalunya.
Violinista Vellsolà 37, Terrassa, E-08222, Spain

1 INTRODUCTION

Accurate modelling of light sources comes to be crucial in any photometric simulation. However, whenever photometric descriptions of source emissions are described, they are based in principles such as conglomerations of point emitters, axially normal subdivisions of the filament or even more complex distributions[1], hard to justify from a physical point of view.

Moreover, such models are usually difficult to implement because of the assumptions made. A procedure to develop a model which allows the calculation of the emission of the source in any direction through simple mathematical methods will be presented. The data needed to develop such a model are either provided by the manufacturer, or easily measured.

2 TECHNICAL DETAILS

The procedure used to develop the source model is presented in three steps, followed by a final step where the developed model is used in a simple photometric simulation.

2.1 Starting data

The polar intensity plots usually provided for any source model (isocandela plots) are an ideal starting data, as they are a description of the luminous flux emission used for a number of different sources. Furtherrmore, they may be measured easily using a goniometer. These data are asumed to represent the average emission properties of the lamp in three orthogonal planes. Eventually, a fourth plane is provided which might be used as a validation of the developed model.

The procedure will be used to develop a model for a typical commercial filament bulb (Philips P21W IncoK®), commonly used in car tail lights and indicators[2]. Although, this procedure has been succesfully applied to other sources types, such as LED and halogen sources. A full description of axes position's relative to the lamp, is presented in Figure 1.

2.2 Curve-fitting

The polar intensity plots will then becurve-fitted in order to obtain analytical expressions for the isocandela plots present in the described XY, YZ and XZ planes. This procedure is not straightforward, as some of the curves may present complex shapes that exceed the capabilities of most of the commercial software available.

Figure 1:Reference system

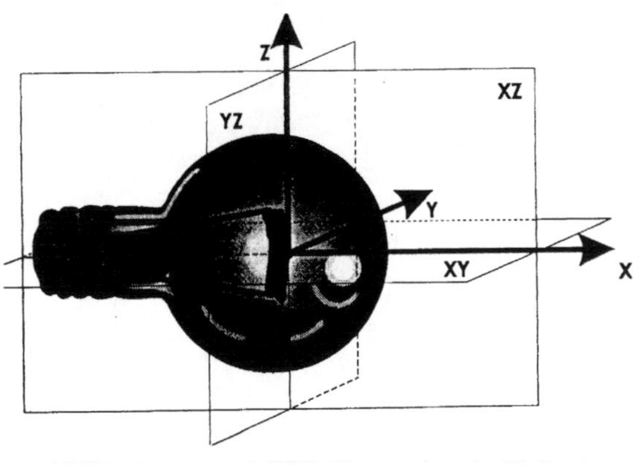

Figure 2: Modelled Philips P21W IncoK®

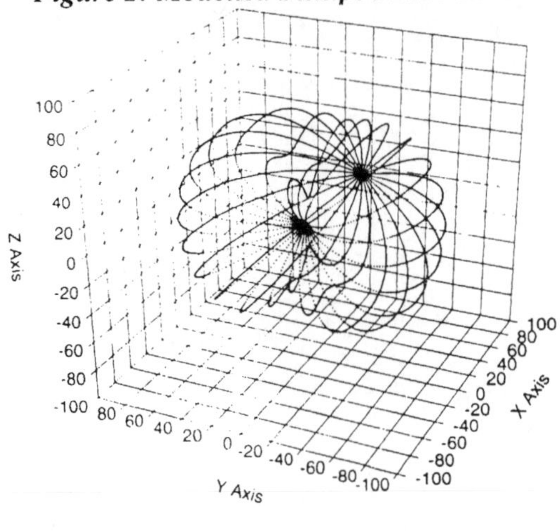

2.3 Interpolation

At this point, analytical expressions of luminous intensity emissions at three orthogonal planes (XY,YZ,XZ) are known. Let's describe the one contained in XY plane as $f_{XY}(\theta_{XY})$, as it is a function in XY plane depending only on the polar angle at that plane. So, $f_{XY}(\theta_{XY})$, $f_{XZ}(\theta_{XZ})$ and $f_{YZ}(\theta_{YZ})$ are known analytical expressions.

Our aim now is to describe an interpolation procedure that allows us to know the light intensity in any desired direction. The chosen procedure has been using the function present in the YZ plane($f_{YZ}(\theta_{YZ})$) as a weighting factor of the functions at each of the other planes. The intensity at all directions on a plane rotated α degrees from XY plane around X axis ($\theta_{YZ}=\alpha$), will be obtained as a weighted average of functions $f_{XY}(\theta_{XY})$ and $f_{XZ}(\theta_{XZ})$, with the weighting factor determined by the value of $f_{YZ}(\alpha)$. Obviously, the cases $\alpha=0°$ and $\alpha=90°$ will reproduce $f_{XY}(\theta_{XY})$ and $f_{XZ}(\theta_{XZ})$. Figure 2 shows the flux emission in space obtained from intensity plots in three orthogonal planes. The model has been validated by comparing the predicted flux emission with data from a polar intensity plot provided at a plane tilted 45° around X axis.

2.4 Results

The developed model has been applied to the calculation of the photometric distribution through a single plano-convex lens at a distant plane. In Figure 3a the optical axis is placed along the described X axis;in Figure 3b the optical axis is placed following the filament, along Z axis. The lack of lateral emission of the filament carrries on different photometric distributions for each of the positions of the source.

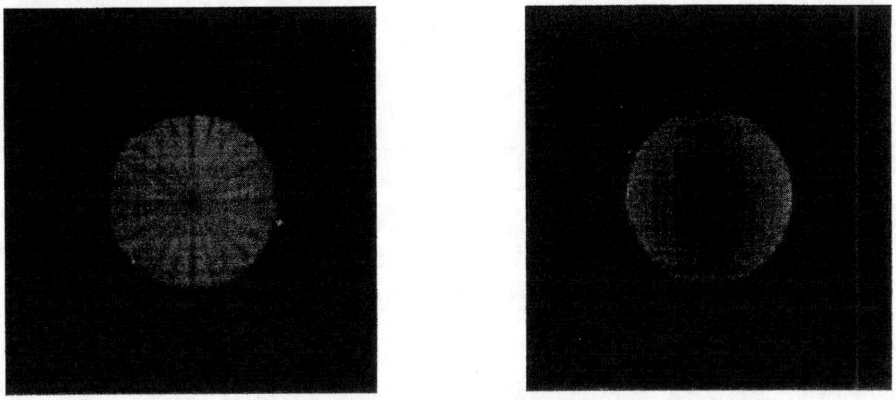

Figure 3: Simulated intensity patterns: a) Left. b)Right.

3 CONCLUSSIONS

A simple procedure for developing light source models has been presented. Starting from the polar intensity plots usually provided by the manufacturer, luminous intensity emission in any desired direction is obtained. A complete model for a typical filament bulb has been developed and validated. Realistic photometric distributions may be performed with this model, which allow to study complex features of the system(the influence of the position of the source) with an easy to obtain model.

4 ACKNOWLEDGEMENTS

The authors wish to thank Cristina Cadevall for her patience, used equally in programming and in her contribution to stablish new links of understanding between physicists and programmers.

5 REFERENCES

[1] I.Powell, and E. Brewsher, Opt.Eng. **33** 1678-1683 (1994)

[2] J.Arasa, S.Royo, J.Martínez and J.Arandia *Proc.3rd Int.Cong. on Project Engineering* ,Vol.2, p.446-453 edited by 3rd Int.Cong.on Project Engineering Secretary. Barcelona, 1996.

M 10

TOWARDS A GENERAL COLLISIONAL RADIATIVE MODEL

J. van Dijk, H.W.P. van der Heijden, A. Hartgers, J. Jonkers and J.A.M. van der Mullen
Department of Applied Physics, Eindhoven University of Technology,
PO Box 513, 5600 MB Eindhoven, The Netherlands

1. INTRODUCTION

Collisional Radiative Models (CRMs) are a widely used tool in the modelling of plasmas. The results of such models appear as source terms in the particle and energy balances of plasma transport models. Usually such models calculate a partial Quasi Steady State Solution (QSSS). This means that the spatial and temporal relaxation of most excited state densities is neglected, which is valid for a broad variety of plasmas.

Recently a new CRM has been developed which combines two interesting properties:

- validity for a broad range of electron densities, as stepwise processes are taken into account;
- the number of non-QSSS levels can be chosen arbitrarily, generalising the concept of the so-called *coefficients of net ionisation and recombination.*

As a result of these properties it is possible to fully describe radiative transfer, while simultaneously the model is valid for higher electron densities, as step processes can be taken into account. An application of this model is discussed in a second contribution to this conference [1]. Other research groups are kindly invited to participate in this research.

2. THE MODEL

In case the change of internal energy of the heavy particles is due to electron collisions and radiative processes the Atomic State Distribution Function (ASDF) can be calculated from plasma parameters, such as the electron density and temperature by solving a set of continuity equations. If the electron density is low only a few excited levels have to be taken into account. This holds for e.g. the conventional fluorescent lamps where, due to an electron density of around 5×10^{17} m^{-3}, a model with only four to six excited atomic mercury levels gives an accurate description of the discharge [2]. If the electron density rises more levels must be taken into account because of the onset of the stepwise excitation and ionisation processes [3].

In the model both numerical and analytical techniques have been combined. The lowest excited states, which are in Corona Balance (CB) have to be calculated numerically. The highest states are in Partial Local Saha Equilibrium (pLSE). In the case of an ionising plasma, the intermediate levels are in a so-called Excitation Saturation Balance (ESB). It was shown both numerically by Fujimoto [4] and analytically and experimentally by Van der Mullen [3,6] that the deviation from Saha equilibrium of the levels in ESB depends on their effective principal quantum number in a known way. Also these levels can therefore be treated analytically and the number of levels for which the continuity equation has to be solved can be further reduced. This truncation technique minimises the numerical effort which is needed and works in the case of recombining plasmas as well [5].

3. NUMERICAL IMPLEMENTATION

The program is written in C++ and has been ported to Windows and UNIX-based systems. For MS-Windows a graphical interface has been developed (see figure 1 for a sample).

The input parameters for the model are the cross sections for the collisional processes, the effective radiative decay frequencies, the electron and heavy-particle temperatures. Further the electron density and the densities of the non-QSSS states must be specified. The program calculates the rate coefficients of the collisional processes from the cross sections by integrating over the velocity distribution. At this moment the Maxwellian equilibrium distribution is assumed, but the implementation of other velocity distributions is possible. From these data, the program calculates the ASDF, the radiation production and the ionisation and recombination coefficients.

Further the program provides default values for the cross sections of the collisional processes involving the higher excited states. These values are calculated from a hydrogen approximation. This approximation is of course not valid for the lower states, for which the cross sections can be entered by specifying parameters of a chosen fitting function. *See [1] for a specific application.*

4. ACKNOWLEDGEMENTS

This work is part of the research programme of the Dutch Technology Foundation STW and is done in corporation with Philips Lighting, Eindhoven.

5. REFERENCES

1. J. van Dijk, R.M.J. Haanraads and J.A.M. van der Mullen: "Radiative transfer in High-Current Ar-Hg Discharges", this conference.
2. J.F. Waymouth: *"Electrical Discharge Lamps"*, MIT Press, Cambridge MA (1971).
3. J.A.M. van der Mullen: Phys. Rep. **191** (1990) 109.
4. T. Fujimoto: J. Phys. Soc. Jap. **47** (1979) 265.
5. D.A. Benoy, J.A.M. van der Mullen and D.C. Schram: J. Quant. Spectrosc. Radiat. Transf. **46** (1991) 195.
6. J.A.M. van der Mullen, B. van der Sijde and D.C. Schram: Phys. Lett. **79A** (1980) 51.

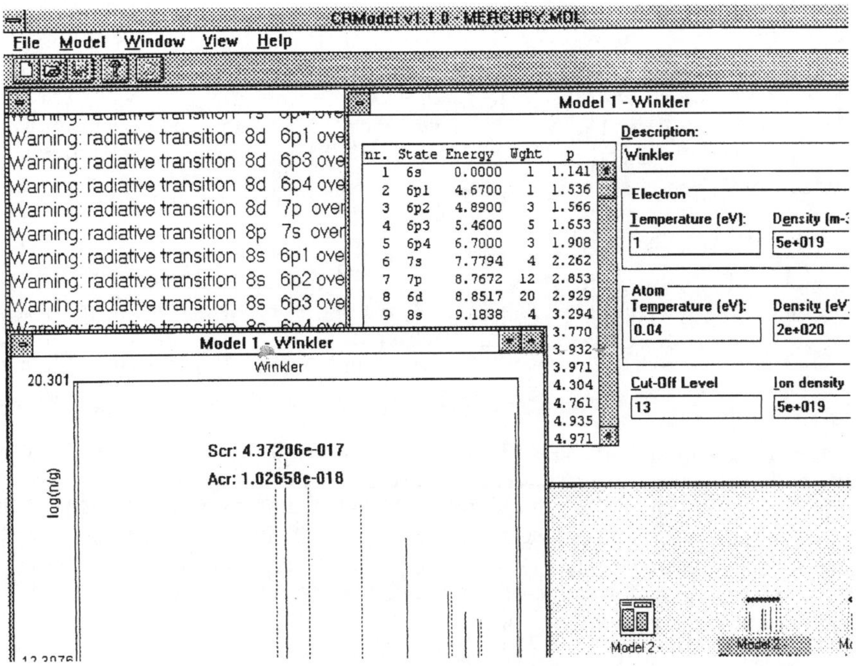

Fig. 1: User interface of the CRM program.

M 11

ON THE STUDY OF THE THERMAL EQILIBRIUM IN THE HIGH-PRESSURE MERCURY PLASMAS UNDER AC OPERATING MODE

Kamel Charrada[1], Georges Zissis[2], Hatem Elloumi[2] and Mongi Stambouli[3]
[1] IPEIM, rte de Kairaoun, 5000 Monastir, Tunisia
[2] CPAT, 118 rte de Narbonne 31062 Toulouse, France
[3] ESSTT, 5 av Taha Hussein, 1008 Tunis, Tunisia

1. INTRODUCTION

In our previous work [1], we studied the effect of the transport phenomena on the thermal equilibrium in the high pressure mercury plasmas. At this stage we examined the dc discharge only. In this work, we will discuss the influence of the discharge excitation mode on the thermal equilibrium state.

2. MODEL DESCRIPTION

2.1 Model assumptions

It was assumed that electron and heavy particle have a Maxwellian distribution function at electron and gas temperatures respectively. We suppose that the occupation of the internal energy states of the heavy particles is governed by Boltzman's law at electron temperature. As ionisation equilibrium is assumed the electron density can be calculated by using the Saha law at electron temperature. Furthermore, as a cylindrical symmetry is considered, the electric field is assumed to be purely axial. In our case as magnetic forces are negligible, thus terms $\mathbf{j}\times\mathbf{B}$ are discarded.

2.2 Model equations

In this model we solve the following equations :

- masse conservation equation
- electron energy equation
- heavy particles energy equation
- circuit equation

These equations are coupled with the perfect gas' law, the Saha's law, Ohm's law and Dalton's law. As far as the numerical techniques are concerned, differentials equations were treated by the element finite method and with a semi-implicite scheme. At this study we use a 1D time dependent model.

3.RESULTS AND DISCUSSIONS

In this section we will give some results showing the evolution of departures from thermal equilibrium for a mercury plasma supplied by ac current. We define the departure from thermal equilibrium as the relative difference between electron and gas temperature: $(T_e\text{-}T_g)/Tg$. Discharge parameters are: Inter electrode length: 72 mm ; Internal diameter: 20mm ; Mercury mass : 5 mg.

Figure 1 shows the variation of the relative departure from equilibrium for mercury plasma as function of time, the electrical field variation is also plotted. The arc current variation was shown on figure 2. We can see that the departure from equilibrium is important at the instants when the electrical field is important and the arc current is low. In figure 3 we have plotted the variation of the electron density as function as time and radial position. The variation departure from equilibrium is given on figure 4. It is shown that the deviation is important in the region near the wall and at when arc current is low.

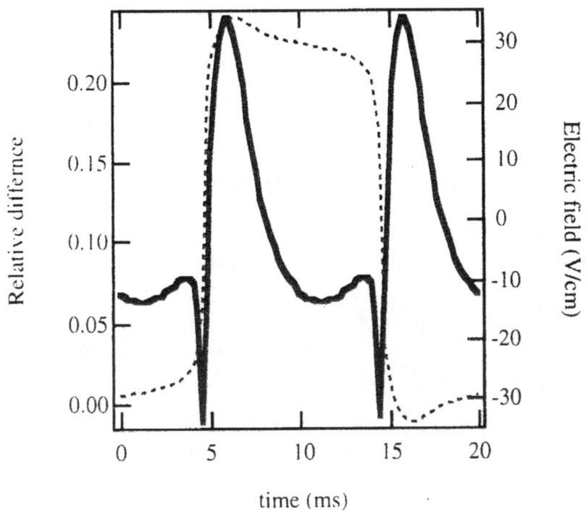

Figure 1

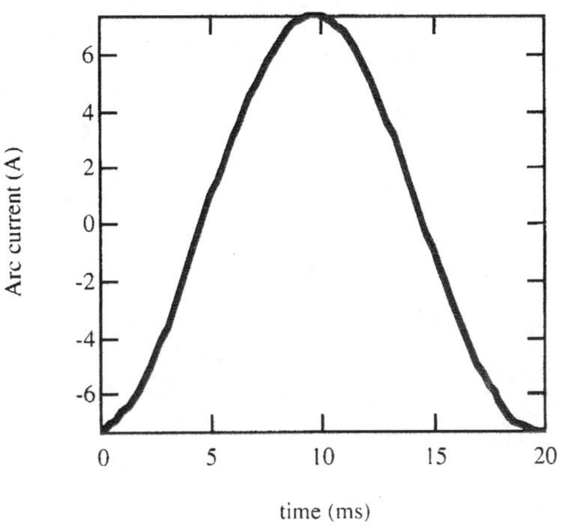

Figure 2

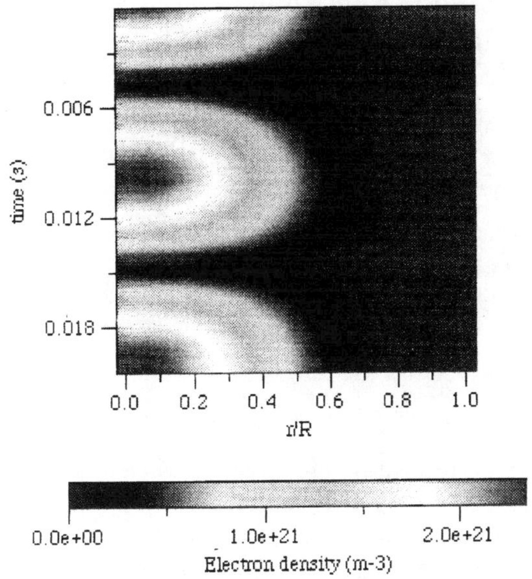

Figure 3

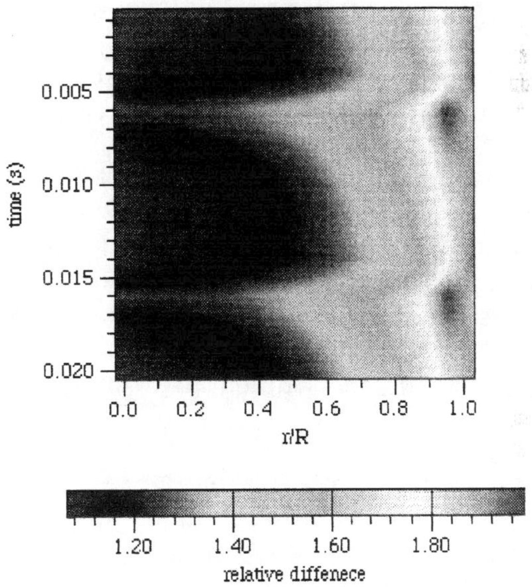

Figure 4

4. CONCLUSION

In this paper we have shown some typical results that can be predicted by our model. Therefore, it was found that important deviation from equilibrium occurs in the case of important electric fields and low currents. Our model can use also in order to obtain other the evolution of several physical quantities in the discharge. This work will completed in the near future by simulation of pulsed operating mode with several duty cycles.

5. REFERENCES

[1] K. Charrada, G. Zissis, M. Aubès, J. Phys D**29**, 2432 (1196)

M 12

NUMERICAL STUDY OF RADIAL TEMPERATURE PROFILES OF HIGH INTENSITY DISCHARGE LAMPS

Seishiro Hashiguchi, Shigeyuki Mori[1] and Kunihide Tachibana[2]
Department of Electronics and Information Science,
Kyoto Institute of Technology, Matugasaki, Sakyo-ku, Kyoto 606, Japan
[1]Japan Storage Battery, Co., Ltd., Nishinosho, Kisshoin, Minami-ku, Kyoto 601, Japan
[2]Graduate School of Engineering, Kyoto University, Yoshida Honmachi, Sakyo-ku, Kyoto 606-01, Japan

1 INTRODUCTION

High-pressure mercury arcs have a wall stabilized radial temperature profile; smaller temperature gradients in the central region and larger gradients at outside region in the tube. Addition of some kind of metal halide makes the temperature profile a constricted type. This occurs when the energy of the upper levels of spectral lines of additives is relatively low compared with the ionization energy. Cause of formation of the constricted temperature profile was explained by Waymouth[1]. In a previous work, the authors found that the radial temperature profile of mercury arcs depended on the radius of the tube[2]; for a smaller tube radius, the heat conduction balanced the input power even in the central region and the temperature profile resembled the constricted type. If a thick tube is used, on the other hand, the constricted temperature profile may be improved due to the increase in the tube diameter. In this work, the radial temperature profile of a Hg-Sc-Na system was studied as a function of the tube radius.

2 METHOD OF CALCULATION

We study a slice of plasma in the central region between electrodes of a standing arc, where physical quantities are axially uniform. Hence, the incoming energy due to convection on the lower base cancels the outgoing one on the upper base; we can neglect convection in the energy equation. Then the energy equation is expressed as

$$\frac{\partial}{\partial t}(r\rho h) = \frac{\partial}{\partial r}(rk\frac{\partial T}{\partial r}) + r(\sigma E^2 - U_{rad})$$

Here, h is the specific enthalpy of the gas mixture, ρ is mass density, k is thermal conductivity, T is temperature, σ is electrical conductivity, E is electric field, and U_{rad} is radiation loss including free-free continuum [2]. We treat 11 strong mercury lines and sodium resonance lines as optically thick. Details of calculation of optically thick lines and transport coefficients are described in [2]. We took into account other 71 mercury lines, 61 sodium lines, 184 scandium lines and 64 ionic scandium lines to calculate the radiation energy. These lines were assumed to be optically thin. As a boundary condition, pressures of NaI and ScI_3 were given at the wall; the wall temperature was assumed to be 1000K. We took into account Hg,

Na and Sc atoms and ions, electrons and I atoms as well as NaI and ScI_3 molecules. Partial pressure of these species in the arc were determined by assuming thermal equilibrium [3]. We deduce a finite difference equation in an implicit form from the energy equation and a steady solution was obtained by repeating numerical calculations.

3 RESULTS

A standard discharge condition was following: radius was 0.9cm, discharge current was 3amp, mercury pressure was about 6000kPa, pressures of NaI and ScI_3 were 133Pa at the wall. Calculations were also made virtually without NaI or without ScI_3 or without both. When the radius was changed the discharge current was changed to keep the maximum temperature at the center unchanged.

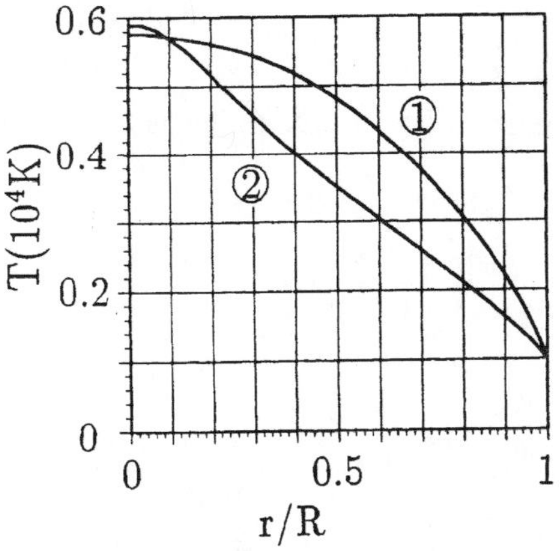

Fig.1 Radial temperature profile of pure mercury arc for the radius R=0.9cm① and R=0.45cm②.

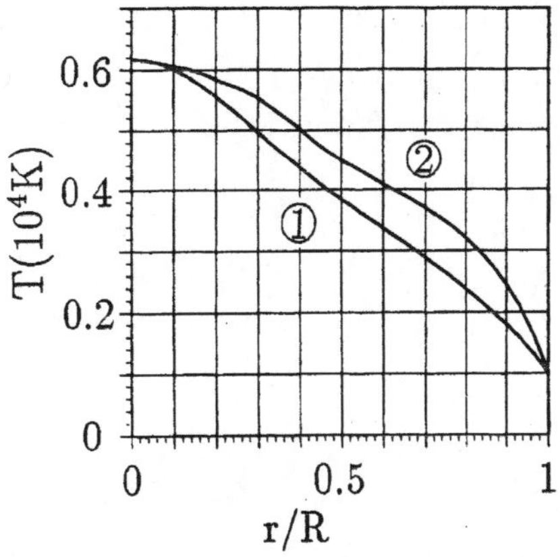

Fig.2 Radial temperature profile profile of Hg-Sc arc for the radius R=0.9cm① and R=1.8cm②.

Figure 1 shows the temperature profile of virtually pure mercury for two values of tube radius. The temperature profile is wall-stabilized type when the tube radius is 0.9cm but it changes to constricted type when the radius is decreased to half. Figure 2 shows the temperature profile of Hg-ScI_3 system. It is constricted type for the tube radius of 0.9cm. Although the profile becomes more flat near the center for the radius of 1.8cm, it still has negative value of divergence of thermal conduction flux in a middle region.

Present results show that the temperature profile depends on the tube radius but the change in the profile is different for different composition of working gas.

REFERENCES

[1] J.F.Waymouth: Electric Discharge Lamps(MIT, Cambridge, 1971), chap.11.

[2] S.Hashiguchi, S.Mori and K.Tachibana: Jpn. J. Appl. Phys. **36**, 6533 (1997).

[3] T.Ishigami: J. Illum. Eng. Inst. Japan **81**, 1010(1997) [in Japanese].

M 13

RIGOROUS DERIVATION OF A NON-LTE ELECTRICAL CONDUCTIVITY FOR ELECTRODE REGIONS OF HIGH PRESSURE DISCHARGE LAMPS

Peter Flesch and Manfred Neiger

Lighting Technology Institute (LTI), University of Karlsruhe, Box 6980, 76128 Karlsruhe, FRG,
E-mail: Peter.Flesch@etec.uni-karlsruhe.de

1 INTRODUCTION

In *Philips Journal of Research* [1] Fischer proposed a time-independent high pressure model for unifying the treatment of a local thermal equilibrium (LTE) plasma and the non-LTE plasma close to the electrodes. Unfortunately, no rigorous derivation of the model equations was published. In our poster we give a derivation which is partly different from [1] starting from fundamental diffusion equations [2], clarifying the different approximations. A length parameter is derived with which the applicability of the model can be judged.

2 DERIVATION OF NON-LTE ELECTRICAL CONDUCTIVITY

We start with an expression for the mass flux in the rigorous kinetic theory of n-component gas mixtures [2]:

$$\sum_{\substack{k=1\\k\neq j}}^{\nu} \frac{n_j n_k}{n^2 D_{jk}}(V_k - V_j) = d_j - \mathrm{grad}(\ln T) \sum_{\substack{k=1\\k\neq j}}^{\nu} \frac{n_j n_k}{n^2 D_{jk}}\left(\frac{D_k^T}{n_k m_k} - \frac{D_j^T}{n_j m_j}\right) \tag{1}$$

$$d_j = \mathrm{grad}\left(\frac{n_j}{n}\right) + \left(\frac{n_j}{n} - \frac{n_j m_j}{\rho}\right)\mathrm{grad}(\ln p) - \frac{n_j m_j}{p\rho}\left(\frac{\rho}{m_j}X_j - \sum_{s=1}^{\nu} n_s X_s\right) \tag{2}$$

with ν : number of plasma components, n_j : number densities (j = „e" : electrons, j = „i" : ions, j = „n" : neutral atoms), n : overall number density, ρ : overall mass density, D_{jk} : binary diffusion coefficients, V_k : velocity, T : temperature, D_k^T : multicomponent thermal diffusion coefficients, m_k : mass, p : pressure, X_j : external forces (due to an electric field in our case). For high pressure discharge lamps we set the pressure p constant. Whether or not we can neglect thermal diffusion in equation (1) depends on the ratio k_j^T of thermal diffusion and diffusion due to the concentration gradients:

$$k_j^T = \frac{\mathrm{grad}(\ln T)}{\mathrm{grad}\left(\frac{n_j}{n}\right)} \sum_{\substack{k=1\\k\neq j}}^{\nu} \frac{n_j n_k}{n^2 D_{jk}}\left(\frac{D_k^T}{n_k m_k} - \frac{D_j^T}{n_j m_j}\right). \tag{3}$$

As an example, for a D1-type model lamp with 4MPa Hg filling (input current I = 0.75A, work function of cathode: 2.6eV) k_j^T is less than 4% for electrons and ions. In high pressure mercury and xenon lamps we can furthermore neglect the number densities of electrons and ions with respect to the total number density. With an error less than 7% (for the example above) we can also neglect the terms containing the binary electron-ion diffusion coefficient $D_{e,i}$. Furthermore we neglect $\mathrm{grad}(n_e/n)$ compared to $(n_i/n)\mathrm{grad}(n_e/n_i)$ with an error less than 10%. Writing j_i for the particle fluxes we consider the production and annihilation of ions (4)[1]

$$\mathrm{div}\, j_{t,i} = R^+ - R^- = q^I n_e n_n - q^R n_e n_i - q^C n_e n_i n_n \tag{4}$$

where q^I, q^R, q^C are the temperature-dependent rate coefficients for ionisation (I), radiative recombination (R) and 3-body collisional recombination (C). In the part of the plasma which is in

LTE (but not in front of the electrodes) the ion production rate is equal to the ion annihilation rate so that we can express q^R and q^C with the help of q^I [1]:

$$q^I \frac{n_{n,LTE}}{n_{i,LTE}} = q^R + q^C n_{n,LTE} \tag{5}$$

with $n_{j,LTE}$ number densities in LTE. Using the relation $j_n = -j_i$ for a time-independent plasma and assuming that n_e/n is proportional to the electrical conductivity σ [1] we get

$$\Delta\sigma = -\frac{q^I n}{2D_{i,n}\alpha}\left(1 - \alpha\frac{\sigma}{\sigma_{LTE}}\right)\sigma \tag{6}$$

where $\alpha = n_i/n_e$ and σ_{LTE} is the electrical conductivity in LTE. The boundary conditions for σ at the cathode [3] and the anode [1] are as follows

$$\sigma_{cathode} = \frac{4j_{e,cathode}\sigma_{LTE}}{ev_{th}n_{e,LTE}} \quad \text{and} \quad \mathrm{grad}\sigma_{anode} = -\frac{e\sigma_{anode}}{2kT}\mathrm{grad}V \tag{7}$$

with $j_{e,cathode}$: electron emission current, v_{th} : thermal velocity of the electrons, T : temperature, V : potential. The rigorous derivation of these equations will be presented in the poster at the conference.

3 CHARACTERISTIC LENGTH Λ_σ

An important parameter is the „characteristic length" $\Lambda_\sigma = (2D_{i,n}/q^I n)^{0.5}$ (see Fig. 1) measuring the distance necessary for the non-LTE σ to reach its LTE value (see Fig. 2). If Λ_σ is too large, there is no smooth approximation of σ to σ_{LTE}, indicating that the model assumptions are no longer valid.

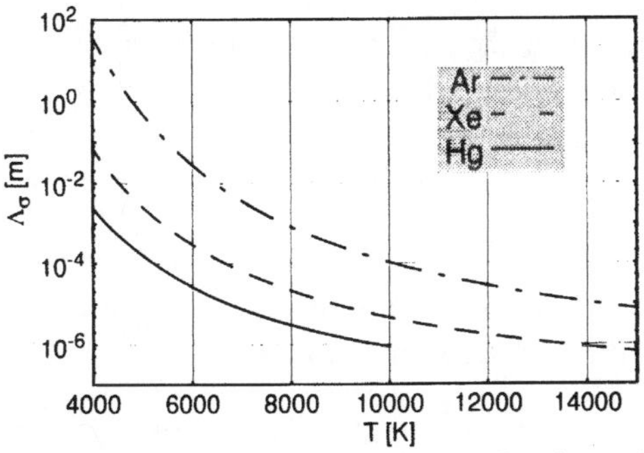

Fig 1: „Characteristic length" Λ_σ for Ar, Xe and Hg for a pressure of 4MPa.

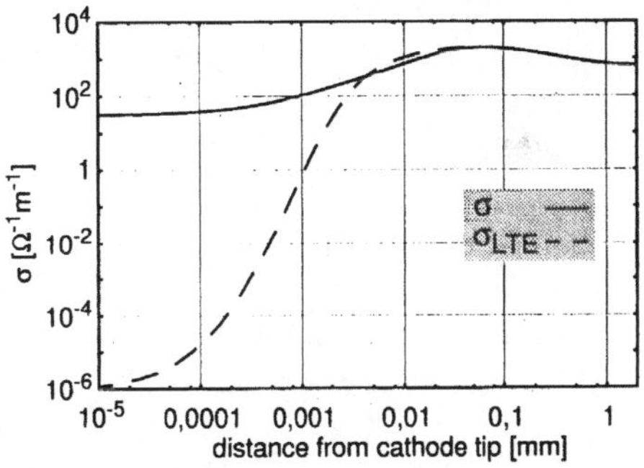

Fig. 2: σ approaches σ_{LTE} within a certain distance from the cathode (non-LTE region). Near the cathode σ is very large compared to σ_{LTE} (Xe, 4MPa).

4 ACKNOWLEDGEMENTS

This work was supported by the BMBF, FKZ 13N7107/0

5 REFERENCES

[1] E. Fischer, *Modelling of low-power high-pressure discharge lamps*, Philips J. Res. 42, pp. 58-85 (1987)

[2] J. O. Hirschfelder, C. F. Curtiss, R. B. Bird, *Molecular Theory of Gases and Liquids*, John Wiley & Sons (1964)

[3] H. Wiesmann, *Zweidimensionale Simulation von Hochdruckgasentladungslampen unter Berücksichigung der Elektrodenbereiche mit der Finite-Elemente-Methode*, PhD dissertation, University of Karlsruhe (1997), ISBN 3-8265-2632-5

M 14

RADIATIVE TRANSFER IN HIGH-CURRENT Ar-Hg DISCHARGES

J. van Dijk, R.M.J. Haanraads and J.A.M. van der Mullen
Department of Applied Physics, Eindhoven University of Technology,
PO Box 513, 5600 MB Eindhoven, The Netherlands

1. INTRODUCTION

Mixtures of noble gases and mercury are widely used in lighting applications, e.g. fluorescent lamps. To optimise the performance of these light sources, numerical modelling has been carried out with the help of PLASIMO, a computer program which is being developed at the Eindhoven University of Technology.

The most important process in efficient light sources is also the one which is least understood: the production and transport of resonant radiation. Attempts were made to provide an accurate description of radiative transfer in the plasmas under consideration. Various solutions to the Holstein equation will be presented and discussed. In order to incorporate these solutions into the fluid model PLASIMO, a novel Collisional Radiative Model has been developed. It is discussed in a second contribution to this conference [1].

2. THEORY

The treatment of the non-local radiation processes (absorption), can be dealt with by introducing an escape-factor Λ. This factor corrects the transition probability A(p,q) of a transition to account for absorption of radiation produced elsewhere, yielding an effective transition probability according to

$$A^{eff}(p,q) = \Lambda(p,q)A(p,q).$$

In the following equation due to Holstein [2], Λ is defined in terms of the density of the upper level n_p and the probability G of capturing photons produced in other regions, which is a function of the absorption coefficient k and distance between emitter and absorber,

$$\Lambda(r) = 1 - \frac{1}{n_p(r)} \int G(r,r')n_p(r')d^3r'.$$

The function G further depends on the line shape; throughout this work pure Doppler broadening has been assumed, although this does not impose any restriction on the general recipe described here.

3. RESULTS

3.1 Λ on the axis of an infinite cylinder

Holstein found a solution for the escape factor which is valid on the axis of an infinite cylinder with uniform radial absorption, a parabolic radial distribution of the excited state density and Doppler-broadening of the resonance line. The figure shows both Holstein's result ands the full numerical solution for the escape factor as a function of the plasma optical depth k_0R. We see how Holstein's derivation, which is based upon the assumption of large optical depth, fails for small tube radii or absorption coefficients (thin line).

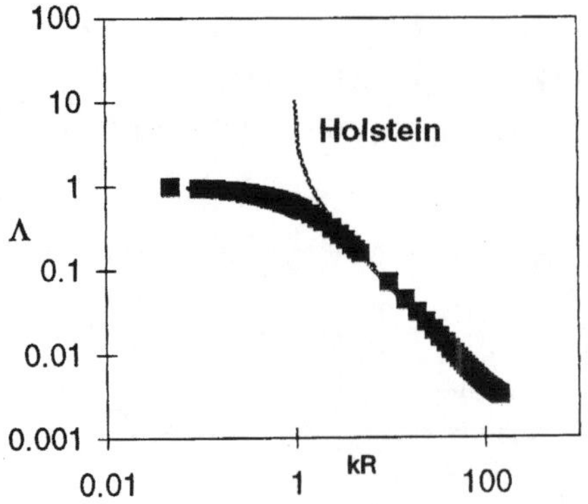

Secondly, the theory has been applied to describe the mercury 254 nm transition in a high-current argon-mercury fluorescent lamp. Results have been obtained with the fluid-dynamics code PLASIMO which is being developed at Eindhoven University. Use has been made of a collisional radiative model which takes stepwise excitation into account [1].

3.2 Off-axis values

The calculation of the escape factor has been incorporated in the PLASIMO simulation program. This allows a calculation which is self-consistent with the computed fluid dynamics data (densities, temperatures). The calculation which is shown in the figure has been obtained for a fluorescent lamp with cylindrical geometry. A strongly non-uniform self-absorption is observed: photons which are produced near the plasma walls are more likely to escape from the lamp.

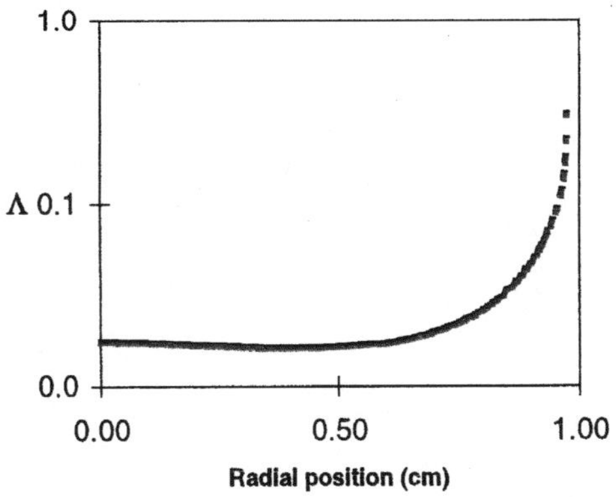

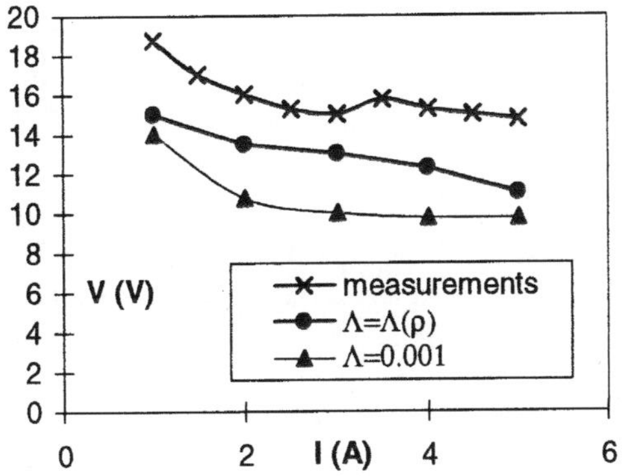

3.3 Influence on the VI-characteristic

The voltage fall over the positive column of a high-current fluorescent lamp has been calculated with the PLASIMO model. Various models for the escape factor have been tested, the results have been compared with measurements. One observes that for higher escape factors the voltages increase, as there is a larger 'leak' in the ionisation channel. For higher currents the assumption of uniform absorption becomes dubious and the results should be interpreted with care.

4. CONCLUSIONS

Holstein's recipe to calculate the escape factor for resonant radiation has been used to obtain values for the escape factor of the mercury 254 nm transition for a plasma with cylindrical geometry. This calculation is incorporated in a fluid-dynamics code, revealing that the converged result features a strongly non-uniform net radiation. The results are consistent with Holstein's asymptotic solution in the latter range of applicability.

5. ACKNOWLEDGEMENTS

This work is part of the research programme of the Dutch Technology Foundation STW and is done in corporation with Philips Lighting, Eindhoven.

6. REFERENCES

1. J. van Dijk, H.W.P. van der Heijden, J. Jonkers and J.A.M. van der Mullen: "Towards a general Collisional Radiative Model", this conference.
2. T. Holstein, Phys. Rev 72(12), 1212 (1947)

INDEX